Riemann–Hilbert Problems, Their Numerical Solution, and the Computation of Nonlinear Special Functions

Riemann–Hilbert Problems, Their Numerical Solution, and the Computation of Nonlinear Special Functions

Thomas Trogdon
New York University
New York, New York

Sheehan Olver
The University of Sydney
New South Wales, Australia

Society for Industrial and Applied Mathematics
Philadelphia

10 9 8 7 6 5 4 3 2 1

Publisher	David Marshall
Acquisitions Editor	Elizabeth Greenspan
Developmental Editor	Gina Rinelli
Managing Editor	Kelly Thomas
Production Editor	David Riegelhaupt
Copy Editor	Nicola Howcroft
Production Manager	Donna Witzleben
Production Coordinator	Cally Shrader
Compositor	Techsetters, Inc.
Graphic Designer	Lois Sellers

Library of Congress Cataloging-in-Publication Data

Trogdon, Thomas D.

Riemann–Hilbert problems, their numerical solution, and the computation of nonlinear special functions / Thomas Trogdon, New York University, New York, New York, Sheehan Olver, The University of Sydney, New South Wales, Australia.

pages cm. -- (Other titles in applied mathematics ; 146)

Includes bibliographical references and index.

ISBN 978-1-611974-19-5

1. Riemann–Hilbert problems. 2. Differentiable dynamical systems. I. Olver, Sheehan. II. Title.

QA379.T754 2016

515'.353--dc23

2015032776

To Karen and Laurel

Contents

Preface

This book grew out of the collaboration of the authors, which began in the Spring of 2010, and the first author's PhD dissertation. The second author developed much of the theory in Part II during his Junior Research Fellowship at St. John's College in Oxford, applying it to the Painlevé II equation in the nonasymptotic regime. The authors, together with Bernard Deconinck, then developed the methodology presented in Chapter 8 for the Korteweg–de Vries (KdV) equation. The accuracy that is observed begged for an explanation, leading to the framework in Chapter 7. Around the same time, the approaches for the nonlinear Schrödinger (NLS) equations, the Painlevé II transcendents, and the finite-genus solutions of the KdV equation were developed. These applications, along with the original KdV methodology, make up Part III. Motivated by the difficulty in finding a comprehensive, beginning graduate-level reference for Riemann–Hilbert problems (RH problems) that included the L^2 theory of singular integrals, the first author compiled much of Part I during his PhD studies at the University of Washington.

Central to the philosophy of this book is the question, What does it mean to "solve" an equation? The most basic answer is to establish *existence* — the equation has at least one solution — and *uniqueness* — the equation has one and only one solution. A more concrete answer is that an equation is solved if its solution can be evaluated, typically by approximation in a reliable and efficient way. This can be accomplished via asymptotics: the solution to the equation is given by an approximation that improves with accuracy in certain parameter regimes. Otherwise, the solution can be evaluated by numerics, a sequence of approximations that converge to the true solution. In the case of linear partial differential equations (PDEs), standard solutions are given as integral representations obtained via, say, Fourier series or Green's functions. Integral representations are preferable to the original PDE because they satisfy all the properties of a "solution" to the equation:

1. *Existence* and *uniqueness* generally follow directly from the well-understood integration theory.

2. *Asymptotics* are achievable via the *method of stationary phase* or the *method of steepest descent.*

3. *Numerics* for integrals have a well-developed theory, and the integral representations can be readily evaluated via quadrature.

In place of integral representations, fundamental integrable nonlinear ordinary differential equations (ODEs) and PDEs have an RH problem representation. RH problems are boundary-value problems for piecewise (or sectionally) analytic functions in the complex plane. Our goal is to solve these integrable nonlinear ODEs and PDEs by utilizing their RH problem representations in a manner analogous to integral representations. In some cases, we use RH problems to establish existence and uniqueness, as well as derive asymptotics. But most importantly, we want to be able to accurately evaluate the solutions inside and outside of asymptotic regimes with a unified numerical approach.

The stringent requirements we put into our definition of a "solution" force all solutions we find to be in a very particular category: *nonlinear special functions*. A *special function* is shorthand for a mathematical function which arises in many physical, biological, or computational contexts or in a variety of mathematical settings. A nonlinear special function is a special function arising from a fundamentally nonlinear setting. For centuries, mathematicians have been studying special functions. An important feature that separates special functions from other elementary functions is that they generically take a *transcendental*[1] form.

The catenary, discovered by Leibniz, Huygens, and Bernoulli in the 1600s, describes the shape of a freely hanging rope in terms of the hyperbolic cousin of the cosine function. The study of special transcendental functions continued with the discovery of the Airy and Bessel functions which share similar but more complicated series representations when compared to the hyperbolic cosine function. These series representations are often derived using a differential equation that is satisfied by the given function. Such a derivation succeeds in many cases when the differential equation is linear.

The 19th century was a golden age for special function theory. Techniques from the field of complex analysis were invoked to study the so-called elliptic functions. These functions are of a fundamentally nonlinear nature: elliptic functions are solutions of nonlinear differential equations. The early 20th century marked the work of Paul Painlevé and his collaborators in identifying the so-called *Painlevé transcendents*. The Painlevé transcendents are solutions of nonlinear differential equations that possess important properties in the complex plane. Independent of their mathematical properties, which are described at length in [52], the Painlevé transcendents have found use in the asymptotic study of water wave models [4, 32, 35] and in statistical mechanics and random matrix theory [109, 120].

Through the study of RH problems, we discuss classical special functions (the Airy function, elliptic functions, and the error function), canonical nonlinear special functions (elliptic functions and the Painlevé II transcendents), and some noncanonical special functions (solutions of integrable nonlinear PDEs) which we advocate for inclusion in the pantheon of nonlinear special functions based on the structure we describe.

We now present the layout of the book to guide the reader. A comprehensive table of notations is given after the Preface, and optional sections are marked with an asterisk.

- Part I contains an introduction to the applied theory of RH problems. Chapter 1 contains a survey of applications in which RH problems arise. Then Chapter 2 describes the classical development of the theory of Cauchy integrals of Hölder continuous functions. This theory is used to explicitly solve many scalar RH problems. Lebesgue and Sobolev spaces are used to develop the theory of singular integral equations in order to deal with the matrix, or noncommutative, case. Some of these results are new, while many others are compiled from a multitude of references. Finally, the method of nonlinear steepest descent developed by P. Deift and X. Zhou is reviewed in a simplified form in Chapter 3. On first reading, many of the proofs in this part can be omitted.

- Part II contains a detailed development of the numerical methodology used to approximate the solutions of RH problems. While there is certainly some dependence of Part II on Part I, for the more numerically inclined, it can be read mostly

[1] In this context, transcendental means that the function cannot be expressed as a finite number of algebraic steps, including rational powers, applied to a variable or variables [63].

independently because the dependencies are made clear. Many aspects of computational/numerical complex analysis are discussed in detail in Chapter 4, including convergence of trigonometric, Laurent, and Chebyshev interpolation. The computation of Cauchy transforms is treated in Chapter 5. This numerical approach to Cauchy transforms is utilized in Chapter 6 to construct a collocation method for solving RH problems numerically. Finally, a uniform approximation theory that allows one to estimate the derivatives of solutions of RH problems is presented in Chapter 7.

- Part III contains applications of the theory of Part II to specific integrable equations. Each of the chapters in Part III depends significantly on the material in Part I, specifically Chapter 3, and on nearly all of Part II. Part III is written in a way that is appropriate for a reader interested only in numerical results wishing to understand the scope and applications of the methodology. As mentioned above, the applications are to the KdV equation (Chapter 8), the NLS equations (Chapter 9), the Painlevé II transcendents (Chapter 10), the finite-genus solutions of the KdV equation (Chapter 11), and the so-called dressing method (Chapter 12) applied to both the KdV and NLS equations to nonlinearly superimpose rapidly decaying solutions and periodic and quasi-periodic finite-genus solutions.

We would like to acknowledge the important contributions to this work of the first author's PhD advisor, Bernard Deconinck. We would also like to acknowledge the encouragement and input of Deniz Bilman, Percy Deift, Ioana Dumitriu, Anne Greenbaum, Randy LeVeque, Chris Lustri, Katie Oliveras, Bob O'Malley, Irina Nenciu, Ben Segal, Natalie Sheils, Chris Swierczewski, Olga Trichtchenko, Vishal Vasan, Geoff Vasil and Xin Yang.

Mathematica code for the applications discussed in Chapter 1 along with code for the more elaborate examples in Part II can be found on the book's website (`www.siam.org/books/ot146`). We sincerely hope the reader finds this to be a valuable resource.

Thomas Trogdon and Sheehan Olver

Notation and Abbreviations

Symbol	Description	Page
(f)	The divisor of a meromorphic function f	272
$[\diamond,\diamond]$	The standard matrix commutator: $[M,N]=MN-NM$	88
$[G;\Gamma]$	A homogeneous L^2 RH problem whose solution tends to the identity at infinity	63
$Z(\Gamma)$	Functions that satisfy the absolutely converging zero-sum condition	152
$Z_n(\Gamma)$	A finite-dimensional subspace of $Z(\Gamma)$	160
$\mathrm{Ai}(z)$	The Airy function	9
$\arg z$	The argument of complex number, $\arg z \in (-\pi,\pi]$	
đs	đ$s = 1/(2\pi \mathrm{i})\mathrm{d}s$	25
$\mathscr{C}$	Equivalent to $\mathscr{C}_{\mathbb{U}}$	128
$\check{f}_k$	The kth Chebyshev coefficient of f	116
$\operatorname{codim} X$	The dimension of the quotient space Y/X where Y is clear from the context	330
$\deg D$	The degree of a divisor D	272
$\dim X$	The dimension of a vector space X	
$\mathrm{e}^{\alpha\hat{\sigma}_3}A$	$\mathrm{e}^{\alpha\sigma_3}A\mathrm{e}^{-\alpha\sigma_3}$	88
ℓ^p	The Banach space of p-summable complex sequences. Context determines if sequences run over $\mathbb{N}$ or $\mathbb{Z}$.	328
$\ell^{(\lambda,R),p}$	p-summable complex sequences with algebraic and exponential decay	329
$\ell^{\lambda,p}$	p-summable complex sequences with algebraic decay	329
$\ell^{(\lambda,R),p}_{\pm}$	$\ell^{(\lambda,R),p}$ with a zero-sum condition	329
$\operatorname{erfc} z$	The complementary error function	5
$\mathscr{F}^{-\mathrm{z}}$	The map to vanishing basis coefficients on the line	121
$\mathtt{F}_{\alpha,\beta}$	The DFT matrix operator	111
$\mathscr{T}$	The operator that maps a function to its Chebyshev coefficients	116
$\mathscr{T}^{\pm\mathrm{z}}$	The map to vanishing Chebyshev basis coefficients	122
$\mathtt{T}_n$	The DCT matrix operator	117
$\Gamma(\zeta)$	The Gamma function	103
Γ^+	The region above a Lipschitz graph Γ	48
$\Gamma^\dagger$	The Schwarz conjugate of contour Γ: $\Gamma^\dagger = \{\bar{z} : z \in \Gamma\}$	77

$\mathscr{K}(X,Y)$	The subspace of compact operators from X to Y 330
$\mathscr{L}(X,Y)$	The Banach space of bounded linear operators from X to Y 330
$\mathscr{S}(\mathbb{R})$	The Schwartz class on $\mathbb{R}$: $\mathscr{S}(\mathbb{R})=\{f\in C^\infty(\mathbb{R}):\sup_x\lvert x^j f^{(k)}(x)\rvert<\infty \text{ for all } j,k\geq 0\}$
$\mathscr{S}_\delta(\mathbb{R})$	The Schwartz class with exponential decay: $\mathscr{S}_\delta(\mathbb{R})=\{f\in\mathscr{S}(\mathbb{R}):\sup_x e^{\delta\lvert x\rvert}\lvert f(x)\rvert<\infty\}$
$\Omega_\pm$	Components to the left/right of a complete contour 24
∂D	The boundary of a set D 45
$\phi(z,b,a)$	The Lerch transcendent 139
$\Phi^\pm(s)=\Phi_\pm(s)$	The boundary values of Φ from the left/right 24
$\operatorname{Res}_{z=a} f(z)$	The residue of a function $f(z)$ at a point $z=a$ 3
$\operatorname{sgn} x$	The sign of real number x, $\operatorname{sgn} 0=0$
σ_1	$\begin{bmatrix} 0 & 1\\ 1 & 0\end{bmatrix}$ 202
σ_3	$\operatorname{diag}(1,-1)$ 88
Σ_∞	The class of Jordan curves tending to straight lines at infinity 46
$\sigma_{\mathrm{B}}(q)$	The Bloch spectrum of q 274
$\Subset$	$D\Subset A$ when D is a connected component of A 58
θ^m	Evenly spaced points on the periodic interval 110
$\tilde{H}^k_\pm(\Gamma)$	$H^k_\pm(\Gamma)\oplus\mathbb{C}^{n\times n}$ when Γ is unbounded 67
$\triangleq$	Used for inline definitions 46
$\boldsymbol{x}^n$	The Chebyshev points 116
$\{s_1,s_2,s_3\}$	Stokes constants 253
$^\top$	Transpose of a matrix or vector
$A(D)$	The Abel map of a divisor D 272
$B(x,\delta)$	The ball centered at x of radius δ
$B_{\theta,\phi}(x,\delta)$	$e^{i\phi}(\{y\in B(x,\delta):\lvert\operatorname{Im}(x-y)\rvert/\lvert x-y\rvert>\sin\theta\}-x)+x$ 27
C	A generic constant
$C^k(A)$	The Banach space of k-times continuously differentiable functions 324
$C^{0,\alpha}(\Gamma)$	The Banach space of α-Hölder continuous functions 32
$C^k_c(A)$	The space of k-times continuously differentiable functions with compact support 324
$C_n[G;\Gamma]$	The collocation matrix for $\mathscr{C}[G;\Gamma]$ 161
D^{-1}	$\{z^{-1}:z\in D\}$ 47
$D_\nu(\zeta)$	The parabolic cylinder function 102
Df	The weak differentiation operator applied to f 59
$f\vert_A$	The restriction of a function $f:D\to R$ to a set $A\subset D$
$f^\dagger$	The Schwarz conjugate of a function f: $f^\dagger(z)=\overline{f(\overline{z})}^\top$ 77

$H^k(\Gamma)$	The kth-order Sobolev space on a self-intersecting admissible contour Γ	59
$H^k_\pm(\Gamma)$	The Sobolev spaces of Zhou	62
$H^k_z(\Gamma)$	$H^k(\Gamma)$ with the $(k-1)$th-order zero-sum condition	60
I	The identity matrix/operator with dimensionality implied	
$L^p(\Gamma)$	The Lebesgue space on a self-intersecting curve Γ	324
n_i	The number of collocation points on Γ_i, $\Gamma = \Gamma_1 \cup \cdots \cup \Gamma_L$	160
$P_{\mathrm{II}}(s_1, s_2, s_3; x)$	The solution of the Painlevé II ODE with Stokes constants s_1, s_2, s_3	253
$q_n(x,t)$	An approximation to the solution of the PDE with n collocation points per contour	206
$R_\pm(\Gamma)$	Functions that are a.e. rational on each component of $\Omega_\pm$	68
$T_k(x)$	The Chebyshev polynomials	340
$T_k^{\pm z}(x)$	The vanishing Chebyshev basis	122
U_k	The Chebyshev U polynomials	342
$\mathcal{H}_\Gamma f(z)$	The Hilbert transform of a function f	126
$\mathcal{T}$	The Chebyshev transform	340
$\mathcal{V}[f]$	The total variation of f	323
$\mathrm{FP}\mathcal{C}_\Gamma f(z)$	The finite-part Cauchy transform of f	146
${}_2F_1$	The Gauss hypergeometric function	140
RH problem	Riemann–Hilbert problem	3
BA	Baker–Akhiezer	276
DCT	Discrete cosine transform	117
DFT	Discrete Fourier transform	110
IST	Inverse scattering transform	87
KdV	Korteweg–de Vries	193
NLS	Nonlinear Schrödinger	231
ODE	Ordinary differential equation	253
PDE	Partial differential equation	193
SIE	Singular integral equation	63

Part I

Riemann–Hilbert Problems

Chapter 1

Classical Applications of Riemann–Hilbert Problems

A fundamental theme of complex analysis is that an analytic function can be uniquely expressed in terms of its behavior at the boundary of its domain of analyticity. A basic example of this phenomena is the partial fraction expansion of a rational function, which states that any rational function can be expressed as a sum of (finite) Laurent series around its poles. Consider the case of a rational function with only simple poles that is bounded at ∞. We can reinterpret the partial fraction result as a solution $r(z)$ to the following problem.

Problem 1.0.1. *Given distinct poles* $\{z_1,\ldots,z_m\}$*, a normalization point* $z_0 \in \mathbb{C} \cup \{\infty\}$ *satisfying* $z_0 \notin \{z_1,\ldots,z_m\}$*, a normalization constant* $c_0 \in \mathbb{C}$*, and residues* $\{A_1,\ldots,A_m\}$*, find* $r : \mathbb{C} \cup \{\infty\} \setminus \{z_1,\ldots,z_m\} \to \mathbb{C}$ *that satisfies the following:*

1. $r(z)$ *is analytic off* $\{z_1,\ldots,z_m\}$*,*

2. $r(z)$ *is bounded at* ∞ *and has at most simple pole singularities throughout the complex plane*
$$\lim_{\tau\to z} |(\tau - z)r(z)| < \infty \textit{ for all } z \in \mathbb{C},$$

3. $r(z_0) = c_0$*, and*

4. $r(z)$ *satisfies the residue conditions*
$$\mathrm{Res}_{z=z_k}\, r(z) = A_k \quad \textit{for} \quad k = 1,\ldots,m.$$

The conditions of this problem are sufficient to uniquely determine r: if there existed another solution, then their difference is bounded and analytic (by Theorem C.3), vanishes at z_0, and thence is zero by Liouville's theorem (Theorem C.2).

If an analytic function is not rational, then it necessarily has more exotic singularities, typically branch cuts and essential singularities. However, in many instances we can build up analogues of Problem 1.0.1 using *Riemann–Hilbert (RH) problems* to uniquely determine a function by its behavior at such singularities. Loosely speaking, an RH problem consists of constructing a *sectionally analytic function* dictated by its behavior at the boundary of regions of analyticity. For s on some contour Γ, we use the notation $\Phi^+(s)$ and $\Phi^-(s)$ to refer to the boundary values that a function $\Phi(z)$ takes as z approaches s

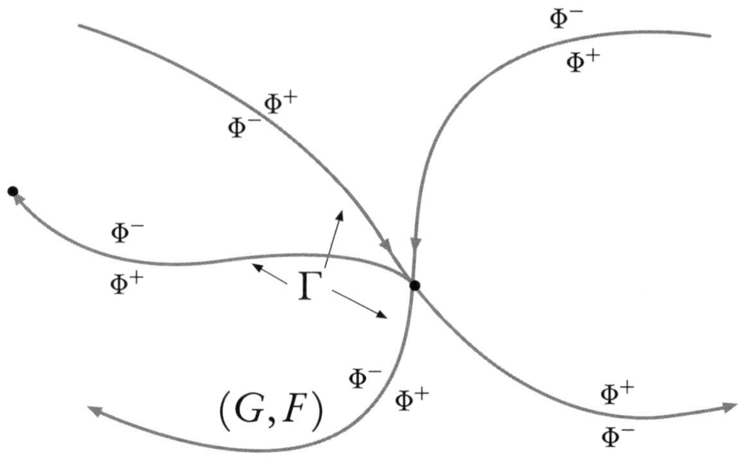

Figure 1.1. *A generic oriented jump contour* Γ *with the pair of jump functions* (G,F). *When* $F=0$ *we specify the jump with* G *alone. The black dots signify (finite) endpoints of the contours.*

from the left and right of Γ, respectively, provided these limits exist. The prototypical RH problem takes the following form.[2]

Problem 1.0.2. *Given a* contour Γ, *a* normalization point $z_0 \in \mathbb{C}\cup\{\infty\}$, *a* normalization constant $C_0 \in \mathbb{C}^{n\times m}$, *and* jump functions $G:\Gamma \to \mathbb{C}^{m\times m}$ *(also called the* jump matrix*) and* $F:\Gamma \to \mathbb{C}^{n\times m}$, *find* $\Phi : \mathbb{C}\setminus\Gamma \to \mathbb{C}^{n\times m}$ *that satisfies the following:*

1. $\Phi(z)$ *is analytic off* Γ *(in each component of* $\mathbb{C}\setminus\Gamma$*),*
2. $\Phi(z)$ *is bounded at* ∞ *and has weaker than pole singularities throughout the complex plane,*
3. $\Phi(z_0) = C_0$, *and*
4. $\Phi(z)$ *satisfies the* jump condition

$$\Phi^+(s) = \Phi^-(s)G(s) + F(s) \quad \textit{for} \quad s \in \Gamma.$$

See Figure 1.1 *for a schematic of a generic RH problem.*

Remark 1.0.1. *When* $F(s)$ *is zero, we use the notation* $[G;\Gamma]$ *to refer to this RH problem, with appropriate additional regularity properties; see Section 2.7.*

Remark 1.0.2. *When the RH problem is scalar* ($m=n=1$)*, we often use lowercase variables: e.g.,* ϕ *satisfies the jump condition*

$$\phi^+(s) = \phi^-(s)g(s) + f(s).$$

In this chapter, we invoke the philosophy of Its [66] and demonstrate how several classical problems can be reduced to RH problems. These fall under three fundamental categories:

[2]The definition of an RH problem is made precise in Definition 2.44.

1. Integral representations: we obtain simple, constant coefficient scalar RH problems for error functions and elliptic integrals.

2. Differential equations: we obtain matrix RH problems that encode the Airy function and the solution to inverse monodromy problems.

3. Spectral analysis of operators: we construct matrix RH problems that encode the potential of a Schrödinger operator and the entries of a Jacobi operator in terms of their spectral data.

Unlike the partial fraction case, existence and uniqueness of such problems is a delicate question, which we defer until Chapter 2. However, for the purposes of this chapter a simple rule of thumb suffices: if G and F are bounded and piecewise continuous, and the winding number[3] of $\det G(s)$ is zero, we expect the RH problem to be solvable.[4]

1.1 ▪ Error function: From integral representation to Riemann–Hilbert problem

The *complementary error function* is defined by

$$\operatorname{erfc} z \triangleq \frac{2}{\sqrt{\pi}} \int_z^\infty e^{-s^2} ds.$$

This is an entire function in z, and behaves like a smoothed step function on the real axis, with $\operatorname{erfc}(\infty) = 0$ and

$$\operatorname{erfc}(-\infty) = \frac{2}{\sqrt{\pi}} \int_{-\infty}^\infty e^{-x^2} dx = 2.$$

The error function is important in statistics: the cumulative density function (CDF) of a standard Gaussian random variable is given by $1 - \frac{1}{2}\operatorname{erfc} x$. It is also a critical tool in describing asymptotics of solutions to differential equations near essential singularities [91].

One can certainly calculate error functions via quadrature: for a quadrature rule

$$\int_0^\infty f(x) dx \approx \sum_{k=1}^n w_k f(x_k),$$

approximate

$$\operatorname{erfc} z \approx \sum_{k=1}^n w_k e^{-(x_k+z)^2}.$$

However, the integrand becomes increasingly oscillatory as the imaginary part of z becomes large so that an increasing number of quadrature points are required to resolve the oscillations.

[3]The winding number is the number of times an image of a function wraps around the origin. Technically speaking, a zero winding number in this context is not sufficient for existence and uniqueness, but we defer these details to Chapter 2; see also Section 8.2.4.

[4]There may be simple additional conditions required to ensure that the solution is unique.

An effective alternative to quadrature rules applied to the integral representation is to apply numerical methods to an RH problem. To this end, we reduce the integral representation to a simple RH problem. The first stage is to manipulate $\operatorname{erfc} z$ so that it has nice behavior at $+\infty$. We can determine its asymptotic behavior via integration by parts:

$$\begin{aligned}\int_z^\infty e^{-s^2}\,ds &= -\int_z^\infty \left(\frac{d}{ds}\left[\frac{e^{-s^2}}{2s}\right] + \frac{e^{-s^2}}{2s^2}\right) ds\\ &= \frac{e^{-z^2}}{2z} - \int_z^\infty \frac{e^{-s^2}}{2s^2}\,ds.\end{aligned}$$

The second integral satisfies, for $\operatorname{Re} z \geq 0$ and $|z| > 1$,

$$\left| e^{z^2}\int_z^\infty \frac{e^{-s^2}}{2s^2}\,ds\right| = \left|\int_0^\infty \frac{e^{-2zx-x^2}}{2(x+z)^2}\,dx\right| \leq \frac{1}{2|z|^2}\int_0^\infty e^{-x^2}\,dx = \frac{\sqrt{\pi}}{4|z|^2}.$$

It follows that $e^{z^2}\operatorname{erfc} z$ is bounded in the right-half plane (boundedness for $|z| \leq 1$ follows from analyticity) and decays as $z \to +\infty$. Similarly, we construct a companion function by integrating from $-\infty$ and normalizing: the function

$$e^{z^2}(2 - \operatorname{erfc} z) = \frac{2}{\sqrt{\pi}} e^{z^2}\int_{-\infty}^z e^{-s^2}\,ds$$

is bounded in the left-half plane and decays as $z \to -\infty$.

Combining these two functions, we can construct a sectionally analytic function

$$\psi(z) \triangleq \begin{cases} e^{z^2}(2 - \operatorname{erfc} z) & \text{if } \operatorname{Re} z < 0,\\ -e^{z^2}\operatorname{erfc} z & \text{if } \operatorname{Re} z > 0.\end{cases}$$

This has a discontinuity along the imaginary axis, and we can encode the jump along this discontinuity via the following scalar RH problem.

Problem 1.1.1. *Find $\psi : \mathbb{C}\setminus i\mathbb{R} \to \mathbb{C}$ that satisfies the following:*

1. *$\psi(z)$ is analytic off the imaginary axis and continuous up to the imaginary axis,*
2. *$\psi(z)$ is bounded throughout the complex plane,*
3. *$\psi(z)$ decays at ∞, and*
4. *for $y \in i\mathbb{R}$ (oriented from $-i\infty$ to $i\infty$),*

$$\psi^+(y) - \psi^-(y) = \frac{2}{\sqrt{\pi}} e^{y^2}\int_{-\infty}^\infty e^{-x^2}\,dx = 2e^{y^2},$$

where $\psi^+(y) = \lim_{\epsilon\downarrow 0}\psi(y-\epsilon)$ and $\psi^-(y) = \lim_{\epsilon\downarrow 0}\psi(y+\epsilon)$ are the limits from the left and right, respectively.

Figure 1.2 depicts the jump contours and jump functions. Figure 1.3 depicts the solution.

This RH problem has a unique solution (Problem 2.3.1), and solving the RH problem *globally throughout the complex plane* can be accomplished via a single fast Fourier transform (Section 5.3).

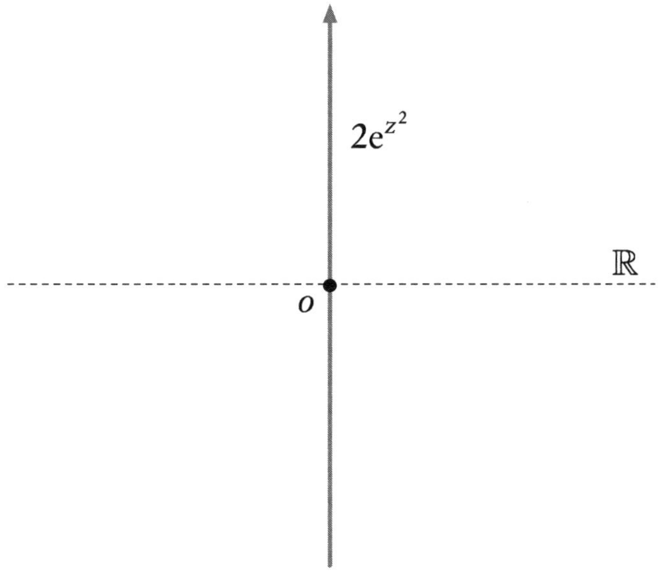

Figure 1.2. *The jump contour and jump function* $G(z) = 2e^{z^2}$ *for the complementary error function.*

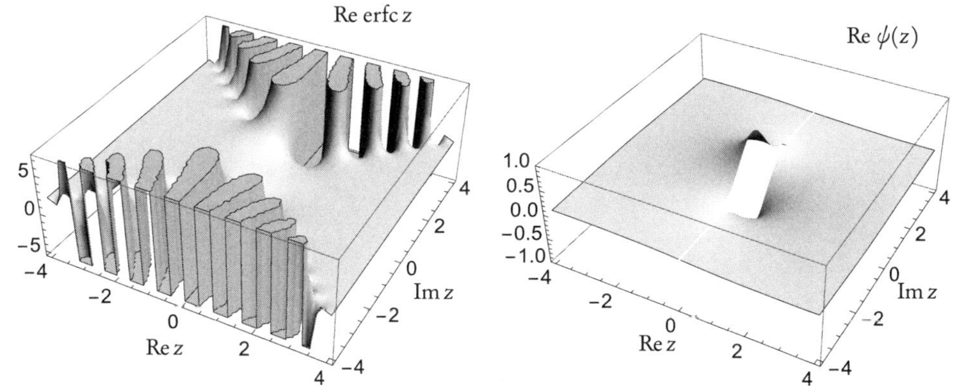

Figure 1.3. *The real part of* erfc z *(left) and the real part of* ψ *(right).*

1.2 ▪ Elliptic integrals

Define the *elliptic integral of the first kind*[5] by

$$g(z;a) \triangleq \int_0^z \frac{1}{\sqrt{1-s^2}\sqrt{a^2-s^2}}\mathrm{d}s$$

for $a > 1$. Here the integration path can be taken to be a straight line from the origin to z, and we introduce branch cuts on $[1,\infty)$ and $(-\infty,-1]$.

Elliptic integrals initially arose in the descriptions of the arc lengths of ellipses. They form a fundamental tool in building functions on Riemann surfaces, along with their

[5]The usual convention is to define Legendre's elliptic integral of the first kind $F(\phi,m) \triangleq m^{-1}g(\sin\phi;m^{-1})$. However, the definition as g leads naturally to an RH problem.

functional inverses, the *Jacobi elliptic functions*. Like error functions, calculating elliptic integrals with quadrature has several issues, including the difficulty introduced by the singularities of the integrand at ± 1 and $\pm a$. Rephrasing g as a solution to an RH problem allows for the resolution of these difficulties.

To determine an RH problem for g, we first construct an RH problem for the integrand

$$\psi(z) = \frac{1}{\sqrt{1-z^2}\sqrt{a^2-z^2}} = \frac{1}{\sqrt{1-z}\sqrt{1+z}\sqrt{a-z}\sqrt{a+z}}.$$

The square root (with the principal branch) satisfies a multiplicative jump for $x<0$ of

$$\sqrt{x}^{+} = \mathrm{i}\sqrt{|x|} = -\sqrt{x}^{-}.$$

This implies that $\psi^+(x) = \psi^-(x)$ for $x > a$ and $x < -a$, i.e., $\psi(x)$ is continuous and hence analytic (by Theorem C.11) on $[a,\infty)$ and $(-\infty,-a]$. On the remaining branch cuts along $[-a,-1]$ and $[1,a]$, it satisfies the jump

$$\psi_+(x)+\psi_-(x)=0 \quad \text{for} \quad -a<x<-1 \quad \text{and} \quad 1<x<a. \tag{1.1}$$

Returning to g, since the singularities of ψ are integrable, for $x \geq 0$

$$g_\pm(x) = \int_0^x \psi_\pm(s)\mathrm{d}s = \begin{cases} \int_0^x \psi(s)\mathrm{d}s & \text{if } 0\leq x\leq 1,\\ \int_0^1 \psi(s)\mathrm{d}s + \int_1^x \psi_\pm(s)\mathrm{d}s & \text{if } 1\leq x\leq a,\\ \int_0^1 \psi(s)\mathrm{d}s + \int_1^a \psi_\pm(s)\mathrm{d}s + \int_a^x \psi(s)\mathrm{d}s & \text{if } a\leq x. \end{cases}$$

Define the *complete elliptic integral*[6]

$$K(\tau) \triangleq \int_0^1 \frac{1}{\sqrt{1-x^2}\sqrt{1-\tau^2x^2}}\mathrm{d}x \quad \text{for} \quad -1<\tau<1. \tag{1.2}$$

We use the cancellation of the integrand to determine, for $1\leq x\leq a$,

$$g_+(x)+g_-(x) = \int_0^x [\psi_+(s)+\psi_-(s)]\mathrm{d}s = 2\int_0^1 \psi(x)\mathrm{d}x = 2\frac{K(a^{-1})}{a}.$$

Similarly, the analyticity of ψ between $[0,1]$ and $[1,\infty)$ tells us, for $x>a$,

$$g_+(x)-g_-(x) = \int_0^x [\psi_+(s)-\psi_-(s)]\mathrm{d}s = 2\int_1^a \psi_+(x)\mathrm{d}x = 2\mathrm{i}\frac{K(\sqrt{1-a^{-2}})}{a},$$

where the final expression follows from applying the change of variables $x = \frac{1}{\sqrt{1-(1-a^{-2})t^2}}$ and simplifying. From the symmetry relationship $g(-z) = -g(z)$, we determine the equivalent jumps on the negative real axis. We arrive at an RH problem.

Problem 1.2.1. *Find g that satisfies the following:*

1. *$g(z)$ is analytic off $(-\infty,-1]\cup[1,\infty)$ and continuous up to the contours,*
2. *$g(z)$ is bounded throughout the complex plane,*

[6]We use the convention of [91, Chapter 19], which differs from *Mathematica*'s `EllipticK` routine: `EllitpicK`$(\tau^2)\equiv K(\tau)$.

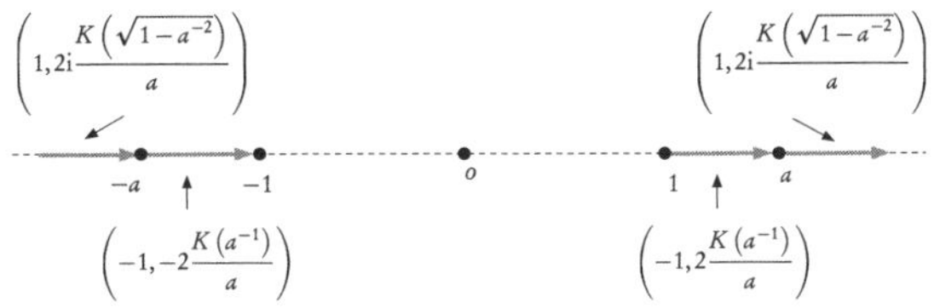

Figure 1.4. *The jump contours and jump functions for the Jacobi elliptic integral RH problem.*

3. $g(0) = 0$, *and*

4. *on the branch cuts it satisfies the jumps*

$$g_+(x) + g_-(x) = 2\frac{K(a^{-1})}{a} \quad \textit{for} \quad 1 < x < a,$$
$$g_+(x) + g_-(x) = -2\frac{K(a^{-1})}{a} \quad \textit{for} \quad -a < x < -1,$$
$$g_+(x) - g_-(x) = 2i\frac{K(\sqrt{1-a^{-2}})}{a} \quad \textit{for} \quad a < x,$$
$$g_+(x) - g_-(x) = 2i\frac{K(\sqrt{1-a^{-2}})}{a} \quad \textit{for} \quad x < -a.$$

See Figure 1.4 *for the jump contours and jump functions.*

We solve this RH problem numerically, giving an approximation to the elliptic integral that is accurate throughout the complex plane; see Section 5.5.

1.3 ▪ Airy function: From differential equation to Riemann–Hilbert problem

RH problems can also be seen as a way to recover solutions to differential equations from their asymptotic behavior. The canonical example is the *Airy equation*

$$Y''(z) = zY(z).$$

The Liouville–Green approximation (or Wentzel–Kramers–Brillouin approximation) informs us that along any given direction for which z approaches ∞, there are two linearly independent solutions that have the asymptotic behavior $z^{-1/4}e^{\pm\frac{2}{3}z^{3/2}}$. However, the well-known *Stokes phenomenon* says that the asymptotic behavior of a single solution changes depending on the sector of the complex plane in which z approaches ∞.

A particular canonical solution is the *Airy function*, which satisfies

$$\mathrm{Ai}(z) \sim \frac{1}{2\sqrt{\pi}} z^{-1/4} e^{-\frac{2}{3}z^{3/2}} \quad \text{for} \quad -\frac{2\pi}{3} \le \arg z \le \frac{2\pi}{3}$$

uniformly[7] as $z \to \infty$. By plugging $\mathrm{Ai}(\omega z)$ and $\mathrm{Ai}(\omega^2 z)$ into the ODE, we see that they are also solutions to the Airy equation for $\omega = e^{-\frac{2i\pi}{3}}$. We can deduce the asymptotics of

[7] This asymptotic behavior can be extended to $|\arg z| \le \pi - \delta$ for any δ; however, we only use the asymptotics in the stated sector.

these solutions from the asymptotics of the Airy function, with a bit of care taken due to the branch cut of the asymptotic formula:

$$\omega \mathrm{Ai}(\omega z) \sim -\frac{z^{-1/4}}{2\sqrt{\pi}} \begin{cases} \mathrm{i}\mathrm{e}^{\frac{2}{3}z^{3/2}} & \text{if } 0 \le \arg z \le \pi, \\ \mathrm{e}^{-\frac{2}{3}z^{3/2}} & \text{if } -\pi \le \arg z \le -\frac{2\pi}{3}, \end{cases}$$

$$\omega^2 \mathrm{Ai}(\omega^2 z) \sim \frac{z^{-1/4}}{2\sqrt{\pi}} \begin{cases} -\mathrm{e}^{-\frac{2}{3}z^{3/2}} & \text{if } \frac{2\pi}{3} \le \arg z \le \pi, \\ \mathrm{i}\mathrm{e}^{\frac{2}{3}z^{3/2}} & \text{if } -\pi \le \arg z \le 0. \end{cases}$$

We choose two linearly independent solutions in each sector to construct a sectionally analytic function

$$y(z) \triangleq \begin{cases} \left[-\omega \mathrm{Ai}(\omega z),\ -\mathrm{i}\omega^2 \mathrm{Ai}(\omega^2 z)\right] & \text{if } -\pi < \arg z < -\frac{2\pi}{3}, \\ \left[\mathrm{Ai}(z),\ -\mathrm{i}\omega^2 \mathrm{Ai}(\omega^2 z)\right] & \text{if } -\frac{2\pi}{3} < \arg z < 0, \\ \left[\mathrm{Ai}(z),\ \mathrm{i}\omega \mathrm{Ai}(\omega z)\right] & \text{if } 0 < \arg z < \frac{2\pi}{3}, \\ \left[-\omega^2 \mathrm{Ai}(\omega^2 z),\ \mathrm{i}\omega \mathrm{Ai}(\omega z)\right] & \text{if } \frac{2\pi}{3} < \arg z < \pi \end{cases}$$

so that

$$y(z) \sim \frac{1}{2\sqrt{\pi}} z^{-1/4} \left[\mathrm{e}^{-\frac{2}{3}z^{3/2}},\ \mathrm{e}^{\frac{2}{3}z^{3/2}}\right]$$

uniformly as $z \to \infty$ throughout the complex plane.

We now determine the jumps of y using the symmetry relationship

$$\mathrm{Ai}(z) + \omega \mathrm{Ai}(\omega z) + \omega^2 \mathrm{Ai}(\omega^2 z) = 0.$$

This relationship holds since $\mathrm{Ai}(z) + \omega \mathrm{Ai}(\omega z) + \omega^2 \mathrm{Ai}(\omega^2 z)$ also solves the Airy equation with zero initial conditions:

$$\begin{aligned} (1 + \omega + \omega^2)\mathrm{Ai}(0) &= \left(1 + 2\cos\frac{2\pi}{3}\right)\mathrm{Ai}(0) = 0, \\ (1 + \omega^2 + \omega^4)\mathrm{Ai}'(0) &= 0. \end{aligned}$$

It follows that y satisfies the following jumps:

$$\begin{aligned} y_+(x) &= y_-(x) \begin{bmatrix} 1 & -\mathrm{i} \\ 0 & 1 \end{bmatrix} && \text{for} \quad x \in (0, \infty), \\ y_+(s) &= y_-(s) \begin{bmatrix} 1 & 0 \\ -\mathrm{i} & 1 \end{bmatrix} && \text{for} \quad s \in (0, \mathrm{e}^{\frac{2\pi \mathrm{i}}{3}} \infty), \\ y_+(s) &= y_-(s) \begin{bmatrix} 1 & 0 \\ -\mathrm{i} & 1 \end{bmatrix} && \text{for} \quad s \in (0, \mathrm{e}^{-\frac{2\pi \mathrm{i}}{3}} \infty), \\ y_+(x) &= y_-(x) \begin{bmatrix} 0 & -\mathrm{i} \\ -\mathrm{i} & 0 \end{bmatrix} && \text{for} \quad x \in (-\infty, 0). \end{aligned}$$

We now remove the asymptotic behavior

$$\phi(z) \triangleq 2\sqrt{\pi} z^{1/4} y(z) \begin{bmatrix} \mathrm{e}^{\frac{2}{3}z^{3/2}} & 0 \\ 0 & \mathrm{e}^{-\frac{2}{3}z^{3/2}} \end{bmatrix}$$

so that $\phi(z) \sim [1,\ 1]$ as $z \to \infty$. We obtain the following vector RH problem.

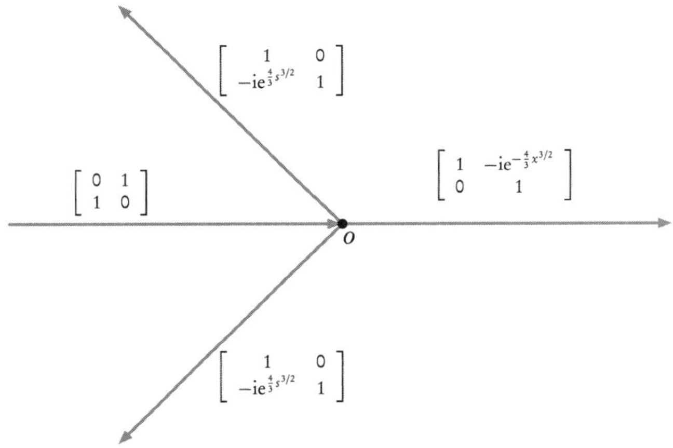

Figure 1.5. *The jump contour and jump functions for the Airy function RH problem.*

Problem 1.3.1. *Find* $\phi(z): \mathbb{C}\backslash\Gamma \to \mathbb{C}^{1\times 2}$, $\Gamma = [0,\infty)\cup(-\infty,0]\cup[0,e^{\frac{2i\pi}{3}}\infty)\cup[0,e^{-\frac{2i\pi}{3}}\infty)$, *that satisfies the following:*

1. $\phi(z)$ *is analytic off* Γ *and continuous up to* $\Gamma\setminus\{0\}$,
2. $\phi(z)$ *has weaker than pole singularities throughout the complex plane,*
3. $\phi(z)\sim[1,\ 1]$ *at* ∞, *and*
4. *on* Γ, ϕ *satisfies the jumps*

$$\phi_+(x) = \phi_-(x)\begin{bmatrix} 1 & -ie^{-\frac{4}{3}x^{3/2}} \\ 0 & 1 \end{bmatrix} \quad \textit{for} \quad x\in(0,\infty),$$
$$\phi_+(s) = \phi_-(s)\begin{bmatrix} 1 & 0 \\ -ie^{\frac{4}{3}s^{3/2}} & 1 \end{bmatrix} \quad \textit{for} \quad s\in(0,e^{\frac{2\pi i}{3}}\infty),$$
$$\phi_+(s) = \phi_-(s)\begin{bmatrix} 1 & 0 \\ -ie^{\frac{4}{3}s^{3/2}} & 1 \end{bmatrix} \quad \textit{for} \quad s\in(0,e^{-\frac{2\pi i}{3}}\infty),$$
$$\phi_+(x) = \phi_-(x)\begin{bmatrix} 0 & 1 \\ 1 & 0 \end{bmatrix} \quad \textit{for} \quad x\in(-\infty,0),$$

see Figure 1.5.

This vector RH problem is solvable numerically, giving a numerical method for calculating the Airy function that is accurate throughout the complex plane; see Section 6.3.

1.4 ▪ Monodromy

Suppose we are given a second-order *Fuchsian ODE*, i.e., $Y:\mathbb{C}\backslash\{z_1,\ldots,z_r\}\to\mathbb{C}^{2\times 2}$ satisfying

$$Y'(z) = A(z)Y(z) = \sum_{k=1}^{r}\frac{A_k}{z-z_k}Y(z),$$

where z_k are distinct and $A_k\in\mathbb{C}^{2\times 2}$. As a concrete example, we consider the case of three singular points $z_1=1$, $z_2=2$, and $z_3=3$ and take $Y(0)=I$. By integrating the differential

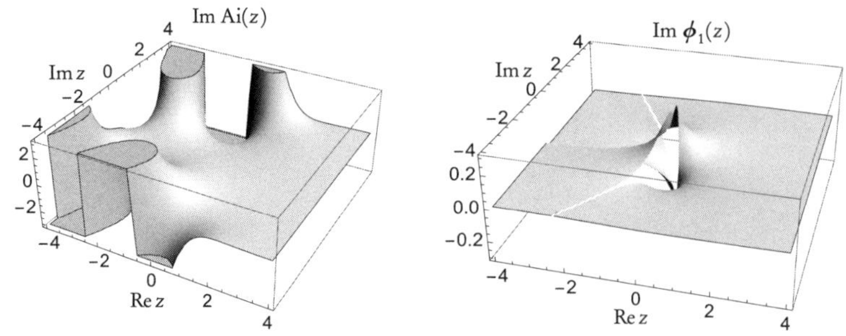

Figure 1.6. *The imaginary part of* Ai(z) *(left) and the imaginary part of* $\phi_1(z)$ *(right).*

equation along a contour that avoids the singularities $\{z_k\}$, we obtain a solution $Y(z)$ that is analytic off the contour $(1,\infty)$. For any x not equal to a singular point z_k, we define the limits from above and below:

$$Y_+(x) = \lim_{\epsilon\downarrow 0} Y(x+\mathrm{i}\epsilon) \qquad \text{and} \qquad Y_-(x) = \lim_{\epsilon\downarrow 0} Y(x-\mathrm{i}\epsilon).$$

These functions can also be considered as solutions to the differential equation obtained by integrating along arcs that avoid the singular points.

Note that $Y_+(x)$ and $Y_-(x)$ satisfy the same ODE as Y and have nonvanishing determinants; hence we know that the columns of $Y_+(x)$ are a linear combination of the columns of $Y_-(x)$. In other words, there is an invertible *monodromy matrix* $M_1 \in \mathbb{C}^{2\times 2}$ so that $Y_+(x) = Y_-(x)M_1$ for any $x \in (1,2)$. By analytic continuation, this relationship is satisfied for all $x \in (1,2)$. Similarly, for all $x \in (2,3)$ we have $Y_+(x) = Y_-(x)M_2$ and for all $x \in (3,\infty)$ we have $Y_+(x) = Y_-(x)M_3$.

Consider the special case where

$$\sum_{k=1}^{r} A_k = 0.$$

We find that $Y(z)$ is analytic at ∞, as verified by performing a change of variables $U(z) = Y(1/z)$ so that

$$\begin{aligned} U'(z) &= -\frac{1}{z^2}Y'(1/z) = -\frac{1}{z^2}\sum_{k=1}^{r}\frac{A_k}{z^{-1}-z_k}Y(1/z) = \sum_{k=1}^{r}\frac{A_k z^{-1}}{1-z_k z}U(z) \\ &= \sum_{k=1}^{r} A_k z^{-1}(1+z_k z+\mathcal{O}(z^2))U(z) = \sum_{k=1}^{r} A_k(z_k+\mathcal{O}(z))U(z) \end{aligned}$$

has a normal point at the origin. For our special case of three singular points, this implies that $Y_+(x) = Y_-(x)$ for all $x \in (3,\infty)$, and hence $M_3 = I$.

The map from $\{A_k\}$ to the monodromy matrices $\{M_k\}$ is known as the *Riemann–Hilbert correspondence*. We now consider the inverse map: recovering $\{A_k\}$ from $\{M_k\}$. Observe that the solution Y can be described by an RH problem given in terms of the following monodromy matrices.

Problem 1.4.1. *Find* $Y(z): \mathbb{C}\setminus[1,3] \to \mathbb{C}^{2\times 2}$ *that satisfies the following:*

1. $Y(z)$ *is analytic off* $[1,3]$ *and continuous up to* $(1,2)\cup(2,3)$,

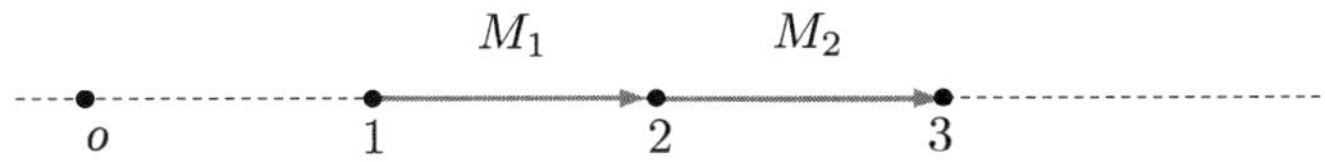

Figure 1.7. *The jump contour and jump functions for the inverse monodromy RH problem.*

2. *$Y(z)$ has weaker than pole singularities throughout the complex plane,*
3. *$Y(0)=I$, and*
4. *Y satisfies the jump*

$$\begin{aligned} Y_+(x) &= Y_-(x)M_1 \quad \textit{for} \quad x\in(1,2),\\ Y_+(x) &= Y_-(x)M_2 \quad \textit{for} \quad x\in(2,3), \end{aligned}$$

see Figure 1.7.

Hilbert's twenty-first problem essentially posed the question of whether one can uniquely recover $\{A_k\}$ from the monodromy matrices $\{M_k\}$. If we can solve this RH problem, then we know that

$$A(z)=Y'(z)Y(z)^{-1}.$$

If this is rational, then we can recover A_k from the Laurent series of A around each singular point. We solve this problem numerically in Section 6.4.

Remark 1.4.1. *Whether the RH problem is always solvable with $A(z)$ rational is a delicate question. For the case of three or fewer singular points, the problem is solvable in terms of hypergeometric functions. The case of four or more singular points is substantially more difficult to solve directly and requires* Painlevé transcendents; *see* [52, *p.* 80]. *In the general case, it was originally answered in the affirmative by Plemelj* [101]; *however, there was a flaw in the argument restricting its applicability to the case where at least one M_k is diagonalizable. Counterexamples were found by Bolibrukh* [15]; *see also* [6]. *On the other hand, setting up and numerically solving an RH problem is straightforward regardless of the number of singular points.*

1.5 ▪ Jacobi operators and orthogonal polynomials

RH problems can be used for *inverse spectral problems.* Our first example is the family of *Jacobi operators*

$$J=\begin{bmatrix} a_0 & \sqrt{b_1} & & & \\ \sqrt{b_1} & a_1 & \sqrt{b_2} & & \\ & \sqrt{b_2} & a_2 & \sqrt{b_3} & \\ & & \ddots & \ddots & \ddots \\ & & & & \end{bmatrix}. \tag{1.3}$$

We assume that J is endowed with a domain space so that it is self-adjoint with respect to the ℓ^2 inner product.[8] Self-adjointness ensures that $(J-z)^{-1}$ is bounded for all z off

[8] If J is bounded on ℓ^2, then J is self-adjoint on ℓ^2. In the unbounded case, J can be viewed as the self-adjoint extension of J_0, the restriction of the Jacobi operator to acting on the space of vectors with a finite number of nonzero entries.

the real axis, and the spectral theorem guarantees the existence of a *spectral measure* μ for J, a probability measure that satisfies

$$\mathbf{e}_0^\top (J-z)^{-1}\mathbf{e}_0 = \int_{\mathbb{R}} \frac{\mathrm{d}\mu(x)}{x-z} \quad \text{for} \quad z \notin \operatorname{supp}\mu \equiv \sigma(J),$$

where $\mathbf{e}_0 = [1,0,0,\ldots]^\top$ and $\sigma(J)$ denotes the spectrum of J. The *spectral map* is the map from the operator J to its spectral measure μ.

We now consider the *inverse spectral map*: recovering the operator J given its spectral measure. The key to this inverse spectral problem is the polynomials orthogonal with respect to the weight $\mathrm{d}\mu(x)$, i.e., orthogonal with respect to the inner product

$$\langle f, g\rangle_\mu = \int_{\mathbb{R}} f(x)g(x)\mathrm{d}\mu(x).$$

We see below that the entries of the operator J are embedded in the three-term recurrence relationship that the orthogonal polynomials satisfy.

Applying the Gram–Schmidt procedure to the sequence $\{1, x, x^2, \ldots\}$ produces a sequence of (monic) orthogonal polynomials[9]

$$\begin{aligned}\pi_{-1}(x) &= 0,\\ \pi_0(x) &= 1,\\ \pi_{n+1}(x) &= (x-\alpha_n)\pi_n(x) - \beta_n\pi_{n-1}(x).\end{aligned}$$

Here the coefficients α_n, β_n are those in the usual *three-term recurrence relation*, given by

$$\alpha_n = \langle x\pi_n, \pi_n\rangle_\mu \gamma_n \text{ and } \beta_n = \langle x\pi_n, \pi_{n-1}\rangle_\mu \gamma_{n-1}$$

for

$$\gamma_n = \langle \pi_n, \pi_n\rangle_\mu^{-1}.$$

It is easy to see that such coefficients exist using $\langle f, xg\rangle_\mu = \langle xf, g\rangle_\mu$. A remarkable fact, which we do not demonstrate here (see [27, p. 31]), is that $\alpha_n = a_n$ and $\beta_n = b_n$ for a large class of Jacobi operators.

Another remarkable fact is that these coefficients can be expressed in terms of a solution of an RH problem. Let $w(x)\mathrm{d}x = \mathrm{d}\mu(x)$, assuming the spectral measure has a continuous density. Then, consider the function

$$\begin{aligned}\Psi_n(z) &= \begin{bmatrix} \pi_n(z) & \mathcal{C}_{\mathbb{R}}[\pi_n w](z) \\ -2\pi\mathrm{i}\gamma_{n-1}\pi_{n-1}(z) & -2\pi\mathrm{i}\gamma_{n-1}\mathcal{C}_{\mathbb{R}}[\pi_{n-1}w](z)\end{bmatrix},\\ \mathcal{C}_{\mathbb{R}}f(z) &= \frac{1}{2\pi\mathrm{i}}\int_{\mathbb{R}} \frac{f(x)}{x-z}\mathrm{d}x,\end{aligned}$$

i.e., $\mathcal{C}_{\mathbb{R}}f(z)$ is the *Cauchy integral* of f. For f with sufficient smoothness, the Cauchy integral satisfies $\mathcal{C}^+_{\mathbb{R}}f(x) - \mathcal{C}^-_{\mathbb{R}}f(x) = f(x)$ for all $x \in \mathbb{R}$, where $\mathcal{C}^\pm_{\mathbb{R}}f(x)$ is the limit of $\mathcal{C}_{\mathbb{R}}f(z)$ from above and below, as described in Lemma 2.7 below. It follows immediately that $\Psi_n(z)$ satisfies the following jump on the real axis:

$$\Psi_n^+(x) = \Psi_n^-(x)\begin{bmatrix} 1 & w(x) \\ 0 & 1\end{bmatrix} \quad \text{for} \quad x \in \mathbb{R}.$$

[9]It can be shown that $\mathbf{e}_0^\top J^k \mathbf{e}_0 = \int_{\mathbb{R}} x^k \mathrm{d}\mu(x)$, implying that all moments are finite and guaranteeing the validity of the Gram–Schmidt procedure.

We can further use the orthogonality of π_k with every lower-degree polynomial to determine the asymptotic behavior as $z \to \infty$:

$$
\begin{aligned}
\mathscr{C}_{\mathbb{R}}[\pi_k w](z) &= \frac{1}{2\pi \mathrm{i}} \int_{\mathbb{R}} \frac{\pi_k(x)}{x-z} \mathrm{d}\mu(x) = -\frac{1}{2\pi \mathrm{i} z} \int_{\mathbb{R}} \pi_k(x) \left(1 + \frac{x}{z} + \frac{x^2}{z^2} + \cdots \right) \mathrm{d}\mu(x) \\
&= -\frac{1}{2\pi \mathrm{i} z^{k+1}} \int_{\mathbb{R}} \pi_k(x) x^k \left(1 + \frac{x}{z} + \frac{x^2}{z^2} + \cdots \right) \mathrm{d}\mu(x) \\
&\sim -\frac{\int_{\mathbb{R}} \pi_k(x) x^k \mathrm{d}\mu(x)}{2\pi \mathrm{i} z^{k+1}} = -\frac{\langle \pi_k, \pi_k \rangle_\mu}{2\pi \mathrm{i} z^{k+1}} = -\frac{1}{2\pi \mathrm{i} \gamma_k z^{k+1}}.
\end{aligned} \tag{1.4}
$$

We thus find that $\Psi_n(z)$ has the following asymptotic behavior:

$$
\Psi_n(z) \begin{bmatrix} z^{-n} & 0 \\ 0 & z^n \end{bmatrix} = I + \mathcal{O}(z^{-1}) \text{ as } |z| \to \infty.
$$

We can now state the inverse spectral problem in terms of the solution of an RH problem.

Problem 1.5.1. *Find* $\Psi \equiv \Psi_n : \mathbb{C} \setminus \mathbb{R} \to \mathbb{C}^{2\times 2}$ *that satisfies the following:*

1. $\Psi(z)$ *is analytic off* $\mathbb{R}$,
2. $\Psi(z)$ *has the asymptotic behavior*

$$
\Psi(z) \begin{bmatrix} z^{-n} & 0 \\ 0 & z^n \end{bmatrix} = I + \mathcal{O}(z^{-1}) \quad \text{as} \quad |z| \to \infty, \quad \text{and}
$$

3. $\Psi(z)$ *satisfies the jump*

$$
\Psi^+(x) = \Psi^-(x) \begin{bmatrix} 1 & w(x) \\ 0 & 1 \end{bmatrix}, \quad x \in \mathbb{R},
$$

see Figure 1.8.

With some technical assumptions on $w(x)$, this RH problem is uniquely solvable and we can recover the Jacobi operator J which has spectral measure $\mathrm{d}\mu(x) = w(x)\mathrm{d}x$. This is accomplished by expanding the solution near infinity:

$$
\Psi(z) = \left(I + \frac{Y_1}{z} + \frac{Y_2}{z^2} + \mathcal{O}(z^{-3})\right) \begin{bmatrix} z^n & 0 \\ 0 & z^{-n} \end{bmatrix}.
$$

The terms in this expansion can be determined from (1.4), from which we compute

$$
(Y_1)_{12}(Y_1)_{21} = \frac{\gamma_{n-1}}{\gamma_n} = \langle \pi_n, \pi_n \rangle_\mu \gamma_{n-1} = \langle \pi_n, x\pi_{n-1} \rangle_\mu \gamma_{n-1} = \beta_n
$$

and

$$
\begin{aligned}
\frac{(Y_2)_{12}}{(Y_1)_{12}} - (Y_1)_{22} &= \gamma_n \langle x^{n+1}, \pi_n \rangle_\mu - \gamma_{n-1} \langle x^n, \pi_{n-1} \rangle_\mu \\
&= \gamma_n \langle x^n, \pi_{n+1} + \alpha_n \pi_n + \beta_n \pi_{n-1} \rangle_\mu - \gamma_{n-1} \langle x^n, \pi_{n-1} \rangle_\mu \\
&= \alpha_n + \gamma_n \beta_n \langle x^n, \pi_{n-1} \rangle_\mu - \gamma_{n-1} \langle x^n, \pi_{n-1} \rangle_\mu = \alpha_n.
\end{aligned}
$$

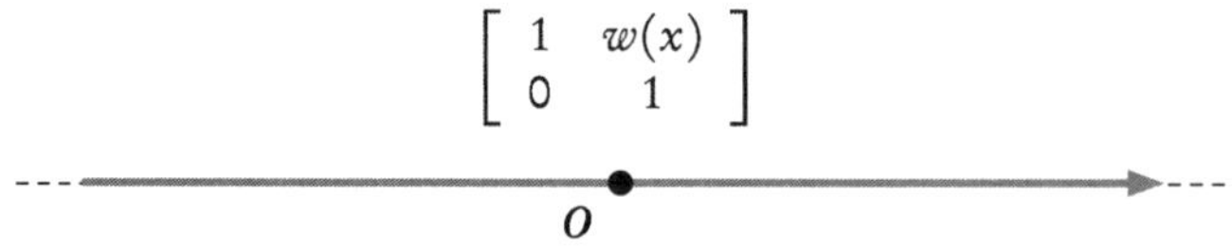

Figure 1.8. *The jump contour and jump functions for the orthogonal polynomial RH problem.*

While this RH problem has differing asymptotic conditions from those of our model RH problem (Problem 1.0.2), it can be reduced to the required form using the so-called *equilibrium measure* arising in potential theory; see [27].

Remark 1.5.1. *The special case of solving this RH problem when $n = 0$ is considered in Example* 2.19.

Remark 1.5.2. *The numerical Riemann–Hilbert approach developed in this book can be adapted to this setting to compute this inverse spectral map in specific cases* [97, 116].

1.6 ▪ Spectral analysis of Schrödinger operators

Our second example of an inverse spectral problem is for the time-independent *Schrödinger operator*

$$Lu = -\frac{d^2 u}{dx^2} - V(x)u, \tag{1.5}$$

where V is, for simplicity, smooth with exponential decay.[10] One initially considers L as acting on smooth, rapidly decaying functions on $\mathbb{R}$. With an appropriate extension, as in the case of Jacobi operators, L becomes self-adjoint: one replaces strong with weak differentiation and has L act on the Sobolev space $H^2(\mathbb{R})$. This particular example is of great importance for integrable systems.

We see that, similar to the spectral map for Jacobi operators, there exists a spectral map from the Schrödinger operator to *spectral data* or *scattering data*. RH problems can then be used for the *inverse spectral map*, recovering the potential V from the operators' spectral data. Details on the spectrum of Schrödinger operators can be found in [69, Chapter 5] (see also [31]).

The determination of the spectral data starts with the free Schrödinger operator $-d^2/dx^2$, whose (essential) spectrum is the positive real axis $[0,\infty)$.[11] The operator L is a relatively compact perturbation of $-d^2/d^2x$; hence the spectrum of L must only differ from $-d^2/d^2x$ by the possible addition of discrete eigenvalues lying in $(-\infty, 0)$. The assumptions made on V are sufficient to ensure that there are only a finite number of eigenvalues $\{\lambda_j\}_{j=1}^n$ [31].

The associated spectral data is defined on the spectrum, by considering the solutions to equation $Lu = \lambda u$.[12] The solution space of the free equation is spanned by $e^{\pm isx}$, $s^2 = \lambda$. Thus, because V decays rapidly, we expect the solutions of (1.5) to be approximately equal

[10] Assuming that $|V(x)|$ is integrable and its first moment is finite is sufficient for most of the results we state here [31].

[11] This can be verified by considering the Fourier transform of the operator.

[12] Generally speaking, to rigorously establish the facts we state here one proceeds with the analysis of an integral equation [2, Chapter 2] (see also Chapter 3 of this book for a detailed analysis of similar integral equations).

to a linear combination of these free solutions asymptotically. Define four solutions of $Lu = \lambda u$ for $s \in \mathbb{R}$ by

$$\psi_{\mathrm{p}}(x;s) \sim e^{+isx}(1+o(1)) \text{ as } x \to +\infty,$$
$$\phi_{\mathrm{p}}(x;s) \sim e^{+isx}(1+o(1)) \text{ as } x \to -\infty,$$
$$\psi_{\mathrm{m}}(x;s) \sim e^{-isx}(1+o(1)) \text{ as } x \to +\infty,$$
$$\phi_{\mathrm{m}}(x;s) \sim e^{-isx}(1+o(1)) \text{ as } x \to -\infty.$$

As there can only be two linearly independent solutions of a second-order differential equation, there must exist an x-independent *scattering matrix*[13] $S(s)$ such that

$$\begin{bmatrix} \psi_{\mathrm{p}}(x;s) & \psi_{\mathrm{p}}'(x;s) \\ \psi_{\mathrm{m}}(x;s) & \psi_{\mathrm{m}}'(x;s) \end{bmatrix} = S(s) \begin{bmatrix} \phi_{\mathrm{p}}(x;s) & \phi_{\mathrm{p}}'(x;s) \\ \phi_{\mathrm{m}}(x;s) & \phi_{\mathrm{m}}'(x;s) \end{bmatrix}. \tag{1.6}$$

Because V is real, we have two further properties, similar to the symmetries of the Fourier transform of a real-valued function, that will prove useful. Complex conjugacy commutes with L, and hence we have a conjugacy relationship between the solutions:

$$\psi_{\mathrm{p}}(x;s) = \overline{\psi_{\mathrm{m}}(x;s)} \quad \text{and} \quad \phi_{\mathrm{p}}(x;s) = \overline{\phi_{\mathrm{m}}(x;s)}.$$

Furthermore, by direct substitution into the differential equation we have the symmetry relationship with respect to negating the spectral variable:

$$\psi_{\mathrm{p}}(x;s) = \psi_{\mathrm{m}}(x;-s) \quad \text{and} \quad \phi_{\mathrm{p}}(x;s) = \phi_{\mathrm{m}}(x;-s).$$

The determinants of the matrices in (1.6) can be expressed in terms of the *Wronskian* $W(f,g) = fg' - gf'$, which we use to simplify the definition of S. Indeed, *Abel's formula* indicates that the Wronskian of any two solutions must be independent of x. From the asymptotic behavior of the solutions, assuming the expansion can be differentiated, one can deduce $W(\psi_{\mathrm{p}}, \psi_{\mathrm{m}}) = -2is = W(\phi_{\mathrm{p}}, \phi_{\mathrm{m}})$. Then $S(s)$ is expressed as

$$S(s) = \begin{bmatrix} A(s) & B(s) \\ b(s) & a(s) \end{bmatrix} = -\frac{1}{2is} \begin{bmatrix} \psi_{\mathrm{p}}(x;s) & \psi_{\mathrm{p}}'(x;s) \\ \psi_{\mathrm{m}}(x;s) & \psi_{\mathrm{m}}'(x;s) \end{bmatrix} \begin{bmatrix} \phi_{\mathrm{m}}'(x;s) & -\phi_{\mathrm{p}}'(x;s) \\ -\phi_{\mathrm{m}}(x;s) & \phi_{\mathrm{p}}(x;s) \end{bmatrix} \tag{1.7}$$

so that each element is again a Wronskian divided by $-2is$. The scattering matrix encodes the spectral data associated with the continuous spectrum of L; in particular, we have the *reflection coefficient* $\rho(s) \triangleq b(s)/a(s)$. The conjugacy relationship of the solutions ensures that $A(s) = \overline{a}(s)$ and $B(s) = \overline{b}(s)$; therefore $\overline{\rho}(s) = \frac{B(s)}{A(s)}$.

We must also determine the spectral data associated with the discrete spectrum. Consider the $(2,2)$ element of $S(s)$:

$$a(s) = -\frac{W(\psi_{\mathrm{m}}, \phi_{\mathrm{p}})}{2is}.$$

Rephrasing these solutions to a second-order differential equation in terms of solutions to Volterra integral equations, one shows that ψ_{p} and ϕ_{m} can be analytically continued

[13]Technically speaking, this matrix only exists if the matrices in (1.6) have nonzero determinants. This follows from Wronksian considerations below.

in the s variable throughout the lower-half plane while ψ_{m} and ϕ_{p} can be analytically continued in the s variable throughout the upper-half plane. This latter fact indicates that $a(z)$ is analytic in the upper-half plane. Now, assume that for discrete values $z_1,\ldots,z_n$ in the upper-half plane, ψ_{p} is a multiple of ϕ_{m}, both of which must decay exponentially with respect to x in their respective directions of definition. The conclusion is that ψ_{p} must be an $L^2(\mathbb{R})$ eigenfunction of L and $z_j^2=\lambda_j$ is an eigenvalue of L. Furthermore, $a(z)$ must vanish at each such $L^2(\mathbb{R})$ eigenvalue and these zeros can be shown to be simple [2, p. 78]. The discrete spectral data are then given by the *norming constants* $c_j=b(z_j)/a'(z_j)$.

Remark 1.6.1. *It is important to note that b is defined on $\mathbb{R}$. It may or may not have an analytic continuation into the upper-half plane. When we define $b(z_j)$ we are not referring to the function on the real axis or its analytic continuation. We are referring to the proportionality constant: $\psi_{\mathrm{p}}(x;z_j)=b(z_j)\phi_{\mathrm{m}}(x;z_j)$. In this way, $b(z_j)$ could be viewed as an abuse of notation because one could have $b=0$ on $\mathbb{R}$ but $b(z_j)\neq 0$.*

We now consider the problem of recovering the potential V from the *spectral data* ρ, $\{c_j\}$, and $\{\lambda_j\}$. We accomplish this task through the analyticity properties of $\psi_{\mathrm{p/m}}(x;s)$ and $\phi_{\mathrm{p/m}}(x;s)$ with respect to the spectral variable s. Before deriving the correct construction, we investigate an approach that fails: rearranging (1.6) only in terms of the analyticity properties of the solutions. This gives us the jump

$$\begin{aligned}\begin{bmatrix}\psi_{\mathrm{m}}(x;s) & \phi_{\mathrm{p}}(x;s)\end{bmatrix} &= \begin{bmatrix}\phi_{\mathrm{m}}(x;s) & \psi_{\mathrm{p}}(x;s)\end{bmatrix}\tilde{G}(s), \quad s\in\mathbb{R},\\ \text{for}\quad \tilde{G}(s) &= \begin{bmatrix} a(s)-\dfrac{B(s)b(s)}{A(s)} & -\dfrac{B(s)}{A(s)}\\ \dfrac{b(s)}{A(s)} & \dfrac{1}{A(s)}\end{bmatrix}.\end{aligned} \tag{1.8}$$

The left-hand side of this equation is analytic in the upper-half plane while the vector on the right-hand side is analytic in the lower-half plane. For fixed x, it can be shown that

$$\begin{aligned}\left[\psi_{\mathrm{m}}(x;z),\ \phi_{\mathrm{p}}(x;z)\right]\begin{bmatrix}e^{izx} & 0\\ 0 & e^{-izx}\end{bmatrix} &= [1,\ 1]+\mathcal{O}(1/z),\\ \left[\phi_{\mathrm{m}}(x;z),\ \psi_{\mathrm{p}}(x;z)\right]\begin{bmatrix}e^{izx} & 0\\ 0 & e^{-izx}\end{bmatrix} &= [1,\ 1]+\mathcal{O}(1/z)\end{aligned}$$

as $|z|\to\infty$ in their respective domains of analyticity. This is close to a valid RH problem. The main issue here is that $\det\tilde{G}(s)=a(s)/A(s)$ does not generically have a zero winding number[14] which violates our rule of thumb, and therefore we do not expect the problem to have a (unique) solution.

Instead, consider

$$\left[\frac{\psi_{\mathrm{m}}(x;s)}{a(s)},\ \phi_{\mathrm{p}}(x;s)\right]=\left[\phi_{\mathrm{m}}(x;s),\ \frac{\psi_{\mathrm{p}}(x;s)}{A(s)}\right]G(s), \quad s\in\mathbb{R}, \tag{1.9}$$

[14]This can be demonstrated by considering that $a(s)$ and $A(s)$ are analytic functions in the upper- and lower-half planes, respectively. Furthermore, $a(s)$ has zeros at $\{z_j\}$ and $A(s)$ at $\{\bar{z}_j\}$. Both functions have poles at $s=0$, and this must be taken into account when computing the winding number with the argument principle.

and our task is to determine G. Straightforward algebra, combined with the definition of ρ, demonstrates

$$G(s) = \begin{bmatrix} 1 - \dfrac{b(s)}{a(s)}\dfrac{B(s)}{A(s)} & -\dfrac{B(s)}{A(s)} \\ \dfrac{b(s)}{a(s)} & 1 \end{bmatrix} = \begin{bmatrix} 1-\rho(s)\bar{\rho}(s) & -\bar{\rho}(s) \\ \rho(s) & 1 \end{bmatrix},$$

which has a unit determinant, and therefore no winding number issues are present for $\det G(s)$. The division by $a(s)$ and $A(s)$ forces us to consider a sectionally *meromorphic* vector-valued function, depending parametrically on x:

$$\Phi(z) \equiv \Phi(x;z) \triangleq \begin{cases} \begin{bmatrix} \dfrac{\psi_{\mathrm{m}}(x;z)}{a(z)}, \ \phi_{\mathrm{p}}(x;z) \end{bmatrix} \begin{bmatrix} e^{izx} & 0 \\ 0 & e^{-izx} \end{bmatrix} & \text{if } z \in \mathbb{C}^+, \\ \begin{bmatrix} \phi_{\mathrm{m}}(x;z), \ \dfrac{\psi_{\mathrm{p}}(x;z)}{A(z)} \end{bmatrix} \begin{bmatrix} e^{izx} & 0 \\ 0 & e^{-izx} \end{bmatrix} & \text{if } z \in \mathbb{C}^- \end{cases} \tag{1.10}$$

so that for $s \in \mathbb{R}$

$$\Phi^+(s) = \Phi^-(s) \begin{bmatrix} e^{-isx} & 0 \\ 0 & e^{isx} \end{bmatrix} G(s) \begin{bmatrix} e^{isx} & 0 \\ 0 & e^{-isx} \end{bmatrix}, \quad \Phi(\infty) = [1, \ 1]. \tag{1.11}$$

Because Φ has poles on the imaginary axis corresponding to $L^2(\mathbb{R})$ eigenvalues of L (the points where a vanishes) we must impose residue conditions, as we must dictate the behavior of Φ at *all* of its singularities. Given $z_j \in i\mathbb{R}^+$ corresponding to an eigenvalue ($z_j^2 = \lambda_j$), we compute

$$\begin{aligned} \operatorname{Res}_{z=z_j} \Phi(z) &= \begin{bmatrix} \dfrac{\psi_{\mathrm{m}}(x;z_j)}{a'(z_j)} e^{iz_j x}, \ 0 \end{bmatrix} = \begin{bmatrix} \dfrac{b(z_j)\phi_{\mathrm{p}}(x;z_j)}{a'(z_j)} e^{iz_j x}, \ 0 \end{bmatrix} \\ &= \lim_{z \to z_j} \Phi(z) \begin{bmatrix} 0 & 0 \\ c_j e^{2iz_j x} & 0 \end{bmatrix}, \end{aligned} \tag{1.12}$$

where c_j are precisely the norming constants of the spectral data. Because $\psi_{\mathrm{p/m}}(x;-s) = \psi_{\mathrm{m/p}}(x;s)$ and $\phi_{\mathrm{p/m}}(x;-s) = \phi_{\mathrm{m/p}}(x;s)$, we have $a(-z) = A(z)$, and hence $\Phi(z)$ satisfies

$$\Phi(-z) = \Phi(z) \begin{bmatrix} 0 & 1 \\ 1 & 0 \end{bmatrix}.$$

Residue conditions may be obtained in the lower-half plane from this relationship.

We arrive at the following inverse spectral problem.[15]

Problem 1.6.1. *Given data* $\rho(s), \{z_j\}_{j=1}^n$, *and* $\{c_j\}_{j=1}^n$, *find* $\Phi : \mathbb{C} \setminus (\mathbb{R} \cup \{z_j\}_{j=1}^n \cup \{\bar{z}_j\}_{j=1}^n) \to \mathbb{C}^{1\times 2}$ *that satisfies the following:*

1. $\Phi(z)$ *is meromorphic in* $\mathbb{C} \setminus \mathbb{R}$,

$$\Phi^+(s) = \Phi^-(s) \begin{bmatrix} 1-\rho(s)\bar{\rho}(s) & -\bar{\rho}(s)e^{-2isx} \\ \rho(s)e^{2isx} & 1 \end{bmatrix}, \quad s \in \mathbb{R}, \quad \Phi(\infty) = [1, \ 1],$$

for where ρ *satisfies* $\rho(-s) = \bar{\rho}(s)$, $|\rho(s)| < 1$ *for* $s \neq 0$,

[15] In the language of integrable systems, solving this RH problem is called *inverse scattering*.

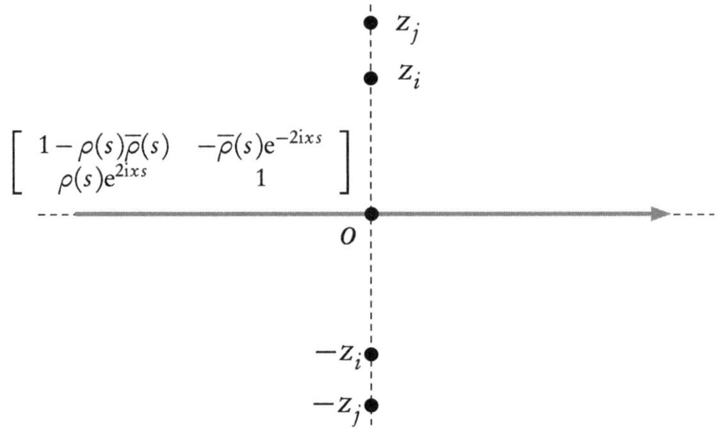

Figure 1.9. *The jump contour and jump functions for the Schrödinger inverse spectral RH problem.*

2. $\Phi(z)$ *has a finite number of specified poles* $\{z_j\}_{j=1}^n$, $\{-z_j\}_{j=1}^n$, *for* $z_j = \mathrm{i}\sqrt{-\lambda_j}$, $\lambda_j < 0$, *on the imaginary axis where it satisfies the given residue conditions*

$$\operatorname{Res}_{z=z_j} \Phi(z) = \lim_{z\to z_j} \Phi(z)\begin{bmatrix} 0 & 0 \\ c_j e^{2\mathrm{i}z_j x} & 0 \end{bmatrix},$$
$$\operatorname{Res}_{z=-z_j} \Phi(z) = \lim_{z\to -z_j} \Phi(z)\begin{bmatrix} 0 & -c_j e^{2\mathrm{i}z_j x} \\ 0 & 0 \end{bmatrix}, \quad c_j \in \mathrm{i}\mathbb{R}^+,$$

and

3. $\Phi(z)$ *satisfies the (essential) symmetry condition*

$$\Phi(-z) = \Phi(z)\begin{bmatrix} 0 & 1 \\ 1 & 0 \end{bmatrix}.$$

See Figure 1.9 *for this RH problem.*

If we can solve this RH problem, we can recover the potential $V(x)$ by

$$V(x) = 2\mathrm{i} \lim_{z\to\infty} z\partial_x \Phi_1(x;z). \tag{1.13}$$

Indeed, for $z \in \mathbb{C}^+$

$$\Phi_1^-(x;z) = \phi_{\mathrm{m}}(x;z)e^{\mathrm{i}zx} = 1 - \frac{1}{2\mathrm{i}z}\int_{-\infty}^{x} \left[1 - e^{2\mathrm{i}z(x-\tau)}\right] V(\tau)\Phi_1^-(\tau;z)\mathrm{d}\tau,$$

where the integral representation can be verified by substitution of $\phi_{\mathrm{m}}(x;z) = \Phi_1^-(x;z)e^{-\mathrm{i}zx}$ into the Schrödinger equation, $Lu = z^2 u$. Therefore,

$$\partial_x \Phi^-(z)_1 = -e^{2\mathrm{i}zx}\int_{-\infty}^{x} e^{-2\mathrm{i}z\tau} V(\tau)\Phi_1^-(\tau;z)\mathrm{d}\tau \sim \frac{\Phi_1^-(x;z)V(x)}{2\mathrm{i}z} \sim \frac{V(x)}{2\mathrm{i}z} \quad \text{as} \quad z\to\infty,$$

using integration by parts and the fact that $\Phi_1^-(x;z) \to 1$ as $z \to \infty$. Similar relations follow for Φ^+.

It can be shown that $\Phi(z)$ is uniquely specified by the above RH problem. Most importantly, we have characterized the operator $-\mathrm{d}^2/\mathrm{d}^2x - V(x)$ uniquely in terms of $\rho(z)$ defined on the essential spectrum and norming constants $\{c_j\}$ defined on the discrete spectrum $\{\lambda_j\}$. In this sense ρ, $\{c_j\}$, and $\{\lambda_j\}$ constitute the spectral data in the spectral analysis of Schrödinger operators. This procedure is critical in the solution of the Korteweg–de Vries equation with the inverse scattering transform, as is discussed in great detail in Chapter 8.

Remark 1.6.2. *In the presentation we have ignored some technical details. One of these is the boundary behavior of $\psi_{\mathrm{m}}(x;z)/a(z)$ as z approaches the real axis. Specifically, one needs to show that $\Phi(z)$ as defined in* (1.10) *is in an appropriate Hardy space (see Section* 2.5*). Additionally, we used implicitly that $a(z)$ and $A(z)$ limit to unity for large z. These details can be established rigorously; see* [31].

Chapter 2

Riemann–Hilbert Problems

This chapter is a comprehensive introduction to the applied theory of RH problems. Other standard references on the subject include [3, 18, 27, 77, 123] and the recent paper of Lenells [75]. On first reading, many proofs may be skipped.

In the next section, we make the definition of an RH problem precise. We then discuss a fundamental tool in the study of RH problems: Cauchy integrals of Hölder continuous functions. This provides a class of functions for which the Cauchy integral can be defined in the entire complex plane, including limits to the contour of integration. The importance of this class is that it allows for the explicit solution of scalar RH problems. From here we describe a restricted class of matrix RH problems that can be solved explicitly. The class of RH problems that we can handle explicitly is much too restrictive to address the RH problems that appear in Chapter 1 and Part III. For this reason, we turn to a general theory of solvability that begins with a description of Hardy spaces and general properties of the Cauchy integral. Our full development requires piecewise-smooth Lipschitz contours. The theory can be developed in a much more general setting, but we restrict our attention so that nearly all the developments can be made explicit here.

We build up to singular integral equations on the so-called zero-sum Sobolev spaces and describe precise conditions that can be checked to determine the unique solvability of RH problems. We concentrate on results concerning the smoothness of solutions because this is of critical importance in the numerical analysis of such problems. Because it is necessary for using RH problems in applications, the chapter concludes with techniques for the deformation of RH problems.

2.1 ▪ Precise statement of a Riemann–Hilbert problem

Recall the preliminary formulation of an RH problem in Problem 1.0.2, which poses the problem of finding a piecewise-analytic function $\Phi : \mathbb{C}\backslash\Gamma \to \mathbb{C}^{n\times m}$, which we often refer to as *sectionally analytic*, that is discontinuous across a contour $\Gamma \subset \mathbb{C}$ with the *jump condition*

$$\Phi^+(s) = \Phi^-(s)G(s) + F(s) \quad \text{for} \quad s \in \Gamma,$$

where Φ^+ denotes the limit from the left and Φ^- denotes the limit from the right. We make the definition of $\Phi^\pm$ precise by assuming that Γ is a complete contour.[16] Together the pair (G, F) are the *jump functions* and G is called the *jump matrix*. We call the RH problem *homogeneous* when $F = 0$.

[16] We remark that there is a fundamental dependence of this definition on the Jordan curve theorem.

Definition 2.1. *Γ is said to be a* complete contour *if Γ can be oriented in such a way that $\mathbb{C}\setminus\Gamma$ can be decomposed into left and right components: $\mathbb{C}\setminus\Gamma = \Omega_+ \cup \Omega_-$, $\Omega_+ \cap \Omega_- = \varnothing$, and Ω_+ (Ω_-) lies to the left (right) of Γ; see Figure 2.1.*

Often, it is beneficial to decompose $\Gamma = \Gamma_1 \cup \cdots \cup \Gamma_L$ so that each Γ_i is smooth and non-self-intersecting. The incomplete contours we encounter can be augmented with additional contours so that they become complete without modifying the solution to the RH problem; see Section 2.8.2.

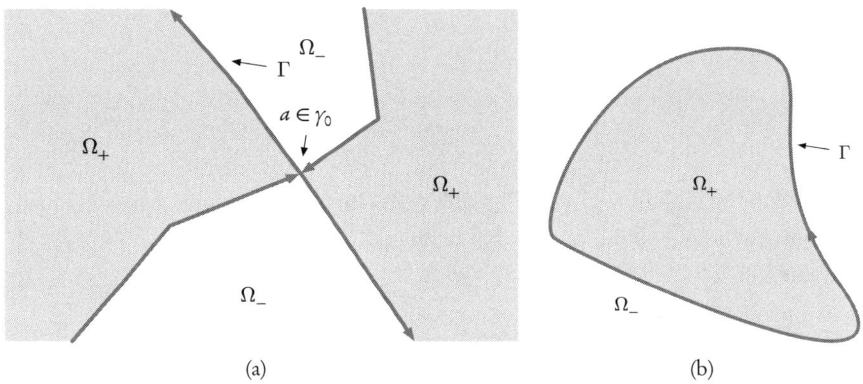

Figure 2.1. (a) *A complete self-intersecting contour Γ with $\Omega_\pm$ labeled.* (b) *A complete non-self-intersecting contour.*

We use the disjoint components $\Omega_\pm$ to define the left and right boundary values pointwise:

$$\Phi_+(s) \equiv \Phi^+(s) \triangleq \lim_{\substack{z\to s \\ z\in\Omega_+}} \Phi(z), \quad \Phi_-(s) \equiv \Phi^-(s) \triangleq \lim_{\substack{z\to s \\ z\in\Omega_-}} \Phi(z).$$

We always assume that z approaches s along a curve that is not tangential to Γ, though this condition can be relaxed whenever $\Phi^\pm$ are continuous at s.

To make the statement of the jump condition

$$\Phi^+(s) = \Phi^-(s)G(s) + F(s)$$

precise, one must specify what is meant by $\Phi^\pm$ as functions on Γ: the pointwise definition is insufficient, as it may not be defined for all $s \in \Gamma$. There are many analogies between RH problems and differential equations, and this here is analogous to when one must specify classical versus weak derivatives. Two common requirements for $\Phi^\pm$ are that

1. $\Phi^\pm(s)$ should exist at every interior point of the contour and be continuous functions except at endpoints of Γ where they should be locally integrable, or

2. $\Phi^\pm(s)$ should exist almost everywhere (with respect to Lebesgue arc length measure) and be in an appropriate L^p space.

The first case is that of a *continuous RH problem* and the second is that of an *L^p RH problem*. The definition used might affect the possibility of solving a specific RH problem. In practice, many difficult problems are reduced to RH problems and requisite conditions on $\Phi^\pm$ fall out of the reduction process.

In addition to boundary behavior, we often specify the behavior of Φ at some point in the complex plane, usually at ∞. A function Φ is of *finite degree* at infinity if

$$\limsup_{|z|\to\infty} |z|^{-n}|\Phi(z)| < \infty \quad \text{for some } n.$$

We require that for a function to solve an RH problem it must be of finite degree at infinity. For matrix RH problems, $|\diamond|$ denotes a matrix norm; see Appendix A.3.

In practice, for continuous RH problems, we specify some terms in the asymptotic series[17] of Φ at ∞, say $\Phi(z) = I + o(1)$, or $\Phi(z) = I + C/z + o(z^{-1})$. In the former case we write $\Phi(\infty) = I$. These conditions share an analogy with boundary/initial conditions in the context of differential equations. We use uniform convergence, i.e.,

$$\Phi(z) = p(z) + o(z^{-n}) \text{ as } z \to \infty \Leftrightarrow \lim_{R\to\infty} \sup_{|z|=R} |\Phi(z) - p(z)||z|^n = 0.$$

Certainly when Γ is bounded, and in many other cases, we are able to write $\Phi(z) = I + \mathcal{O}(z^{-1})$, or $\Phi(z) = I + C/z + \mathcal{O}(z^{-2})$, where

$$\Phi(z) = p(z) + \mathcal{O}(z^{-n}) \text{ as } z \to \infty \Leftrightarrow \limsup_{R\to\infty} \sup_{|z|=R} |\Phi(z) - p(z)||z|^n < \infty.$$

When considering an L^2 RH problem we write $\Phi(\infty) = I$ if $\Phi - I$ is in an appropriate Hardy space; see Section 2.5.

Remark 2.1.1. *In what follows we often pose an RH problem by only stating its jump condition and normalization. The region of analyticity is assumed to be the complement of the jump contour.*

2.2 ▪ Hölder theory of Cauchy integrals

The fundamental object of study in the theory of RH problems is the Cauchy integral. Given an oriented contour Γ and a function $f : \Gamma \to \mathbb{C}$, the *Cauchy integral* is defined by

$$\mathcal{C}_\Gamma f(z) = \int_\Gamma \frac{f(s)}{s-z} \text{đ}s, \quad \text{đ}s = \frac{\mathrm{d}s}{2\pi \mathrm{i}}. \tag{2.1}$$

The Cauchy integral maps functions on a contour to analytic functions off the contour. We shall see later that under specific regularity conditions these functions can be put into a one-to-one correspondence. In this way, Cauchy integrals are critical in the solution of RH problems from both a numerical and an analytical perspective.

As in the precise statement of an RH problem, we must understand the limiting values of (2.1), specifically issues related to existence and regularity. We describe a class of functions for which the Cauchy integral has nice properties.

Definition 2.2. *Given $\Omega \subset \mathbb{C}$, a function $f : \Omega \to \mathbb{C}$ is α-Hölder continuous on Ω if for each $s^* \in \Omega$, there exist $\Lambda(s^*)$, $\delta(s^*) > 0$ such that*

$$|f(s) - f(s^*)| \le \Lambda(s^*)|s - s^*|^\alpha, \textit{ for } |s - s^*| < \delta(s^*) \textit{ and } s \in \Omega.$$

Note that this definition is useful when $\alpha \in (0, 1]$. If $\alpha = 1$, f is Lipschitz and if $\alpha > 1$, f must be constant.

[17] We also impose the following entrywise if Φ is a matrix-valued function.

Definition 2.3. *A function $f : \Omega \to \mathbb{C}$ is uniformly α-Hölder continuous on a set Ω if Λ and δ can be chosen independently of s^*.*

A matrix- or vector-valued function is said to be α-Hölder continuous if it is entrywise α-Hölder continuous.

Lemma 2.4. *Each uniformly α-Hölder continuous function on a bounded curve Γ with corresponding constants δ and Λ satisfies*

$$\sup_{s_1 \neq s_2,\, s_1, s_2 \in \Gamma} \left\{ \frac{|f(s_1)-f(s_2)|}{|s_1 - s_2|^\alpha} \right\} < C\Lambda < \infty,$$

where C depends only on δ and Γ.

Proof. It suffices to show that $|f(s_1)-f(s_2)| \leq C\Lambda|s_1 - s_2|^\alpha$ for any choice of s_1, s_2. We select a uniform grid on Γ: $\{p_1, p_2, \ldots, p_N\}$ such that $|p_i - p_j| \geq \delta/2$ for $i \neq j$ and $|p_i - p_{i+1}| = \delta/2$. We assume Γ is oriented from s_1 to s_2. Let p_i be the first element of the partition after s_1 and p_j be the first before s_2. We assume $|s_1 - s_2| \geq \delta$; then

$$\begin{aligned}\frac{|f(s_1)-f(s_2)|}{|s_1 - s_2|^\alpha} &\leq \frac{|f(s_1)-f(p_i)|}{|s_1 - s_2|^\alpha} + \frac{|f(p_i)-f(p_{i+1})|}{|s_1 - s_2|^\alpha} + \cdots + \frac{|f(p_j)-f(s_2)|}{|s_1 - s_2|^\alpha} \\ &\leq \frac{|f(s_1)-f(p_i)|}{|s_1 - p_i|^\alpha} + \frac{|f(p_i)-f(p_{i+1})|}{|p_i - p_{i+1}|^\alpha} + \cdots + \frac{|f(p_j)-f(s_2)|}{|p_j - s_2|^\alpha} \leq N\Lambda. \qquad \square\end{aligned}$$

Remark 2.2.1. *We use a definition of uniformly Hölder functions that differs slightly from the classical definition. Classically, the conclusion of Lemma 2.4 is used as a definition. The results stated below are necessarily local, which makes this definition more convenient.*

For differentiable functions $f(t)$ and $g(t)$ with the same domain, we say that $|\mathrm{d}g| \leq |\mathrm{d}f|$ if $|f'(t)| \leq |g'(t)|$ for all t in the domain of f and g. We also say a curve is smooth if it is parameterized by an infinitely differentiable function with nonvanishing derivative. It will be clear that a finite amount of differentiability will suffice for our results. We continue with a number of technical lemmas.

Lemma 2.5. *Let Γ be a bounded, smooth curve and for $s, s^* \in \Gamma$ define $r = |s - s^*|$. Then there exists $\delta > 0$ such that for every $s^* \in \Gamma$ and $s \in B(s^*, \delta)$, $|\mathrm{d}s| \leq C|\mathrm{d}r|$, where s is a function of the arc length variable.*

Proof. Introduce the arc length parameterization of Γ, $s(t) = \alpha(t) + \mathrm{i}\beta(t)$. Let t^* be such that $s(t^*) = s^* = a + \mathrm{i}b$. Then

$$r = [(\alpha(t)-a)^2 + (\beta(t)-b)^2]^{1/2}, \quad \mathrm{d}r = r^{-1}(t)[(\alpha(t)-a)\alpha'(t) + (\beta(t)-b)\beta'(t)]\mathrm{d}t.$$

Near $t = t^*$, we use Taylor's theorem to write

$$\begin{aligned}&(\alpha(t)-a)\alpha'(t) + (\beta(t)-b)\beta'(t) \\ &\qquad = (t-t^*)\left[\alpha'(t^*)(\alpha'(t^*) + \mathcal{O}(t-t^*)) + \beta'(t^*)(\beta'(t^*) + \mathcal{O}(t-t^*))\right] \\ &\qquad = (t-t^*)[(\alpha'(t^*))^2 + (\beta'(t^*))^2] + \mathcal{O}((t-t^*)^2).\end{aligned}$$

Additionally,

$$r(t) = |t-t^*|[(\alpha'(t^*))^2 + (\beta'(t^*))^2]^{1/2}(1 + \mathcal{O}(|t-t^*|)).$$

From boundedness and smoothness we know that there exists a constant $C > 1$ such that

$$\frac{1}{C} \leq [(\alpha'(t))^2 + (\beta'(t))^2]^{1/2} \leq C.$$

It follows that

$$\left| \frac{(\alpha(t)-a)\alpha'(t) + (\beta(t)-b)\beta'(t)}{r(t)} \right| \geq C^{-2} + \mathcal{O}(|t-t^*|).$$

Therefore for $t \in (t^* - \gamma, t^* + \gamma)$, $\gamma > 0$,

$$|\mathrm{d}t| \leq (1+C)|\mathrm{d}r|,$$

where γ depends only on the magnitude of the second derivatives of α and β and can be made small enough so that it does not depend on t^* (or s^*). The smoothness of $s(t)$ and the nonvanishing of $s'(t)$ gives $|\mathrm{d}s| \leq C|\mathrm{d}r|$ for a new constant. This is valid in $B(s^*, \delta)$ where $\delta = \sup_{t \in (t^*-\gamma, t^*+\gamma)} |s(t) - s^*|$. $\square$

2.2.1 ▪ Boundary values

We use the notation $B_\theta(x,\delta) = \{y \in B(x,\delta) : |\mathrm{Im}(x-y)|/|x-y| > \sin\theta\}$ for $0 < \theta < \pi$. This is a ball with two cones subtracted; see Figure 2.2. To allow for rotations define $B_{\theta,\phi}(x,\delta) = (B_\theta(x,\delta) - x)\mathrm{e}^{\mathrm{i}\phi} + x$. The following technical lemma begins to illustrate the importance of the class of Hölder continuous functions.

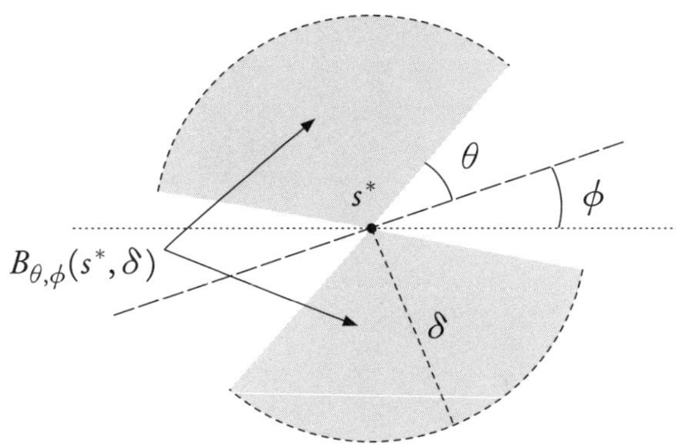

Figure 2.2. *A representation of $B_{\theta,\phi}(s^*,\delta)$.*

Lemma 2.6. *Let Γ be a bounded, smooth curve. Let $f : \Gamma \to \mathbb{C}$ be α-Hölder continuous and let $f(s^*) = 0$ for some $s^* \in \Gamma$. Then*

1. *$\mathcal{C}_\Gamma f(s^*)$ exists, and*
2. *for each $\theta > 0$ there exists $\delta > 0$ such that $\mathcal{C}_\Gamma f(z)$ is continuous in $B_{\theta,\phi}(s^*,\delta)$ where ϕ is the angle the tangent of Γ at s^* makes with the horizontal.*

Proof. We prove each part separately.

1. This follows from the Hölder condition on f. The only unboundedness of the integrand behaves like $|s - s^*|^{\alpha-1}$, which is integrable.

2. Examine

$$I(z) = \int_\Gamma \frac{f(s)}{s-z} \text{đ}s - \int_\Gamma \frac{f(s)}{s-s^*} \text{đ}s = \int_\Gamma f(s)\left(\frac{1}{s-z} - \frac{1}{s-s^*}\right) \text{đ}s$$
$$= \int_\Gamma \frac{(f(s)-f(s^*))(z-s^*)}{(s-z)(s-s^*)} \text{đ}s.$$

We decompose $\Gamma = \Gamma_\delta \cup \Gamma_\delta^c$, where $\Gamma_\delta = \Gamma \cap B(s^*, \delta)$ and Γ_δ^c is the complement relative to Γ. We assume $z \in B(s^*, \delta/2)$ and set up some elementary inequalities. For $s \in \Gamma_\delta^c$

- $|z - s^*| < \delta/2$,
- $|z - s| > \delta/2$, and
- $|s - s^*| > \delta$.

For $s \in \Gamma_\delta$ we use the law of sines (see Figure 2.3) to obtain $|z-s^*|/|z-s| \leq 1/\sin\theta_1$, where θ_1 is the angle between the line from z to s^* and that from s^* to s. Note that $0 < \theta_1 < \pi$ is bounded away from zero (and π) provided δ is sufficiently small and $z \in B_{\theta,\phi}(s^*, \delta)$, $0 < \theta < \pi$.

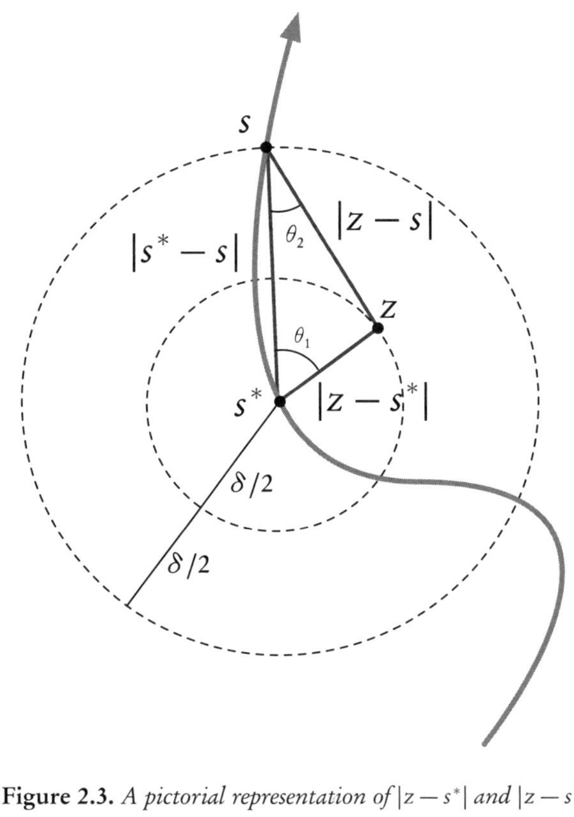

Figure 2.3. *A pictorial representation of* $|z - s^*|$ *and* $|z - s|$.

For $\delta > 0$, there exists $C > 0$ such that

$$\left|\int_{\Gamma_\delta^c} \frac{(f(s)-f(s^*))(z-s^*)}{(s-z)(s-s^*)} \text{đ}s\right| \leq 2|z-s^*|\delta^{-2} \int_{\Gamma_\delta^c} |f(s)-f(s^*)||\text{đ}s| \leq C|z-s^*|\delta^{-2}.$$

The right-hand side tends to zero as $z \to s^*$. We estimate the remaining terms.

$$\left| \int_{\Gamma_\delta} \frac{(f(s)-f(s^*))(z-s^*)}{(s-z)(s-s^*)} \, đs \right| \leq \frac{\Lambda(s^*)}{\sin\theta_1} \int_{\Gamma_\delta} |s-s^*|^{\alpha-1} |đs|.$$

Set $r = |s-s^*|$. The final estimate we need is that for δ sufficiently small, $|đs| \leq C|dr|$ (see Lemma 2.5). Thus by modifying C,

$$|I(z)| \leq C\left(2\frac{\Lambda(s^*)}{\alpha \sin\theta_1}\delta^\alpha + |z-s^*|\delta^{-2}\right). \tag{2.2}$$

For any $\epsilon > 0$, we choose δ so that the first term is less than ϵ and let $z \to s^*$. This proves that $I(z) \to 0$ as $z \to s^*$. □

Remark 2.2.2. *If f is uniformly α-Hölder continuous, then the right-hand side of (2.2) depends on s^* just through $|z-s^*|$ and this fact is important for proving Corollary 2.9 below.*

Now we discuss the limiting values of $\mathscr{C}_\Gamma f$ when f is α-Hölder continuous but does not vanish at the limiting point. We assume that Γ is a bounded, positively oriented, closed curve. We denote the region lying to the left of Γ by Ω_+ and that to the right by Ω_-. For any point $s^* \in \Gamma$ consider

$$\mathscr{C}_\Gamma f(z) = \int_\Gamma \frac{f(s)-f(s^*)}{s-z} \, đs + f(s^*) \int_\Gamma \frac{đs}{s-z}.$$

Due to Cauchy's theorem, we have

$$\int_\Gamma \frac{đs}{s-z} = \begin{cases} 1 & \text{if } z \in \Omega^+, \\ 0 & \text{otherwise.} \end{cases}$$

Combining this with Lemma 2.6 we can determine the left and right limits of the Cauchy integral:

$$\mathscr{C}_\Gamma^+ f(s^*) \triangleq \lim_{\substack{z\to s^* \\ z\in\Omega_+}} \mathscr{C}_\Gamma f(s) = f(s^*) + \int_\Gamma \frac{f(s)-f(s^*)}{s-s^*} \, đs, \tag{2.3}$$

$$\mathscr{C}_\Gamma^- f(s^*) \triangleq \lim_{\substack{z\to s^* \\ z\in\Omega_-}} \mathscr{C}_\Gamma f(s) = \int_\Gamma \frac{f(s)-f(s^*)}{s-s^*} \, đs, \tag{2.4}$$

where the limits are again taken nontangentially. We rewrite the integral appearing in this formula. For $s^* \in \Gamma$ we define the Cauchy principal value integral

$$⨍_\Gamma \frac{f(s)}{s-s^*} \, đs \triangleq \lim_{\delta\downarrow 0} \int_{\Gamma\setminus B(s^*,\delta)} \frac{f(s)}{s-s^*} \, đs.$$

Again, let $\Gamma_\delta = \Gamma \cap B(s^*, \delta)$:

$$\begin{aligned} \int_\Gamma \frac{f(s)-f(s^*)}{s-s^*} \, đs &= \lim_{\delta\downarrow 0} \int_{\Gamma\setminus\Gamma_\delta} \frac{f(s)-f(s^*)}{s-s^*} \, đs \\ &= \lim_{\delta\downarrow 0} \int_{\Gamma\setminus\Gamma_\delta} \frac{f(s)}{s-s^*} \, đs - \lim_{\delta\downarrow 0} \int_{\Gamma\setminus\Gamma_\delta} \frac{đs}{s-s^*}. \end{aligned}$$

The existence of the second limit shows the existence of the first:

$$\lim_{\delta \downarrow 0} \int_{\Gamma_\delta} \frac{\text{đ}s}{s - s^*} = \frac{1}{2\pi \mathrm{i}} \lim_{\delta \downarrow 0} \log\left(\frac{s^* - s_+}{s^* - s_-}\right) = -\frac{1}{2},$$

where $s_\pm$ are the endpoints of the arc Γ_δ. This fact is made precise within the proof of Lemma 2.49, in Appendix A.1.3.

Remark 2.2.3. *The curve Γ need not be closed. We can close Γ and define $f = 0$ on the added portion. These results still follow provided we stay away from the endpoints of Γ.*

We arrive at the following lemma.

Lemma 2.7 (Plemelj). *Let Γ be a smooth arc from a to b and let f be α-Hölder continuous on Γ. Then for $s^* \in \Gamma \setminus \{a, b\}$*

$$\mathscr{C}_\Gamma^+ f(s^*) = \frac{1}{2} f(s^*) + ⨍ \frac{f(s)}{s - s^*} \text{đ}s, \tag{2.5}$$

$$\mathscr{C}_\Gamma^- f(s^*) = -\frac{1}{2} f(s^*) + ⨍ \frac{f(s)}{s - s^*} \text{đ}s \tag{2.6}$$

and

$$\mathscr{C}_\Gamma^+ f(s^*) - \mathscr{C}_\Gamma^- f(s^*) = f(s^*),$$

$$\mathscr{C}_\Gamma^+ f(s^*) + \mathscr{C}_\Gamma^- f(s^*) = 2 ⨍ \frac{f(s)}{s - s^*} \text{đ}s.$$

Remark 2.2.4. *In Problem 2.3.1, we interpret this lemma as stating the Cauchy integral is the (unique) solution to a simple scalar RH problem.*

2.2.2 ▪ Regularity and singularity behavior of boundary values

We take up the issue of understanding the continuity properties of the functions $\mathscr{C}_\Gamma^\pm f$ when f is α-Hölder continuous.

Lemma 2.8. *Let Γ be a bounded, smooth arc and let f be uniformly α-Hölder continuous on Γ. Let $\Gamma' \subset \Gamma$ be an arc with endpoints lying a finite distance from the endpoints a and b of Γ. Then $\mathscr{C}_\Gamma^\pm f$ is uniformly α-Hölder continuous on Γ'.*

Proof. Let $s_1, s_2 \in \Gamma'$, and we write

$$\mathscr{C}_\Gamma^\pm f(s_1) - \mathscr{C}_\Gamma^\pm f(s_2) = f(s_1)\mathscr{C}_\Gamma^\pm 1(s_1) - f(s_2)\mathscr{C}_\Gamma^\pm 1(s_s) + I(s_1, s_2),$$

$$I(s_1, s_2) = \int_\Gamma \left(\frac{f(s) - f(s_1)}{s - s_1} - \frac{f(s) - f(s_2)}{s - s_2} \right) \text{đ}s.$$

By a simple contour deformation argument, it can be seen that

$$\mathscr{C}_\Gamma 1(z) = \int_\Gamma \frac{\text{đ}s}{s - z}$$

has infinitely smooth boundary values on Γ'. Thus the study of Hölder continuity for both $\mathscr{C}_\Gamma^\pm f$ is reduced to the study of $I(s_1,s_2)$. Define $\Gamma'_\delta = \Gamma' \cap B(s_1,\delta)$ for $\delta > 0$ such that $s_2 \in \partial B(s_1,\delta/2)$; see Figure 2.4. Separate $I(s_1,s_2) = I_0(s_1,s_2) + I_1(s_1,s_2)$, where I_0 contains an integral over Γ'_δ and I_1, over the complement, relative to Γ, of Γ'_δ. For I_0, using the Hölder condition, we obtain

$$|I_0(s_1,s_2)| \leq \Lambda \int_{\Gamma'_\delta} |s-s_1|^{\alpha-1}|đs| + \Lambda \int_{\Gamma'_\delta} |s-s_2|^{\alpha-1}|đs|.$$

Define $r_1 = |s-s_1|$ and $r_2 = |s-s_2|$. For sufficiently small $\delta > 0$, depending only on Γ (see Lemma 2.5), there exists a constant C that depends only on Γ such that $|đs| \leq C|\mathrm{d}r_1|$ and $|đs| \leq C|\mathrm{d}r_2|$. Thus

$$\begin{aligned} |I_0(s_1,s_2)| &\leq C\Lambda 2 \int_0^\delta r^{\alpha-1}\mathrm{d}r + C\Lambda \int_0^{\delta/2} r^{\alpha-1}\mathrm{d}r + C\Lambda \int_0^{3\delta/2} r^{\alpha-1}\mathrm{d}r \\ &\leq 5C\Lambda\alpha^{-1}\delta^\alpha. \end{aligned}$$

For I_1, we write

$$I_1(s_1,s_2) = (f(s_2)-f(s_1)) \int_{(\Gamma'_\delta)^c} \frac{đs}{s-s_1} + \int_{(\Gamma'_\delta)^c} (f(s)-f(s_2)) \left(\frac{1}{s-s_1} - \frac{1}{s-s_2} \right) đs. \quad (2.7)$$

The first integral is bounded, showing that the first term satisfies a uniform α-Hölder condition. We simplify the second integral

$$\begin{aligned} I_2(s_1,s_2) &= \int_{(\Gamma'_\delta)^c} (f(s)-f(s_2)) \left(\frac{1}{s-s_1} - \frac{1}{s-s_2} \right) đs \\ &= \int_{(\Gamma'_\delta)^c} (f(s)-f(s_2)) \frac{s_1-s_2}{(s-s_1)(s-s_2)} đs. \end{aligned}$$

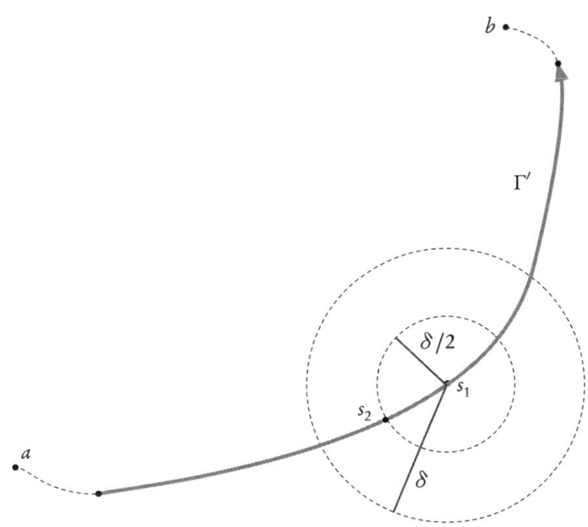

Figure 2.4. *The positioning of s_1 and s_2 on Γ.*

We find that for $s \in (\Gamma'_\delta)^c$

$$\frac{|s-s_2|}{|s-s_1|} \geq \frac{|s-s_2|}{|s-s_2|+|s_2-s_1|} \geq \frac{2}{2+\delta/|s-s_2|} \geq \frac{1}{2}.$$

We estimate

$$|I_2(s_1,s_2)| \leq \Lambda |s_1-s_2| \int_{(\Gamma'_\delta)^c} \frac{|s-s_2|^{\alpha-1}}{|s-s_1|} |ds| \leq 2^{1-\alpha}\Lambda|s_1-s_2| \int_{(\Gamma'_\delta)^c} |s-s_1|^{\alpha-2}|ds|.$$

This integral is easily bounded:

$$|I_2(s_1,s_2)| \leq 2^{1-\alpha}\Lambda\delta^{\alpha-2}|(\Gamma'_\delta)^c||s_1-s_2|.$$

This proves the lemma. $\square$

Define the space $C^{0,\alpha}(\Gamma)$, $0<\alpha\leq 1$, for Γ smooth, bounded, and closed, consisting of uniformly α-Hölder continuous functions. We introduce the seminorm

$$|f|_{0,\alpha} = \sup_{s_1\neq s_2,\ s_1,s_2\in\Gamma} \frac{|f(s_1)-f(s_2)|}{|s_1-s_2|^\alpha},$$

which is finite for every function in $C^{0,\alpha}(\Gamma)$ by Lemma 2.4. $C^{0,\alpha}(\Gamma)$ is a Banach space when equipped with the norm [49, p. 254]

$$\|f\|_{0,\alpha} = \sup_{s\in\Gamma}|f(s)| + |f|_{0,\alpha}.$$

Corollary 2.9. *The Cauchy integral operators $\mathcal{C}_\Gamma^\pm$ are bounded linear operators on $C^{0,\alpha}(\Gamma)$ when Γ is smooth, bounded, and closed.*

Proof. The bounds in the previous lemma and Lemma 2.4 depend only on δ, Λ, and C from Lemma 2.5. This shows

$$|\mathcal{C}_\Gamma^\pm f|_{0,\alpha} \leq C_1\|f\|_{0,\alpha}, \quad C_1>0.$$

It remains to show $\sup_{s\in\Gamma}|\mathcal{C}_\Gamma^\pm f(s)| \leq C_2\|f\|_{0,\alpha}$, $C_2>0$. For $s^*\in\Gamma$, consider Γ_δ and Γ_δ^c as above. We write

$$\mathcal{C}_\Gamma^+ f(s^*) = \int_{\Gamma_\delta^c} \frac{f(s)}{s-s^*} đs + f(s^*) + \int_{\Gamma_\delta} \frac{f(s)-f(s^*)}{s-s^*} đs.$$

Thus

$$|\mathcal{C}_\Gamma^+ f(s^*)| \leq |f(s^*)| + \sup_{s\in\Gamma}|f(s)|\delta^{-1}\int_{\Gamma_\delta^c}|đs| + |f|_{0,\alpha}\int_{\Gamma_\delta}|s-s^*|^{\alpha-1}|đs| \leq C_2\|f\|_{0,\alpha},$$

for some $C_2>0$, by previous arguments. Taking a supremum proves the corollary for $\mathcal{C}_\Gamma^+$. The result for $\mathcal{C}_\Gamma^-$ can be inferred from Lemma 2.7. $\square$

Definition 2.10. *A function f satisfies an (α,γ)-Hölder condition on a curve Γ if f is α-Hölder away from the endpoints of Γ and if at each endpoint c, f satisfies*

$$f(s) = \frac{\tilde{f}(s)}{(s-c)^\gamma}, \quad \gamma = a+ib, \quad 0\leq a<1,$$

and $\tilde{f}$ is α-Hölder in a neighborhood of c.

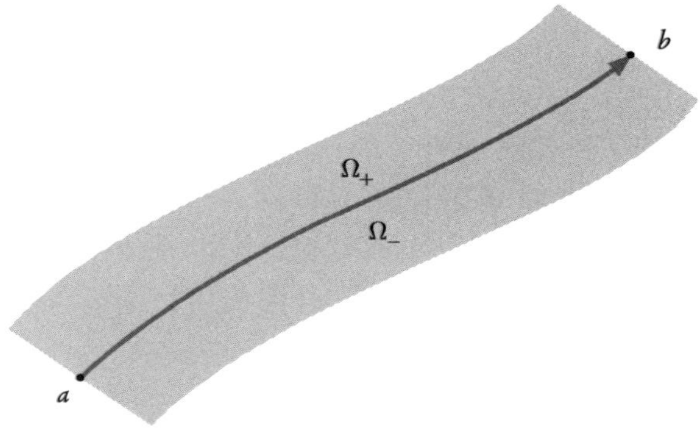

Figure 2.5. *An arc with* $\Omega_\pm$ *labeled.*

Now, we discuss the important features of Cauchy integrals near endpoints of the arc of integration through the following lemma. This lemma is compiled from the results in [3, p. 521] and [86, p. 73].

Lemma 2.11. *Let* Γ *be a smooth arc from* a *to* b *with* Ω_+ (Ω_-) *defined as regions lying directly to the left (right) of* Γ*; see Figure 2.5. Assume* f *satisfies an* (α,γ)*-Hölder condition. The following hold for any endpoint* $c = a$ *or* b*:*

1. *If* $\gamma = 0$*, then we have the following:*

   (a) *As* $z \to c$, $z \in \Omega_\pm$,

$$\mathscr{C}_\Gamma f(z) = \sigma \frac{f(c)}{2\pi i} \log \frac{1}{z-c} + F_0(z;\sigma).$$

   (b) *As* $s \to c$, $s \in \Gamma$,

$$\mathscr{C}_\Gamma^\pm f(s) = \sigma \frac{f(c)}{2\pi i} \log \frac{1}{|s-c|} + H_0(s;\sigma).$$

   Here H_0 *and* F_0 *both tend to definite limits at* c *and* $\sigma = -1$ *if* $c = b$ *and* $\sigma = +1$ *if* $c = a$*. The branch cut for the logarithm is taken along* Γ.

2. *If* $\gamma \neq 0$*, then we have the following:*

   (a) *As* $z \to c$, $z \in \Omega_\pm$,

$$\mathscr{C}_\Gamma f(z) = \sigma \frac{e^{i\sigma\gamma\pi}}{2i\sin(\gamma\pi)} \frac{\tilde{f}(c)}{(z-c)^\gamma} + F_0(z;\sigma).$$

   (b) *As* $s \to c$, $s \in \Gamma$,

$$\mathscr{C}_\Gamma^\pm f(s) = \sigma \frac{\cot(\gamma\pi)}{2i} \frac{\tilde{f}(c)}{(s-c)^\gamma} + H_0(s;\sigma).$$

   Here H_0 *and* F_0 *both tend to definite limits at* c *and* $\sigma = -1$ *if* $c = b$ *and* $\sigma = +1$ *if* $c = a$*. The branch cut for* $(z-c)^\gamma$ *is taken to be along the arc* Γ*. If* $\operatorname{Re}\gamma > 0$*, then for some* $0 < \alpha^* < \operatorname{Re}\gamma$*, and constants* A *and* B*,*

$$|F_0(z)| < \frac{A}{|z-c|^{\alpha^*}}, \quad |H_0(s,\sigma)| < \frac{B}{|s-c|^{\alpha^*}}.$$

Proof. We prove part 1 here. For the proof of part 2 see [86, p. 73]. Write

$$\int_a^b \frac{f(s)}{s-z} \text{đ}s = f(c)\int_a^b \frac{\text{đ}s}{s-z} + \int_a^b \frac{f(s)-f(c)}{s-z}\text{đ}s.$$

Thus

$$f(c)\int_a^b \frac{\text{đ}s}{s-z} = \sigma\frac{f(c)}{2\pi \mathrm{i}}\log\frac{1}{z-c} + V_0(z),$$

where V_0 is analytic near $z=c$. Thus

$$F_0(z;\sigma) = V_0(z) + \int_a^b \frac{f(s)-f(c)}{s-z}\text{đ}s,$$

which tends to a definite limit as $z \to c$ by Lemma 2.6. The limit $z \to s \in \Gamma$, $z \in \Omega_\pm \setminus B(c,\delta)$, $\delta > 0$, of $F_0(z)$ is $H_0(s;\sigma)$, which also tends to a definite limit as $s \to c$. □

Remark 2.2.5. *An important consequence of this result is that for functions bounded at the end of an arc, a singularity is introduced. For functions singular at the end of an arc, the singularity structure is preserved. More precisely, if f is (α,γ)-Hölder with $\gamma \neq 0$, then $\mathscr{C}_\Gamma^\pm f$ is (α',γ)-Hölder for some $\alpha' > 0$.*

2.3 ▪ The solution of scalar Riemann–Hilbert problems

We have presented a fairly wide class of functions, the α-Hölder continuous functions, for which the limits of Cauchy integrals are well-defined and regular. We continue with the solution of the simplest RH problem on smooth, closed, and bounded curves.

2.3.1 ▪ Smooth, closed, and bounded curves

Problem 2.3.1. *Find ϕ that solves the continuous RH problem*

$$\phi^+(s) - \phi^-(s) = f(s), \quad s \in \Gamma, \quad \phi(\infty) = 0, \quad f \in C^{0,\alpha}(\Gamma), \tag{2.8}$$

where Γ is a smooth, bounded, and closed curve.

This problem is solved directly by the Cauchy integral $\phi(z) = \mathscr{C}_\Gamma f(z)$. Indeed, Lemma 2.7 gives

$$\phi^+(s) - \phi^-(s) = \mathscr{C}_\Gamma^+ f(s) - \mathscr{C}_\Gamma^- f(s) = f(s), \quad s \in \Gamma. \tag{2.9}$$

To show $\phi(\infty) = 0$ we use the following lemmas, which provide more precise details. Note that Γ need not be bounded in these results.

Lemma 2.12. *If $|s|^j f(s) \in L^1(\Gamma)$ for $j = 0,\dots,n$, then*

$$\int_\Gamma \frac{f(s)}{s-z}\text{đ}s = \sum_{j=1}^{n-1} c_j z^{-j} + \mathcal{O}(z^{-n}) \quad \textit{as} \quad |z| \to \infty$$

for

$$c_j = -\int_\Gamma s^{j-1} f(s) \text{đ}s$$

if z, *sufficiently large, satisfies* $\inf_{s\in\Gamma} |z-s| \geq c > 0$.

Proof. The asymptotic series is obtained from a geometric series expansion for $1/(s-z)$ for s fixed and $|z|$ large:

$$\frac{1}{s-z} = -\frac{1}{z}\sum_{j=0}^{n-1}\left(\frac{s}{z}\right)^j + \frac{(s/z)^n}{s-z}.$$

Define $P_n(s,z) = (s/z)^n/(z-s)$. It is clear that

$$|P_n(s,z)| \leq c^{-1}\left|\frac{s}{z}\right|^n.$$

From this estimate

$$\int_\Gamma |f(s)||P_n(s,z)||\text{đ}s| = \mathcal{O}(z^{-n}) \quad \text{for} \quad \inf_{s\in\Gamma}|z-s| \geq c > 0. \qquad \square$$

The assumptions in this lemma can be relaxed at the expense of weaker asymptotics using the *dominated convergence theorem* (Theorem A.3).

Lemma 2.13. *For* $f \in L^1(\Gamma)$,

$$\lim_{z\to\infty} z\int_\Gamma \frac{f(s)}{s-z}\text{đ}s = -\int_\Gamma f(s)\text{đ}s,$$

where the limit is taken in a direction that is not tangential to Γ.

We have addressed existence in a constructive way. Now we address uniqueness. Let $\psi(z)$ be another solution of Problem 2.3.1. The function $D(z) = \psi(z) - \phi(z)$ satisfies

$$D^+(s) - D^-(s) = 0, \quad s \in \Gamma, \quad D(\infty) = 0.$$

The trivial jump $D^+(s) = D^-(s)$ is equivalent to stating that D is continuous up to Γ. It follows that D is analytic at every point on Γ (Theorem C.11), and hence D is entire. By Liouville's theorem, it must be identically zero. This shows that the Cauchy integral of f is the unique solution to Problem 2.3.1.

We defer the extension of these results to unbounded contours such as $\mathbb{R}$ until Section 2.5.4. This case is dealt with in a more straightforward way using L^p and Sobolev spaces. All solution formulae hold with slight changes in interpretation.

We move to the simplest case of a scalar RH problem with a multiplicative jump.

Problem 2.3.2. *Find* ϕ *that solves the homogeneous*[18] *continuous RH problem*

$$\phi^+(s) = \phi^-(s)g(s), \quad s\in\Gamma, \quad \phi(\infty) = 1, \quad g \in C^{0,\alpha}(\Gamma), \tag{2.10}$$

where Γ *is a smooth, bounded, and closed curve, and* $g(s) \neq 0$.

[18] We use the term homogeneous to refer to the jump condition, not the behavior at infinity.

Formally, this problem can be solved via the logarithm. Consider the RH problem solved by $X(z) = \log \phi(z)$:

$$X^+(s) = X^-(s) + G(s) \Leftrightarrow X^+(s) - X^-(s) = G(s), \quad G(s) = \log g(s).$$

If $\log g(s)$ is well-defined and Hölder continuous, the solution is given by

$$\phi(z) = \exp(\mathcal{C}_\Gamma G(z)). \tag{2.11}$$

Furthermore, because $|\mathcal{C}_\Gamma G(z)| < \infty$ for all $z \in \mathbb{C} \setminus \Gamma$, and it is continuous up to Γ, we have $|\mathcal{C}_\Gamma G(z)| \leq C$ for some C. This implies that $\phi(z)$ and $1/\phi(z)$ are both bounded, continuous functions on $\mathbb{C} \setminus \Gamma$. To see uniqueness, let $\psi(z)$ be another solution and consider $R(z) = \psi(z)/\phi(z)$. Then $R^+(s) = R^-(s)$ on Γ, and hence $R(z)$ is entire (Theorem C.11). $R(z)$ is uniformly bounded on $\mathbb{C} \setminus \Gamma$, and thus $R(z) \equiv 1$, or $\psi(z) = \phi(z)$.

For a general Hölder continuous function g, $\log g$ may not be well-defined. Indeed, even if one fixes the branch of the logarithm, $\log g$ generically suffers from discontinuities. To rectify this issue we define the *index* of a function g with respect to traversing Γ in the positive direction to be the normalized increment of its argument:

$$\operatorname{ind}_\Gamma g(s) \triangleq \frac{1}{2\pi}[\arg g(s)]_\Gamma = \frac{1}{2\pi i}[\log g(s)]_\Gamma = \int_\Gamma \mathrm{d} \log g(s). \tag{2.12}$$

First, if g is α-Hölder continuous and $\operatorname{ind}_\Gamma g(s) = 0$, then $\log g(s)$ is also α-Hölder continuous. In this case the branch cut of $\log s$ can be taken so that it is Lipschitz continuous in an open set containing $\{g(s) : s \in \Gamma\}$. If $\operatorname{ind}_\Gamma g(s) = \kappa \neq 0$ and (without loss of generality) $z = 0$ is in Ω_+, the region to the left of Γ, then $\operatorname{ind} s^{-\kappa} g(s) = 0$. Thus we can uniquely solve the problem

$$\psi^+(z) = \psi^-(s) s^{-\kappa} g(s), \quad s \in \Gamma, \quad \psi(\infty) = 1,$$

with the expression (2.11) because $s^{-\kappa} g(s)$ has index zero. There are two cases:

- If $\kappa > 0$, then

  $$\phi(z) = P(z) \begin{cases} \psi(z) & \text{if } z \in \Omega_+, \\ \psi(z) z^{-\kappa} & \text{if } z \in \Omega_-, \end{cases}$$

  where P is a polynomial of degree κ with leading coefficient 1, solves Problem 2.3.2. Next, we show that all solutions are of this form. Let $\gamma(z)$ be another solution of Problem 2.3.2 and consider the ratio

  $$R(z) = \gamma(z)/\phi(z) \Rightarrow R^+(s) = R^-(s), \quad s \in \Gamma, \quad R(\infty) = 1.$$

  It follows from Theorem C.11 that R has poles at the zeros of $\phi(z)$ and zeros at the zeros of $\gamma(z)$. From previous considerations $|\psi(z)|$ is bounded above and below, globally, and $R(z)$ has at most κ poles, counting multiplicities. Thus from the asymptotic behavior of R, $R(z) = P_1(z)/P_2(z)$, where P_1 and P_2 are polynomials of degree κ.

- If $\kappa < 0$, then

  $$\phi(z) = P(z) \begin{cases} \psi(z) & \text{if } z \in \Omega_+, \\ \psi(z) z^{-\kappa} & \text{if } z \in \Omega_- \end{cases}$$

cannot satisfy $\phi(\infty)=1$ for any polynomial P. Thus the only sectionally analytic function that satisfies the jump of Problem 2.3.2 and is bounded at ∞ is the zero solution. Indeed, assume γ is such a function, and take $P(z)=1$. Then

$$R(z)=\gamma(z)/\phi(z) \Rightarrow R^+(s)=R^-(s),\ R(\infty)=0,$$

and R is entire because $|\phi(z)|$ is bounded below. Liouville's theorem implies $R\equiv 0$ and $\gamma\equiv 0$.

In both cases above, when $P=1$ we call the function ϕ the *fundamental solution*.

We move to consider inhomogeneous scalar RH problems. We see a direct parallel between the methods presented here and the method of variation of parameters for ODEs.

Problem 2.3.3. *Find ϕ that solves the inhomogeneous, continuous RH problem*

$$\phi^+(s)=\phi^-(s)g(s)+f(s),\ s\in\Gamma,\ \phi(\infty)=0,\ g,f\in C^{0,\alpha}(\Gamma), \tag{2.13}$$

where Γ is a smooth, bounded, and closed curve, and $g\neq 0$.

To solve this problem we first find the fundamental solution of the homogeneous problem. Just as in the case of variation of parameters for differential equations the solution of the homogeneous problem allows us to solve the inhomogeneous problem. We use $\nu(z)$ to denote the fundamental solution. Assume $\operatorname{ind}_\Gamma g(s)=\kappa$. The Hölder continuity of $s^{-\kappa}g(s)$ shows us that ν does not vanish in the finite plane. Dividing ϕ by ν and using $\nu^+(s)=\nu^-(s)g(s)$ in (2.13) we find that

$$\frac{\phi^+(s)}{\nu^+(s)}=\frac{\phi^-(s)}{\nu^-(s)}+\frac{f(s)}{\nu^+(s)},\ \frac{\phi(z)}{\nu(z)}=\mathcal{O}(z^{\kappa-1}) \text{ as } z\to\infty.$$

Again, there are two cases:

- If $\kappa\geq 0$, then we obtain an expression for $\phi(z)/\nu(z)$ using Plemelj's lemma (Lemma 2.7):

$$\frac{\phi(z)}{\nu(z)}=\int_\Gamma \frac{f(s)}{\nu^+(s)(s-z)}đs=\mathcal{O}(z^{\kappa-1}) \text{ as } z\to\infty,\ \nu(z)=\mathcal{O}(z^{-\kappa}). \tag{2.14}$$

- If $\kappa<0$, then the asymptotic condition here requires higher-order decay of a Cauchy integral. The solution formula is still (2.14), but we use Lemma 2.12 to find the *moment conditions*

$$\int_\Gamma s^n\frac{f(s)}{\nu^+(s)}đs=0,\ n=0,\ldots,\kappa-1. \tag{2.15}$$

  If any of these conditions is not satisfied, then no solution of Problem 2.3.3 that vanishes at infinity exists.

Once a valid expression for $\phi(z)/\nu(z)$ is obtained, the general solution is given by

$$\phi(z)=\nu(z)\left(\int_\Gamma\frac{f(s)}{\nu^+(s)(s-z)}đs+P(z)\right),$$

where $P(z)$ is a polynomial of degree $<\kappa$ if $\kappa>0$; otherwise P is zero.

Remark 2.3.1. *The definition of the index in* (2.12) *corresponds directly with that of the Fredholm index; see Definition* A.18. *As we can see below, a solution of the homogeneous problem vanishing at infinity corresponds to an element of the kernel of an integral operator. Furthermore,* (2.15) *are the conditions for f to lie in the range of the same integral operator.*

Example 2.14. Once an inhomogeneous RH problem can be solved, other combinations of jump conditions and asymptotic behavior can be discussed. For example, a small modification is

$$\phi^+(s) = \phi^-(s)g(s) + f(s), \quad s \in \Gamma, \quad \phi(\infty) = 1, \quad g, f \in C^{0,\alpha}(\Gamma).$$

Let $\psi(z) = \phi(z) - 1$, and we obtain

$$\psi^+(s) = \psi^-(s)g(s) + g(s) - 1 + f(s), \quad s \in \Gamma, \quad \psi(\infty) = 0, \quad g, f \in C^{0,\alpha}(\Gamma),$$

a new inhomogeneous RH problem. ■

2.3.2 ▪ Smooth, bounded, and open curves

The solution procedure for scalar RH problems is not much more difficult in practice when the curve Γ is not closed. A complication comes from the fact that in the case of arcs, additional solutions are introduced. To highlight this, consider the following continuous RH problem.

Problem 2.3.4. *Find ϕ that solves the homogeneous continuous RH problem*

$$\phi^+(s) = \phi^-(s)g(s), \quad s \in \Gamma, \quad \phi(\infty) = 1, \quad g \in C^{0,\alpha}(\Gamma), \tag{2.16}$$

where Γ is a smooth, bounded, and open curve extending from $z = a$ to $z = b$, and $g(s) \neq 0$.

If $\phi(z)$ satisfies the jump condition, then so does $\frac{1}{2}(1 + \frac{z-a}{z-b})^k \phi(z)$, away from a, b for any integer k. We impose *local integrability* (see Definition C.5) in our definition of the solution of a continuous RH problem; otherwise we would have an infinite number of solutions.

Before we solve Problem 2.3.4 we discuss the solution of Problem 2.3.1 in the case that Γ has open endpoints and f is (α, γ)-Hölder. One solution is certainly $\mathcal{C}_\Gamma f(z)$ (see Lemma 2.11). To see uniqueness, let ψ be another solution. Then $\mathcal{C}_\Gamma f(z) - \psi(z)$ is an analytic function away from the endpoints of Γ that decays at infinity. The local integrability condition precludes the existence of poles at these endpoints. Again, Liouville's theorem shows $\mathcal{C}_\Gamma f(z) = \psi(z)$.

We proceed to solve Problem 2.3.4 as if Γ is closed. Define $G(s) = \log g(s)$ taking a branch of the logarithm so that $G(s)$ varies continuously over the curve. Taking any other branch would modify $G(s)$ by $2\pi i n$ for some n. It can be shown that the choices of n_a and n_b below remove the dependence on n. Define

$$\psi(z) = \exp(\mathcal{C}_\Gamma G(z)).$$

A straightforward calculation shows that ψ satisfies the appropriate jump condition. We must determine whether ψ is locally integrable and, if it is not, modify. Using Lemma 2.11 we see that for an arc from a to b

$$\psi(s) = (s - b)^{\zeta_b} H_b(s), \quad \psi(s) = (s - a)^{\zeta_a} H_a(s),$$

where $\zeta_a = \frac{\log g(a)}{2\pi i}$ and $\zeta_b = -\frac{\log g(b)}{2\pi i}$. Here $H_a(s), H_b(s)$ are functions that tend to definite limits as $s \to a$ and b, respectively. Let $\zeta_c = \lambda_c + i\mu_c$ and let n_c be an integer such that $-1 < n_c + \lambda_c < 1$ for $c = a, b$. It follows that

$$\nu(z) = (z-a)^{n_a}(z-b)^{n_b}\psi(z)$$

is a solution of (2.16) since it is locally integrable. Note that if $\lambda_c \in \mathbb{Z}$, then n_c is uniquely specified. Specifically, if g takes only positive values, then the solution is unique. In the case of arcs, any locally integrable $\nu(z)$ is called a fundamental solution.

We follow the same procedure as above to solve Problem 2.3.3 in the case of Γ being an arc.

Problem 2.3.5. *Find ϕ that solves the inhomogeneous continuous RH problem*

$$\phi^+(s) = \phi^-(s)g(s) + f(s), \quad s \in \Gamma, \quad \phi(\infty) = 0, \quad f, g \in C^{0,\alpha}(\Gamma), \tag{2.17}$$

where Γ is a smooth, bounded arc extending from $z = a$ to $z = b$, and $g(s) \neq 0$.

Divide (2.17) by ν and write

$$\frac{\phi^+(s)}{\nu^+(s)} = \frac{\phi^-(s)}{\nu^-(s)} + \frac{f(s)}{\nu^+(s)}, \quad \frac{\phi(z)}{\nu(z)} = \mathcal{O}(z^{n_a+n_b-1}) \text{ as } z \to \infty. \tag{2.18}$$

We assume f satisfies an α-Hölder condition. Thus $f(s)/\nu^+(s)$ satisfies an $(\alpha, -\zeta_a)$-Hölder condition near $z = a$ and a similar condition near $z = b$. A solution of (2.18), assuming possible moment conditions (2.15) are satisfied, is given by

$$\phi(z) = \nu(z) \int_\Gamma \frac{f(s)}{\nu^+(s)(s-z)} ds.$$

By Remark 2.2.5 we see that $\phi(z)$ has bounded singularities at the endpoints of Γ whenever $\zeta_c \neq 0$; otherwise there is a logarithmic singularity present. As before,

$$\phi(z) = \nu(z)\left(\int_\Gamma \frac{f(s)}{\nu^+(s)(s-z)} ds + P(z)\right)$$

is the solution, where $P(z)$ is a polynomial of degree less than $-(n_a + n_b)$ if $n_a + n_b < 0$; otherwise $P = 0$ and we have $n_a + n_b$ orthogonality conditions for a solution to exist (see (2.15) with $\kappa = n_a + n_b$). Note that different fundamental solutions can be chosen when the fundamental solution is not unique. This gives rise to solutions with different boundedness properties. This is illustrated in the following example.

Example 2.15. Consider the RH problem

$$\phi^+(s) + \phi^-(s) = 1 - s, \quad s \in (-\ell, \ell), \quad \ell > 0, \quad \phi(\infty) = 0.$$

Assume we want ϕ to be uniformly bounded in the plane. We first find the fundamental solution, ν, taking any branch of the logarithm:

$$\log g(s) = -i\pi,$$

$$\exp\left(\int_{-\ell}^{\ell} \frac{\log g(s)}{s-z} ds\right) = \sqrt{\frac{z+\ell}{z-\ell}}, \quad -\zeta_{-\ell} = \zeta_\ell = -\frac{1}{2}.$$

For ν to be bounded, $n_\ell = 1$ and $n_{-\ell} = 0$ so that $\mathrm{Re}(\zeta_{\pm\ell}) + n_{\pm\ell} > 0$. Thus

$$\nu(z) = \sqrt{(z-\ell)(z+\ell)}.$$

The solution ϕ is given by ($P = 0$)

$$\phi(z) = \nu(z)\left(\int_{-\ell}^{\ell} \frac{1-s}{s-z} \frac{đs}{\sqrt{(s-\ell)(s+\ell)}^+}\right),$$

provided that

$$\int_{-\ell}^{\ell} (1-s) \frac{đs}{\sqrt{(s-\ell)(s+\ell)}^+} = -\frac{i}{2}\left(\ell^2 - 2\right) = 0.$$

The solution exists if $\ell = \sqrt{2}$. ∎

We conclude this section with another fairly simple example that is of use later.

Example 2.16. Consider the RH problem

$$\phi^+(s) = \alpha\phi^-(s), \quad s \in (a,b), \quad \phi \text{ of finite degree at } \infty.$$

Set $\alpha = |\alpha| e^{i\theta}$ so that $\log \alpha = \log|\alpha| + i\theta$. We find a solution

$$\psi(z) = \exp\left(\frac{\log|\alpha| + i\theta}{2\pi i} \log\left(\frac{z-b}{z-a}\right)\right) = \left(\frac{z-b}{z-a}\right)^{-i\log|\alpha|/(2\pi) + \theta/(2\pi)}.$$

The general form of a solution is

$$\phi(z) = (z-a)^{n_a}(z-b)^{n_b}\psi(z),$$

with integers n_a, n_b chosen so that $-1 < n_a - \theta/(2\pi) < 1$ and $-1 < n_b + \theta/(2\pi) < 1$. ∎

2.4 ▪ The solution of some matrix Riemann–Hilbert problems

The general form for the jump condition of a matrix RH problem defined for a contour Γ is

$$\Phi^+(s) = \Phi^-(s) G(s) + F(s), \quad s \in \Gamma,$$

where $\Phi : \mathbb{C} \setminus \Gamma \to \mathbb{C}^{m\times n}$, $G : \Gamma \to \mathbb{C}^{n\times n}$, and $F : \Gamma \to \mathbb{C}^{m\times n}$. Most often in our applications, $m = n = 2$. We assume $m = n$ in this section. Unlike scalar RH problems, matrix RH problems cannot, in general, be solved in closed form. Issues related to existence and uniqueness are also more delicate. The general theory involves the analysis of singular integral operators. Specifically, it involves questions related to their invertibility. We address this in Section 2.7. Here we take a constructive approach and describe a procedure for solving three types of RH problems: diagonal problems, constant jump matrix problems, and triangular problems.

All solution techniques in this section rely on the reduction of the matrix problem to a sequence of scalar problems. When these techniques fail we must develop a completely new theory that is in some sense independent of dimensionality. This theory is developed in the remaining sections of this chapter.

2.4.1 ▪ Diagonal Riemann–Hilbert problems

Problem 2.4.1. *Find Φ that solves the homogeneous, diagonal, and continuous RH problem*

$$\Phi^+(s) = \Phi^-(s)D(s), \quad s \in \Gamma, \quad \Phi(\infty) = I, \quad \Phi : \mathbb{C} \setminus \Gamma \to \mathbb{C}^{n\times n}, \quad D \in C^{0,\alpha}(\Gamma), \tag{2.19}$$

and $D(s) = \operatorname{diag}(d_1(s), \ldots, d_n(s))$ *with* $\det D(s) \neq 0$. *Assume that* $\log d_i(s) \in C^{0,\alpha}(\Gamma)$ *for each i and some $\alpha > 0$.*

This problem decouples into n scalar RH problems:

$$\phi_i^+(s) = \phi_i^-(s)d_i(s), \quad s \in \Gamma, \quad \phi_i(\infty) = 1, \quad i = 1, \ldots, n. \tag{2.20}$$

Each of these has a solution

$$\phi_i(z) = \exp\left(\int_\Gamma \frac{\log d_i(s)}{s - z} \mathrm{d}s\right)$$

because $\operatorname{ind} \log d_i(s) = 0$ from continuity. A solution of (2.19) is given by

$$\Phi(z) = \operatorname{diag}(\phi_1(z), \ldots, \phi_n(z)).$$

If we restrict to smooth, closed, and bounded curves, the solution is unique. To see this let Ψ be another solution. It is clear that $\Phi^{-1}(z)$ exists for all $z \in \mathbb{C} \setminus \Gamma$. Thus

$$\Psi^+(s)\Phi_+^{-1}(s) = \Psi^-(s)D(s)D^{-1}(s)\Phi_-^{-1}(s) = \Psi^-(s)\Phi_-^{-1}(s),$$

Liouville's theorem applied to each element shows that $\Psi(s)\Phi^{-1}(s)$ is constant. The condition at infinity implies Ψ and Φ are the same function. As one would imagine, the theory for diagonal matrix RH problems on arcs mimics that of scalar problems on arcs. This is explored further in the following section.

2.4.2 ▪ Constant jump matrix problems

Problem 2.4.2. *Find Φ that solves the continuous RH problem*

$$\Phi^+(s) = \Phi^-(s)A, \quad s \in (a, b), \quad \Phi \text{ of finite degree at } \infty,$$

where A is an invertible, diagonalizable matrix $A = U\Lambda U^{-1}$, $\Lambda = \operatorname{diag}(\lambda_1, \ldots, \lambda_n)$.

The diagonalizable condition reduces this to a diagonal RH problem for $D(z) = U^{-1}\Phi(z)U$:

$$D^+(s) = D^-(s)\begin{bmatrix} \lambda_1 & & & \\ & \lambda_2 & & \\ & & \ddots & \\ & & & \lambda_n \end{bmatrix}, \quad D(\infty) = I.$$

We decouple this as we did for (2.19):

$$D_i^+(s) = D_i^-(s)\lambda_i.$$

Example 2.16 gives us the form of all the possible solutions of this problem. Thus $\Phi(z) = U \operatorname{diag}(D_1(z), \ldots, D_n(z))U^{-1}$ is a solution. Note that if $\lambda_i = 0$ for some i, the solution procedure fails.

Example 2.17. Consider the RH problem

$$\Phi^+(s)=\Phi^-(s)\begin{bmatrix} 0 & c \\ 1/c & 0 \end{bmatrix}, \quad s\in(a,b), \quad c\neq 0, \quad \Phi \text{ of finite degree at } \infty.$$

First, diagonalize

$$\begin{bmatrix} 0 & 1/c \\ c & 0 \end{bmatrix}=U\begin{bmatrix} -1 & 0 \\ 0 & 1 \end{bmatrix}U^{-1}, \quad U^{-1}=\begin{bmatrix} -1/c & 1 \\ 1/c & 1 \end{bmatrix}.$$

We solve the two auxiliary problems

$$h_1^+(z)=-h_1^-(z), \quad h_2^+(z)=h_2^-(z).$$

It is clear that $h_2(z)=1$ and

$$h_1(z)=(z-a)^{n_a}(z-b)^{n_b}\sqrt{\frac{z-b}{z-a}}, \quad n_a=0,1, \quad n_b=-1,0,$$

are the corresponding fundamental solutions. The solution is

$$\Phi(z)=U\begin{bmatrix} -h_1(z)/c & h_1(z) \\ h_2(z)/c & h_2(z) \end{bmatrix}.$$

We can multiply Φ on the left by any matrix of polynomials to obtain another solution of finite degree. ∎

We include one more example that is used later.

Example 2.18. Consider the RH problem

$$\Phi^+(s)=\Phi^-(s)\begin{bmatrix} 0 & 1 \\ -1 & 0 \end{bmatrix}, \quad s\in(a,b), \quad \Phi \text{ of finite degree at } \infty.$$

First, diagonalize

$$\begin{bmatrix} 0 & 1 \\ -1 & 0 \end{bmatrix}=U\begin{bmatrix} \mathrm{i} & 0 \\ 0 & -\mathrm{i} \end{bmatrix}U^{-1}, \quad U^{-1}=\begin{bmatrix} -\mathrm{i} & 1 \\ \mathrm{i} & 1 \end{bmatrix}.$$

We solve the two auxiliary problems

$$\begin{aligned} h_1^+(z)&=\mathrm{i}h_1^-(z), \\ h_2^+(z)&=-\mathrm{i}h_2^-(z). \end{aligned}$$

We find that

$$\begin{aligned} h_1(z)&=\exp\left(\int_a^b \frac{\log \mathrm{i}}{s-z}\,\mathrm{d}s\right)=\left(\frac{z-b}{z-a}\right)^{1/4}, \\ h_2(z)&=\exp\left(\int_a^b \frac{\log -\mathrm{i}}{s-z}\,\mathrm{d}s\right)=\left(\frac{z-b}{z-a}\right)^{-1/4} \end{aligned}$$

are the corresponding fundamental solutions. A solution is

$$\Phi(z)=U\begin{bmatrix} -\mathrm{i}h_1(z) & h_1(z) \\ \mathrm{i}h_2(z) & h_2(z) \end{bmatrix}, \quad \Phi(\infty)=I.$$

Again, we can multiply Φ on the left by any matrix of polynomials to obtain another solution of finite degree. ∎

2.4.3 ▪ Triangular Riemann–Hilbert problems

We restrict to smooth, closed, and bounded curves.

Problem 2.4.3. *Find Φ that solves the homogeneous, upper-triangular, and continuous RH problem*

$$\Phi^+(s) = \Phi^-(s)U(s), \quad s \in \Gamma, \quad \Phi(\infty) = I, \quad U \in C^{0,\alpha}(\Gamma),$$

where $U(s)$ is upper triangular, $U(s) = (U_{i,j}(s))_{1\leq i,j\leq n}$, $U_{i,j}(s) = 0$ if $i < j$.

To ensure unique solvability, assume $\operatorname{ind}_\Gamma U_{i,i}(s) = 0$ for $i = 1, \ldots, n$.[19] In essence, this problem is solved by solving successive scalar RH problems, akin to forward/backward substitution for solving triangular linear systems. It is important to note that each row can be found independently of other rows. The first row of the solution is determined by the following scalar problems:

$$\begin{aligned}
&\Phi^+_{1,1}(s) = \Phi^-_{1,1}(s)U_{1,1}(s), && \Phi_{1,1}(\infty) = 1,\\
&\Phi^+_{1,2}(s) = \Phi^-_{1,2}(s)U_{2,2}(s) + \Phi^-_{1,1}(s)U_{1,2}(s), && \Phi_{1,2}(\infty) = 0,\\
&\Phi^+_{1,3}(s) = \Phi^-_{1,3}(s)U_{3,3}(s) + \Phi^-_{1,2}(s)U_{2,3}(s) + \Phi^-_{1,1}(s)U_{1,3}(s), && \Phi_{1,3}(\infty) = 0,\\
&\quad\vdots
\end{aligned}$$

Note that $\Phi^-_{1,1}(s)U_{1,2}(s)$ in the second equation above can be considered an inhomogeneous term since $\Phi_{1,1}$ is known from the first equation. For the second row,

$$\begin{aligned}
&\Phi^+_{2,1}(s) = \Phi^-_{2,1}(s)U_{1,1}(s), && \Phi_{2,1}(\infty) = 0,\\
&\Phi^+_{2,2}(s) = \Phi^-_{2,2}(s)U_{2,2}(s) + \Phi^-_{2,1}(s)U_{1,2}(s), && \Phi_{2,2}(\infty) = 1,\\
&\Phi^+_{2,3}(s) = \Phi^-_{2,3}(s)U_{3,3}(s) + \Phi^-_{2,2}(s)U_{2,3}(s) + \Phi^-_{2,1}(s)U_{1,3}(s), && \Phi_{2,3}(\infty) = 0,\\
&\quad\vdots
\end{aligned}$$

From the condition $\operatorname{ind} U_{1,1}(s) = 0$ we know, $\Phi_{2,1}(z) = 0$, which means that the RH problem for $\Phi_{2,2}$ is homogeneous and the condition at infinity can be satisfied. In general, for row j, the first $j-1$ entries vanish identically. We present the general procedure.

1. Solve

$$\Phi^+_{j,j}(s) = \Phi^-_{j,j}(s)U_{j,j}(s), \quad \Phi_{j,j}(\infty) = 1, \quad j = 1, \ldots, n.$$

   All of these solutions exist and are unique by the imposed index conditions.

2. For each $j = 1, \ldots, n$, solve for $i = 1, \ldots, n-j$

$$\Phi^+_{j,j+i}(s) = \Phi^-_{j,j+i}(s)U_{j+i,j+i}(s) + F_{i,j}(s),$$
$$F_{i,j}(s) = \sum_{k=0}^{i-1} \Phi^-_{j,j+k}(s)U_{j+k,j+i}(s).$$

[19] This can be generalized by using the techniques introduced for scalar RH problems.

The resulting solution is unique. This can be shown by the same argument used at the end of Section 2.4.1.

Remark 2.4.1. *For a general problem of the form*

$$\Phi^+(s) = \Phi^-(s)A,$$

where A is a constant, possibly nondiagonalizable, matrix, we proceed by finding its Jordan normal form and applying the approach for upper triangular RH problems. Alternatively, one can use the Schur decomposition $A = QUQ^*$, where Q is a unitary matrix and U is an upper triangular matrix, to reduce the problem to upper triangular form. Jordan normal form is unstable under perturbations in A; hence the Schur decomposition is preferable for numerics.

We end this section with an important example that connects matrix RH problems with scalar RH problems.

Example 2.19. Consider the continuous RH problem

$$\Phi^+(s) = \Phi^-(s)\begin{bmatrix} 1 & f(s) \\ 0 & 1 \end{bmatrix}, \quad f \in C^{0,\alpha}(\Gamma), \quad \Phi(\infty) = I.$$

This arose in Problem 1.5.1 (with $n = 0$) for the inverse spectral problem associated with the Jacobi operator. We follow the general procedure. First solve

$$\Phi_{1,1}^+(s) = \Phi_{1,1}^-(s), \quad \Phi_{1,1}(\infty) = 1,$$
$$\Phi_{2,2}^+(s) = \Phi_{2,2}^-(s), \quad \Phi_{2,2}(\infty) = 1.$$

It is clear that $\Phi_{1,1} = \Phi_{2,2} = 1$. It remains to find $\Phi_{1,2}$:

$$\Phi_{1,2}^+(s) = \Phi_{1,2}^-(s) + f(s), \quad \Phi_{1,2}(\infty) = 0.$$

Therefore $\Phi_{1,2}(z) = \mathscr{C}_\Gamma f(z)$ and

$$\Phi(z) = \begin{bmatrix} 1 & \mathscr{C}_\Gamma f(z) \\ 0 & 1 \end{bmatrix}. \quad \blacksquare$$

2.5 ▪ Hardy spaces and Cauchy integrals

In this section we discuss Cauchy integrals of functions analytic off a contour which have boundary values on that contour that live in an appropriate space. This is a natural setting in which to study RH problems. Consider an L^2 RH problem for a function Φ that tends to the identity at infinity. We write $\Phi(z) = I + \mathscr{C}_\Gamma u(z)$ for $u \in L^2(\Gamma)$ to reduce the RH problem to the problem of finding u, as described in Section 2.7 below. To justify this representation of Φ, we must carefully study the convergence of $\mathscr{C}_\Gamma u(z)$ as z approaches Γ.

Essentially, the theory of Hardy spaces allows for the extension of the Cauchy integral formula to a larger class of functions. It also allows precise properties of the Cauchy integral to be established. The following results are closely tied to L^p spaces. All results are proved for $p = 2$. When the generality does not distract from the end goal we state results for general p.

2.5.1 ▪ Hardy spaces on the unit disk

Let $f(z)$ be analytic for $z \in \mathbb{D} \triangleq \{z \in \mathbb{C} : |z| < 1\}$. For $r < 1$ and $0 < p \leq \infty$ we define the quantity

$$M_p(f, r) = \int_{|z|=r} |f(z)|^p |dz|.$$

Definition 2.20. *We say that a function is of Hardy class $\mathcal{H}^p$ if*

$$\sup_{r<1} M_p(f, r) < \infty.$$

We state an essential result; see [47, p. 169].

Theorem 2.21. *If $f \in \mathcal{H}^p$ for $0 < p < \infty$, then $f(z) = \lim_{r\to 1^-} f(rz)$ exists for a.e. $z \in \partial\mathbb{D}$. Also*

$$\lim_{r\to 1^-} M_p(f, r) = M_p(f, 1)$$

and

$$\lim_{r\to 1^-} \int_{\partial\mathbb{D}} |f(rz) - f(z)|^p |dz| = 0.$$

Another result that is of fundamental importance for what follows is also found in [47, p. 170].

Theorem 2.22. *Every function $f \in \mathcal{H}^1$ can be expressed as the Cauchy integral of its boundary function. In fact,*

$$f^{(k)}(z) = k! \int_{\partial\mathbb{D}} \frac{f(\zeta)}{(\zeta - z)^{k+1}} d\zeta, \quad |z| < 1, \quad k \subset \mathbb{N},$$

and this integral vanishes identically for $|z| > 1$.

Since $\partial\mathbb{D}$ has finite measure, $\mathcal{H}^p \subset \mathcal{H}^1$ for all $p > 1$.

2.5.2 ▪ Hardy spaces on general domains

The theory for the unit circle can be extended to other domains. We preliminarily consider bounded domains, following [47, Chapter 10]. Assume $D \subset \mathbb{C}$ is a simply connected open set and ∂D is a rectifiable (i.e., has bounded variation) *Jordan curve* (i.e., a non-self-intersecting, continuous closed curve).

Definition 2.23. *A function $f(z)$ analytic in D is of class $\mathcal{E}^p(D)$ if there exists a sequence C_n, $n = 1, 2, \ldots$, of rectifiable curves in D tending to ∂D in the sense that C_n eventually surrounds every compact subset of D such that*

$$\sup_n \int_{C_n} |f(z)|^p |dz| < \infty.$$

We summarize some results from [47, pp. 169–170] in the following theorems.

Theorem 2.24. *For $0 < p < \infty$, every function $f \in \mathscr{E}^p(D)$ has a non-tangential limit at the boundary of D, the boundary function $f \in L^p(\partial D)$ cannot vanish on a set of positive measure unless $f \equiv 0$, and for $p \geq 1$, the Cauchy integral formula holds:*

$$f(z) = \int_{\partial D} \frac{f(s)}{s - z} đs, \quad z \in D.$$

Theorem 2.25. *Let $\phi(w)$ map $\{|w| < 1\}$ conformally onto D, and let Γ_r be the image of $\{|w| = r\}$ under ϕ. Then the following are equivalent:*

- $\sup_{r<1} \int_{\Gamma_r} |f(z)|^p |dz| < \infty$,
- $f(z) \in \mathscr{E}^p(D)$,
- $f(z) = \mathscr{C}_{\partial D} g(z)$ *for some* $g \in L^p(\partial D)$,
- $F(w) = f(\varphi(w))[\varphi'(w)]^{1/p} \in \mathscr{H}^p$ *for some conformal mapping $\varphi(w)$ of $\{|w| < 1\}$ onto D, and*
- $F(w) \in \mathscr{H}^p$ *for all such mappings* φ.

Furthermore, as we state below, Theorem 2.22 holds with $\partial \mathbb{D}$ replaced by ∂D. This shows that Definition 2.23 is a consistent extension of Definition 2.20.

Remark 2.5.1. *If there exists $C > 1$ such that*

$$\frac{1}{C} < |\varphi'(w)| < C,$$

then $\mathscr{E}^p(D)$ and $\mathscr{H}^p$ are isomorphic.

Define the space $\mathscr{E}^p(D_-)$ which is the class of functions f analytic on $\mathbb{C} \setminus \overline{D}$ with finite L^p norms as curves approach ∂D in analogy with Definition 2.23. For the Cauchy integral formula to hold we need to impose the restriction that $f(\infty) = 0$.

We now make the extension to unbounded domains. One approach is to conformally map the unbounded domain to a bounded domain. A canonical example is the fractional transformation

$$x = \mathrm{i}\frac{z+1}{z-1},$$

to map the unit circle in the z-plane to the real axis in the x-plane. If this is used as a change of variables in an integral, then

$$dz \mapsto \frac{2}{\mathrm{i}(x-\mathrm{i})^2} dx.$$

Proceeding this way, the Hardy space on the line does not share a nice relationship with $L^p(\mathbb{R})$. As we see below, $L^p(\mathbb{R})$ is critical for the study of the Cauchy integral operators.[20]

[20] There is an alternate approach that relies on modifying the Cauchy kernel [124].

Definition 2.26. *Denote the class of Jordan curves which tend to straight lines at infinity by Σ_∞. More precisely, Σ_∞ consists of the set of Jordan curves where $\Sigma_\infty^{-1} \triangleq \{z^{-1} : z \in \Sigma_\infty\}$ has a transverse, and rectifiable, intersection at the origin.*

We now use all of these ideas to deal with a simply connected, unbounded domain with $\partial D \in \Sigma_\infty$. This is always assumed unless specified otherwise. The remainder of this section proceeds as follows.

Definition 2.27. *When D is unbounded, a function f analytic in D is of class $\mathcal{E}^p(D)$, $1 < p < \infty$, if there exists a sequence C_n, $n = 1,2,\dots$, of bounded, closed curves in D satisfying*

$$\sup_n \int_{C_n} \frac{|dz|}{|z-a|^q} < \infty, \quad \textit{for some} \quad a \in \mathbb{C} \setminus \overline{D},$$
$$1/p + 1/q = 1,$$

that tend to ∂D in the sense that C_n eventually surrounds every compact subset of D such that

$$\sup_{n \geq 1} \int_{C_n} |f(z)|^p |dz| < \infty.$$

We show below that an explicit sequence C_n can be taken, in general, when ∂D satisfies some additional assumptions. The following is adapted from [124].

Theorem 2.28. *For $\partial D \in \Sigma_\infty$ and $f \in \mathcal{E}^p(D)$, $1 \leq p < \infty$, nontangential boundary values exist a.e., the boundary function $f \in L^p(\partial D)$, and the Cauchy integral formula holds.*

Proof. We assume that $a = 0 \in \mathbb{C} \setminus \overline{D}$ in Definition 2.27. Consider the conformal map $z \mapsto 1/z$. It is clear that the Cauchy integral formula holds if and only if

$$\frac{1}{z} f\left(\frac{1}{z}\right) = \int_{\partial D^{-1}} \frac{\frac{1}{\zeta} f\left(\frac{1}{\zeta}\right)}{\zeta - z} d\zeta, \quad z \in D^{-1},$$

where $D^{-1} = \{z : 1/z \in D\}$. Let $\Gamma \in \Sigma_\infty$ be a curve in D. Then Γ^{-1} is a rectifiable curve in $\overline{D^{-1}}$ and

$$\int_{\Gamma^{-1}} |1/z f(1/z)||dz| = \int_\Gamma |1/z f(z)||dz| \leq \|1/\diamond\|_{L^q(\Gamma)} \|f\|_{L^p(\Gamma)}, \quad 1/p + 1/q = 1.$$

Replacing Γ with C_n we have that $1/\diamond f(1/\diamond) \in \mathcal{E}^1(D^{-1})$ since $\sup_n \|1/\diamond\|_{L^q(C_n)} < \infty$. Therefore the Cauchy integral formula holds and the nontangential limits exist a.e. by Theorem 2.24.

To examine the boundary function, for positive integers k, m, consider

$$F(z) = z^{-2k} f(1/z)^m,$$

which is analytic in D^{-1}. For $\Gamma \in \Sigma_\infty$, $\Gamma \subset D$, consider

$$\int_{\Gamma^{-1}} |F(z)|^{p/m} |dz| = \int_\Gamma |z|^{2kp/m-2} |f(z)|^p |dz| \leq \sup_{z \in \Gamma} |z|^{2kp/m-2} \int_\Gamma |f(z)|^p |dz|.$$

Again, replacing Γ with C_n we see that provided $kp/m \leq 1$, $F \in \mathscr{E}^{p/m}(D^{-1})$ and the boundary function satisfies (see [47, Theorems 10.1 and 10.3])

$$\int_{\partial D^{-1}} |F(z)|^{p/m} |\mathrm{d}z| = \int_{\partial D} |z|^{2kp/m-2} |f(z)|^p |\mathrm{d}z| \leq \sup_{z \in \partial D} |z|^{2kp/m-2} \limsup_n \int_{C_n} |f(z)|^p |\mathrm{d}z|.$$

Now choose k and m such that $kp/m \uparrow 1$, and the dominated convergence theorem gives that

$$\int_{\partial D} |f(z)|^p |\mathrm{d}z| \leq \limsup_n \int_{C_n} |f(z)|^p |\mathrm{d}z|. \qquad \square$$

We look to provide additional characterization of $\mathscr{E}^p(D)$, and to do this we need to restrict the class of curves.[21] We assume ∂D is Lipschitz and in Σ_∞, i.e., after possible rotation,

$$\partial D = \{x + \mathrm{i}\nu(x) : x \in \mathbb{R}\},$$

where ν is real-valued and $\|\nu'\|_\infty < \infty$. Such a curve is referred to as a *Lipschitz graph*. Note that we require a Lipschitz graph to also lie in Σ_∞. With this restriction, the remainder of this section proceeds as follows. We have shown for $f \in \mathscr{E}^p(D)$, $1 \leq p \leq \infty$, that the boundary function $f \in L^p(\partial D)$. We define a simple seminorm on a subspace of $\mathscr{E}^p(D)$. Once we obtain an estimate on the operator norm of the Cauchy integral operator directly in terms of Lipschitz constants we use that $f \in L^p(\partial D)$ to show that this subspace is actually all of $\mathscr{E}^p(D)$ and the seminorm is a bona fide norm. From here, we further simplify the characterization of $\mathscr{E}^p(D)$.

Consider the sequence $(L_n)_{n\geq 1}$, given by a shift of ∂D:

$$L_n = \{x + \mathrm{i}\nu(x) + \mathrm{i}/n : x \in \mathbb{R}\}.$$

Note that this is not a bounded, closed curve and it does not satisfy the necessary properties given above for C_n. But, in particular, the curves L_n are Lipschitz with a uniformly bounded Lipschitz constant. Define the region above[22] L_n by $L_n^+ \triangleq \{z : \operatorname{Im} z > \nu(\operatorname{Re} z) + 1/n\}$ and similarly $\Gamma^+ \triangleq \{z : \operatorname{Im} z > \nu(\operatorname{Re} z)\}$. Define the distance to the curve

$$d_\Gamma(z) = \inf_{a \in \Gamma} |a - z|.$$

It is clear that

$$|d_\Gamma(z) - d_{L_n}(z)| < 1/n \quad \forall z \in L_n^+.$$

The next lemma follows in much the same way as Theorem 2.28.

Lemma 2.29. *The map* $f \mapsto \|f\|_{\mathscr{E}^p(D)} \triangleq \limsup_{n\to\infty} \|f\|_{L^p(L_n)}$ *satisfies*

$$\|f\|_{L^p(\partial D)} \leq \|f\|_{\mathscr{E}^p(D)}. \tag{2.21}$$

Additionally, $\|f\|_{\mathscr{E}^p(D)}$ *defines a norm on* $\check{\mathscr{E}}^p(D) \triangleq \{f \in \mathscr{E}^p(D) : \|f\|_{\mathscr{E}^p(D)} < \infty\}$.

[21] We impose a restriction only to simplify the presentation.

[22] Note that Γ^+ coincides with Ω^+ where $\mathbb{C} \setminus \Gamma = \Omega^+ \cup \Omega^-$ as in Definition 2.1. We use this alternate notation in this section for convenience.

Proof. It is clear that $||\diamond||_{\mathcal{E}^p(D)}$ defines a seminorm. Since the Cauchy integral formula holds for each $f \in \mathcal{E}^p(D)$, (2.21) shows it is a norm. Now we prove (2.21). It follows from previous results that $z^{-1}f(z^{-1}) \in \mathcal{E}^p(D^{-1})$ and has limits a.e. on ∂D^{-1}. This implies

$$f(z+i/n) \to f(z) \text{ a.e.}$$

We have for $n > 1$

$$\int_{L_n} |f(z)|^p |dz| = \int_\Gamma |f(z+i/n)|^p |dz|.$$

Fatou's lemma (see [54, p. 52]) gives

$$\int_\Gamma |f(z)|^p|dz| \leq \liminf_{n\to\infty} \int_\Gamma |f(z+i/n)|^p|dz| \leq \limsup_{n\to\infty} \int_{L_n} |f(z)|^p|dz|,$$

and the lemma is proved. □

We now work towards mapping properties of the Cauchy integral operator and its relation to Hardy spaces and show, in the process, that $\check{\mathcal{E}}^p(D) = \mathcal{E}^p(D)$.[23] There are many ways to proceed, but a quite direct way is outlined in [21]. The proofs of these intermediate results require estimates from harmonic analysis that are beyond the scope of this book. Given a Lipschitz graph Γ, define the weighted *Bergman space* by

$$\mathcal{B}^+(\Gamma) = \{f \text{ holomorphic in } \Gamma^+ : ||f||_{\mathcal{B}^+(\Gamma)} < \infty\},$$

with the inner product defined by

$$\langle g, f\rangle_{\mathcal{B}^+(\Gamma)} = \iint_{\Gamma^+} g(z)\overline{f(z)} d_\Gamma(z) dx dy, \quad z = x + iy,$$

so that $\mathcal{B}^+(\Gamma)$ is a Hilbert space [21].

We pause for some technical lemmas that are required in what follows. Note that the following results make explicit use of the Hilbert space structure. Therefore we restrict ourselves to $p = 2$. The main results here, Lemma 2.32 and Theorem 2.35, hold for $1 < p < \infty$ [82, p. 184].

Lemma 2.30 (See [21]). *Suppose Γ is a Lipschitz graph and F is analytic in Γ^+ and decays to zero at infinity. Then*

$$||F||_{L^2(\Gamma)} \leq C(1 + ||\nu'||_\infty)||F'||_{\mathcal{B}^+(\Gamma)}.$$

Lemma 2.31 (See [21]). *Let Γ be a Lipschitz graph and let $f \in \mathcal{E}^2(\Gamma^+)$. Define*

$$\mathcal{J}f(\zeta) = \iint_{\Gamma^+} \frac{f(z)d_\Gamma(z)}{(z-\zeta)^2} dx dy, \quad \zeta \in \Gamma.$$

Then $||\mathcal{J}f||_{L^2(\Gamma)} \leq C(1 + ||\nu'||_\infty)||f||_{\mathcal{B}^+(\Gamma)}$.

[23] This holds for any $1 < p < \infty$, but we only prove it for $p = 2$.

Lemma 2.32. *Let Γ be a Lipschitz graph and $\Gamma_1 = \{x + \mathrm{i}\nu_1(x) : x \in \mathbb{R}\}$ be a Lipschitz graph in $\overline{\Gamma^+}$. Then for $f \in L^2(\Gamma)$*

$$\|\mathscr{C}_\Gamma^+ f\|_{L^2(\Gamma_1)} \leq C(1+\|\nu_1'\|_\infty)(1+\|\nu'\|_\infty)\|f\|_{L^2(\Gamma)}.$$

Proof. To prove this we follow [21]. Let

$$B = \{h \in \mathscr{E}^2(\Gamma^+) : \|h\|_{\mathscr{E}^2(\Gamma^+)} \leq 1,\ h \text{ compactly supported in } \Gamma^+\}.$$

Then we know that for any $f \in \mathscr{E}^2(\Gamma^+)$, $\|f\|_{\mathscr{E}^2(\Gamma^+)} = \sup_{h\in B} |\langle f, h\rangle_{\mathscr{E}^2(\Gamma^+)}|$. Also, denote $\mathscr{C}_\Gamma^{+\prime} f(z) = \mathrm{d}/\mathrm{d}z \mathscr{C}_\Gamma^+ f(z)$. Lemma 2.30 applied to Γ_1 and Γ_1^+ along with Fubini's theorem gives

$$\begin{aligned}\|\mathscr{C}_\Gamma^+ f\|_{L^2(\Gamma_1)} &\leq C(1+\|\nu_1'\|_\infty)\|\mathscr{C}_\Gamma^{+\prime} f\|_{\mathscr{B}^+(\Gamma_1)} \\ &= C(1+\|\nu_1'\|_\infty)\left(\iint_{\Gamma_1^+} |\mathscr{C}_\Gamma^{+\prime} f(z)|^2 d_{\Gamma_1}(z)\mathrm{d}x\mathrm{d}y\right)^{1/2}.\end{aligned}$$

From the choice of Γ_1 we have that $d_{\Gamma_1}(z) \leq d_\Gamma(z)$ so that

$$\begin{aligned}\iint_{\Gamma_1^+} |\mathscr{C}_\Gamma^{+\prime} f(z)|^2 d_{\Gamma_1}(z)\mathrm{d}x\mathrm{d}y &\leq \iint_{\Gamma^+} |\mathscr{C}_\Gamma^{+\prime} f(z)|^2 d_\Gamma(z)\mathrm{d}x\mathrm{d}y \\ &= \sup_{h\in B} |\langle \mathscr{C}_\Gamma^{+\prime} f, h\rangle_{\mathscr{B}^+(\Gamma)}|^2 \\ &= \sup_{h\in B}\left|\iint_{\Gamma^+}\left(-\int_\Gamma \frac{f(\zeta)\mathrm{d}\zeta}{(z-\zeta)^2}\right)\overline{h(z)} d_\Gamma(z)\mathrm{d}x\mathrm{d}y\right|^2 \\ &= C\sup_{h\in B}\left|\int_\Gamma f(\zeta)\mathscr{J}(\overline{h})(\zeta)\mathrm{d}\zeta\right|^2 \\ &\leq C\|f\|_{L^2(\Gamma)}^2 \sup_{h\in B}\|\mathscr{J}(\overline{h})\|_{L^2(\Gamma)}^2 \\ &\leq C(1+\|\nu'\|_\infty)^2\|f\|_{L^2(\Gamma)}^2.\end{aligned}$$

This proves the lemma. □

Before the next theorem, we prove two technical lemmas.

Lemma 2.33. *Let $\Gamma = \{x + \mathrm{i}\nu(x) : x \in \mathbb{R}\}$ be a Lipschitz graph and let $S_\rho = \{x + \mathrm{i}\nu(x) + \mathrm{i}\rho : x \in \mathbb{R}\}$. Then for $z \in S_\rho$*

$$\int_\Gamma \frac{|\mathrm{d}s|}{|s-z|^2} \leq 8(1+\|\nu'\|_\infty)^2/\rho.$$

Proof. For $y \in \mathbb{R}$ we must consider

$$F(y) = \int_{\mathbb{R}} \frac{|1+\mathrm{i}\nu'(x)|\mathrm{d}x}{(x-y)^2 + (\nu(x)-\nu(y)-\rho)^2}.$$

Then

$$\frac{|1+\mathrm{i}\nu'(x)|}{(x-y)^2+(\nu(x)-\nu(y)-\rho)^2} \le (1+\|\nu'\|_\infty)\begin{cases} 4/\rho^2 & \text{if } |x-y| \le \frac{\rho}{2}(1+\|\nu'\|_\infty)^{-1}, \\ |x-y|^{-2} & \text{otherwise.} \end{cases}$$

By explicit integration

$$F(y) \le (4+4(1+\|\nu'\|_\infty)^2)/\rho,$$

and the claim follows. □

Lemma 2.34. *Let $\Gamma = \{x+\mathrm{i}\nu(x) : x \in \mathbb{R}\}$ be a Lipschitz graph and let $S_\rho = \{x+\mathrm{i}\nu(x)+\mathrm{i}\rho : x \in \mathbb{R}\}$. If $f \in L^2(\Gamma)$, then for any $\rho > 0$*

$$\lim_{R\to\infty} \sup_{|z|=R,\ z\in S_\rho^+} |\mathscr{C}_\Gamma f(z)| = 0,$$

i.e., $\mathscr{C}_\Gamma f(z) = o(1)$ in S_ρ^+ as $z \to \infty$.

Proof. Let $\Gamma_R = \Gamma \cap B(0,R/2)$, and we consider

$$\mathscr{C}_\Gamma f(z) = \mathscr{C}_{\Gamma_R} f(z) + \mathscr{C}_{\Gamma\setminus\Gamma_R} f(z).$$

For $|z| = R$ we have

$$|\mathscr{C}_{\Gamma_R} f(z)| \le \frac{1}{\pi R}\|f\|_{L^2(\Gamma)}|\Gamma_R|^{1/2} = \mathcal{O}(R^{-1/2})$$

as $R \to \infty$. Then for $z \in S_\rho^+$, using Lemma 2.33,

$$|\mathscr{C}_{\Gamma\setminus\Gamma_R} f(z)| \le \|f\|_{L^2(\Gamma\setminus\Gamma_R)} \sup_{z\in S_\rho^+}\left(\int_\Gamma \frac{|\mathrm{d}s|}{|s-z|^2}\right)^{1/2} \le \frac{\sqrt{2}}{\pi\rho^{1/2}}\|f\|_{L^2(\Gamma\setminus\Gamma_R)}(1+\|\nu'\|_\infty) = o(1)$$

as $R \to \infty$. □

Theorem 2.35. *Assume Γ is a Lipschitz graph given by $\Gamma = \{x+\mathrm{i}\nu(x) : x \in \mathbb{R}\}$. Then $\mathscr{C}_\Gamma^+ g \in \check{\mathscr{E}}^2(\Gamma^+)$ whenever $g \in L^2(\Gamma)$, and hence $\check{\mathscr{E}}^2(\Gamma^+) = \mathscr{E}^2(\Gamma^+)$. Furthermore, $\mathscr{E}^2(\Gamma^+)$ consists of functions f, analytic in Γ^+ such that*

$$\sup_{0<\rho<\infty} \int_{S_\rho} |f(z)|^2|\mathrm{d}z| < \infty, \quad S_\rho = \{x+\mathrm{i}\nu(x)+\mathrm{i}\rho : x \in \mathbb{R}\}. \tag{2.22}$$

Proof. For the first claim it suffices to show that $\mathscr{C}_\Gamma^+$ is a bounded operator from $L^2(\Gamma)$ to $L^2(L_n)$ and the bound can be taken to be uniform in n. Applying Lemma 2.32 to the sequence $\{L_n\}$ demonstrates this. It is clear that $\check{\mathscr{E}}^2(\Gamma^+) \subset \mathscr{E}^2(\Gamma^+)$. For $g \in \mathscr{E}^2(\Gamma^+)$ we know that the Cauchy integral formula holds: $g = \mathscr{C}_\Gamma g$ and $g \in L^2(\Gamma)$, and hence $g \in \check{\mathscr{E}}^2(\Gamma^+)$.

It remains to prove the second claim. Any $f \in \mathscr{E}^2(\Gamma^+)$ satisfies (2.22) by Lemma 2.32. We need to show that if (2.22) is satisfied, then $f \in \mathscr{E}^2(\Gamma^+)$. Under this assumption it is first shown that for $\rho_1 > 0$

$$f(z) = \int_{S_{\rho_1}} \frac{f(s)}{s-z} \text{đ}s, \quad \nu(\operatorname{Re} z) + \rho_1 < \operatorname{Im} z. \tag{2.23}$$

We use ideas introduced by Titchmarsh [107] to demonstrate this. Consider the contour $S_{\rho_1,\rho_2,R}$ defined to be the boundary of the region enclosed by S_{ρ_1}, S_{ρ_2}, and $I_R = \{z : \operatorname{Re} z = \pm R\}$. Then for fixed z in the region and $M > 1$

$$f(z) = \int_M^{M+1} \int_{S_{\rho_1,\rho_2,R}} \frac{f(s)}{s-z} \text{đ}s \, \mathrm{d}R.$$

It follows for $j = 1,2$ that

$$\int_{S_{\rho_j} \setminus \{-R < \operatorname{Re} z < R\}} \frac{f(s)}{s-z} \mathrm{d}s \to 0$$

as $R \to \infty$. Then

$$\begin{aligned}
&\int_M^{M+1} \int_{S_{\rho_1,\rho_2,R} \cap I_R} \frac{f(s)}{s-z} \text{đ}s \, \mathrm{d}R \\
&\quad = \int_{\rho_1}^{\rho_2} \int_{S_t \cap \{M < \operatorname{Re} z < M+1\}} \frac{f(s)}{s-z} \text{đ}s \, \mathrm{d}t - \int_{\rho_1}^{\rho_2} \int_{S_t \cap \{-M-1 < \operatorname{Re} z < -M\}} \frac{f(s)}{s-z} \text{đ}s \, \mathrm{d}t.
\end{aligned}$$

We estimate with the Cauchy–Schwarz inequality

$$\left| \int_{\rho_1}^{\rho_2} \int_{S_t \cap \{M < \operatorname{Re} z < M+1\}} \frac{f(s)}{s-z} \mathrm{d}s \, \mathrm{d}t \right| \leq C^{1/2} \int_{\rho_1}^{\rho_2} \left(\int_{S_t \cap \{M < \operatorname{Re} z < M+1\}} \frac{|\mathrm{d}s|}{|s-z|^2} \right)^{1/2} \mathrm{d}t.$$

This tends to zero as $M \to \infty$. A similar estimate follows for the integral over $[-M-1, M]$. Then from

$$f(z) = \lim_{M \to \infty} \int_M^{M+1} \int_{S_{\rho_1,\rho_2,R}} \frac{f(s)}{s-z} \text{đ}s \, \mathrm{d}R,$$

and then taking $\rho_2 \to \infty$, (2.23) follows. Now, let $\rho_1 = 1/n$, $\rho_2 = n$, and we determine $R(n) \to \infty$ so that

$$\sup_{n>1} \int_{S_{1/n,n,R(n)}} |f(z)|^2 |\mathrm{d}z| < \infty. \tag{2.24}$$

It remains to estimate the norm on the vertical segments of the contour. Such an $R(n)$ must exist by Lemma 2.34 because $f(z) = \mathscr{C}_{S_{1/(2n)}} f(z)$, where $f \in L^2(S_{1/(2n)})$. This shows (2.24). □

The following corollary is our fundamental result concerning L^2-based Hardy spaces.

Corollary 2.36. *The following are consequences of Theorem* 2.35. *Assume that* $\Gamma \in \Sigma_\infty$ *is a Lipschitz graph given by* $\{x + \mathrm{i}\nu(x) : x \in \mathbb{R}\}$ *after rotation; then*

- $\mathscr{E}^2(\Gamma^+) = \{\mathscr{C}_\Gamma f : f \in L^2(\Gamma)\}$,
- $\|\diamond\|_{L^2(\Gamma)}$ *and* $\|\diamond\|_{\mathscr{E}^2(\Gamma^+)}$ *define equivalent norms on* $\mathscr{E}^2(\Gamma^+)$,
- $\mathscr{E}^2(\Gamma^+)$ *is a Hilbert space, and*
- $\mathscr{C}_\Gamma^\pm$ *is well-defined and bounded from* $L^2(\Gamma)$ *to itself with nontangential boundary values a.e.*

Since the Cauchy integral formula holds when applied to $\mathscr{E}^2(\Gamma^+)$ functions, for any $f \in \mathscr{E}^2(\Gamma^+)$ we have that $\mathscr{C}_\Gamma^+ f - f = 0$ a.e. on Γ for $f \in \mathscr{E}^2(\Gamma^+)$. It follows from Theorem 2.35 that

$$\left\| \frac{\mathrm{d}}{\mathrm{d}z} \mathscr{C}_\Gamma(\mathscr{C}^+{}_\Gamma f - f) \right\|_{\mathscr{B}^+(\Gamma)} \leq \|\mathscr{C}_\Gamma^+ f - f\|_{L^2(\Gamma)} = 0.$$

Therefore $\mathscr{C}_\Gamma^+(\mathscr{C}_\Gamma^+ f - f)$ is constant in all of Γ^+. That constant must be zero from the $z \to \infty$ limit. We obtain

$$\mathscr{C}_\Gamma^+ \mathscr{C}_\Gamma^+ f = \mathscr{C}_\Gamma^+ f, \tag{2.25}$$

and $\mathscr{C}_\Gamma^+$ is a projection on $L^2(\Gamma)$. Also, the Plemelj formula holds:

$$\mathscr{C}_\Gamma^+ f - \mathscr{C}_\Gamma^- f = f \tag{2.26}$$

for $f \in L^2(\Gamma)$. This can be seen by the fact that for $f \in L^2(\Gamma)$, $F(x) = f(\ell(x)) \in L^2(\mathbb{R})$, $\ell(x) = x + i\nu(x)$. Approximate F by step functions F_n in $L^2(\mathbb{R})$. Then $f_n = F_n \circ \ell^{-1}$ converges to f in $L^2(\Gamma)$. It is straightforward to check from previous results that $\mathscr{C}_\Gamma^+ f_n(s) - \mathscr{C}_\Gamma^- f_n(s) = f_n(s)$ a.e. This implies the Plemelj formula for all $L^2(\Gamma)$ functions. Alternatively, the result follows from Theorem D.6.

We know that if $f \in \mathscr{E}^2(\Gamma^+)$, then $\mathscr{C}_\Gamma^+ f = f$. This implies that

$$\mathscr{C}_\Gamma^- f = 0, \quad f \in \mathscr{E}^2(\Gamma^+). \tag{2.27}$$

If we consider $\mathscr{E}^2(\Gamma^-)$, $\Gamma^- \triangleq \{z : \operatorname{Im} z < \nu(\operatorname{Re} z)\}$, then we replace $+$ with $-$ in (2.27) and (2.25) becomes

$$\mathscr{C}_\Gamma^- \mathscr{C}_\Gamma^- f = -\mathscr{C}_\Gamma^- f, \tag{2.28}$$

when we take orientation into account.

2.5.3 ▪ An application of Hardy spaces to PDEs

In this section we discuss a derivation of the classical solution formula for linear, constant-coefficient PDEs on the line via RH problems. This is an important and elementary application of the theory of Hardy spaces. The critical observation here can be found in the work of Fokas and Gel'fand [51] in which they derived so-called *Lax pairs* (or Lax representations) for linear, constant-coefficient evolution PDEs. The Lax pair allows for an algorithmic derivation of a transform to solve the given PDE.

The tools used in this approach are of fundamental importance because they generalize in two ways. First, they are able to handle boundary-value problems on the half-line and finite interval [50] (see also [26]). The second is that the nonlinearization of this transform is the inverse scattering transform as presented in Chapter 3. It is worth noting that nonlinear boundary-value problems can also be solved with the Lax representation.

Consider solving

$$\begin{aligned} \mathrm{i}q_t + \omega(-\mathrm{i}\partial_x)q = 0, \quad x \in \mathbb{R},\ t \in (0,T], \\ q(\diamond,0) = q_0 \in H^n(\mathbb{R}), \end{aligned} \tag{2.29}$$

where subscripts denote partial differentiation and $\omega(z)$ is a polynomial of degree n. We assume global $H^n(\mathbb{R})$ existence of the solution. The well-known solution formula for this problem is

$$q(x,t) = \frac{1}{2\pi}\int_{\mathbb{R}} \mathrm{e}^{\mathrm{i}zx-\mathrm{i}\omega(z)t}\hat{q}_0(z)\mathrm{d}z, \quad \hat{q}_0(z) \triangleq \int_{\mathbb{R}} \mathrm{e}^{-\mathrm{i}zx} q_0(x)\mathrm{d}x. \tag{2.30}$$

We now derive this formula from first principles.

We require that the imaginary part of $\omega(z)$, $z \in \mathbb{R}$, be bounded from above. From the work of Fokas and Gel'fand there exists a system of ODEs[24]

$$\mu_x - \mathrm{i}z\mu = q, \quad z \in \mathbb{R}, \tag{2.31}$$

$$\mu_t + \mathrm{i}\omega(z)\mu = \sum_{j=0}^{n-1} c_j(z)\partial_x^j q, \tag{2.32}$$

where $\{c_j(z)\}$ is determined through the equality

$$\mathrm{i}\left(\frac{\omega(z)-\omega(l)}{z-l}\right)\bigg|_{l=-\mathrm{i}\partial_x} = c_j(z)\partial_x^j,$$

because the quantity in parentheses is a polynomial in z and l. An important formal calculation is that $\mu_{xt} = \mu_{tx} \Leftrightarrow q$ solves (2.29), ignoring the initial condition. The conditions (2.31) and (2.32) are referred to as a Lax pair.

We obtain two simultaneous solutions of (2.31) and (2.32) given by

$$\mu^+(x,t;z) = \int_{-\infty}^{x} \mathrm{e}^{\mathrm{i}z(x-s)} q(s,t)\mathrm{d}s, \quad \mu^-(x,t;z) = -\int_{x}^{\infty} \mathrm{e}^{-\mathrm{i}z(s-x)} q(s,t)\mathrm{d}s.$$

Consider the difference

$$\nu(x,t;z) = \mu^+(x,t;z) - \mu^-(x,t;z) = \int_{-\infty}^{\infty} \mathrm{e}^{\mathrm{i}z(x-s)} q(s,t)\mathrm{d}s. \tag{2.33}$$

It follows that ν satisfies

$$\begin{aligned} \nu_x - \mathrm{i}z\nu = 0, \\ \nu_t + \mathrm{i}\omega(z)\nu = 0. \end{aligned}$$

The first equation implies that $\nu(x,t;z) = A(t;z)\mathrm{e}^{\mathrm{i}zx}$ and the second that $\nu(x,t;z) = B(x;z)\mathrm{e}^{-\mathrm{i}\omega(z)t}$ for some functions A and B. Evaluating at $t=0$, $A(0;z)\mathrm{e}^{\mathrm{i}zx} = B(x;z)$, and therefore $\nu(x,t;z) = A(0;z)\mathrm{e}^{\mathrm{i}zx-\mathrm{i}\omega(z)t}$. Evaluating this at $x=t=0$, we find $A(0;z) = \hat{q}_0(z)$. We arrive at the relation

$$\mu^+(x,t;z) - \mu^-(x,t;z) = \mathrm{e}^{\mathrm{i}zx-\mathrm{i}\omega(z)t}\hat{q}_0(z). \tag{2.34}$$

[24]The assumptions on q_0 can be relaxed to $H^1(\mathbb{R})$ if we allow for distributional solutions of these ODEs. We use subscripts to denote partial differentiation.

We note that $\mu^{\pm}$ is analytic for z in $\mathbb{C}^{\pm}$ and we have established a jump condition. To finish setting up an RH problem we must prove that $\mu^{\pm}$ are in appropriate Hardy spaces. Furthermore, to make the method useful we must determine a way to extract $q(x,t)$ from $\mu^{\pm}$. We have two claims:

- $\mu^{\pm}(x,t;\diamond) \in \mathscr{E}^2(\mathbb{C}^{\pm})$, and
- $q(x,t) = \mathrm{i} \lim_{|z|\to\infty} z\mu^{\pm}(x,t;z)$.

Then (2.30) follows since[25] $\hat{q}_0 \in L^1(\mathbb{R})$ (see Lemma 2.13) and

$$\mu^{\pm}(x,t;z) = \mathscr{C}_{\mathbb{R}}\left[\mathrm{e}^{\mathrm{i}x\diamond - \mathrm{i}\omega(\diamond)t}\hat{q}_0(\diamond)\right](z), \quad z \in \mathbb{C}^{\pm}.$$

We show the claims only for μ^+ since they follow for μ^- in precisely the same way. Let $z = z_r + \mathrm{i}z_i$, $z_i > 0$, and it is clear that

$$\left\| \int_{-\infty}^{x} \mathrm{e}^{\mathrm{i}(\diamond + \mathrm{i}z_i)(x-s)} q(s,t)\mathrm{d}s \right\|_{L^2(\mathbb{R})} \leq \sqrt{2\pi}\|q(\diamond,t)\|_{L^2((-\infty,x))} \leq \sqrt{2\pi}\|q(\diamond,t)\|_{L^2(\mathbb{R})} \quad (2.35)$$

since the Fourier transform, appropriately defined, is unitary on $L^2(\mathbb{R})$. This shows that $\mu^+(x,t;\diamond) \in \mathscr{E}^2(\mathbb{C}^+)$. To show the second claim it suffices to show that $\mu_x^+(x,t;z) = o(1)$ as $|z| \to \infty$ for z bounded away from the real axis. We integrate the expression for μ_x^+ by parts:

$$\mu_x^+(x,t;z) = q(x,t) + \mathrm{i}z \int_{-\infty}^{x} \mathrm{e}^{\mathrm{i}z(x-s)} q(s,t)\mathrm{d}s = \int_{-\infty}^{x} \mathrm{e}^{\mathrm{i}z(x-s)} q_x(s,t)\mathrm{d}s.$$

Replacing q with q_x in (2.35) shows that $\mu_x^+(x,t;\diamond) \in \mathscr{E}^2(\mathbb{C}^+)$. To prove the last claim we appeal to Lemma 2.13.

It may seem unsatisfactory that we have used properties of the Fourier transform in this derivation. One often thinks of the Lax pair as a replacement for guessing the necessary transform. Invoking properties of the Fourier transform becomes necessary as one reduces regularity. If we used $q_0 \in \mathscr{S}(\mathbb{R})$ instead, i.e., a Schwartz class initial condition, we could avoid such issues. This section should be compared with its nonlinearization in Section 3.1.

2.5.4 ▪ Cauchy integrals on self-intersecting contours

As seen above, the theory of Cauchy integrals on unbounded contours is arguably more natural on L^p spaces. This development extends in a straightforward way to contours with self-intersections, provided that the contour can be separated into Lipschitz components.[26] We concentrate on $p = 2$, but the results extend to $1 < p < \infty$. We have shown that if Γ is a Lipschitz graph, then

$$\|\mathscr{C}_\Gamma^{\pm} f\|_{L^2(\Gamma)} \leq C(\Gamma)\|f\|_{L^2(\Gamma)},$$

where $C(\Gamma)$ depends continuously on the Lipschitz constant of Γ.

Corollary 2.37. *Assume Γ is a Lipschitz graph defined by $x + \mathrm{i}\nu(x)$, $x \in \mathbb{R}$, after possible rotation. Let $\Gamma_{a,b} = \{x + \mathrm{i}\nu(x) : x \in (a,b)\}$ for $a < b$ in the extended real numbers. Then for $f \in L^2(\Gamma_{a,b})$ and $c < d$*

---

[25] $q_0 \in H^1(\mathbb{R}) \Rightarrow (1 + |\diamond|)\hat{q}_0(\diamond) \in L^2(\mathbb{R}) \Rightarrow \hat{q}_0 \in L^1(\mathbb{R})$.

[26] Generally speaking, it is known that $\mathscr{C}_\Gamma^{\pm}$ is bounded if and only if Γ is a so-called *Carleson curve*. See [82, Chapter 12] and [75] for this general theory.

1. $\|\mathcal{C}^{\pm}_{\Gamma_{a,b}} f\|_{L^2(\Gamma_{c,d})} \leq C(\Gamma)\|f\|_{L^2(\Gamma_{a,b})}$, *and specifically,*
2. $\|\mathcal{C}^{\pm}_{\Gamma_{a,b}} f\|_{L^2(\Gamma_{a,b})} \leq C(\Gamma)\|f\|_{L^2(\Gamma_{a,b})}$,

with the same constant $C(\Gamma)$ *as in Lemma 2.32 with* $\nu_1 = \nu$.

Proof. Let f be defined on $\Gamma_{a,b}$. Extend it to Γ by $f = 0$ on $\Gamma \setminus \Gamma_{a,b}$. Then

$$\|\mathcal{C}^{\pm}_{\Gamma_{a,b}} f\|_{L^2(\Gamma_{c,d})} = \|\mathcal{C}^{\pm}_{\Gamma} f\|_{L^2(\Gamma_{c,d})} \leq \|\mathcal{C}^{\pm}_{\Gamma} f\|_{L^2(\Gamma)} \leq C(\Gamma)\|f\|_{L^2(\Gamma)} = C(\Gamma)\|f\|_{L^2(\Gamma_{a,b})}.$$

Set $c = a$, $d = b$ to obtain the second claim. □

Remark 2.5.2. *This corollary states that if one can extend a finite or semi-infinite contour* Γ *to be a Lipschitz graph by adding contours, then* $\mathcal{C}^{\pm}_{\Gamma}$ *is a bounded operator on* $L^2(\Gamma)$. *This can be done by adding contours with constant slope that do not increase the Lipschitz constant.*

Lemma 2.38. *Let* $\Gamma = \Gamma_1 \cup \cdots \cup \Gamma_L$. *Suppose there exist constants* $C_{ij} > 0$ *such that for* $f \in L^2(\Gamma_j)$, $\|\mathcal{C}^{\pm}_{\Gamma_j} f\|_{L^2(\Gamma_i)} \leq C_{ji}\|f\|_{L^2(\Gamma_j)}$. *Then*

$$\|\mathcal{C}^{\pm}_{\Gamma} f\|_{L^2(\Gamma)} \leq C\|f\|_{L^2(\Gamma)}, \quad C = \max_{1\leq j\leq n} \left(\sum_{i=1}^{n} C_{ji}^2\right)^{1/2}.$$

Proof. Define χ_i to be the characteristic function of Γ_i. We write

$$\begin{aligned}
\|\mathcal{C}^{\pm}_{\Gamma} f\|_{L^2(\Gamma)} &= \left\|\sum_{i=1}^{n} \chi_i \mathcal{C}^{\pm}_{\Gamma}\left(\sum_{j=1}^{n} \chi_j f\right)\right\|_{L^2(\Gamma)} = \left\|\sum_{i=1}^{n} \chi_i \sum_{j=1}^{n} \mathcal{C}^{\pm}_{\Gamma_j} f\right\|_{L^2(\Gamma)} \\
&\leq \sum_{i=1}^{n}\sum_{j=1}^{n} \left\|\chi_i \mathcal{C}^{\pm}_{\Gamma_j} f\right\|_{L^2(\Gamma)} = \sum_{i=1}^{n}\sum_{j=1}^{n} \left\|\mathcal{C}^{\pm}_{\Gamma_j} f\right\|_{L^2(\Gamma_i)} \\
&\leq \sum_{j=1}^{n}\sum_{i=1}^{n} C_{ji}\|f\|_{L^2(\Gamma_j)} \leq \max_j \left(\sum_{i=1}^{n} C_{ji}^2\right)^{1/2} \left(\sum_{j=1}^{n} \|f\|^2_{L^2(\Gamma_j)}\right)^{1/2}.
\end{aligned}$$

The final inequality follows from the Cauchy–Schwarz inequality. □

Lemma 2.39. *Suppose* $\Gamma = \{e^{i\phi}(x + i\nu(x)) + z_1 : c \leq x \leq d\}$ *and* $\Lambda = \{e^{i\theta}(x + i\gamma(x)) + z_2 : a \leq x \leq b\}$ *are segments of shifted Lipschitz graphs such that if* Λ *intersects* Γ, *then* $\Delta = \Gamma \cap \Lambda$ *has a finite number of connected components. If* $\|\nu'\|_\infty, \|\gamma'\|_\infty < 1$, *then* $\mathcal{C}^{\pm}_{\Gamma\cup\Lambda}$ *is bounded on* $L^2(\Gamma \cup \Lambda)$ *and*

$$\|\mathcal{C}^{\pm}_{\Gamma\cup\Lambda}\|_{\mathcal{L}(L^2(\Gamma\cup\Lambda))} \leq C\left(1 + \frac{1+\|\nu'\|_\infty}{1-\|\nu'\|_\infty} + \frac{1+\|\gamma'\|_\infty}{1-\|\gamma'\|_\infty}\right)^2, \tag{2.36}$$

where $C > 0$ *is a constant, independent of* Γ *and* Λ.

Proof. By shifting the contours and applying a rotation, we assume $\phi = z_1 = 0$. Then a reparameterization of Λ allows for $z_2 = 0$. By reversing orientation it suffices to assume

$|\theta| \leq \pi/2$. Extend $\nu(x) = \nu(d)$ for $x > d$ and $\nu(x) = \nu(c)$ for $x < c$, and, similarly for γ, $\gamma(x) = \gamma(b)$ for $x > b$ and $\gamma(x) = \gamma(a)$ for $x < a$. We assume the extended contours either intersect with a connected intersection or are disjoint. The general case is easily treated with these ideas. Next, we rotate the plane by the angle $-\theta/2$ so that we can consider the rotated contours[27]

$$\tilde{\Gamma} = \{e^{-i\theta/2}(x + i\nu(x)) : x \in \mathbb{R}\}, \quad \tilde{\Lambda} = \{e^{i\theta/2}(x + i\gamma(x)) : x \in \mathbb{R}\}.$$

It follows that $\tilde{\Gamma} = \{y + i\tilde{\nu}(y) : y \in \mathbb{R}\}$ for a Lipschitz function $\tilde{\nu}$ provided that $\|\nu'\|_\infty < |\cot\theta/2|$. To see this, write

$$e^{-i\theta/2}(x + i\nu(x)) = x\cos\frac{\theta}{2} - \nu(x)\sin\frac{\theta}{2} + i\left(x\sin\frac{\theta}{2} + \nu(x)\cos\frac{\theta}{2}\right) = y + i\tilde{\nu}(y).$$

Using the implicit function theorem we require $\|\nu'\|_\infty < |\cot\theta/2|$ so that $y = x\cos\frac{\theta}{2} - \nu(x)\sin\frac{\theta}{2}$ is uniquely solvable for $x(y)$. We also find

$$|x'(y)| \leq \frac{1}{\left|\cos\frac{\theta}{2} - \nu'(x(y))\sin\frac{\theta}{2}\right|},$$

$$|\tilde{\nu}'(y)| \leq \frac{\left|\tan\frac{\theta}{2}\right| + \|\nu'\|_\infty}{1 - \left|\tan\frac{\theta}{2}\right| \|\nu'\|_\infty} \leq \frac{1 + \|\nu'\|_\infty}{1 - \|\nu'\|_\infty}.$$

The same restriction is put on γ so that $\tilde{\Gamma} = \{y + i\tilde{\gamma}(y) : y \in \mathbb{R}\}$ for a Lipschitz function $\tilde{\gamma}$ and a similar estimate follows for $\tilde{\gamma}'$. Note that $|\cot\frac{\theta}{2}| \geq 1$ for the values of θ considered.

We turn to the estimation of the operator norm. Assume we have a connected intersection. Let Γ_1 and Γ_2 be the components of Γ before and after Δ, respectively, using orientation. Define Λ_1 and Λ_2 in an analogous manner. Consider the contour $\Gamma' = \Gamma_1 \cup \Delta \cup \Lambda_2$. It is straightforward to see that the Lipschitz constant of this contour is bounded by

$$1 + \frac{\left|\tan\frac{\theta}{2}\right| + \|\nu'\|_\infty}{1 - \left|\tan\frac{\theta}{2}\right| \|\nu'\|_\infty} + \frac{\left|\tan\frac{\theta}{2}\right| + \|\gamma'\|_\infty}{1 - \left|\tan\frac{\theta}{2}\right| \|\gamma'\|_\infty}.$$

From Lemma 2.32 the operators

$$\|\mathcal{C}_\Gamma^\pm\|_{\mathcal{L}(L^2(\Gamma),L^2(\Gamma'))}, \ \|\mathcal{C}_\Lambda^\pm\|_{\mathcal{L}(L^2(\Gamma),L^2(\Gamma'))}, \|\mathcal{C}_{\Gamma'}^\pm\|_{\mathcal{L}(L^2(\Gamma'),L^2(\Gamma))},$$
$$\|\mathcal{C}_{\Gamma'}^\pm\|_{\mathcal{L}(L^2(\Gamma'),L^2(\Gamma))}, \ \|\mathcal{C}_{\Gamma'}^\pm\|_{\mathcal{L}(L^2(\Gamma'))}$$

all satisfy the estimate in (2.36) with a different constant C. Therefore, using the ideas from Corollary 2.37 we have a bound on the L^2 Cauchy integral operator for functions defined on $\Gamma_1 \cup \Delta$ and $\Lambda_2 \cup \Delta$ and evaluated on $\Gamma_1 \cup \Delta$, $\Gamma_2 \cup \Delta$, $\Lambda_1 \cup \Delta$, and $\Lambda_2 \cup \Delta$, and vice versa. Repeating this argument with $\Gamma' = \Gamma_2 \cup \Delta \cup \Lambda_1$ and applying Lemma 2.38 proves this lemma in this case. When the contours have no intersection the lemma follows immediately from Lemma 2.32. □

The condition on ν' in the above lemma may seem like a significant restriction. It is used below in the special case of $\nu \equiv 0$. Furthermore, the hypotheses of this theorem are satisfied locally by any smooth contour.

[27]To obtain sharper estimates a different rotation angle can be used, but this choice suffices here.

Remark 2.5.3. *Note that a relaxed condition for similar estimates as in Lemma 2.39 is* $||\nu'||_\infty < |\cot\frac{\theta}{2}|$ *and* $||\gamma'||_\infty < |\cot\frac{\theta}{2}|$.

Using these ideas, we aim to define Hardy spaces of functions when Γ has intersections. The next three definitions accomplish this. We use the notation $D \Subset A$ when D is a connected component of A.

Definition 2.40. *Γ is said to be an* admissible contour *if*

1. *Γ is complete (see Definition 2.1) with only transverse self-intersections and*
2. *for each $D \Subset \Omega_\pm$, ∂D is piecewise smooth and if ∂D is unbounded, it is a segment of a piecewise-smooth Lipschitz graph in Σ_∞ (see Definition 2.26):*

$$\partial D = \{e^{i\theta}(x + i\nu(x)) + z : -\infty \le a \le x \le b \le \infty\}, \quad \lim_{x\to\pm\infty} \nu'(x) = 0.$$

Theorem 2.41. *The Cauchy integral operators $\mathscr{C}_\Gamma^\pm$ are bounded linear operators on $L^2(\Gamma)$ whenever Γ is admissible. More generally, $\mathscr{C}_\Gamma^\pm$ are bounded linear operators on $L^2(\Gamma)$ whenever part 2 in Definition 2.40 holds for each contour in a finite decomposition of Γ.*

Proof. Let $\Gamma_1 \cup \cdots \cup \Gamma_L$ be a decomposition of Γ into smooth non-self-intersecting components where each Γ_j is a segment of a Lipschitz graph satisfying

$$\Gamma_j = \{e^{i\theta_j}(x + i\nu_j(x)) + z_j : a_j \le x \le b_j\}, \quad ||\nu_j'||_\infty < 1.$$

Note that a finite decomposition exists because contours tend to straight lines at infinity, i.e., $\nu_j(x)$ must tend to constants for $|x|$ large. Then an application of Lemma 2.38 with Lemma 2.39 proves the theorem. □

Corollary 2.42. *If Γ is a (measurable) subset of an admissible contour, then $\mathscr{C}_\Gamma^\pm$ are bounded operators on $L^2(\Gamma)$.*

Proof. Let χ_Γ be the characteristic function of Γ and Γ' be the larger admissible contour. This follows from $||\mathscr{C}_\Gamma^\pm f||_{L^2(\Gamma)} = ||\mathscr{C}_{\Gamma'}^\pm[f\chi_\Gamma]||_{L^2(\Gamma)} \le ||\mathscr{C}_\Gamma^\pm[f\chi_\Gamma]||_{L^2(\Gamma')} \le C||f\chi_\Gamma||_{L^2(\Gamma')} = C||f||_{L^2(\Gamma)}$. □

Definition 2.43. *If Γ is an admissible contour, then $\mathscr{E}^\pm(\Gamma)$ consists of functions F holomorphic in $\Omega_\pm$ such that $F \in \mathscr{E}^2(D_\pm)$ for each $D_\pm \Subset \Omega_\pm$. It is clear that $\mathscr{E}^\pm(\Gamma)$ may be identified with*

$$\bigoplus_{D_\pm \Subset \Omega_\pm} \mathscr{E}^2(D_\pm).$$

Next we introduce a third notion of a solution of an RH problem. It is made clear below that this is not a new definition but a developed definition of an L^2 RH problem.

Definition 2.44. *A function Φ is an $\mathscr{E}^\pm(\Gamma)$ solution of an RH problem if it satisfies the jump condition a.e. and $(\Phi - I)|_{\Omega_\pm} \in \mathscr{E}^\pm(\Gamma)$, i.e., $\Phi - I \in \mathscr{E}^+(\Gamma) \oplus \mathscr{E}^-(\Gamma)$. If $\Phi \in \mathscr{E}^+(\Gamma) \oplus \mathscr{E}^-(\Gamma)$, then we call Φ a* vanishing solution.

Remark 2.5.4. *Note that this definition captures the $\Phi(\infty) = I$ condition and that the Cauchy integral formula holds for these functions; for each $D \Subset \Omega_{\pm}$ we have $\Phi = I + \mathscr{C}_{\partial D} u$ for $u \in \mathscr{E}^2(D)$, implying $u \in L^2(\partial D)$. From above, (2.27) implies that $\mathscr{C}_{\partial D} u$ vanishes identically outside D. By patching together these representations we find that $\Phi = I + \mathscr{C}_\Gamma \hat{u}$ for some $\hat{u} \in L^2(\Gamma)$. Similarly, for a vanishing solution, $\Phi = \mathscr{C}_\Gamma \hat{u}$ for some $\hat{u} \in L^2(\Gamma)$.*

2.6 ▪ Sobolev spaces

In applications stemming from RH problems, particularly in the numerical solution of RH problems, the smoothness of the solutions on the boundary is a necessary object of study. We begin with a discussion of differentiability on oriented contours and self-intersecting contours. Assume $\Gamma = \Gamma_1 \cup \cdots \cup \Gamma_L$, where each Γ_i is non-self-intersecting and C^∞ smooth[28], and define γ_0, the set of *nonsmooth points*, to be the union of the endpoints of the contours Γ_i. We define D, the distributional differentiation operator for functions defined on $\Gamma \setminus \gamma_0$. Let D_i be the distributional differentiation operator on the smooth contour Γ_i as described in Appendix A.1.3. Let $f_i = f|_{\Gamma_i}$ be the restriction of f to Γ_i and define

$$Df(z) = D_i f_i(z) \text{ if } z \in \Gamma_i \setminus \gamma_0,$$

assuming that $D_i f_i$ corresponds to a locally integrable function.

Definition 2.45. *Define*

$$H^k(\Gamma) = \left\{ f \in L^2(\Gamma) : D^j f \in L^2(\Gamma),\ j = 0, \ldots, k \right\},$$

with norm

$$\|f\|^2_{H^k(\Gamma)} = \sum_{j=0}^{k} \|D^{(j)} f\|^2_{L^2(\Gamma)}.$$

It is clear that there is a generalization of this definition to $W^{k,p}(\Gamma)$, the Sobolev space with the L^2 norm replaced with the L^p norm for $1 \le p \le \infty$, as well as to matrix-valued functions; see Appendix A.1.3. All these spaces are Banach spaces and $H^k(\Gamma)$ is a Hilbert space; also see Appendix A.1.3.

We state a well-known result concerning Sobolev spaces on the line. This may be deduced from the results in [49, Chapter 5].

Theorem 2.46. *If $f \in H^k((a,b))$, then*

- *$f^{(j-1)}$ is differentiable a.e. in (a,b) and $f^{(j)}(z) = D^j f(z)$ a.e., $1 \le j \le k$, and*

$$f^{(j)}(x) - f^{(j)}(a) = \int_a^x f^{(j+1)}(s)\mathrm{d}s,\quad 0 \le j < k, \tag{2.37}$$

- *$\sum_{j=0}^{k-1} \|f^{(j)}\|_u \le C\|f\|_{H^k(\Gamma)}$, where $\|\diamond\|_u$ denotes the uniform norm,*
- *$\lim_{x\to a} f^{(j)}(x)$ and $\lim_{x\to b} f^{(j)}(x)$ exist for $0 \le j < k$,*
- *f has a unique $C^{k-1}([a,b])$ representation,*

[28] This can be relaxed to only imposing a finite degree of smoothness.

- *$C^\infty([a,b])$ functions with compact support are dense in $H^k((a,b))$ for all $k \geq 0$,*
- *multiplication is continuous: $\|fg\|_{H^k((a,b))} \leq C\|f\|_{H^k((a,b))}\|g\|_{H^k((a,b))}$, and*
- *$f^{(j)}$ is uniformly $1/2$-Hölder continuous on $[a,b]$ for $0 \leq j < k$.*

If a or b is infinite, the same conclusions follow but with C^{k-1} replaced with C_0^{k-1}, the space of C^{k-1} functions, decaying at infinity. If $\alpha : [a,b] \to \Gamma$ is a bounded C^∞ parameterization of Γ, then it is clear that $H^k((a,b))$ and $H^k(\Gamma)$ are isomorphic for every k and (2.37) holds on Γ.

Remark 2.6.1. *Since $f \in H^k(\Gamma)$ may be arbitrary on a set of measure zero, we always assume we are working with the C^{k-1} extension.*

Given a contour Γ, let γ_0 be the set of non-smooth points, as before. The behavior of functions at these points is important for what follows. In particular, $\mathscr{C}_\Gamma^\pm f$ is not smooth if f is not sufficiently well behaved at intersection points, even if f is smooth. When Γ has open endpoints we include these in the set γ_0: after augmenting the contour so that it is complete, open endpoints effectively become a nonsmooth point; see Section 2.8.2 below.

Definition 2.47. *Assume that $a \in \gamma_0$ and let $\Gamma_{j_1}, \ldots, \Gamma_{j_\ell}$ be a counterclockwise ordering of sub-components of Γ which contain $z = a$ as an endpoint. For $f \in H^k(\Gamma)$, define*

$$f_i^{(j)} = \begin{cases} -\lim_{z\to a} \left(\frac{\mathrm{d}}{\mathrm{d}z}\right)^j f|_{\Gamma_{j_i}}(z) & \text{if } \Gamma_{j_i} \text{ is oriented away from } a, \\ \lim_{z\to a} \left(\frac{\mathrm{d}}{\mathrm{d}z}\right)^j f|_{\Gamma_{j_i}}(z) & \text{if } \Gamma_{j_i} \text{ is oriented toward } a. \end{cases} \tag{2.38}$$

We say that f satisfies the $(k-1)$th-order zero-sum condition at a if

$$\sum_{i=1}^{\ell} f_i^{(j)} = 0 \text{ for } j = 0, \ldots, k-1. \tag{2.39}$$

We say that f satisfies the $(k-1)$th-order zero-sum condition if it satisfies the condition at all $a \in \gamma_0$.

Remark 2.6.2. *These definitions imply $f^{(j)} = 0$, $j = 0, \ldots, k-1$, when f satisfies the zero-sum condition and Γ has an open endpoint (if Γ is not complete and hence not admissible). This is discussed further in Section 2.8.2.*

Definition 2.48. *Define*

$$H_z^k(\Gamma) = \{f \in H^k(\Gamma) : f \text{ satisfies the } (k-1)\text{th-order zero-sum condition}\},$$

with the $H^k(\Gamma)$ norm.

By Theorem 2.46, point evaluation of any of the $k-1$ derivatives of an H^k function is a bounded linear functional. Thus $H_z^k(\Gamma)$ is the kernel of a bounded linear functional and is therefore a closed subspace of $H^k(\Gamma)$. We see below that $H_z^k(\Gamma)$ is a natural Sobolev space on which to consider the operators $\mathscr{C}_\Gamma^\pm$. With this motivation, we begin with a lemma from [83, p. 67] that is proved in Appendix A.1.3.

Lemma 2.49 (See [83]). *Let Γ be a smooth, nonclosed curve oriented from $z=a$ to $z=b$. If $f \in H^1(\Gamma)$, then*

$$D\int_\Gamma \frac{f(\zeta)}{\zeta - s}\mathrm{d}\zeta = \frac{f(a)}{a-s} - \frac{f(b)}{b-s} + \int_\Gamma \frac{Df(\zeta)}{\zeta - s}\mathrm{d}\zeta.$$

The important consequence of this lemma is that D^j, $0 \leq j \leq k$, commutes with the Cauchy integral operators $\mathscr{C}_\Gamma^\pm$ for functions in $H_z^k(\Gamma)$ as the boundary terms cancel. Thus $\mathscr{C}_\Gamma^\pm$ are bounded from $H_z^k(\Gamma)$ to $H^k(\Gamma)$ with the operator norm being the same as the $L^2(\Gamma)$ norm even if Γ is not complete. Indeed, a stronger statement is true.

Theorem 2.50. *Let Γ be an admissible contour so that $\|\mathscr{C}_\Gamma f\|_{L^2(\Gamma)} \leq C\|f\|_{L^2(\Gamma)}$. If $f \in H_z^k(\Gamma)$, then $\mathscr{C}_\Gamma^\pm f \in H_z^k(\Gamma)$ and hence $\|\mathscr{C}_\Gamma^\pm f\|_{H_z^k(\Gamma)} \leq C\|f\|_{H_z^k(\Gamma)}$ with the same constant.*

Before proving this theorem we establish a lemma.

Lemma 2.51 (Uniform Hölder continuity near intersection points). *Let Γ be an admissible contour, let $f \in H_z^1(\Gamma)$, and let $a \in \gamma_0$. There exists $\delta > 0$ such that $\mathscr{C}_\Gamma^\pm f$ is uniformly $1/2$-Hölder continuous in each $D \Subset B(a,\delta)\setminus\Gamma$ and hence in $\overline{D}$.*

Proof. Let the components of $\Gamma' = (\Gamma \cap B(a,\delta))\setminus\{a\}$ be denoted by $\Gamma_1,\ldots,\Gamma_n$, with a counterclockwise ordering; see Figure 2.6. Choose $\delta > 0$ sufficiently small so that the contours satisfy the hypotheses of Lemma 2.39, i.e., that each of the contours is of the form $\{e^{i\varphi}(x + i\nu(x)) : a \leq x \leq b,\ \|\nu'\|_\infty < 1\}$. Let $x_1 \in \Gamma_1$, $x_2 \in \Gamma_2$, and let Λ be the line segment from x_1 to x_2. For $i = 1,2$, define $r_i = |x_i - a|$. Assume $r_i < \delta/2$.

For $f \in H_z^1(\Gamma)$, extend f trivially to vanish on Λ. Set $F = \mathscr{C}_{\Gamma'} f$. Then f still satisfies the zero-sum condition on $\Gamma' \cup \Lambda$ and F is H^1 on Λ because differentiation commutes with the Cauchy integral operator. Then

$$\begin{aligned}
F(z) - F(z') &= \int_z^{z'} F'(s)\mathrm{d}s,\\
|F(z) - F(z')| &\leq \|F'\|_{L^2(\Lambda)}\sqrt{|z-z'|},\\
|F(z) - F(z')| &\leq \|\mathscr{C}_{\Gamma'} f'\|_{L^2(\Lambda)}\sqrt{|z-z'|}.
\end{aligned} \tag{2.40}$$

But we know that $\|\mathscr{C}_{\Gamma'} f'\|_{L^2(\Lambda)} = \|\mathscr{C}_{\Gamma'\cup\Lambda} f'\|_{L^2(\Lambda)} \leq C(\Lambda)\|f\|_{L^2(\Gamma'\cup\Lambda)} = C(\Lambda)\|f\|_{H^1(\Gamma')}$. It is clear from Lemma 2.39 that $C(\Lambda)$ can be taken to be independent of x_1 and x_2. □

Proof. We return to the proof of Theorem 2.50. For $f \in H_z^k(\Gamma)$ we extend f to a function $\tilde{f}$ by defining it to be zero on Λ. Then $\tilde{f} \in H_z^k(\Gamma \cup \Lambda)$ so that $D^j(\mathscr{C}_\Gamma \tilde{f}) = \mathscr{C}_\Gamma D^j \tilde{f}$. Since $D^j \tilde{f} \in H^{k-j}(\Gamma \cup \Lambda)$ for $0 \leq j \leq k$, we apply Lemma 2.51. Then $D^j(\mathscr{C}_\Gamma \tilde{f})$ is uniformly Hölder in the sense of Lemma 2.51. Choose $a \in \gamma_0$. By the orientation of an admissible Γ, for every contour Γ_1 oriented towards a there exists a contour Γ_2 oriented away such that both contours are part of the boundary of the same component of $\Omega_\pm$. By the uniform

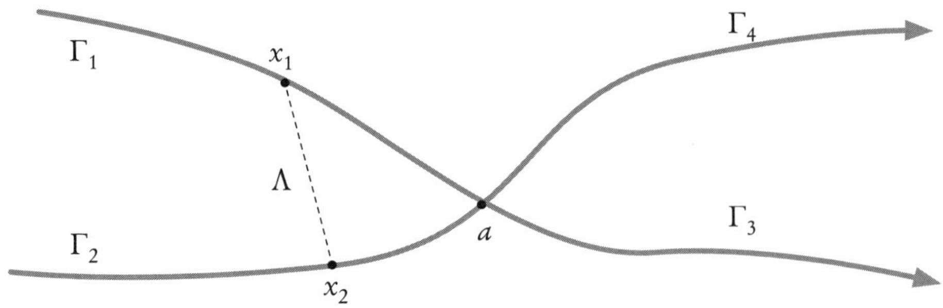

Figure 2.6. *Zooming in on the intersection point a with a counterclockwise ordering of the components of* Γ'.

Hölder continuity of $\mathscr{C}_\Gamma^\pm f$

$$\lim_{z\to a}\left(\frac{\mathrm{d}}{\mathrm{d}z}\right)^j (\mathscr{C}_\Gamma^\pm f)|_{\Gamma_1}(z) = \lim_{z\to a}\left(\frac{\mathrm{d}}{\mathrm{d}z}\right)^j (\mathscr{C}_\Gamma^\pm f)|_{\Gamma_2}(z).$$

This implies the zero-sum condition. □

Note that the previous result implies that $\mathscr{C}_\Gamma f(z)$ is uniformly $1/2$-Hölder continuous in each $D \Subset \Omega_\pm$ whenever Γ is admissible and $f \in H_z^1(\Gamma)$. Furthermore, the Hölder constant Λ is given in terms of $\|f\|_{H_z^1(\Gamma)}$. This combined with Lemma 2.34 results in the following corollary.

Corollary 2.52. *If* $f \in H^1(\Gamma)$ *and* Γ *is admissible, then*

$$\lim_{R\to\infty} \sup_{|z|=R, z\notin\Gamma} |\mathscr{C}_\Gamma f(z)| = 0,$$

i.e., $\mathscr{C}_\Gamma f(z) = o(1)$ *as* $z \to \infty$.

For what follows, we identify $H_z^k(\Gamma)$ with $n \times n$ matrix-valued functions, with each entry being in the scalar $H_z^k(\Gamma)$ space and n (unless otherwise noted) used to denote the dimensions of these matrices.

2.6.1 ▪ The Sobolev spaces of Zhou

The spaces $H_z^k(\Gamma)$ have not appeared in the classical theory of RH problems. In the influential paper of Zhou [123], he considered the two Sobolev spaces (for an admissible Γ)

$$H_\pm^k(\Gamma) = \{f \in L^2(\Gamma) : f \in H_z^k(\partial D) \text{ for every } D \Subset \Omega_\pm\}, \tag{2.41}$$

both of which are equipped with the standard H^k norm.

Remark 2.6.3. *Note that these spaces are not Hardy spaces: they do not require* f *to be the limit of a function analytic in* $D \Subset \Omega_\pm$.

These spaces fit nicely within the framework of the *Gohberg–Krein matrix factorization theory* [18] that is discussed in Section 2.7. It is easy to see that $H_\pm^k(\Gamma) \subset H_z^k(\Gamma)$.

Furthermore, Lemma 2.51 shows that if $f \in H_z^k(\Gamma)$, then $\mathscr{C}_\Gamma^\pm f \in H_\pm^k(\Gamma)$. Since $f = \mathscr{C}_\Gamma^+ f - \mathscr{C}_\Gamma^- f$, we can express

$$H_z^k(\Gamma) = H_+^k(\Gamma) + H_-^k(\Gamma),$$

where a decomposition is given by $\mathscr{C}_\Gamma^+ f$ and $-\mathscr{C}_\Gamma^- f$. The following additional properties follow [123, 124]:

- $\mathscr{C}_\Gamma^\pm : H_\pm^k(\Gamma) \to H_\pm^k(\Gamma)$ and
- $\mathscr{C}_\Gamma^\pm : H_\mp^k(\Gamma) \to H_+^k(\Gamma) \cap H_-^k(\Gamma)$.

Next we prove a proposition that allows us to invert matrix functions with no loss of regularity.

Proposition 2.53. *For $X \in H_\pm^k(\Gamma) + \mathbb{C}^{n\times n}$, $k \geq 1$, if $\det X$ never vanishes, including at infinity, then $X^{-1} \in H_\pm^k(\Gamma) + \mathbb{C}^{n\times n}$.*

Proof. It follows from the classical cofactor expansion that each entry in X^{-1} is a sum of products of the entries of X divided by $\det X$. Since $H_\pm^k(\Gamma) + \mathbb{C}$ (scalar case) is closed under multiplication, it remains to show that $1/\det X \in H_\pm^k(\Gamma) + \mathbb{C}$. Without loss of generality, we assume Γ is unbounded. Let $a = \lim_{|z|\to\infty} \det X(z)$ and consider

$$F = \frac{1}{\det X} - \frac{1}{a} = \frac{a - \det X}{a \det X}.$$

It is clear that $a - \det X \in H_\pm^k(\Gamma)$ and because $1/\det X$ is bounded, $F \in L^2(\Gamma)$. The jth derivative of F produces a ratio with $(\det X)^{j+1}$ as a denominator and a numerator that is a sum of products of the first j derivatives of a and $\det X$. If $j < k$, then each term in the numerator is a product of functions that are at least as regular as $H_\pm^1(\Gamma)$. Therefore $F^{(j)} \in L^2(\Gamma)$. Finally, the only time the kth derivative of $\det X$ appears in $F^{(k)}$ is when it is divided by $-a \det X$ or multiplied by $(-)^k(a - \det X)(\det X)^{-k-1}$, and therefore each term in the numerator of $F^{(k)}$ is an $L^2(\Gamma)$ function. Thus $F \in H_\pm^k(\Gamma) + \mathbb{C}^{n\times n}$. ☐

2.7 ▪ Singular integral equations

We discuss the theory of *singular integral equations* (SIEs), focusing on the solution of RH problems through SIEs. Consider the solution of an L^2 RH problem of the form

$$\Phi^+(s) = \Phi^-(s)G(s) + F(s), \quad s \in \Gamma, \quad \Gamma \text{ admissible}, \quad \Phi(\infty) = I. \tag{2.42}$$

Assume for simplicity that all functions above are $n \times n$ matrix-valued. We again use the pair $[G;\Gamma]$ to refer to this RH problem when $F = 0$. See Appendix A.3 for a discussion of the extension of function spaces to matrix-valued functions.

2.7.1 ▪ Singular integral equations on $L^2(\Gamma)$

We look for solutions where $\Phi^\pm(s)$ exist a.e. and $\Phi^\pm - I \in L^2(\Gamma)$. It is clear that if Φ is an $\mathscr{E}^\pm(\Gamma)$ solution, then it is a solution in the L^2 sense. It turns out that considering this more restrictive class of $\mathscr{E}^\pm(\Gamma)$ solutions is sufficient.

If we can find an $\mathscr{E}^{\pm}(\Gamma)$ solution, then $\Phi(z)=I+\mathscr{C}_\Gamma u(z)$ for some $u\in L^2(\Gamma)$. Note that $u=\Phi^+-\Phi^-$. We make this substitution in (2.42) and use Lemma 2.7 (more precisely, (2.26)) to find

$$u(s)-\mathscr{C}_\Gamma^- u(s)(G(s)-I)=G(s)-I+F(s),\quad s\in\Gamma. \tag{2.43}$$

This is a singular integral equation for u. We use this equation extensively.

There exists another singular integral equation for Φ^-, which we now derive. By rearranging the jump condition, we have

$$\Phi^+(s)-I=\Phi^-(s)(G(s)-I)+F(s)+\Phi^-(s)-I,\quad s\in\Gamma.$$

After making mild assumptions on G and F (made clear below), we apply $\mathscr{C}_\Gamma^-$, noting that $\mathscr{C}_\Gamma^-(\Phi^+(s)-I)=0$ and $\mathscr{C}_\Gamma^-(\Phi^-(s)-I)=-(\Phi^-(s)-I)$ from (2.25) and (2.27). Rearranging, we obtain

$$\Phi^- - \mathscr{C}_\Gamma^-[\Phi^-(G-I)]=I+F,\quad s\in\Gamma. \tag{2.44}$$

This is the equation introduced by Beals and Coifman [10] and studied by Zhou [123]. It is a fundamental aspect of the method of nonlinear steepest descent [34].

The discussion turns to function spaces and the mapping properties of these operators. To ease notation, define

$$\begin{aligned}\mathscr{C}[G;\Gamma]u &= u-\mathscr{C}_\Gamma^- u(G-I),\\ \mathscr{C}'[G;\Gamma]\Phi^- &= \Phi^- - \mathscr{C}_\Gamma^-[\Phi^-(G-I)].\end{aligned}$$

It is clear that $\mathscr{C}[G;\Gamma]$ is bounded on $L^2(\Gamma)$, provided that $G\in L^\infty(\Gamma)$. We impose that the right-hand side of (2.43) is an $L^2(\Gamma)$ function. Thus, we require that $G-I,F\in L^2(\Gamma)$. In the second case, since $\Phi^-=f+I$, $f\in L^2(\Gamma)$. Thus

$$\Phi^-(G-I)=f(G-I)+G-I,$$

which is found to be in $L^2(\Gamma)$, provided that the same conditions on G hold.

2.7.2 ▪ Singular integral equations on Sobolev spaces

When Γ has self-intersections, $\mathscr{C}_\Gamma^- u(s)(G(s)-I)$ does not generically satisfy the zero-sum condition, even for $u\in H_z^k(\Gamma)$. This is the case no matter how smooth G is. If one wishes to find smooth solutions of (2.43), a regularity condition for G must be satisfied at each intersection point, in addition to G being smooth away from intersection points.

Definition 2.54. *A jump matrix G defined on Γ is k-regular if Γ is admissible and G has an algebraic factorization*

$$G(s)=X_-^{-1}(s)X_+(s),$$

where $X_\pm - I, X_\pm^{-1}-I\in H_\pm^k(\Gamma)$.

It is clear that this requires $G-I,G^{-1}-I\in H^k(\Gamma)$. If Γ is not complete, then $G=I$ at an isolated endpoint of Γ.

Definition 2.55. *Assume $a \in \gamma_0$, the set of nonsmooth points of Γ. Let $\Gamma_{j_1},\ldots,\Gamma_{j_\ell}$ be a counterclockwise ordering of subcomponents of Γ which contain $z = a$ as an endpoint. For $G \in H^k(\Gamma)$ we define $\hat{G}_i$ by $G|_{\Gamma_{j_i}}$ if Γ_{j_i} is oriented outwards and by $(G|_{\Gamma_{j_i}})^{-1}$ otherwise. We say that G satisfies the $(k-1)$th-order product condition if, using the $(k-1)$th-order Taylor expansion of each $\hat{G}_i$, we have*

$$\prod_{i=1}^{\ell} \hat{G}_i = I + \mathcal{O}\left(|z-a|^{k-1/2}\right) \ \forall a \in \gamma_0. \tag{2.45}$$

The product condition is precisely the condition that produces smooth solutions. Viewing an RH problem as a solution to the Cauchy–Riemann equations, the classical theory on elliptic PDEs would indicate that singularities should be present at every corner of the domain. A kth-order product condition will indicate that the singularity still possesses some differentiability. In the case that each $\hat{G}_i$ is analytic in a neighborhood of an intersection point, we will see that there is no singularity and solutions are infinitely smooth (actually, analytic).

Remark 2.7.1. *In analogy with the zero-sum condition, if Γ is not complete, the product condition implies that $\partial_z^{(j)}(G-I) = 0$, $j = 0,\ldots,k-1$, at each isolated endpoint. See Section 2.8.2 for a deeper discussion of this.*

Theorem 2.56 (See [124]). *$G-I, G^{-1}-I \in H^k(\Gamma)$, and G satisfies the $(k-1)$th-order product condition if and only if G is k-regular.*

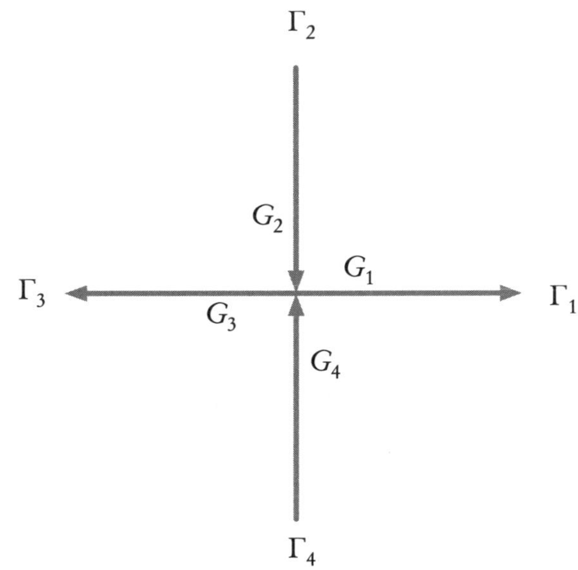

Figure 2.7. *A sample self-intersection with consistent orientation and G_i labeled.*

Sketch of proof. The proof of this theorem is straightforward when viewed from the correct angle, although it is lengthy. We only prove it for $\Gamma = \mathbb{R} \cup i\mathbb{R}$; see Figure 2.7. To ensure we have the orientation correct we define

$$\Gamma_1 = (0,\infty), \quad \Gamma_2 = \mathrm{i}(\infty,0),$$
$$\Gamma_3 = (0,-\infty), \quad \Gamma_4 = \mathrm{i}(-\infty,0).$$

Also, define

$$G_i(s) = \tilde{G}_i(s) + \mathcal{O}(|s|^{k-1/2}), \quad \tilde{G}_i(s) = \sum_{j=0}^{k-1} A_{ij}s^j, \quad s \in \Gamma_i, \quad i = 1,2,3,4.$$

The power $k-1/2$ comes from the integral form of the remainder in Taylor's theorem.

First, assume that $G_i = \tilde{G}_i$ and we do not care about large s behavior. We must find the proper $X_\pm$ factorization of the full jump G. We start with Γ_1. Define $X_-|_{\Gamma_1} = I$ so that $X_+|_{\Gamma_1} = \tilde{G}_1$. We move counterclockwise around Figure 2.7. Set $X_+|_{\Gamma_2} = \tilde{G}_1$, which requires $X_-|_{\Gamma_2} = \tilde{G}_1\tilde{G}_2^{-1}$. Set $X_-|_{\Gamma_3} = \tilde{G}_1\tilde{G}_2^{-1}$, which requires $X_+|_{\Gamma_3} = \tilde{G}_1\tilde{G}_2^{-1}\tilde{G}_3$. Finally, $X_+|_{\Gamma_4} = \tilde{G}_1\tilde{G}_2^{-1}\tilde{G}_3$ and $X_-|_{\Gamma_4} = \tilde{G}_1\tilde{G}_2^{-1}\tilde{G}_3\tilde{G}_4^{-1}$. Under the assumption that the jump satisfies the $(k-1)$th-order product condition, we see that $X_-|_{\Gamma_4} = I + \mathcal{O}(|s|^{k-1/2})$.

We move to the general case where $G_i = \tilde{G}_i + \mathcal{O}(s^k)$. Let $\phi \in C^\infty(\Gamma)$ satisfy

$$\phi(s) = \begin{cases} I & \text{if } |s| < r, \\ 0 & \text{if } |s| > 2r. \end{cases}$$

We note that if we can find a local factorization in a small neighborhood of $s = 0$, $G = \tilde{X}_-^{-1}X_+$ such that $\tilde{X}_-$ is invertible in this neighborhood, then $\tilde{X}_-\phi + (I-\phi)$ is a good candidate for X_- provided that its determinant does not vanish. We must show that ϕ can be chosen so that $(\tilde{X}_- - I)\phi + I$ is non-singular. This only needs to be discussed for $r \le |s| \le 2r$. For simplicity, we choose $\phi(s) = t(|s|)I$, where $t(s)$ is scalar-valued. A little thought shows that the problem on Γ_1 reduces to choosing $t(s) : [r,2r] \to \mathbb{C}$ so that $t(r) = 1$, $t(2r) = 0$, and the eigenvalues of $(\tilde{X}_- - I)\phi$ are never -1. The eigenvalues $\{\lambda_i\}$ of $\tilde{X}_- - I$ for $s \in [r,2r]$ are located in a collection of balls $\{B(\lambda_i, \delta(r))\}$ in the complex plane with radius $\delta(r)$ that tends to zero as $r \to 0$. For sufficiently small r, because there are a finite number of such eigenvalues, there must exist θ such that the ray connecting 0 and $e^{i\theta}$ and an arc connecting -1 and $e^{i\theta}$ do not intersect any ball in $\{B(\lambda_i,\delta(r))\}$. Let $h(s)$ be a smooth and monotonic function on $\mathbb{R}^+$ that satisfies $h(x) = 1$ for $x < \epsilon$ and $h(x) = 0$, $x \ge r - \epsilon$. Let $M = \max_i |\lambda_i| + \delta(r)$ and define $r < s^* < 2r$ by $h(s^*) = 1/(2M)$ (if $1/(2M) \ge h(s)$ for all s, then set $s^* = r$); then

$$t(s) = \begin{cases} e^{i\theta h(2r(s-r)/(s^*-r))} & \text{if } s < s^*, \\ e^{i\theta} h(2r(s-s^*)/(2r-s^*)) & \text{if } s \ge s^*. \end{cases}$$

With this choice $X_- = (\tilde{X}_- - I)\phi + I$ and $X_+ = X_- G$. The definition of t, in general, differs on each contour Γ_i. We now construct $\tilde{X}_\pm$. Define

$$\begin{aligned}
\tilde{X}_-|_{\Gamma_1} &= I, \\
\tilde{X}_+|_{\Gamma_1} &= G_1 = \tilde{G}_1 + (G_1 - \tilde{G}_1), \\
\tilde{X}_+|_{\Gamma_2} &= \tilde{G}_1, \\
\tilde{X}_-|_{\Gamma_2} &= \tilde{G}_1 G_2^{-1} = \tilde{G}_1\tilde{G}_2^{-1} + \tilde{G}_1(G_2^{-1} - \tilde{G}_2^{-1}), \\
\tilde{X}_-|_{\Gamma_3} &= \tilde{G}_1\tilde{G}_2^{-1},
\end{aligned}$$

$$\tilde{X}_+|_{\Gamma_3} = \tilde{G}_1\tilde{G}_2^{-1}G_3 = \tilde{G}_1\tilde{G}_2^{-1}\tilde{G}_3 + \tilde{G}_1\tilde{G}_2^{-1}(G_3 - \tilde{G}_3),$$
$$\tilde{X}_+|_{\Gamma_4} = \tilde{G}_1\tilde{G}_2^{-1}\tilde{G}_3,$$
$$\tilde{X}_-|_{\Gamma_4} = \tilde{G}_1\tilde{G}_2^{-1}\tilde{G}_3G_4^{-1} = \tilde{G}_1\tilde{G}_2^{-1}\tilde{G}_3\tilde{G}_4^{-1} + \tilde{G}_1\tilde{G}_2^{-1}\tilde{G}_3(G_4^{-1} - \tilde{G}_4^{-1}).$$

It is easy to see that $\tilde{X}_-^{-1}\tilde{X}_+ = G$ and $\tilde{X}_\pm$ agrees with $\tilde{X}_\pm$ on the next contour up to $\mathcal{O}(|s|^{k-1/2})$ because the terms in parentheses are small. This property carries over to $X_\pm$.

Conversely, if G is k-regular, then, for example on the boundary of the fourth quadrant, $X_-|_{\Gamma_1} = X_-|_{\Gamma_4} + \mathcal{O}(|s|^{k-1/2})$ or $X_-|_{\Gamma_4}^{-1}X_-|_{\Gamma_1} = I + \mathcal{O}(|s|^{k-1/2})$. Similar relationships hold for the three other quadrants, and the product condition follows. □

The factorization $G = X_-^{-1}X_+$ is now applied to (2.43).

$$\begin{aligned}
u - \mathcal{C}_\Gamma^- u(X_-^{-1}X_+ - I) &= X_-^{-1}X_+ - I + F,\\
uX_+^{-1} - \mathcal{C}_\Gamma^- u(X_-^{-1} - X_+^{-1}) &= X_-^{-1} - X_+^{-1} + FX_+^{-1},\\
\mathcal{C}_\Gamma^+ uX_+^{-1} - \mathcal{C}_\Gamma^- uX_+^{-1} - \mathcal{C}_\Gamma^- u(X_-^{-1} - X_+^{-1}) &= X_-^{-1} - X_+^{-1} + FX_+^{-1},\\
\mathcal{C}_\Gamma^+ uX_+^{-1} - \mathcal{C}_\Gamma^- uX_-^{-1} &= X_-^{-1} - X_+^{-1} + FX_+^{-1}.
\end{aligned} \tag{2.46}$$

Note that if $f \in H_\pm^k(\Gamma)$ and $g \in H_z^k(\Gamma)$, then $f\mathcal{C}_\Gamma^\pm g \in H_\pm^k(\Gamma) \subset H_z^k(\Gamma)$. This operation is continuous since multiplication is continuous in $H^k(\Gamma)$. Thus

$$\mathcal{C}[X_+, X_-; \Gamma]u = \mathcal{C}_\Gamma^+ uX_+^{-1} - \mathcal{C}_\Gamma^- uX_-^{-1}$$

is an operator on $H_z^k(\Gamma)$. This is especially convenient since this modification of the operator will not affect u. For the right-hand side to be an $H_z^k(\Gamma)$ function, we require that $F \in H_+^k(\Gamma)$.

We turn our attention to (2.44):

$$\Phi^- - \mathcal{C}_\Gamma^-[\Phi^-(X_-^{-1}X_+ - I)] = I + F.$$

Define $\Psi = \Phi^- X_-^{-1}$. Then

$$\begin{aligned}
\Psi X_- - \mathcal{C}_\Gamma^-[\Psi(X_+ - X_-)] &= I + F,\\
\mathcal{C}_\Gamma^+[\Psi X_-] - \mathcal{C}_\Gamma^-[\Psi X_-] - \mathcal{C}_\Gamma^-[\Psi(X_+ - X_-)] &= I + F,\\
\mathcal{C}_\Gamma^+[\Psi X_-] - \mathcal{C}_\Gamma^-[\Psi X_+] &= I + F.
\end{aligned} \tag{2.47}$$

To properly interpret this equation we define $\tilde{H}_\pm^k(\Gamma) \triangleq H_\pm^k(\Gamma) \oplus \mathbb{C}^{n\times n}$ when Γ is unbounded and $\tilde{H}_\pm^k(\Gamma) \triangleq H_\pm^k(\Gamma)$ when Γ is bounded. For $A \in \mathbb{C}^{n\times n}$ we define $\mathcal{C}_\Gamma^+ A = A$ and $\mathcal{C}_\Gamma^- A = 0$ so that Lemma 2.7 holds. For $\Psi \in \tilde{H}_-^k(\Gamma) \cap \tilde{H}_+^k(\Gamma)$, $\Psi X_\pm \in \tilde{H}_\pm^k(\Gamma)$. Thus using $\mathcal{C}_\Gamma^\pm : H_\pm^k(\Gamma) \to H_\pm^k(\Gamma)$ we find that $\mathcal{C}_\Gamma^\pm[\Psi X_\mp] \in \tilde{H}_-^k(\Gamma) \cap \tilde{H}_+^k(\Gamma)$. The operator

$$\mathcal{C}'[X_+, X_-; \Gamma]\Psi = \mathcal{C}_\Gamma^+[\Psi X_-] - \mathcal{C}_\Gamma^-[\Psi X_+]$$

is bounded on $\tilde{H}_-^k(\Gamma) \cap \tilde{H}_+^k(\Gamma)$, and we require $F \in \tilde{H}_-^k(\Gamma) \cap \tilde{H}_+^k(\Gamma)$.

Fredholm properties: We first state our main rational approximation theorem that is proved in Appendix D.

Definition 2.57. *When Γ is admissible, let $R_\pm(\Gamma)$ be the space of functions r such that $r|_{\partial D}$ is equal a.e. to a rational function for every $D \Subset \Omega_\pm$.*

Remark 2.7.2. *When $\Omega_\pm$ has more than one component (on the Riemann sphere), a function $f \in R_\pm(\Gamma)$ does not have to agree a.e. on Γ with a rational function. For example, let Γ be the contour in Figure 2.7. The function*

$$f(z) = \begin{cases} (z - e^{i5\pi/4})^{-n} & \text{if } z \in \Gamma_1 \cup \Gamma_2, \\ (z - e^{i\pi/4})^{-n} & \text{if } z \in \Gamma_3 \cup \Gamma_4 \end{cases}$$

is an $R_+(\Gamma)$ function, but it is not a true rational function.

Theorem 2.58. *If Γ is admissible, then $R_\pm(\Gamma) \cap L^2(\Gamma)$ is dense in $H^k_\pm(\Gamma)$ and $L^2(\Gamma) \cap R_\pm(\Gamma) + \mathbb{C}^{n\times n}$ is dense in $\tilde{H}^k_\pm(\Gamma)$.*

Remark 2.7.3. *In light of these results there is a simple way to define the above spaces. $H^k_\pm(\Gamma)$ is the closure of functions f such that $f|_{\partial D}$, $D \subset \Omega_\pm$, is rational in the $H^k(\Gamma)$ norm. Furthermore, $H^k_z(\Gamma) = H^k_+(\Gamma) + H^k_-(\Gamma)$ with $\pm\mathscr{C}^\pm_\Gamma$ providing a decomposition. Note that these definitions coincide only when Γ is sufficiently regular (Γ being admissible is sufficient).*

Throughout this section we assume that Γ is admissible and that the jump matrix $G = X_-^{-1}X_+$ is k-regular for $k \geq 1$.

Theorem 2.59. *The following operators are Fredholm (see Definition A.16):*

- *$\mathscr{C}[X_+, X_-; \Gamma]$ on $H^k_z(\Gamma)$ and*
- *$\mathscr{C}[G; \Gamma]$ on $L^2(\Gamma)$.*

Proof. We begin the proof with a lemma.

Lemma 2.60. *If $g \in \tilde{H}^k_\mp(\Gamma)$ for $k \geq 1$, the operator $L^\pm_g$ defined by*

$$L^\pm_g u = \mathscr{C}^\pm_\Gamma[[\mathscr{C}^\mp_\Gamma u]g]$$

is compact on $H^k_z(\Gamma)$, $H^k_+(\Gamma) \cap H^k_-(\Gamma)$, and $L^2(\Gamma)$.

Proof. We prove this for the upper sign. The result follows analogously for the lower sign. Assume $g \in \tilde{H}^k_-(\Gamma)$ and let g_n be a sequence of $R_-(\Gamma)$ functions that converge to g in $H^k_-(\Gamma)$. Let $D \Subset \Omega_-$, $F = \mathscr{C}^-_\Gamma u$, and let $\tilde{g}_n$ be the rational function that agrees with $g_n|_{\partial D}$ a.e. We note that

$$u \mapsto \operatorname{Res}_{s=\zeta} \frac{F(s)\tilde{g}_n(s)}{s - z}, \tag{2.48}$$

when ζ is a jth-order pole of g_n, is a bounded, (at most) rank j operator on all Sobolev spaces we consider here: taking a residue at ζ produces a rational function of z in inverse powers of $(\zeta - z)$ where the coefficients depend on j inner products of u with j L^2 functions. We show that $\mathscr{C}_\Gamma[[\mathscr{C}^-_\Gamma u]g_n]$ is a finite sum of such residues for $z \in \Omega_+$. For $z \in \Omega_+$

we express

$$\mathscr{C}_\Gamma[[\mathscr{C}_\Gamma^- u]g_n](z) = -\sum_{D\Subset\Omega_-} \mathscr{C}_{\partial D}[[\mathscr{C}_\Gamma^- u]g_n](z),$$

where the minus sign comes from the orientation. On ∂D for all $D \Subset \Omega_-$, $[\mathscr{C}_\Gamma^- u]g_n$ has a meromorphic extension in D, and therefore $\mathscr{C}_{\partial D}[[\mathscr{C}_\Gamma^- u]g_n]$ must be a finite sum of residues of the form (2.48) plus (potentially) a residue at $s = z$, when considering $\mathscr{C}_\Gamma^+$, $z \notin D$. Summing over every component of Ω_- we see that the operator is finite rank. Since $\|L_{g_n}^\pm - L_g^\pm\|_{\mathscr{L}(H_z^k(\Gamma))} \to 0$ as $n \to \infty$, we conclude that $L_g^\pm$ is compact because the limit of finite-rank operators is compact. Note that the operators are compact on $L^2(\Gamma)$ provided $g \in \tilde{H}_\pm^1(\Gamma)$ to allow (uniform) rational approximation.

We return to the proof of the theorem. We show that $\mathscr{C}[X_+^{-1}, X_-^{-1};\Gamma]$ is a Fredholm regulator for $\mathscr{C}[X_+, X_-;\Gamma]$.

$$\begin{aligned}
\mathscr{H}u &\triangleq \mathscr{C}[X_+,X_-;\Gamma][\mathscr{C}[X_+^{-1},X_-^{-1};\Gamma]u] \\
&= \mathscr{C}_\Gamma^+[u + \mathscr{C}_\Gamma^+ u(X_+^{-1}-I) - \mathscr{C}_\Gamma^- u(X_-^{-1}-I)]X_+ \\
&\quad - \mathscr{C}_\Gamma^-[u + \mathscr{C}_\Gamma^+ u(X_+^{-1}-I) - \mathscr{C}_\Gamma^- u(X_-^{-1}-I)]X_- \\
&= \mathscr{C}_\Gamma^+[u + \mathscr{C}_\Gamma^+ u(X_+^{-1}-I)]X_+ - \mathscr{C}_\Gamma^-[u - \mathscr{C}_\Gamma^- u(X_-^{-1}-I)]X_- \\
&\quad - \mathscr{C}_\Gamma^+[\mathscr{C}_\Gamma^- u(X_-^{-1}-I)]X_+ - \mathscr{C}_\Gamma^-[\mathscr{C}_\Gamma^+ u(X_+^{-1}-I)]X_-.
\end{aligned}$$

We concentrate on the first two terms since the last two are compact by Lemma 2.60.

$$\begin{aligned}
\mathscr{C}_\Gamma^+[u + \mathscr{C}_\Gamma^+ u(X_+^{-1}-I)]X_+ &= \mathscr{C}_\Gamma^+[u + \mathscr{C}_\Gamma^+ uX_+^{-1} - \mathscr{C}_\Gamma^+ u]X_+ \\
&= \mathscr{C}_\Gamma^+[\mathscr{C}_\Gamma^+ uX_+^{-1} - \mathscr{C}_\Gamma^-]X_+ = \mathscr{C}_\Gamma^+[\mathscr{C}_\Gamma^+ uX_+^{-1}]X_+, \\
\mathscr{C}_\Gamma^-[u - \mathscr{C}_\Gamma^- u(X_-^{-1}-I)]X_- &= -\mathscr{C}_\Gamma^-[\mathscr{C}_\Gamma^- uX_-^{-1}]X_-.
\end{aligned}$$

Then,

$$\begin{aligned}
\mathscr{C}_\Gamma^+[u + \mathscr{C}_\Gamma^+ u(X_+^{-1}-I)]X_+ &- \mathscr{C}_\Gamma^-[u - \mathscr{C}_\Gamma^- u(X_-^{-1}-I)]X_- \\
&= \mathscr{C}_\Gamma^+[\mathscr{C}_\Gamma^+ uX_+^{-1}]X_+ + \mathscr{C}_\Gamma^-[\mathscr{C}_\Gamma^- uX_-^{-1}]X_- \\
&= \mathscr{C}_\Gamma^+ u + \mathscr{C}_\Gamma^-[\mathscr{C}_\Gamma^+ uX_+^{-1}]X_+ - \mathscr{C}_\Gamma^- u + \mathscr{C}_\Gamma^+[\mathscr{C}_\Gamma^- uX_-^{-1}]X_- \\
&= u + \mathscr{C}_\Gamma^-[\mathscr{C}_\Gamma^+ uX_+^{-1}]X_+ + \mathscr{C}_\Gamma^+[\mathscr{C}_\Gamma^- uX_-^{-1}]X_-.
\end{aligned}$$

Adding this all together we obtain

$$\begin{aligned}
\mathscr{H}u &= u + \mathscr{C}_\Gamma^-[\mathscr{C}_\Gamma^+ uX_+^{-1}]X_+ + \mathscr{C}_\Gamma^+[\mathscr{C}_\Gamma^- uX_-^{-1}]X_- \\
&\quad - \mathscr{C}_\Gamma^+[\mathscr{C}_\Gamma^- u(X_-^{-1}-I)]X_+ - \mathscr{C}_\Gamma^-[\mathscr{C}_\Gamma^+ u(X_+^{-1}-I)]X_- \\
&= u + \mathscr{T}u,
\end{aligned}$$

where $\mathscr{T}$ is compact and therefore the operator is right Fredholm. Changing the roles of $X_\pm$ and $X_\pm^{-1}$ we obtain that the operator is indeed Fredholm.

In order to see that $\mathscr{C}[G;\Gamma]$ is Fredholm on $L^2(\Gamma)$ note that $\mathscr{C}[X_+,X_-;\Gamma]u \cdot X_+ = \mathscr{C}[G;\Gamma]u$. Thus the operator defined by $u \mapsto \mathscr{C}[X_+^{-1},X_-^{-1};\Gamma](uX_+^{-1})$ is a regulator for $\mathscr{C}[G;\Gamma]$. □

Similar calculations show that $\mathscr{C}'[X_+,X_-;\Gamma]$ and $\mathscr{C}'[G;\Gamma]$ are Fredholm.

Theorem 2.61 (See [123]). *The following operators are Fredholm:*

- $\mathcal{C}'[X_+,X_-;\Gamma]$ *on* $\tilde{H}_+^k(\Gamma)\cap\tilde{H}_-^k(\Gamma)$, *and*
- $\mathcal{C}'[G;\Gamma]$ *on* $L^2(\Gamma)$.

2.7.3 ▪ Determining the Fredholm index and kernel

We have shown that singular integral operators associated with RH problems are Fredholm under some precise conditions on the jump matrix. We aim to determine conditions under which the associated integral operators are invertible and hence the associated RH problem is uniquely solvable. To do this, we must briefly discuss the *Gohberg–Krein (GK) matrix factorization theory* in a simplified form. This theory may be discussed in great depth and detail. It is an elegant theory with profound results. In view of the density of rational functions, we use the GK theory directly to discuss only those RH problems with jump matrices that are the product of an $R_-(\Gamma)$ and an $R_+(\Gamma)$ function.

GK factorization of rational matrix functions: Let Γ be a rectifiable Jordan curve. A matrix function $A:\Gamma\to\mathbb{C}^{n\times n}$ with each entry being a rational function (i.e., a rational matrix function) is said to admit a *right-standard factorization relative to* Γ if

$$A=A_-\theta A_+,$$

where $A_\pm$ are bounded rational functions with all poles in $\Omega_\mp$ such that $\det A_\pm\neq 0$ in $\overline{\Omega}_\pm$. Furthermore,

$$\theta(z)=\operatorname{diag}\left(\left(\frac{z-z_+}{z-z_-}\right)^{\kappa_1},\ldots,\left(\frac{z-z_+}{z-z_-}\right)^{\kappa_n}\right)$$

for $z_\pm\in\Omega_\pm$. These conditions imply that $A_\pm^{-1}$ has no poles in $\overline{\Omega}_\pm$. The integers $\kappa_1\geq\cdots\geq\kappa_n$ are the right-partial indices, or just *partial indices*, and $\kappa=\sum_{i=1}^n\kappa_i$ is the *total index*. We show that κ has a direct relationship to the Fredholm index of the associated singular integral operators. The following theorem is a fundamental result from the first chapter of [18].

Theorem 2.62. *Let A be a rational matrix function with a nonvanishing determinant on Γ. Then A admits a right-standard factorization relative to Γ for any $z_\pm\in\Omega_\pm$ and the partial indices are determined uniquely.*

We extend this result to intersecting contours. We abuse notation and call the matrix $A=X_-^{-1}X^+$, where $X_\pm\in R_\pm(\Gamma)$ a rational matrix function. It is important to note that A may not itself be a true rational function.

Theorem 2.63. *Assume that Γ is admissible and that A is a rational matrix function. Then A admits a right-standard factorization relative to Γ.*

Proof. We follow [123] and perform induction on the number of components of $\Omega_\pm$ as subsets of the Riemann sphere. First, we assume $X_+=I$ so that $A=X_-^{-1}\in R_-(\Gamma)$. The general case is reduced to this case below.

If Ω_- has one component, Theorem 2.62 implies the result. Now assume $A=X_-^{-1}$ admits a right-standard factorization if Ω_- has $m<n$ components. We show it admits a factorization when Ω_- has n components. Assume Ω_- has n components. Let $D\subset\Omega_-$ be

a component. Then $\hat{A} = A|_{\overline{\Gamma\setminus\partial D}} \in R_-(\overline{\Gamma\setminus\partial D})$ admits a GK factorization since $\overline{\Gamma\setminus\partial D}$ is admissible and $\mathbb{C}\setminus(\overline{\Gamma\setminus\partial D}) = \Omega'_+ \cup \Omega'_-$, a disjoint union, similar to $\Omega_\pm$ with $n-1$ components. Thus

$$\hat{A} = M_{1-}\theta_1 M_{1+}, \quad \theta_1(z) = \operatorname{diag}\left(\left(\frac{z-z_+}{z-z_-}\right)^{\kappa_1}, \ldots, \left(\frac{z-z_+}{z-z_-}\right)^{\kappa_n}\right),$$

$z_\pm \in \Omega_\pm$. We use M_1 to denote either function $M_{1\pm}$ away from $(\overline{\Gamma\setminus\partial D})$.

This factorization is used to, in effect, reduce the factorization problem to one on ∂D. Fix $z'_- \in \Omega'_+$ and define

$$\hat{\theta}_1(z) = \operatorname{diag}\left(\left(\frac{z-z_-}{z-z'_-}\right)^{\kappa_1}, \ldots, \left(\frac{z-z_-}{z-z'_-}\right)^{\kappa_n}\right).$$

Next, define $\tilde{A} = M_{1-}\check{A}M_{1+}^{-1}\hat{\theta}_1$ for $\check{A} = A|_{\partial D}$. Since $\tilde{A}$ is a rational function on a rectifiable Jordan curve, Theorem 2.62 gives that $\tilde{A}$ admits a right-standard factorization relative to ∂D according to Theorem 2.63:

$$\tilde{A} = M_{2-}\theta_2 M_{2+}, \quad \theta_2(z) = \operatorname{diag}\left(\left(\frac{z-z_+}{z-z'_-}\right)^{\nu_1}, \ldots, \left(\frac{z-z_+}{z-z'_-}\right)^{\nu_n}\right).$$

To be consistent we must also decompose $\mathbb{C}\setminus\partial D = \Omega''_+ \cup \Omega''_-$. Define

$$\theta(z) = \operatorname{diag}\left(\left(\frac{z-z_-}{z-z'_-}\right)^{\nu_1}, \ldots, \left(\frac{z-z_-}{z-z'_-}\right)^{\nu_n}\right),$$

and $\hat{\theta}_2 = \theta_2\theta^{-1}$. We construct a sectionally analytic function

$$M(z) = \begin{cases} \hat{\theta}_2(z)M_2(z)\hat{\theta}_1^{-1}(z)M_1(z) & \text{if } z \in \Omega_+, \\ M_1(z)\theta_1(z)\hat{\theta}_1(z)M_2^{-1}(z)\theta_2^{-1}(z) & \text{if } z \in \Omega'_- \cap \Omega_-, \\ M_1^{-1}(z)M_2(z) & \text{if } z \in \Omega''_- \cap \Omega_-. \end{cases}$$

For $z \in \Gamma\setminus\partial D$

$$\begin{aligned} M_-(z)\theta(z)M_+(z) &= M_{1-}(z)\theta_1(z)\hat{\theta}_1(z)M_{2-}^{-1}(z)\theta_2^{-1}(z)\theta(z)\hat{\theta}_2(z)M_{2+}(z)\hat{\theta}_1^{-1}(z)M_{1+}(z) \\ &= M_{1-}(z)\theta_1(z)M_{1+}(z) = \hat{A}(z) \end{aligned}$$

since $M_{2-} = M_{2+}$ on $\Gamma\setminus\partial D$. For $z \in \Gamma\setminus\partial D$ we have

$$M_-\theta M_+ = M_{1-}^{-1}\check{A}\theta_1^{-1}M_{1+} = A|_{\partial D}.$$

Thus $M_-\theta M_+$ is the desired factorization. Switching $-$ with $+$ we see that we may find a right-standard factorization if $A \in R_+(\Gamma)$. For the general case $A = X_-^{-1}X_+$ we find $X_+ = M_{1-}\theta_1 M_{1+}$. We factor $X_-^{-1}M_{1-}\theta_1 = M_{2-}\theta_2 M_{2+}$. Thus $A = M_{2-}\theta_2(M_{2+}M_{1+})$. This is the desired factorization. □

Theorem 2.64 (See [123]). *The partial indices $\kappa_1 \leq \cdots \leq \kappa_n$ are unique.*

Proof. Assume two factorizations of $A = M_{1-}\theta_1 M_{1+} = M_{2-}\theta_2 M_{2+}$ with

$$\theta_1(z) = \operatorname{diag}(z^{\kappa_1},\ldots,z^{\kappa_n}), \quad \theta_2(z) = \operatorname{diag}(z^{\nu_1},\ldots,z^{\nu_n}),$$
$$\kappa_i \le \kappa_{i+1}, \qquad \nu_i \le \nu_{i+1}.$$

We may assume θ_1, θ_2 are of this simple form by using a Möbius transformation that takes rational functions to rational functions. After this transformation assume $0 \in \Omega_+$, $\infty \in \Omega_-$. We obtain

$$N_+\theta_1 = \theta_2 N_-, \quad N_- \triangleq M_{2-}^{-1}M_{1-}, \quad N_+ \triangleq M_{2+}M_{1+}^{-1}. \tag{2.49}$$

Set $N_- = (N_{ij}^-)_{1\le i,j\le n}$ and $N_+ = (N_{ij}^+)_{1\le i,j\le n}$. Assume that $\nu_l < \kappa_l$ for some l. Then

$$\nu_1 \le \cdots \le \nu_l < \kappa_l \Rightarrow \nu_i - \kappa_j < 0, \;\; j \ge l, \;\; i \le l.$$

From (2.49)

$$N_{ij}^+(z) z^{\kappa_j - \nu_i} = N_{ij}^-(z).$$

If $\kappa_j - \nu_i > 0$, then $N_{ij}^-(z) z^{-(\kappa_j - \nu_i)}$ is an entire function that decays at infinity: it must be identically zero.

Thus $N_{ij}^-(z) = 0$ for $j \ge l, \;\; i \le l$, i.e., N_- has an $l \times (n-l)+1$ block of zeros. The span of the last $(n-l)+1$ columns of n_- is at most $n-l$ (the dimension of the nonzero block below the zero block). Thus the rank of N_- is at most $l-1+n-l = n-1$, contradicting the fact that N_- is invertible. The case $\kappa_l < \nu_l$ can be treated with a slightly modified version of this argument. □

Now that we have established that the partial indices are unique we relate them to the Fredholm index of an operator. We begin with a lemma.

Lemma 2.65. *Assume G is k-regular. If $X_\pm$ and $X_\pm'$ represent two (algebraic) $H^k(\Gamma) + \mathbb{C}^{n\times n}$ factorizations of G, then*

$$u \mapsto u X_\pm (X_\pm')^{-1}$$

is an invertible operator on $H_z^k(\Gamma)$.

Proof. It is clear that $X_\pm(X_\pm')^{-1}$ is bounded and invertible on $H^k(\Gamma)$. We need to show it preserves the zero-sum conditions. With

$$X_-^{-1}X_+ = (X_-')^{-1}X_+',$$

we find

$$X_+(X_+')^{-1} = X_-(X_-')^{-1} \in H_-^k(\Gamma) \cap H_+^k(\Gamma).$$

Since a function in $H_-^k(\Gamma) \cap H_+^k(\Gamma)$ has limits that agree from every direction at an intersection point, it is clear that the zero-sum condition is preserved under multiplication by such functions. □

Theorem 2.66. *Let Γ be an admissible contour and $G : \Gamma \to \mathbb{C}^{n\times n}$ be a rational matrix function. Assume G has partial indices $\kappa_1 \le \cdots \le \kappa_n$. Then for any (algebraic) factorization of $G = X_-^{-1}X_+$ we have*

- $\dim\ker \mathscr{C}[X_+,X_-;\Gamma] = n\sum_{\kappa_i>0}|\kappa_i|$,
- $\operatorname{codim}\operatorname{ran}\mathscr{C}[X_+,X_-;\Gamma] = n\sum_{\kappa_i<0}|\kappa_i|$, *and*
- $\operatorname{ind}\mathscr{C}[X_+,X_-;\Gamma] = -n\sum_{i=1}^{n}\kappa_i$.

Proof. Let $G = R_-^{-1}\theta R_+$ be a right-standard factorization of G. We claim that

$$[\mathscr{C}[I,\theta;\Gamma]\mathscr{C}[R_+,R_-;\Gamma]u]R_+X_+^{-1} = \mathscr{C}[X_+,X_-;\Gamma].$$

This can be verified using the projection properties of the Cauchy integral operator. We digress for an important lemma. See Remark 2.7.6 below for a method to check whether $(\Phi_\pm^{-1} - I) \in \mathscr{E}^\pm(\Gamma)$.

Lemma 2.67. *If a k-regular RH problem with jump matrix G has an $\mathscr{E}^\pm(\Gamma)$ solution Φ with $(\Phi_\pm^{-1} - I) \in \mathscr{E}^\pm(\Gamma)$ and $\Phi_\pm - I, \Phi_\pm^{-1} - I \in H_\pm^k(\Gamma)$, then*

$$\mathscr{C}[X_+,X_-;\Gamma]^{-1}u = \mathscr{C}_\Gamma^+[uX_+\Phi_+^{-1}]\Phi_+ - \mathscr{C}_\Gamma^-[uX_+\Phi_+^{-1}]\Phi_-.$$

Proof. Let $\mathscr{T}u = \mathscr{C}_\Gamma^+[uX_+\Phi_+^{-1}]\Phi_+ - \mathscr{C}_\Gamma^-[uX_+\Phi_+^{-1}]\Phi_-$. First, we find that

$$(\mathscr{C}[X_+,X_-;\Gamma]u)X_+\Phi_+^{-1} = \mathscr{C}_\Gamma^+ u\Phi_+^{-1} - \mathscr{C}_\Gamma^- u\Phi_-^{-1}.$$

The projection properties of $\mathscr{C}_\Gamma^\pm$ show that $\mathscr{C}_\Gamma^+ u\Phi_+^{-1} - \mathscr{C}_\Gamma^- u\Phi_-^{-1}$ and $\mathscr{C}_\Gamma^+ u\Phi_+ - \mathscr{C}_\Gamma^- u\Phi_-$ are inverses. Thus $\mathscr{T}\circ\mathscr{C}[X_+,X_-;\Gamma] = I$. We show the other composition. We find

$$\begin{aligned}\mathscr{C}[X_+,X_-;\Gamma]\circ\mathscr{T}u &= \mathscr{C}_\Gamma^+[uX_+\Phi_+^{-1}]\Phi_+X_+^{-1} - \mathscr{C}_\Gamma^-[uX_+\Phi_+^{-1}]\Phi_-X_-^{-1}\\ &= u + \mathscr{C}_\Gamma^-[uX_+\Phi_+^{-1}]\Phi_+X_+^{-1} - \mathscr{C}_\Gamma^-[uX_+\Phi_+^{-1}]\Phi_-X_-^{-1}.\end{aligned}$$

It is clear from previous calculations that $\Phi_+X_+^{-1} = \Phi_-X_-^{-1}$, proving the identity. □

We return to the proof of the theorem. It is clear that $u \mapsto uR_+X_+^{-1}$ defines an invertible operator. The lemma shows that $\mathscr{C}[R_+,R_-;\Gamma]$ is also invertible. Thus any deficiency of $\mathscr{C}[X_+,X_-;\Gamma]$ must come from

$$\mathscr{C}[I,\theta;\Gamma] = \mathscr{C}_\Gamma^+ \diamond - \mathscr{C}_\Gamma^-[\diamond]\theta.$$

Recall that

$$\theta(z) = \operatorname{diag}\left(\left(\frac{z-z_+}{z-z_-}\right)^{\kappa_1}, \ldots, \left(\frac{z-z_+}{z-z_-}\right)^{\kappa_n}\right).$$

We convert this to an equivalent RH problem. We look for solutions of

$$\Phi^+(s) = \Phi^-(s)\theta(s), \quad s\in\Gamma, \quad \Phi(\infty) = 0.$$

This is a diagonal RH problem. We use the methods of Section 2.4.1, determining n scalar problems

$$\phi_i^+(s) = \phi_i^-(s)M(s)^{\kappa_i}, \quad M(s) = \frac{s-z_+}{s-z_-}, \quad s\in\Gamma, \quad \phi_i(\infty) = 0, \quad i = 1,\ldots,n. \tag{2.50}$$

It is easy to check that if $\kappa_i > 0$, then

$$\phi_i^+(z) = \sum_{l=0}^{\kappa_i} \alpha_l M(z)^l, \quad \phi_i^-(z) = \sum_{l=0}^{\kappa_i} \alpha_l M(z)^{-\kappa_i + l}$$

is a solution of (2.50) for any choice of $\{\alpha_l\}$ provided α_0 is chosen so that $\phi_i^-(\infty) = 0$. Therefore $\Phi^\pm(z) = \operatorname{diag}(\phi_1^\pm(z), \ldots, \phi_n^\pm(z))$ and $u(s) = \Phi^+(s) - \Phi^-(s)$ is an element of the kernel of $\mathscr{C}[I, \theta; \Gamma]$. For $\kappa_i \leq 0$, solutions must be of this same form, but we may change the range for l,

$$\phi_i^+(z) = \sum_{l < \kappa_i} \alpha_l M(z)^l, \quad \phi_i^-(z) = \sum_{l < \kappa_i} \alpha_l M(z)^{-\kappa_i + l},$$

since all exponents on ϕ_i^- must be negative. Considering ϕ_i^+ we see that for it to be analytic, $\alpha_l = 0$ for all l. We have characterized the kernel by $\sum_{\kappa_i > 0} \kappa_i$ elements.

To analyze the range we set $u = u_+ + u_-$, where $u_\pm \in \mathscr{C}_\Gamma^\pm H_z^k(\Gamma)$ is a vector function. If we choose $u_- = 0$, we find $\mathscr{C}[I, \theta; \Gamma] u_+ = u_+$. Therefore $\mathscr{C}_\Gamma^+ H_z^k(\Gamma)$ is in the range. Now choose $u_+ = 0$. We find $\mathscr{C}[I, \theta; \Gamma] u = \mathscr{C}_\Gamma^- u\_\theta$. We break this into the entries u_i^- of u_- and consider

$$\mathscr{C}_\Gamma^- u_i^- M^{\kappa_i}.$$

If $\kappa_i \geq 0$ and $u \in \mathscr{E}^-(\Gamma)$, then $M^{-\kappa_i} u \in \mathscr{E}^-(\Gamma)$, implying that

$$\mathscr{C}_\Gamma^- [M^{-\kappa_i} u] M^{\kappa_i} = u.$$

Thus this maps onto $\mathscr{C}_\Gamma^- H_z^k(\Gamma)$. Now if $\kappa_i < 0$, then $\mathscr{C}_\Gamma^- u_i^- M^{\kappa_i}$ always has a $(-\kappa_i)$th-order zero at $z = z_-$, implying that if $g \in \operatorname{ran} \mathscr{C}_\Gamma^- \cdot M^{\kappa_i}$, then $\frac{\mathrm{d}}{\mathrm{d}z} \mathscr{C}_\Gamma^- g(z_-) = 0$ for $j = 0, \ldots, -\kappa_i - 1$. This amounts to $\sum_{\kappa_i < 0} |\kappa_i|$ orthogonality conditions. Thus we have characterized the deficiency of the range by $\sum_{\kappa_i < 0} |\kappa_i|$ bounded linear functionals. We have constructed the kernel and identified the deficiency of the range when $\mathscr{C}[I, \theta; \Gamma]$ acts on vectors. Since each row is acted on independently, the dimension of the kernel and codimension of the range is multiplied by the number of rows. □

Remark 2.7.4. *If we apply $\mathscr{C}[X_+, X_-; \Gamma]$ to $m \times n$-matrix-valued functions (instead of $n \times n$), then*

- $\dim \ker \mathscr{C}[X_+, X_-; \Gamma] = m \sum_{\kappa_i > 0} |\kappa_i|$,
- $\operatorname{codim} \operatorname{ran} \mathscr{C}[X_+, X_-; \Gamma] = m \sum_{\kappa_i < 0} |\kappa_i|$, *and*
- $\operatorname{ind} \mathscr{C}[X_+, X_-; \Gamma] = -m \sum_{i=1}^n \kappa_i$.

We connect the total index κ to something easily computable. If A is a rational matrix function with a right-standard factorization, then

$$A = A_- \theta A_+ \Rightarrow \det A = \det A_- \left(\frac{z - z_+}{z - z_-} \right)^\kappa \det A_+.$$

It follows that

$$
\begin{aligned}
\operatorname{ind}_{\partial D} \det A_\pm &= 0, \quad D \Subset \Omega_\pm,\\
\operatorname{ind}_{\partial D}\left[\left(\frac{z-z_+}{z-z_-}\right)^{\kappa}\right] &= \begin{cases} \kappa & \text{if } z_+ \in D,\\ 0 & \text{otherwise,}\end{cases}
\end{aligned}
\tag{2.51}
$$

using the argument principle. Orientation is taken into account in the second expression (2.51). Summing over components,

$$
\sum_{D\subset\Omega_+} \int_{\partial D} \text{đ}\log\left(\det A_+ \left(\frac{z-z_+}{z-z_-}\right)^{\kappa}\right) = \kappa, \quad \sum_{D\subset\Omega_-} \int_{\partial D} \text{đ}\log\det A_- = 0.
$$

We decompose $\Gamma = \Gamma_1 \cup \cdots \cup \Gamma_l$, with Γ_l smooth and non-self-intersecting. Then

$$
\kappa = \sum_{D\subset\Omega_+} \int_{\partial D} \text{đ}\log\left(\det A_+ \left(\frac{z-z_+}{z-z_-}\right)^{\kappa}\right) - \sum_{D\subset\Omega_-} \int_{\partial D} \text{đ}\log\det A_- = \sum_{j=1}^{l} \int_{\Gamma_j} \text{đ}\log\det A.
$$

We obtain the following lemma.

Lemma 2.68. *Let Γ be admissible and G be a rational matrix function with an algebraic factorization $X_\pm - I \in H_\pm^k(\Gamma)$; then*

$$
\operatorname{ind} \mathscr{C}[X_+, X_-; \Gamma] = n \sum_{j=1}^{l} \int_{\Gamma_l} \text{đ}\log\det G
$$

as an operator on $H_z^k(\Gamma)$. Recall that this operator is applied to $n \times n$ matrix-valued functions.

Theorem 2.69. *Lemma* 2.68 *holds when G is just k-regular.*

Proof. Let $G = X_-^{-1} X_+$ and approximate $X_\pm$ with a $H_\pm^k(\Gamma) + \mathbb{C}^{n\times m}$ convergent sequence of rational matrix functions $(X_\pm^n)_{n\geq 0}$. It follows that

$$
\|\mathscr{C}[X_+, X_-; \Gamma] - \mathscr{C}[X_+^n, X_-^n; \Gamma]\|_{\mathscr{L}(H_z^k(\Gamma))} \to 0 \text{ as } n \to \infty.
$$

Thus for n sufficiently large, $\mathscr{C}[\Gamma; X_+^n, X_-^n]$ has the same index as $\mathscr{C}[X_+, X_-; \Gamma]$. We express

$$
\begin{aligned}
\operatorname{ind} \mathscr{C}[X_+^n, X_-^n; \Gamma] &= n \sum_{j=1}^{l} \int_{\Gamma_l} \text{đ}\log\det X_+^n - \text{đ}\log\det X_+^n\\
&= n \sum_{D\subset\Omega_+} \operatorname{ind}_{\partial D} \det X_+^n - \sum_{D\subset\Omega_-} \operatorname{ind}_{\partial D} \det X_-^n.
\end{aligned}
$$

Since $X_\pm^n \to X_\pm$ in $L^\infty(\Gamma)$, we have that for sufficiently large n

$$
\begin{aligned}
\operatorname{ind} \mathscr{C}[X_+^n, X_-^n; \Gamma] &= n \sum_{D\subset\Omega_+} \operatorname{ind}_{\partial D} \det X_+ - n \sum_{D\subset\Omega_-} \operatorname{ind}_{\partial D} \det X_-\\
&= n \sum_{j=1}^{l} \int_{\Gamma_l} \text{đ}\log\det G = \operatorname{ind} \mathscr{C}[X_+, X_-; \Gamma]. \qquad \square
\end{aligned}
$$

Remark 2.7.5. *Note that the index is independent of k and if G is k-regular, then the invertibility of* $\mathscr{C}[X_+,X_-;\Gamma]$ *on* $H_z^j(\Gamma)$ *for one* $j \in \{0,1,2,\ldots,k\}$ *implies invertibility for any* j. *Similar statements follow for* $\mathscr{C}[G;\Gamma]$.

Remark 2.7.6. *Let* Φ *be a bounded* L^2 *solution of the RH problem with jump* G. *Then*

$$\det\Phi^+ = \det\Phi^- \det G, \quad \det\Phi(\infty) = 1. \tag{2.52}$$

Assume that $\operatorname{ind}_\Gamma \det G = 0$ *and that* $\det G$ *is* 1*-regular. The theory presented thus far shows that* $\det\Phi$ *is the only solution to this problem (the only partial index is zero). Indeed, in many cases,* $G = X_-^{-1}X_+$ *so that* $\det G$ *must satisfy the (scalar) product condition. We find a possibly new algebraic factorization* $\det G = Y_+/Y_-$ *for scalar-valued functions* $Y_\pm$. *Enumerate* $\{D_i^+\}$ *and* $\{D_i^-\}$ *where* $D_i^\pm \Subset \Omega_\pm$. *Select points* $z_i^\pm \in D_i^\pm$ *and define the integers*

$$\kappa_i^\pm = \operatorname{ind}_{D_i^\pm} Y_\pm.$$

Let

$$P(z) = \prod_{D_{i,+}\Subset\Omega_+} (z-z_i^+)^{-\kappa_i^+}, \quad M(z) = \prod_{D_{i,-}\Subset\Omega_-} (z-z_i^-)^{-\kappa_i^-}.$$

It follows from $\operatorname{ind}_\Gamma G = 0$ *that* $\lim_{|z|\to\infty} P(z)/M(z) = 1$ *and*

$$\kappa_i^+ = \operatorname{ind}_{D_{i,+}} Y_+P/M = 0, \quad \kappa_i^- = \operatorname{ind}_{D_{i,-}} Y_-M/P = 0.$$

Define $Z_+ = Y_+P/M$ *and* $Z_- = Y_-M/P$ *so that* $G = Z_+/Z_-$. *Again* $Z_\pm$ *are scalar-valued. We find a representation for* $\det\Phi$:

$$\begin{aligned}\log\det\Phi(z) &= \int_\Gamma \frac{\log Z_+(s) - \log Z_-(s)}{s-z} ds \\ &= \sum_{D_{i,+}\Subset\Omega_+} \int_{D_{i,+}} \frac{\log Z_+(s)}{s-z} ds - \sum_{D_{i,-}\Subset\Omega_-} \int_{D_{i,-}} \frac{\log Z_-(s)}{s-z} ds.\end{aligned}$$

From this it is clear that $\det\Phi$ *cannot vanish; both Cauchy integrals are bounded. It is worth noting that if* $\det G = 1$, *then* $\det\Phi = 1$.

Some additional remarks are in order.

Remark 2.7.7. *Theorem* 2.66 *and Lemma* 2.68 *hold for* $\mathscr{C}[G;\Gamma]$ *and* $\mathscr{C}'[G;\Gamma]$ *in* $L^2(\Gamma)$ *provided* $k \geq 1$ *and* $\mathscr{C}'[X_+,X_-;\Gamma]$ *on* $\tilde{H}_+^k(\Gamma) \cap \tilde{H}_-^k(\Gamma)$.

Remark 2.7.8. *In applications it is often the case that* $\det G = 1$. *This immediately shows that the index is zero.*

We have shown there is a one-to-one correspondence between $\mathscr{E}^\pm(\Gamma)$ solutions of an RH problem with jump matrix G and solutions of $\mathscr{C}[X_+,X_-;\Gamma]u = X_+^{-1} - X_-^{-1}$. The question of solvability of an RH problem is reduced to the question of invertibility of a Fredholm singular integral operator. To show unique solvability from this, one usually determines that the singular integral operator has index zero. Additional results need to be used to show that the kernel must be trivial. An appeal to the *open mapping theorem* and one of its well-known corollaries then proves the invertibility of the operator.

Theorem 2.70 (Open mapping). *Let X and Y be Banach spaces. If $T \in \mathcal{L}(X,Y)$ is surjective, then T is open.*

Corollary 2.71. *If X and Y are Banach spaces and $T \in \mathcal{L}(X,Y)$ is bijective, then T is an isomorphism: $T^{-1} \in \mathcal{L}(Y,X)$.*

To get a handle on the kernel we present an incarnation of the so-called *vanishing lemma*. See [53] for other versions of vanishing lemmas.

Definition 2.72. *A contour is said to be Schwarz invariant if $\Gamma^\dagger \triangleq \{\bar{z} : z \in \Gamma\} = \Gamma$, respecting orientation. The Schwarz conjugate of a matrix function F is defined by $F^\dagger(z) = \overline{F(\bar{z})}^\top$.*

Theorem 2.73 (Vanishing lemma [123]). *Let G be defined on a Schwarz invariant, admissible contour Γ. If G is k-regular, $G + G^\dagger$ is positive semidefinite on $\mathbb{R}$, positive definite on a set of positive measure, and $G = G^\dagger$ on $\Gamma \setminus \mathbb{R}$, then* $\dim\ker \mathscr{C}[X_+,X_-;\Gamma] = 0$.

Proof. Let Φ be a vanishing $\mathcal{E}^\pm(\Gamma)$ solution, i.e., a solution of the associated RH problem:

$$\Phi^+(s) = \Phi^-(s)G(s), \quad \Phi(\infty) = 0.$$

The statement that the only vanishing solution is the zero solution is equivalent to the operator having a trivial kernel. So, we assume $\Phi^\pm \in L^2(\Gamma)$ so that $v = \Phi^+ - \Phi^- \in \ker \mathscr{C}[X_+,X_-;\Gamma]$. Let $D \Subset \mathbb{C} \setminus \Gamma$; then $D^\dagger \Subset \mathbb{C} \setminus \Gamma$. Further, $\partial D^\dagger = (\partial D)^\dagger$, respecting orientation. Set $u_1 = \Phi|_{\partial D}$ and $u_2 = \Phi|_{\partial D^\dagger}$. Since $u_1 \in \mathcal{E}^2(D)$, we know that $\int_{\partial D} u_1(z) r(z) dz = 0$ for any rational function r with poles outside D. Since $u_2^\dagger$ may be approximated with rational functions in $L^2(\partial D)$ satisfying this condition, we find

$$\int_{\partial D} u_1(z) u_2^\dagger(z) \mathrm{d}z = 0.$$

Summing over all components in the upper-half plane we find

$$\sum_{D \subset \mathbb{C}^+ \setminus \Gamma} \int_{\pm\partial D} u_1(z) u_2^\dagger(z) \mathrm{d}z = 0. \tag{2.53}$$

We choose $\pm$ to make ∂D positively oriented. Let $\{\Gamma_i\}$ be smooth components of Γ in the open upper-half plane. Then Γ_i is part of the boundary of two components $D_\pm \subset \Omega_\pm$, and Φ has boundary values $u_1^\pm$ from $\Omega_\pm$ on Γ_i. Boundary values for u_2, $u_2^\pm$ exist analogously on $\Gamma_i^\dagger$. Integrals along Γ_i appear twice in (2.53):

$$\int_{\Gamma_i} u_1^+(z)(u_2^-)^\dagger(z) \mathrm{d}z = \int_{\Gamma_i} u_1^-(z) G(z) (u_2^-)^\dagger(z) \mathrm{d}z,$$

$$\int_{-\Gamma_i} u_1^-(z)(u_2^+)^\dagger(z) \mathrm{d}z = \int_{-\Gamma_i} u_1^-(z) G^\dagger(z) (u_2^-)^\dagger(z) \mathrm{d}z.$$

Using $G = G^\dagger$ off the real axis, we see that these integrals cancel. Only integrals on the real axis survive, producing

$$\sum_{D \subset \mathbb{C}^+ \setminus \Gamma} \int_{\pm\partial D} u_1(z) u_2^\dagger(z) \mathrm{d}z = \int_{\mathbb{R}} u_1^+(x)(u_2^-)^\dagger(x) \mathrm{d}x = \int_{\mathbb{R}} u_1^-(x) \tilde{G}(x) (u_1^-)^\dagger(x) \mathrm{d}x,$$

where $\tilde{G} = G$ or $G^\dagger$ depending on how each subset of the real axis is oriented. This equation is added to its Schwarz reflection, providing us with

$$0 = \int_{\mathbb{R}} u_1^-(x)(G(x) + G^\dagger(x))(u_1^-)^\dagger(x)\mathrm{d}x.$$

Thus $u_1^- = 0$ and hence Φ^- is zero a.e. on $\mathbb{R}$ by positive definiteness. It is known that any function in $\mathscr{E}^2(D)$ which vanishes on a set of positive measure on the boundary must be identically zero. If Φ is zero in any component of $\mathbb{C} \setminus \Gamma$, then the jump condition forces Φ to be zero in all neighboring components. We conclude that $\Phi = 0$ and $\dim\ker \mathscr{C}[X_+, X_-; \Gamma] = 0$. □

The notion of partial indices can be generalized to so-called *decomposing algebras*. If these are used, Theorem 2.73 can be strengthened to show invertibility of the operator without appealing to a zero index statement. We do not elaborate on this here but note that for our purposes $\det G$ is constant, or simple. Establishing zero index is then trivial.

Remark 2.7.9. *Note that if a* 1*-regular RH problem is uniquely solvable for an* $\mathscr{E}^\pm(\Gamma)$ *solution, then the solution of the SIE* (2.43) *with* $F \equiv 0$ *is in* $H_z^1(\Gamma)$ *and from Lemma* 2.51 *and Corollary* 2.52 *the RH problem is a uniquely solvable continuous RH problem. For this reason, throughout Parts* II *and* III *of this text, we always assume we are looking for an* $\mathscr{E}^\pm(\Gamma)$ *solution whenever we phrase an RH problem.*

2.8 ▪ Additional considerations

In the remainder of this chapter we present a collection of results that are useful in further applications of RH problems. We describe how to estimate the derivatives of solutions, deform contours in the spirit of contour integrals, and augment contours so that they become admissible.

2.8.1 ▪ Regularity

We expand on the theory above by providing estimates for the derivatives of solutions of SIEs. This provides criteria in terms of closeness of jump matrices to test when the solution of one problem is "close" to the solution of another problem. We continue to associate $[G;\Gamma]$ with the L^2 RH problem

$$\Phi^+(s) = \Phi^-(s)G(s), \quad s \in \Gamma, \quad \Phi(\infty) = I.$$

If G is k-regular, we also say $[G;\Gamma]$ is k-regular.

Theorem 2.74. *Given an RH problem* $[G;\Gamma]$ *which is* k*-regular, assume* $\mathscr{C}[G;\Gamma]$ *is invertible on* $L^2(\Gamma)$*. Then the solution of* (2.43)*,* $u \in H_z^k(\Gamma)$*, satisfies*

$$D^k u = \mathscr{C}[G;\Gamma]^{-1}\Big(D^k(G - I) + D^k(\mathscr{C}_\Gamma^- u \cdot (G - I)) - \mathscr{C}_\Gamma^- D^k u \cdot (G - I)\Big), \tag{2.54}$$

where the right-hand side of (2.54) *after expansion does not depend on* $D^k u$*.*

Proof. Because $u \in H_z^k(\Gamma)$, where u is the solution of $\mathscr{C}[G;\Gamma]u = G - I$, we can (weakly) differentiate this equation k times, each time passing the differentiation operator though the Cauchy integral operator onto u. The first iteration is

$$\begin{aligned} u - \mathscr{C}_\Gamma^- u(G-I) &= G-I,\\ Du - \mathscr{C}_\Gamma^- Du(G-I) &= (I+\mathscr{C}_\Gamma^- u)D(G-I),\\ Du &= \mathscr{C}[G;\Gamma]^{-1}((I+\mathscr{C}_\Gamma^- u)D(G-I)). \end{aligned}$$

The procedure is clear for subsequent iterations. □

Corollary 2.75. *Under the hypotheses of Theorem 2.74, u satisfies an inequality of the form*

$$\|u\|_{H^k(\Gamma)} \leq p_k\Big(\|G-I\|_{W^{k,\infty}(\Gamma)} \|\mathscr{C}[G;\Gamma]^{-1}\|_{\mathscr{L}(L^2(\Gamma))}\Big) \|\mathscr{C}[G;\Gamma]^{-1}\|_{\mathscr{L}(L^2(\Gamma))} \|G-I\|_{H^k(\Gamma)}, \tag{2.55}$$

where p_k is a polynomial of degree k whose coefficients depend on $\|\mathscr{C}_\Gamma^-\|_{\mathscr{L}(L^2(\Gamma))}$.

Proof. Taking the norm of (2.54) gives a bound on the seminorm $\|D^k u\|_{L^2(\Gamma)}$ in terms of $\|u\|_{H^{k-1}(\Gamma)}$. Using (2.54) for $k-1$ gives a bound in terms of $\|u\|_{H^{k-2}(\Gamma)}$. This process produces a bound of the form (2.55). □

Remark 2.8.1. *The expression for the derivative in Theorem 2.74 can be used to bound Sobolev norms of the solution in terms of Sobolev norms of the jump matrix if a bound on the norm of the inverse operator $\mathscr{C}[G;\Gamma]$ is known. When the jump matrix depends on a parameter, and the Sobolev norms of the jump matrix are bounded or decaying, the resulting bounds are useful.*

We often use the following theorem, which is derived from Theorem A.20.

Lemma 2.76. *Consider a sequence of k-regular RH problems $([G_\xi;\Gamma])_{\xi\geq 0}$ on the fixed contour Γ. Assume $G_\xi \to G$ in $H^k(\Gamma)$ if $k \geq 1$ or in $L^2 \cap L^\infty(\Gamma)$ if $k = 0$ as $\xi \to \infty$. The following hold:*

- *If $\mathscr{C}[G;\Gamma]$ is invertible on $L^2(\Gamma)$, then there exists a $T > 0$ such that $\mathscr{C}[G_\xi;\Gamma]$ is also invertible for $\xi > T$.*
- *If Φ_ξ is the solution of $[G_\xi;\Gamma]$, and Φ is the solution of $[G;\Gamma]$, then $\Phi_\xi^\pm - \Phi^\pm \to 0$ in $H^k(\Gamma)$.*
- *$\|\Phi_\xi - \Phi\|_{W^{j,\infty}(S)} \to 0$ for all j whenever S is bounded away from Γ.*

Proof. The first statement follows from the fact that $\mathscr{C}[G_\xi;\Gamma]$ converges to $\mathscr{C}[G;\Gamma]$ in the operator norm. The second property follows from Corollary 2.75. The final property is a consequence of the Cauchy–Schwarz inequality and the fact that $\|u_\xi - u\|_{L^2(\Gamma)} \to 0$. □

2.8.2 ▪ Truncation, augmentation, and practical aspects of RH problems

When RH problems are used in practice the contour involved is often not complete and maybe not even be admissible. Furthermore, we aim to solve RH problems numerically (in Chapter 6), and contours of infinite length may cause problems. In this section we discuss what can be done when a contour is not complete and how for most RH problems on infinite contours there exists a "nearby" problem on a bounded contour.

Augmentation: Throughout much of the above discussion we assumed Γ is complete and admissible. Here we demonstrate that this can be assumed for all the L^2 RH problems we consider. Let Γ be an incomplete but piecewise-smooth contour with only transverse intersections. We also assume that any infinite components tend to straight lines at infinity in the sense that Γ^{-1} has transverse intersections at the origin. We describe a procedure that converts Γ to a complete, admissible contour.

We join all open endpoints back to Γ or to ∞ with smooth contours. Define the resulting contour to be Γ'. Now $\Omega = \mathbb{C} \setminus \Gamma'$ consists of disjoint open sets D such that ∂D is piecewise smooth with only transverse intersections. Note that this is not necessarily true of $\mathbb{C}\setminus\Gamma$. Choose $D_1 \Subset \Omega$ and reorient (if necessary) Γ' so that $\Gamma' \cap \bar{D}_1 = \partial D_1$ (positively oriented) where orientation is accounted for in the equality. Look at all neighboring components $D_2,\ldots,D_m$ of D_1. Either Γ' can be reoriented so that $\Gamma' \cap \bar{D}_2 = -\partial D_2$ (negatively oriented), or it cannot. If not, additional contours are added resulting in Γ'' so that this can be accomplished. We repeat this process until the contour is complete. We demonstrate this procedure in Figure 2.8. Now, in the case that a component $D \Subset \Omega_\pm$ is unbounded and its boundary is not a Lipschitz graph, we can perform an additional augmentation step, shown in Figure 2.9, to make it admissible. We do not prove one can always augment but it is evident in all cases we consider.

For an RH problem $[G;\Gamma]$ we discuss the extension to a new RH problem, $[\tilde{G},\Gamma'']$, and the conditions needed so that $[\tilde{G},\Gamma'']$ is k-regular. First, define $\tilde{G} = I$ on $\Gamma''\setminus\Gamma$. Next if a given component Γ_i of Γ had its orientation reversed as a component of Γ'', define $\tilde{G} = G^{-1}$ on Γ_i. It is easily seen that if G satisfies the $(k-1)$th-order product condition on Γ, then $[\tilde{G},\Gamma'']$ is k-regular. Also, it is clear that the solutions of $[G;\Gamma]$ and $[\tilde{G};\Gamma'']$ are in a one-to-one correspondence.

Indeed, we compare the singular integral equations

$$\mathscr{C}[G,\Gamma]u = G - I \quad \text{on } L^2(\Gamma), \tag{2.56}$$

$$\mathscr{C}[\tilde{G},\Gamma'']\tilde{u} = \tilde{G} - I \quad \text{on } L^2(\Gamma''). \tag{2.57}$$

It follows that (2.57) is equivalent to (2.56). On $\Gamma''\setminus\Gamma$

$$\mathscr{C}[\tilde{G},\Gamma'']\tilde{u} = \tilde{u} - \mathscr{C}^-_{\Gamma''}\tilde{u}\cdot(\tilde{G}-I) = \tilde{u} = 0 \quad \text{on } \Gamma''\setminus\Gamma. \tag{2.58}$$

Furthermore, let $\Gamma_i \subset \Gamma$ have reversed orientation $-\Gamma_i \subset \Gamma''$. Let Γ_i^c be the complement of Γ_i with respect to Γ''. On Γ_i

$$\begin{aligned}
\tilde{u} - \mathscr{C}^-_{\Gamma''}\tilde{u}\cdot(\tilde{G}-I) &= \tilde{G}-I\\
\Leftrightarrow \mathscr{C}^+_{\Gamma''}\tilde{u} - \mathscr{C}^-_{\Gamma''}\tilde{u}\cdot\tilde{G} &= \tilde{G}-I\\
\Leftrightarrow (\mathscr{C}^+_{-\Gamma_i} + \mathscr{C}^+_{\Gamma_i^c})\tilde{u} - (\mathscr{C}^-_{-\Gamma_i} + \mathscr{C}^-_{\Gamma_i^c})\tilde{u}\cdot\tilde{G} &= \tilde{G}-I\\
\Leftrightarrow (\mathscr{C}^+_{-\Gamma_i} + \mathscr{C}^+_{\Gamma_i^c})\tilde{u} - (\mathscr{C}^-_{-\Gamma_i} + \mathscr{C}^-_{\Gamma_i^c})\tilde{u}\cdot G^{-1} &= G^{-1}-I\\
\Leftrightarrow (\mathscr{C}^+_{-\Gamma_i} + \mathscr{C}^+_{\Gamma_i^c})\tilde{u}\cdot G - (\mathscr{C}^-_{-\Gamma_i} + \mathscr{C}^-_{\Gamma_i^c})\tilde{u} &= I-G.
\end{aligned}$$

On Γ_i we have $\mathscr{C}^+_{\Gamma_i^c}\tilde{u} = \mathscr{C}^-_{\Gamma_i^c}\tilde{u}$, and using the orientation $\mathscr{C}^\pm_{-\Gamma_i} = -\mathscr{C}^\mp_{\Gamma_i}$ we obtain

$$(-\mathscr{C}^+_{\Gamma_i} + \mathscr{C}^+_{\Gamma_i^c})\tilde{u} - (-\mathscr{C}^-_{\Gamma_i} + \mathscr{C}^-_{\Gamma_i^c})\tilde{u}\cdot G = G - I.$$

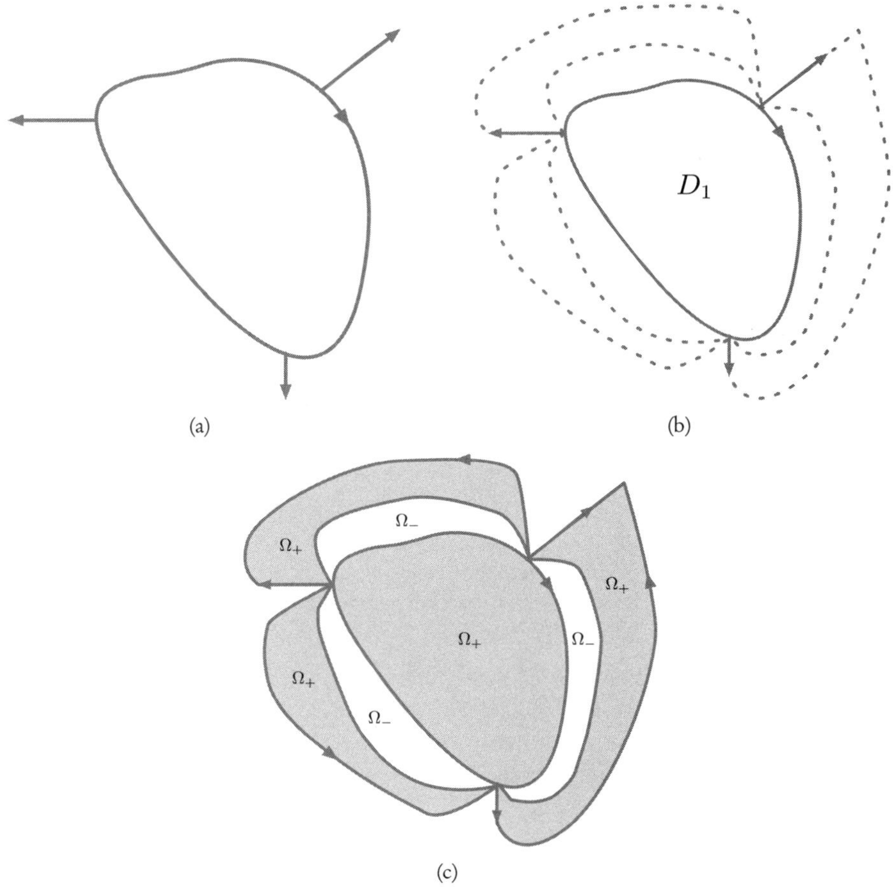

Figure 2.8. *The* augmentation *of an incomplete contour with open endpoints to a complete contour. For this sketch we assume all contours are finite.* (a) *The initial contour.* (b) *Fixing D_1 and adding contours as needed to obtain Γ'.* (c) *The resulting contour Γ'' is now complete with $\Omega_\pm$ clearly defined.*

Now, redefine $\tilde{u} = -\tilde{u}$ on Γ_i. This argument applied to each re-oriented contour combined with (2.58) shows

$$\tilde{u} - \mathcal{C}_\Gamma^- \tilde{u} \cdot (G - I) = G - I \quad \text{on } \Gamma_i,$$

thus proving equivalence. We arrive at the following theorem.

Theorem 2.77. *Let Γ be a possibly incomplete contour and let G satisfy the $(k-1)$th-order product condition on Γ. If for an augmentation Γ'' of Γ and an extension $\tilde{G}$ of G, $\mathcal{C}[\tilde{G};\Gamma'']$ is invertible on $L^2(\Gamma'')$, then $\mathcal{C}[G;\Gamma]$ is invertible on $L^2(\Gamma)$, and $u = \mathcal{C}[G;\Gamma]^{-1}(G-I) \in H_z^k(\Gamma)$.*

Remark 2.8.2. *A theoretical contribution of this chapter is that the Sobolev spaces of Zhou are not defined on incomplete contours and require augmentation. As we have seen, the spaces $H_z^k(\Gamma)$ are ideal when Γ is incomplete. This is one reason for considering $\mathcal{C}[G;\Gamma]$ instead of $\mathcal{C}'[G;\Gamma]$.*

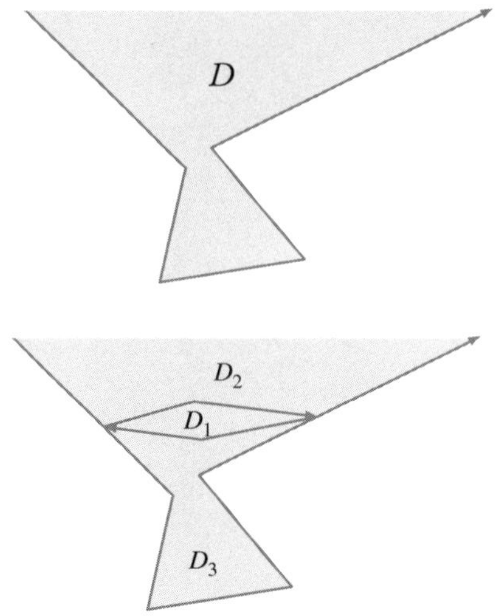

Figure 2.9. *The augmentation of a complete contour to address the fact that ∂D is not a Lipschitz graph. By adding piecewise affine contours we are able to ensure that every component of the complement of the contour is either bounded or has a Lipschitz graph boundary (after possible rotations).*

Truncation: Thus far we have justified augmenting incomplete contours to obtain an equivalent RH problem on complete contours. Often, in practice, we start with an RH problem posed on an unbounded contour and we wish to truncate it to a problem on a finite contour. This is the first of many deformation techniques we present. The following result justifies this process.

Proposition 2.78. *Assume $[G;\Gamma]$ is k-regular. For every $\epsilon > 0$ there exists a function G_ϵ defined on Γ and a bounded contour $\Gamma_\epsilon \subset \Gamma$ such that*

- $G_\epsilon = I$ *on* $\Gamma \setminus \Gamma_\epsilon$,
- $\|G_\epsilon - G\|_{W^{k,\infty}(\Gamma)\cap H^k(\Gamma)} < \epsilon$, *and*
- $[G_\epsilon;\Gamma_\epsilon]$ *is k-regular and*

$$\|\mathscr{C}[G;\Gamma] - \mathscr{C}[G_\epsilon;\Gamma]\|_{\mathscr{L}(L^2(\Gamma))} < \epsilon \|\mathscr{C}_\Gamma^-\|_{\mathscr{L}(L^2(\Gamma))}.$$

Proof. A matrix-valued function f is chosen such that

- $f|_{\Gamma_i} \in C^\infty(\Gamma_i)$,
- $f = I$ in a neighborhood of all intersection points,
- f has compact support, and
- $\|(G-I)f + I - G\|_{W^{k,\infty}(\Gamma)\cap H^k(\Gamma)} < \epsilon$.

Equate $G_\epsilon = (G-I)f + I$. The last property follows immediately. $\square$

This justifies the truncation of infinite contours to bounded contours at the cost of a small error, and it shows this process preserves smoothness. For numerical computations in the next part, we truncate contours when the jump matrix is, to machine precision, the identity matrix.

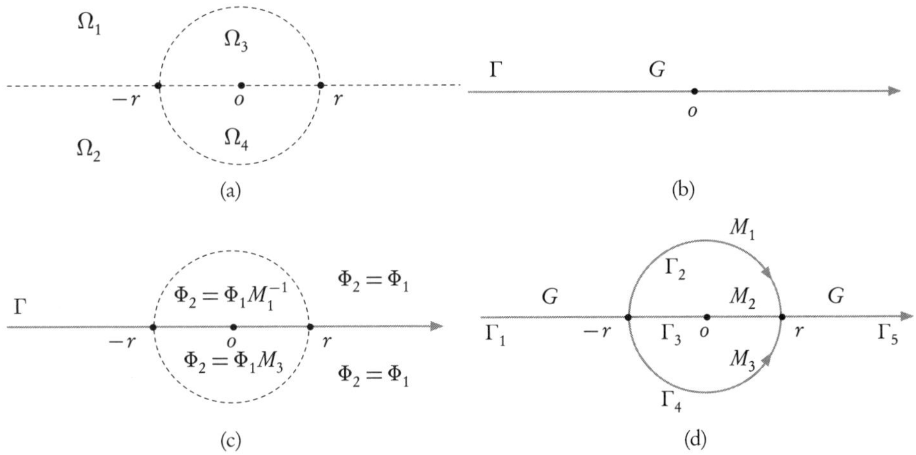

Figure 2.10. *The lensing process.* (a) *Regions in the complex z-plane.* (b) *Contours and jump matrices for the RH problem for* Φ_1. (c) *Definition of* Φ_2. (d) *Contours and jump matrices for the RH problem for* Φ_2.

Lensing: Here we go over, in detail, the process of *lensing* an RH problem. This is one of the many deformation techniques that are useful both in the numerical method presented in the following chapter and for the method of nonlinear steepest descent. We start with an RH problem (Figure 2.10(b))

$$\Phi_1^+(s) = \Phi_1^-(s)G(s), \quad s \in \Gamma \subset \mathbb{R}, \quad \Phi_1(\infty) = I.$$

We choose $\Gamma \subset \mathbb{R}$ for simplicity and assume $0 \in \Gamma$. Assume that G has a factorization

$$G(z) = M_3(z)M_2(z)M_1(z).$$

Assume for simplicity that all matrices are nonsingular and continuous. Fix an $r > 0$ and define the regions Ω_i, $i = 1,2,3,4$ (see Figure 2.10(a)), by

$$\begin{aligned}
\Omega_1 &= \{z \in \mathbb{C} : \operatorname{Im} z > 0 \text{ and } |z| > r\},\\
\Omega_2 &= \{z \in \mathbb{C} : \operatorname{Im} z < 0 \text{ and } |z| > r\},\\
\Omega_3 &= \{z \in \mathbb{C} : \operatorname{Im} z > 0 \text{ and } |z| < r\},\\
\Omega_4 &= \{z \in \mathbb{C} : \operatorname{Im} z < 0 \text{ and } |z| < r\}.
\end{aligned}$$

Further assume that M_3 has an analytic extension in a neighborhood of Ω_4, and that M_1 has an analytic extension in a neighborhood of Ω_3. We wish to change the RH problem on $(-r, r)$ by a lensing process. Define a new function Φ_2 by (Figure 2.10(c))

$$\Phi_2(z) = \begin{cases} \Phi_1(z)M_1^{-1}(z) & \text{if } z \in \Omega_3,\\ \Phi_1(z)M_3(z) & \text{if } z \in \Omega_4,\\ \Phi_1(z) & \text{otherwise.} \end{cases}$$

Define new contours Γ_i, $i = 1,2,3,4,5$, all oriented in the direction of increasing real part, by

$$\begin{aligned}
\Gamma_1 &= \Gamma \cap (-\infty, r),\\
\Gamma_2 &= \{z \in \mathbb{C} : \operatorname{Im} z > 0 \text{ and } |z| = r\},\\
\Gamma_3 &= (-r, r),\\
\Gamma_4 &= \{z \in \mathbb{C} : \operatorname{Im} z < 0 \text{ and } |z| = r\},\\
\Gamma_5 &= \Gamma \cap (r, \infty).
\end{aligned}$$

We compute the jumps of Φ_2 on these contours. As an example consider Γ_2. Set

$$\Phi_2^+(s) = \Phi_2^-(s)U(s), \quad s \in \Gamma_2,$$

for some matrix $U(s)$ to be determined. In this case $\Phi_2^+(s) = \Phi_1^+(s) = \Phi_1^-(s) = \Phi_2^-(s)M_1(s)$. We find that $U(s) = M_1(s)$. Repeating this process on all contours shows that Φ_2 satisfies the following RH problem (Figure 2.10(d)):

$$\Phi_2^+(s) = \begin{cases} \Phi_2^-(s)G(s) & \text{if } s \in \Gamma_1,\\ \Phi_2^-(s)M_1(s) & \text{if } s \in \Gamma_2,\\ \Phi_2^-(s)M_2(s) & \text{if } s \in \Gamma_3,\\ \Phi_2^-(s)M_3(s) & \text{if } s \in \Gamma_4,\\ \Phi_2^-(s)G(s) & \text{if } s \in \Gamma_5, \end{cases} \quad \Phi_2(\infty) = I.$$

It is clear that this generalizes to contours off the line and is only limited by the analyticity properties of the factorization. Furthermore, if $M_1, M_3 \to I$ as $z \to \infty$, in the proper regions the lensing can be employed in infinite regions. Note that one of the matrices M_i could be the identity, in which case that contour is dropped from the RH problem.

Conjugation: *Conjugation* is a simple concept that allows additional deformations of an RH problem. Consider an RH problem $[G;\Gamma]$ with solution Φ. Let $\Psi_\pm, \Psi_\pm^{-1} \in L^\infty(\Gamma)$ be a solution of another problem $[H;\Gamma]$. Then

$$\Phi_+\Psi_+^{-1} = \Phi_-\Psi_-^{-1}\Psi_- G H^{-1}\Psi_-^{-1}.$$

Concisely, $\Phi\Psi^{-1}$ is a solution of $[\Psi_- G H^{-1}\Psi_-^{-1};\Gamma]$. If $H = G$, then the jump becomes the identity. This method is especially useful in conjunction with lensing.

Example 2.79. Assume M_2 (Figure 2.10) is diagonal. The problem $[M_2;\Gamma_3]$ (see Figure 2.10 for Γ_3) can be solved with the techniques in Section 2.4.1. Let Ψ be this solution. By Lemma 2.11 we know that Ψ has singularities at $\pm r$. The function $\Sigma = \Phi_2\Psi^{-1}$ satisfies

$$\begin{aligned}
\Sigma_+ &= \Sigma_-\Psi_- G\Psi_+^{-1} && \text{on } \Gamma_1, \Gamma_5,\\
\Sigma_+ &= \Sigma_-\Psi_- M_1\Psi_+^{-1} && \text{on } \Gamma_2,\\
\Sigma_+ &= \Sigma_-\Psi_- M_3\Psi_+^{-1} && \text{on } \Gamma_4,\\
\Sigma_+ &= \Sigma_- && \text{on } \Gamma_3.
\end{aligned}$$

This process has removed the jump on Γ_3 from the RH problem at the expense of singularities at $\pm r$. This is used in the numerical solution of many RH problems. If the entries of M_2 are real-valued, it can be shown using Lemma 2.11 that the solution of $[M_1;\Gamma_3]$ is bounded in the whole plane. In this case we call Ψ a *parametrix* because its use simplifies the RH problem. To be a "good" parametrix it should make the RH problem easier to solve. Many different parametrices are encountered below.[29] ∎

[29]Often in the Riemann–Hilbert literature the word parametrix is reserved for a function that captures the singularity behavior of the problem and is an approximate solution. We use this term more generally.

Contour scaling and decoupling: The final deformation technique we present in generality is *contour scaling* and decoupling. For simplicity, assume the RH problem $[G^\xi, \Gamma^\xi]$ depends on a single parameter $\xi \geq 0$. Further, assume $\Gamma^\xi = \alpha(\xi)\Lambda + \beta(\xi)$, where Λ is a fixed contour. It follows that the matrix-valued function

$$H^\xi(k) = G^\xi(\alpha(\xi)k + \beta(\xi))$$

is defined on Λ.

We introduce a parameter into the problem to demonstrate a situation in which the RH problems on each disjoint contour decouple from each other in the limit.

Example 2.80. Assume $\Gamma^\xi = \Gamma_1^\xi \cup \Gamma_2^\xi$ and

$$\Gamma_1^\xi = \alpha_1(\xi)\Lambda_1 + \beta_1,$$
$$\Gamma_2^\xi = \alpha_2(\xi)\Lambda_2 + \beta_2, \quad |\beta_1 - \beta_2| > 0.$$

Assume that both Λ_1 and Λ_2 are bounded. We consider the L^2 norm of the Cauchy integral operator applied to a function defined on Γ_1^ξ and evaluated on Γ_2^ξ: $\mathscr{C}_{\Gamma_1^\xi} u(z)|_{\Gamma_2^\xi}$. Explicitly,

$$\mathscr{C}_{\Gamma_1^\xi} u(z) = \int_{\Gamma_1^\xi} \frac{u(s)}{s-z} \mathrm{d}s.$$

This is a *Hilbert–Schmidt operator*, and

$$\|\mathscr{C}_{\Gamma_1^\xi}\|^2_{\mathscr{L}(L^2(\Gamma_1^\xi), L^2(\Gamma_2^\xi))} \leq \int_{\Gamma_1^\xi} \int_{\Gamma_2^\xi} \frac{|\text{đ}s\,\mathrm{d}k|}{|s-k|^2}. \tag{2.59}$$

A simple change of variables shows that

$$\|\mathscr{C}_{\Gamma_1^\xi}\|^2_{\mathscr{L}(L^2(\Gamma_1^\xi), L^2(\Gamma_2^\xi))} \leq |\alpha_1(\xi)\alpha_2(\xi)| \int_{\Lambda_1} \int_{\Lambda_2} \frac{|\text{đ}s\,\mathrm{d}y|}{|\alpha_1(\xi)s - \alpha_2(\xi)y + \beta_1 - \beta_2|^2}. \tag{2.60}$$

Since the denominator in the integral in (2.60) is bounded away from zero and both Λ_i are bounded, the right-hand side tends to zero if either α_1 or α_2 tend to zero. ■

Chapter 3

Inverse Scattering and Nonlinear Steepest Descent

Our main application of RH problems in this text is the solution of integrable PDEs. For this reason, we present a unified discussion of the solution of the *defocusing NLS equation*

$$\begin{aligned} iq_t + q_{xx} + 2\lambda|q|^2 q = 0, \quad \lambda = -1, \\ q(\diamond, 0) = q_0 \in \mathscr{S}_\delta(\mathbb{R}), \end{aligned} \tag{3.1}$$

using the *inverse scattering transform* (IST) and the *method of nonlinear steepest descent*. The IST can be seen as a nonlinearization of classical Fourier analysis on $\mathbb{R}$. The full development we present in this chapter requires $q_0 \in \mathscr{S}_\delta(\mathbb{R})$, i.e., Schwartz class functions with exponential decay. The assumptions on q_0 for each individual result are made explicit. In some cases we prove results in spaces of nonsmooth functions to illustrate the parallels between the IST and the Fourier transform. First, we present the solution of this problem using the IST in the notation of Fokas [50]. The solution of the NLS equation can be recovered from the solution of a matrix RH problem. We apply the Deift and Zhou method of nonlinear steepest descent to this RH problem to extract the long-time asymptotics. This whole process involves invoking much of the theory of Chapter 2.

The foundation for the IST is the spectral analysis of the Lax pair associated with the NLS equation. One should compare the procedure in Section 3.1 below with the linear case in Section 2.5.3 above. Simultaneous solutions of the Lax pair with different boundary conditions are considered. The comparison of these solutions allows one to define the spectral data and gives the forward transform. In the case of Section 2.5.3 the forward transform is just the usual Fourier transform on $\mathbb{R}$ and the spectral data is the Fourier transform of the initial condition. The analyticity of the simultaneous solutions is analyzed so that one can formulate an RH problem where the jump matrix depends on the spectral data; see (2.34) for reference. The final piece one needs is the reconstruction formula, a way to reconstruct $q(x,t)$ from the solution of the RH problem, as in Theorem 3.10 below.

The method of nonlinear steepest descent as presented in Section 3.2.2 should be compared with the classical method of steepest descent for integrals. This procedure allows one to extract the long-time behavior of the solution of the defocusing NLS equation from the RH problem (3.17) below. The process of lensing, combined with conjugation, is used to deform this RH problem, which arises from the IST, so that it localizes in a precise sense near the stationary phase point. Once deformed, an approximate jump matrix can be defined (see, for example, (3.22)). It is then shown that the solution of the RH problem with the approximate jump matrix has two fundamental properties:

- it can be constructed explicitly, and
- it converges to the solution of the original, deformed RH problem as $t \to \infty$.

Then the asymptotics of the solution of the defocusing NLS equation can be extracted from the approximate RH problem.

We emphasize that this presentation of the method of nonlinear steepest descent is important to understand the rationale behind the deformations used for computational purposes in later chapters. Such presentations may seem opaque without an understanding of the goals of the method of nonlinear steepest descent.

3.1 ▪ The inverse scattering transform

We first develop the IST for the defocusing NLS equation. Section 2.5.3 is an important prerequisite for this section. This equation has the following Lax pair representation [50, p. 226]:

$$\mu_x + \mathrm{i}z[\sigma_3, \mu] = Q\mu, \tag{3.2}$$

$$\mu_t + 2\mathrm{i}z^2[\sigma_3, \mu] = \tilde{Q}\mu, \tag{3.3}$$

where

$$Q(x,t) = \begin{bmatrix} 0 & q(x,t) \\ -\lambda\bar{q}(x,t) & 0 \end{bmatrix}, \quad \tilde{Q}(x,t) = 2zQ - \mathrm{i}Q_x\sigma_3 + \mathrm{i}\lambda|q|^2\sigma_3,$$

where $\sigma_3 = \mathrm{diag}(1,-1)$ and $[A,B] = AB - BA$ is the standard matrix commutator. This Lax pair is equivalent to the NLS equation in the sense that $\mu_{xt} = \mu_{tx} \Leftrightarrow q$ solves the NLS equation, where μ_{xt} is determined from the first equation differentiated with respect to t and μ_{tx} is determined from the second equation differentiated with respect to x. More precisely, for a matrix μ to be a simultaneous solution of both (3.2) and (3.3), q must solve the NLS equation. Through the developments in this chapter it will become clear how this fact allows for the effective solution of (3.1).

This Lax pair is also equivalent to the differential form[30]

$$\mathrm{d}\left(\mathrm{e}^{\mathrm{i}(zx+2z^2t)\hat{\sigma}_3}\mu(x,t,z)\right) = \mathrm{e}^{\mathrm{i}(zx+2z^2t)\hat{\sigma}_3}(Q\mu\mathrm{d}x + \tilde{Q}\mu\mathrm{d}t), \quad \mathrm{e}^{\alpha\hat{\sigma}_3}A \triangleq \mathrm{e}^{\alpha\sigma_3}A\mathrm{e}^{-\alpha\sigma_3}. \tag{3.4}$$

Note that $\mathrm{e}^{\alpha\hat{\sigma}_3}A$ is not matrix multiplication but a conjugation operation.

3.1.1 ▪ Solutions of Volterra integral equations

Integrating the differential form over various contours in the (x,t)-plane yields different integral equations and different solutions μ. We spend a significant portion of this section studying properties of these solutions before we can proceed to solving the NLS equation. For any solution μ, it follows that $M = \mu\mathrm{e}^{-\mathrm{i}(zx+2z^2t)\sigma_3}$ solves

$$M_x + \mathrm{i}z\sigma_3 M = QM,$$

$$M_t + 2\mathrm{i}z^2\sigma_3 M = \tilde{Q}M.$$

For any two nonsingular solutions M_1, M_2 of this system, $M_1(x,t,z) = M_2(x,t,z)S(z)$ for some matrix S. Thus for any two solutions μ_1, μ_2 of (3.4),

$$\mu_1(x,t,z) = \mu_2(x,t,z)\mathrm{e}^{-\mathrm{i}(zx+2z^2t)\hat{\sigma}_3}S(z) \tag{3.5}$$

[30] We treat z as a parameter so that $\mathrm{d}f(x,t,z) = f_x(x,t,z)\mathrm{d}x + f_t(x,t,z)\mathrm{d}t$.

for some matrix $S(z)$. We follow [50, p. 227] and define μ_1, μ_2 to be the solutions of

$$\begin{aligned}\mu_1(x,t,z) &= I + \int_{(-\infty,t)}^{(x,t)} e^{-iz(x-\xi)\hat{\sigma}_3}(Q(\xi,\tau)\mu_1(\xi,\tau,z))\,d\xi,\\ \mu_2(x,t,z) &= I + \int_{(\infty,t)}^{(x,t)} e^{-iz(x-\xi)\hat{\sigma}_3}(Q(\xi,\tau)\mu_2(\xi,\tau,z))\,d\xi,\end{aligned} \tag{3.6}$$

where μ_1 is defined by an integral of the differential form over $\{(s,t): -\infty < s < x\}$ and μ_2 by an integral over $\{(s,t): x < s < \infty\}$. Define the operator $\mathscr{T}_{\pm,z}$ by

$$\begin{aligned}\mathscr{T}_{\pm,z}u(x) &= \int_{(\pm\infty,0)}^{(x,0)} e^{-iz(x-\xi)\hat{\sigma}_3}Q(\xi,0)u(\xi)d\xi,\\ &= \int_{\pm\infty}^{x} e^{-iz(x-\xi)\hat{\sigma}_3}Q(\xi,0)u(\xi)d\xi.\end{aligned}$$

Since each component of $Q(x,0)$ is in $L^1(\mathbb{R})$, it is seen that $\mathscr{T}_{\pm,z}$ is a bounded operator on $L^\infty(\mathbb{R})$.

Lemma 3.1. *For $q_0 \in L^1(\mathbb{R})$ and $z \in \mathbb{R}$, $||\mathscr{T}_{\pm,z}^n||_{\mathscr{L}(L^\infty(\mathbb{R}))} \leq C||q_0||_{L^1(\mathbb{R})}^n/n!$.*

Proof. Since $z \in \mathbb{R}$, we consider

$$e^{izx\hat{\sigma}_3}[\mathscr{T}_{\pm,z}u(x)] = \int_{\pm\infty}^{x} e^{iz\xi\hat{\sigma}_3}Q(\xi,0)e^{iz\xi\hat{\sigma}_3}u(\xi)d\xi.$$

To enable us to consider a kernel that is applied through left multiplication, define $\hat{u}(x) \triangleq e^{izx\hat{\sigma}_3}u(x)$ and $\hat{Q}(x,z) \triangleq e^{izx\hat{\sigma}_3}Q(x,0)$. For $z \in \mathbb{R}$, it is clear that the claim follows if we show it for

$$\hat{\mathscr{T}}_{\pm,z}\hat{u}(x) \triangleq \int_{\pm\infty}^{x} \hat{Q}(\xi,z)\hat{u}(\xi)d\xi. \tag{3.7}$$

It is also clear that, after switching the order of integration using Fubini's theorem, the operator $\hat{\mathscr{T}}_{\pm,z}^n$ has the kernel defined by

$$\begin{aligned}\hat{Q}_0(x,\xi,z) &\triangleq \hat{Q}(\xi,z),\\ \hat{Q}_n(x,\xi,z) &\triangleq \hat{Q}(\xi,z)\int_{\xi}^{x} \hat{Q}_{n-1}(x,s,z)ds.\end{aligned}$$

Therefore

$$||\hat{\mathscr{T}}_{\pm,z}^n||_{\mathscr{L}(L^\infty(\mathbb{R}))} \leq \int_{-\infty}^{\infty} |\hat{Q}(\xi,z)|\left(\pm\int_{\pm\infty}^{\xi} |\hat{Q}(s,z)|ds\right)^{n-1} d\xi,$$

where $|\hat{Q}(\xi,z)|$ denotes any consistent submultiplicative matrix norm. This right-hand side simplifies:

$$||\hat{\mathscr{T}}_{\pm,z}^n||_{\mathscr{L}(L^\infty(\mathbb{R}))} \leq \left(\int_{-\infty}^{\infty} |\hat{Q}(\xi,z)|d\xi\right)^n/n!.$$

This proves the result. □

This lemma combined with Theorem A.21 shows the existence of $\mu_1(x,z)\equiv\mu_1(x,0,z)$ and $\mu_1(x,z)\equiv\mu_2(x,0,z)$ for x and z on the real axis. With exponentially decaying data, more is true.

Lemma 3.2. *Assume* $q_0\in\mathscr{S}_\delta(\mathbb{R})$. *Then* $\mu_1(x,z)$ *and* $\mu_2(x,z)$ *are uniquely defined by* (3.6) *for* $|\operatorname{Im} z|<\delta/2$ *and* $\sup_{|\operatorname{Im} z|\le C<\delta/2}\|\mu_j(\diamond,z)\|<\infty$ *for* $j=1,2$.

Proof. We consider μ_1 here. The argument for μ_2 follows in precisely the same way. Define $\hat\mu_1(x,z)=\mathrm{e}^{\mathrm{i}zx\hat\sigma_3}\mu_1(x,z)$. From (3.6) we have

$$(I-\hat{\mathscr{T}}_{-,z})\hat\mu_1=I.$$

This equation is uniquely solvable via iteration for $\hat\mu_1\in L^\infty(\mathbb{R})$ since $\hat{Q}(\diamond,z)\in L^1(\mathbb{R})$. Therefore μ_1 is uniquely defined by (3.6) despite the fact that it contains exponentially growing components for $z\notin\mathbb{R}$ because of the decay of q_0. The uniformity of the norm follows from the fact that the inverse operator $(I-\hat{\mathscr{T}}_{-,z})^{-1}$ has a norm that can be bounded independently of z within the region $|\operatorname{Im} z|\le C<\delta/2$. □

We now discuss the technical aspects concerning analyticity and decay properties of μ_1 and μ_2 with respect to z. We need a technical lemma first because we can no longer reduce to the study of $\hat\mu_j$.

Lemma 3.3. *Assume* $q\in\mathscr{S}_\delta(\mathbb{R})$ *and* $c\ge 0$. *For* $\gamma<\delta/2$ *and* $\pm x\le c$

$$\mp\int_{\pm\infty}^{x}|\xi-x|^k\mathrm{e}^{\pm 2\gamma(x-\xi)}|q(\xi)|\mathrm{d}\xi\le C_{k,\gamma}\mathrm{e}^{\delta c}\sup_{x\in\mathbb{R}}\mathrm{e}^{\delta|x|}|q(x)|<\infty.$$

Proof. We prove the lemma for just one choice of signs. It is clear that a change of variables can be used to reduce the other case to this. Then expanding $|\xi-x|^k\le(|\xi|+|x|)^k$ gives terms of the form

$$\begin{aligned}\left|\int_{-\infty}^{x}|x|^i|\xi|^j\mathrm{e}^{2\gamma(x-\xi)}|q(\xi)|\mathrm{d}\xi\right|&=|x|^i\int_{-\infty}^{x}\mathrm{e}^{2\gamma(x-\xi)+\delta\xi}\mathrm{e}^{-\delta\xi}|\xi|^j|q(\xi)|\mathrm{d}\xi\\&=|x|^i\mathrm{e}^{2\gamma x}\int_{-\infty}^{x}|\xi|^j\mathrm{e}^{(\delta-2\gamma)\xi}\mathrm{e}^{-\delta\xi}|q(\xi)|\mathrm{d}\xi.\end{aligned}$$

It follows that $\sup_{\xi\in\mathbb{R}}\mathrm{e}^{\delta|\xi|}|q(\xi)|<\infty$ and there exists a constant $D_{i,j,\gamma}>0$:

$$|x|^i\int_{-\infty}^{x}|\xi|^j\mathrm{e}^{(\delta-2\gamma)\xi}\mathrm{d}\xi\le D_{i,j,\gamma}\begin{cases}1 & \text{if } x<0,\\ 1+|x|^i & \text{if } x\ge 0.\end{cases}$$

Then for a new constant $\tilde{D}_{i,j,\gamma}$

$$|x|^i\mathrm{e}^{2\gamma x}\int_{-\infty}^{x}|\xi|^j\mathrm{e}^{(\delta-2\gamma)\xi}\mathrm{d}\xi\le\tilde{D}_{i,j,\gamma}\begin{cases}1 & \text{if } x<0,\\ \mathrm{e}^{\delta c} & \text{if } x\ge 0.\end{cases}$$

This proves the lemma. □

For now, we continue to consider $q_0\in\mathscr{S}_\delta(\mathbb{R})$.

Lemma 3.4. *If $q_0 \in \mathcal{S}_\delta(\mathbb{R})$, then $\mu_1(x,z)$ and $\mu_2(x,z)$ are analytic functions of z in the interior of the strip $|\operatorname{Im} z| < \delta/2$ and for $j=1,2$, $n \geq 0$, fixed x and as $|z| \to \infty$, $|\operatorname{Im} z| \leq C < \delta/2$,*

$$\partial_z^n(\mu_j(x,z)-I) = \mathcal{O}(z^{-1}), \quad \partial_x(\mu_j(x,z)-I) = \mathcal{O}(z^{-1}).$$

Proof. Choose z and h so that $|\operatorname{Im} z| < \delta/2$ and $|\operatorname{Im}(z+h)| < \delta/2$ and consider

$$\begin{aligned}
&\frac{\mu_1(x,z+h)-\mu_1(x,z)}{h} \\
&= \frac{1}{h}\int_{-\infty}^{x} \left(e^{-i(z+h)(x-\xi)\hat\sigma_3} Q(\xi,0)\mu_1(\xi,z+h) - e^{-iz(x-\xi)\hat\sigma_3} Q(\xi,0)\mu_1(\xi,z)\right) d\xi \\
&= \int_{-\infty}^{x} e^{-iz(x-\xi)\hat\sigma_3} Q(\xi,0) \frac{\mu_1(\xi,z+h)-\mu_1(\xi,z)}{h} d\xi \\
&+ \int_{-\infty}^{x} \frac{e^{-i(z+h)(x-\xi)\hat\sigma_3} Q(\xi,0) - e^{-iz(x-\xi)\hat\sigma_3} Q(\xi,0)}{h} \mu_1(\xi,z+h) d\xi.
\end{aligned} \tag{3.8}$$

The complex differentiability of μ_1 follows once we show that the last term tends to a definite limit in the $L^\infty((-\infty,c))$, $c < \infty$ norm, implying complex differentiability for each x. We truncate to $(-\infty,c)$ because the z derivative may grow for large, positive x.

Consider $u_h(\xi,z) \triangleq h^{-1}(\mu_1(x,z+h)-\mu_1(x,z))$ so that (3.8) can be written as $u_h(\diamond,z) = (I-\mathcal{T}_{-,z})^{-1}F_h(\diamond,z)$, where $(I-\mathcal{T}_{-,z})^{-1}$ is bounded on $L^\infty((\infty,c))$ by Theorem A.21 using Lemma 3.3. Thus if F_h has a limit in the space, then so does u_h. The proof of Lemma 3.3 with $\gamma = \max\{|\operatorname{Im}(z+h)|, |\operatorname{Im} z|\} < \delta/2$ shows that

$$\left|e^{-i(z+h)(x-\xi)\hat\sigma_3} Q(\xi,0) - e^{-iz(x-\xi)\hat\sigma_3} Q(\xi,0)\right| \leq C2e^{2\gamma c} e^{-\epsilon|\xi|}$$

for some $\epsilon > 0$. Therefore the dominated convergence theorem (Theorem A.3) demonstrates that the integral of the left-hand side tends to zero as $h \to 0$. This proves the continuity of $\mathcal{T}_{\pm,z}$, $(I-\mathcal{T}_{\pm,z})^{-1}$ and hence $\mu_1(x,z)$ with respect to z for $|\operatorname{Im} z| < \delta/2$ in the sense that $\|\mu_1(\diamond,z+h)-\mu_1(\diamond,z)\|_{L^\infty((-\infty,c))} \to 0$ as $h \to 0$. A similar calculation, along with the continuity of $\mu_1(x,z)$, shows that

$$\begin{aligned}
&\int_{-\infty}^{x} \frac{e^{-i(z+h)(x-\xi)\hat\sigma_3} Q(\xi,0) - e^{-iz(x-\xi)\hat\sigma_3} Q(\xi,0)}{h} \mu_1(\xi,z+h) d\xi \\
&\quad \overset{h\to 0}{\to} \int_{-\infty}^{x} i(x-\xi)[\sigma_3, e^{-iz(x-\xi)\hat\sigma_3} Q(\xi,0)\mu_1(\xi,z)] d\xi
\end{aligned}$$

in $L^\infty((-\infty,c))$ using Lemma 3.3; see Theorem A.4. It is clear that similar calculations follow for μ_2 demonstrating analyticity for both μ_1 and μ_2.

Now, define $m = [m_1, m_1]^\top$ to be the first column of μ_1. We present the following arguments for only one column for simplicity. From integration by parts,

$$\begin{aligned}
\begin{bmatrix} m_1(x,z) \\ m_2(x,z) \end{bmatrix} &= \begin{bmatrix} 1 \\ 0 \end{bmatrix} + \int_{-\infty}^{x} \begin{bmatrix} q_0(\xi)m_2(\xi,z) \\ -e^{2i(x-\xi)z} \lambda \bar q_0(\xi) m_1(\xi,z) \end{bmatrix} d\xi \\
&= \begin{bmatrix} 1 \\ (2iz)^{-1}\lambda \bar q_0(x) m_1(x,z) \end{bmatrix} \\
&+ \int_{-\infty}^{x} \begin{bmatrix} q_0(\xi)m_2(\xi,z) \\ -(2iz)^{-1} e^{2i(x-\xi)z} (\lambda \bar q_0'(\xi) m_1(\xi,z) + \lambda |q_0(\xi)|^2 m_2(\xi,z)) \end{bmatrix} d\xi,
\end{aligned} \tag{3.9}$$

where we used (3.2) in the last line to replace $\partial_x m_1(\xi,z)$ with $q_0(\xi)m_2(\xi,z)$. Thus, $\|m(\diamond,z)-[1,\ 0]^\top\|_\infty = \mathcal{O}(z^{-1})$, where the prerequisite is that $m(x,z)$ is bounded uniformly in x as z becomes large; see Lemma 3.2. For fixed x, this property is inherited by $\partial_z^n m(x,z)$ so that $\partial_z^n(m(x,z)-[1,\ 0]^\top) = \mathcal{O}(z^{-1})$ for x fixed and $n>0$. This follows in general by examining the equation $\partial_z^n m(x,z)$ satisfies to show boundedness in z for fixed x and then differentiating (3.9) with respect to z, using Lemma 3.3 for both calculations.

Next, we investigate the x derivative:

$$\partial_x \begin{bmatrix} m_1(x,z) \\ m_2(x,z) \end{bmatrix} - \int_{-\infty}^{x} \begin{bmatrix} 0 \\ e^{2i(x-\xi)z}(\lambda \overline{q}_0'(\xi) m_1(\xi,z) + \lambda |q_0(\xi)|^2 m_2(\xi,z)) \end{bmatrix} d\xi \\ = \mathcal{O}(z^{-1}), \tag{3.10}$$

where the error term is uniform in x. Integration by parts gives

$$\partial_x \begin{bmatrix} m_1(x,z) \\ m_2(x,z) \end{bmatrix} + \frac{1}{2iz}\int_{-\infty}^{x} \begin{bmatrix} 0 \\ e^{2i(x-\xi)z}(\lambda \overline{q}_0'(\xi) \partial_x m_1(\xi,z) + \lambda |q_0(\xi)|^2 \partial_x m_2(\xi,z)) \end{bmatrix} d\xi \\ = \mathcal{O}(z^{-1}).$$

This can be written as

$$(I-\mathcal{K})\partial_x m = \mathcal{O}(z^{-1}),$$
$$\mathcal{K}\partial_x m(x,z) \triangleq -\frac{1}{2iz}\int_{-\infty}^{x} e^{2i(x-\xi)z} \begin{bmatrix} 0 & 0 \\ \lambda \overline{q}_0'(\xi) & \lambda |q_0(\xi)|^2 \end{bmatrix} \partial_x m(\xi,z) d\xi.$$

From this it follows that $I-\mathcal{K} \to I$ in $\mathcal{L}(L^\infty(\mathbb{R}))$ as $|z| \to \infty$ and $\|\partial_x m(\diamond,z)\|_\infty = \mathcal{O}(z^{-1})$ for $|\mathrm{Im}\, z| \le C < \delta/2$ from Lemma 3.3. Estimates for other columns follow in an analogous way. □

Remark 3.1.1. *The estimates in Lemma* 3.4 *can be extended to all regions of analyticity of the entries of* μ_1 *by considering when Lemma* 3.3 *applies. For example, the first column of* μ_1 *satisfies the estimates of Lemma* 3.4 *in the entire upper-half plane because* $e^{2i(x-\xi)z}$ *is bounded in that region for* $\xi \le x$.

We turn to results for q_0 in less restrictive spaces. It follows that each column of μ_1 and μ_2 is determined independently of the other. Furthermore, the only exponential present in the first column of the equation for μ_1 is $e^{2iz(x-\xi)}$, implying that the first column μ_1^+ of μ_1 is analytic in the open upper-half plane. Indeed, μ_1^+ satisfies

$$\mu_1^+(x,z) = \begin{bmatrix} 1 \\ 0 \end{bmatrix} + \int_{-\infty}^{x} Q(\xi,0) \begin{bmatrix} e^{2iz(x-\xi)} & 0 \\ 0 & 1 \end{bmatrix} \mu_1^+(\xi,z) d\xi. \tag{3.11}$$

The methods of Lemma 3.4 applied to this equation prove the desired analyticity properties. Note that this follows assuming only that $q_0 \in L^1(\mathbb{R})$. Similarly, the second column μ_1^- is analytic in the open lower-half plane. For μ_2 the reverse is true. The first (second) column μ_2^- (μ_2^+) of μ_2 is analytic in the open lower- (upper-) half plane. We consider (3.11) in more detail.

Lemma 3.5. *If $q_0 \in L^1 \cap L^2(\mathbb{R})$, $v_1^+ \triangleq \mu_1^+ - [1,\ 0]^\top$ is an element of the Hardy space $\mathscr{E}^+(\mathbb{R})$.*

Proof. Consider the equation satisfied by v_1^+:

$$v_1^+(x,z) - \int_{-\infty}^{x} Q(\xi,0)\begin{bmatrix} e^{2iz(x-\xi)} & 0 \\ 0 & 1 \end{bmatrix} v_1^+(\xi,z)\mathrm{d}\xi = \int_{-\infty}^{x} Q(\xi,0)\begin{bmatrix} e^{2iz(x-\xi)} \\ 0 \end{bmatrix}\mathrm{d}\xi.$$

We have already demonstrated that for $z \in \mathbb{R}$, $q_0 \in L^1(\mathbb{R})$ this equation can be solved by iteration. The kernel is given by

$$\mathscr{K}_0(x,\xi,z) \triangleq Q(\xi,0)\begin{bmatrix} e^{2iz(x-\xi)} & 0 \\ 0 & 1 \end{bmatrix},$$
$$\mathscr{K}_n(x,\xi,z) \triangleq \int_{\xi}^{x} \mathscr{K}_0(x,s,z)\mathscr{K}_{n-1}(s,\xi,z)\mathrm{d}s.$$

Then

$$v_1^+(x,z) = f(x,z) + \sum_{n=0}^{\infty}\int_{-\infty}^{x} \mathscr{K}_n(x,\xi,z)f(\xi,z)\mathrm{d}\xi, \tag{3.12}$$
$$f(x,z) \triangleq \int_{-\infty}^{x} Q(\xi,0)\begin{bmatrix} e^{2iz(x-\xi)} \\ 0 \end{bmatrix}\mathrm{d}\xi.$$

We estimate the $L^2(\mathbb{R})$ norm

$$\left(\int_{-\infty}^{\infty}\left|\int_{-\infty}^{x} \mathscr{K}_n(x,\xi,z)f(\xi,z)\mathrm{d}\xi\right|^2 \mathrm{d}z\right)^{1/2}$$
$$\leq \int_{-\infty}^{x}\left(\int_{-\infty}^{\infty} |\mathscr{K}_n(x,\xi,z)f(\xi,z)|^2\mathrm{d}z\right)^{1/2}$$
$$\leq \int_{-\infty}^{x} \sup_{z\in\mathbb{R}} |\mathscr{K}_n(x,\xi,z)|\mathrm{d}\xi \times \sup_{\xi\in\mathbb{R}} \|f(\xi,\diamond)\|_{L^2(\mathbb{R})},$$

where the first inequality stems from Minkowski's inequality for integrals [54, Theorem 6.19]. We estimate the two factors separately. First,

$$\int_{-\infty}^{x} \sup_{z\in\mathbb{R}} |\mathscr{K}_n(x,\xi,z)|\mathrm{d}\xi \leq \int_{-\infty}^{\infty} |Q(\xi,0)|\left(\int_{\xi}^{\infty} |Q(\xi,0)|\mathrm{d}\xi\right)^{n-1} \leq C\|q_0\|_{L^1(\mathbb{R})}^n/n!. \tag{3.13}$$

A simple change of variables relates f to the Fourier transform

$$f(\xi,z) = \frac{1}{2}\int_{-\infty}^{0} e^{izs}Q(\xi+s/2)\begin{bmatrix} e^{-izs} \\ 0 \end{bmatrix}\mathrm{d}s,$$

and therefore $\|f(\xi,\diamond)\|_{L^2(\mathbb{R})} \leq C\|Q\|_{L^2(\mathbb{R})}$. Since $\sum_n D^n/n!$ converges for any $D \in \mathbb{R}$, we have shown that the L^2 norm of $v_1^+(x,z)$ is bounded for z on the real axis. For $z = s + i\rho$, $\rho > 0$, we note that these same estimates hold, showing that $v_1^+ \in \mathscr{E}^+(\mathbb{R})$; see Definition 2.23. □

Similar arguments apply to μ_1^- and $\mu_2^\pm$ to show that once $[1,\ 0]^\top$ or $[0,\ 1]^\top$ is subtracted they are elements of an appropriate Hardy space. We also look at decay properties of the columns of $\mu_1 - I$ and $\mu_2 - I$.

Lemma 3.6. *For* $q_0 \in L^1(\mathbb{R})$ *and* f *defined in* (3.12), $|f(x,z)| = o(1)$ *as* $|z| \to \infty$ *uniformly for* $\operatorname{Im} z \geq 0$.

Proof. Fix $\epsilon > 0$ and approximate $\lambda\overline{q}$ with a simple function $\sum_j \alpha_j \chi_{A_j}$ so that $\|\lambda\overline{q} - \sum_j \alpha_j \chi_{A_j}\|_{L^1(\mathbb{R})} < \epsilon/2$. Let $|z| > 4\sum_j |\alpha_j||A_j|/\epsilon$. Then for $|z| > 1$

$$|f(x,z)| \leq \left| \int_{-\infty}^{x} e^{iz(x-\xi)} \left((\lambda\overline{q}(\xi) - \sum_j \alpha_j \chi_{A_j}(\xi) \right) d\xi \right| + \left| \int_{-\infty}^{x} e^{iz(x-\xi)} \sum_j \alpha_j \chi_{A_j}(\xi) d\xi \right| \leq \epsilon,$$

by evaluating the final integral explicitly. □

Lemma 3.7. *For* $q_0 \in L^1(\mathbb{R})$ *and* x *fixed,* $\mu_1^+(x,z) = [1,\ 0]^\top + o(1)$ *as* $|z| \to \infty$ *uniformly for* $\operatorname{Im} z \geq 0$.

Proof. Taking a supremum over $z \in \mathbb{R}$ in (3.12) while noting that $f(x,z)$ has the desired decay and the estimate (3.13) can still be applied produces the result. □

Remark 3.1.2. *Statements similar to Lemma* 3.7 *follow for* μ_1^- *in* $\mathbb{C}^-$ *and* $\mu_2^\pm$ *in* $\mathbb{C}^\pm$.

3.1.2 ▪ Properties of the scattering matrix

We move to a discussion of the matrix $S(z)$ that is defined by (3.5). Liouville's formula[31] shows that $\mu_1(x,z)$ and $\mu_2(x,z)$ have determinants that are independent of x. Indeed, both satisfy

$$\left(e^{izx\hat{\sigma}_3}\mu\right)_x = e^{izx\hat{\sigma}_3} Q(e^{izx\hat{\sigma}_3}\mu),$$

and $e^{izx\hat{\sigma}_3}Q$ is traceless so that $\det \mu_1$ and $\det \mu_2$ are independent of x. Then for each z we let $x \to -\infty$ for μ_1 and $x \to \infty$ for μ_2 to see that $\det \mu_1(x,z) = \det \mu_2(x,z) = 1$. Therefore $\det S(z) = 1$. The first column μ_1^+ of μ_1 satisfies

$$\frac{d\mu_1^+}{dx} + iz \begin{bmatrix} 0 & 0 \\ 0 & -2 \end{bmatrix} \mu_1^+ = Q\mu_1^+, \quad \mu_1^+(-\infty) = \begin{bmatrix} 1 \\ 0 \end{bmatrix},$$

while the second column μ_1^- satisfies

$$\frac{d\mu_1^-}{dx} + iz \begin{bmatrix} 2 & 0 \\ 0 & 0 \end{bmatrix} \mu_1^- = Q\mu_1^-, \quad \mu_1^-(-\infty) = \begin{bmatrix} 0 \\ 1 \end{bmatrix}.$$

[31] Liouville's formula states that if $y' = A(x)y$, then $\det y(x) = C_0 \exp(\int \operatorname{tr} A(x) dx)$.

Let $\mu_1^-(x,z)=[w_1(x,z),\ w_2(x,z)]^\top$ and define

$$\mu_1^*(x,z)=\left[\begin{array}{c}\overline{w_2(x,\overline{z})}\\ \overline{\lambda w_1(x,\overline{z})]}\end{array}\right].$$

It follows that $\mu_1^+(x,z)=\mu_1^*(x,z)$ because the solutions to the ODEs are unique. This means that swapping rows and columns, followed by a Schwarz conjugate of each entry and multiplication on the off-diagonal by λ, leaves μ_1 invariant. Similar considerations follow for μ_2 using $\mu_2^\pm$. Let

$$S(z)=\left[\begin{array}{cc}a(z) & \mathscr{B}(z)\\ b(z) & \mathscr{A}(z)\end{array}\right]. \tag{3.14}$$

A tedious but simple calculation shows that $\mathscr{A}(z)=\overline{a(\overline{z})}$ and $\mathscr{B}(z)=\lambda\overline{b(\overline{z})}$, where $\lambda^2=1$ is used explicitly. Furthermore, $\det S(z)=1$ implies

$$|a(z)|^2-\lambda|b(z)|^2=1 \ \text{ for } \ z\in\mathbb{R}. \tag{3.15}$$

Using $\det\mu_1=\det\mu_2=1$ and the fact that each entry of μ_1-I and μ_2-I is in $L^\infty\cap L^2(\mathbb{R})$, we find that $a-1, b\in L^\infty\cap L^2(\mathbb{R})$. If $q_0\in\mathscr{S}_\delta(\mathbb{R})$, it follows that $S(z)$ is analytic in the strip $|\operatorname{Im} z|<\delta/2$. If we only require $q_0\in L^1\cap L^2(\mathbb{R})$, b cannot be, in general, analytically extended off $\mathbb{R}$ but

$$a-1\in\mathscr{E}^+(\mathbb{R}). \tag{3.16}$$

Following Lemma 3.7 for sufficiently large $z\in\overline{\mathbb{C}^+}$, $a(z)$ must be bounded away from zero. We assume here that $a(z)$ does not vanish in the finite plane, implying that $1/a(z)$ is bounded in $\mathbb{C}^+$. This assumption is known to be true when $\lambda=-1$ [10]. Define

$$\Phi=\begin{cases}\Phi^+ & \text{in } \mathbb{C}^+,\\ \Phi^- & \text{in } \mathbb{C}^-,\end{cases}$$

$$\Phi^+(z)=\left[\begin{array}{c}\dfrac{\mu_1^+(z)}{a(z)},\mu_2^+(z)\end{array}\right],\quad \Phi^-(z)=\left[\begin{array}{c}\mu_2^-(z),\dfrac{\mu_1^-(z)}{\overline{a(\overline{z})}}\end{array}\right].$$

It follows that $\Phi(z)=\Phi(x,t,z)$ solves the following RH problem $[G;\mathbb{R}]$ when $t=0$:

$$\Phi^+(s)=\Phi^-(s)G(s),\quad s\in\mathbb{R},\quad \Phi(\infty)=I, \tag{3.17}$$

$$G(z)=G(x,t,z)=\left[\begin{array}{cc}1+\lambda\overline{\rho(\overline{z})}\rho(z) & \lambda\overline{\rho(\overline{z})}e^{-2izx-4iz^2t}\\ \rho(z)e^{2izx+4iz^2t} & 1\end{array}\right]$$

and

$$Q(x,0)=\lim_{|z|\to\infty} i[\sigma_3, z\Phi(x,0,z)]. \tag{3.18}$$

The function $\rho\triangleq b/a$ is the reflection coefficient. When $\lambda=-1$, from (3.15)

$$|\rho(z)|^2=\left|\frac{b(z)}{a(z)}\right|^2=1-\frac{1}{|a(z)|^2}<1. \tag{3.19}$$

The relation (3.18) can then be seen by noting that Φ solves (3.2) and $\Phi_x(z)=\mathscr{O}(1/z)$ as $z\to\infty$ if $q_0\in\mathscr{S}_\delta(\mathbb{R})$; see Remark 3.1.1.

We derive another representation for $a(z)$. Note that

$$M(x,z) = \left[\mu_1^+(x,z),\ \mu_2^+(x,z)\right]$$

is a solution of (3.2). Furthermore, it follows that $a(z) = \det M(x,z)$. Since $\mu_2^+ \to [0,\ 1]^\top$ as $x \to \infty$, we find that $a(z) = \lim_{x\to\infty} \mu_{11}^+(x,z)$, where μ_{11}^+ denotes the first component of μ_1^+. We look at the large x behavior of $f(x,z)$ in (3.12) to determine the x and z dependence of μ_{11}^+. We find

$$|f(x,z)| \le \left|\int_{-\infty}^{\infty} \mathrm{e}^{-2\mathrm{i}z\xi} q_0(\xi)\mathrm{d}\xi\right| + \left|\int_{x}^{\infty} \mathrm{e}^{-2\mathrm{i}z\xi} q_0(\xi)\mathrm{d}\xi\right| \le |\hat{q}_0(2z)| + I(x),$$

where $I(x) \to 0$ as $x \to \infty$. We assume $q_0 \in \mathscr{S}_\delta(\mathbb{R})$ so that the Fourier transform satisfies $\hat{q}_0(z) \in \mathscr{S}(\mathbb{R})$. This implies that $a(z) = 1 + \mathcal{O}(1/z^n)$ as $z \to \infty$ for every $n > 0$. Combining this with (3.15) we see $b(z) = \mathcal{O}(1/z^n)$ for every $n > 0$. Also, following Lemma 3.4 we see that $a - 1, b \in H^k(\mathbb{R})$ for every $k > 0$ whenever $q_0 \in \mathscr{S}_\delta(\mathbb{R})$. These facts suffice for the theory we present here, yet more regularity can be shown: for $q_0 \in \mathscr{S}(\mathbb{R})$ it is known that the reflection coefficient $\rho = b/a \in \mathscr{S}(\mathbb{R})$ [10].

Remark 3.1.3. *The RH problem for Φ has a unique solution. If $q_0 \in \mathscr{S}_\delta(\mathbb{R})$, the RH problem is at least 1-regular and the jump matrix J has a unit determinant such that the Fredholm index of the associated singular integral operator $\mathscr{C}[J;\mathbb{R}]$ is zero (Theorem 2.69). Furthermore, the symmetry of $G(x,t,z)$ is such that the vanishing lemma applies (Theorem 2.73: $G(x,t,z) + G^\dagger(x,t,z)$ is positive definite for $z \in \mathbb{R}$ by Sylvester's criterion when $\lambda = \pm 1$).*

We prove a result that concerns the differentiability of Φ with respect to x or t.

Lemma 3.8. *Consider an L^2 RH problem for a function $\Psi(z) = \Psi(x,z)$:*

$$\Psi^+(s) = \Psi^-(s)K(x,s), \quad s \in \Gamma, \quad \Phi(\infty) = I,$$
$$K(x,s) = H_d(s) + \mathrm{e}^{\mathrm{i}\gamma(s)x\hat{\sigma}_3} H_o(s), \quad x \in \mathbb{R}, \quad \gamma : \Gamma \to \mathbb{R},$$

where H_d contains the diagonal elements of K and $\mathrm{e}^{\gamma(s)x\hat{\sigma}_3} H_o$, the off-diagonal elements. With Γ being admissible, assume $K(x,\diamond) - I \in L^2 \cap L^\infty(\mathbb{R})$ and $\mathscr{C}[K,\Gamma]$ is invertible on $L^2(\Gamma)$ with bounded inverse. If $\gamma(s)H_0(s) \in L^2 \cap L^\infty(\Gamma)$, then $\partial_x \Psi(z)$ exists for every x and $z \in \mathbb{C} \setminus \Gamma$ and $(\partial_x \Psi(z))^\pm \in L^2(\Gamma)$.

Proof. Let u^x be the solution of $u^x - \mathscr{C}_\Gamma^- u^x(K(x,\diamond) - I) = K(x,\diamond) - I$. Consider the difference

$$\frac{u^{x+h} - u^x}{h} - \frac{1}{h}\left(\mathscr{C}_\Gamma^- u^x(K(x,\diamond) - I) - \mathscr{C}_\Gamma^- u^{x+h}(K(x+h,\diamond) - I)\right) = \frac{K(x,\diamond) - K(x+h,\diamond)}{h}.$$

Define $v^{x,h} = (u^{x+h} - u^x)/h$. We find

$$\mathscr{C}[K(x,\diamond),\Gamma]v^{x,h} = (I + \mathscr{C}_\Gamma^- u^{x+h})\frac{K(x,\diamond) - K(x+h,\diamond)}{h}.$$

Based on the assumptions the right-hand side tends to $(I + \mathscr{C}_\Gamma^- u^x)\partial_x K(x,\diamond)$ in $L^2(\Gamma)$. Using the boundedness of the inverse operator we find that $v^x = \lim_{h\to 0} v^{x,h}$ exists in $L^2(\Gamma)$. Therefore $\partial_x \Psi(z) = \mathscr{C}_\Gamma v^x$. ☐

3.1.3 ▪ The dressing method

We state a result from [50] that allows us to complete the solution of the initial-value problem for the NLS equation.

Proposition 3.9 (The dressing method [50]). *Assume Γ is an admissible contour. Let $M(z) = M(x,t,z)$ solve the 2×2 L^2 RH problem*

$$M^+(s) = M^-(s)e^{-i(sx+2s^2t)\hat{\sigma}_3}S(s), \quad s \in \Gamma, \quad M(\infty) = I, \tag{3.20}$$

where $\det S(s) = 1$. *Assume this RH problem has a unique solution that is differentiable in x and t for every $z \notin \Gamma$. Assume further that every derivative vanishes at infinity and has $L^2(\Gamma)$ boundary values. Then M satisfies* (3.2) *and* (3.3) *with*

$$Q(x,t) = \begin{bmatrix} 0 & q_1(x,t) \\ q_2(x,t) & 0 \end{bmatrix}$$

defined by

$$Q(x,t) = i \lim_{z\to\infty} [\sigma_3, zM(x,t,z)].$$

Furthermore, Q satisfies $iQ_t - Q_{xx}\sigma_3 + 2Q^3\sigma_3 = 0$.

Choosing jumps and contours so that $q_2 = -\lambda \bar{q}_1$ produces solutions of NLS equations. To do this, we impose the symmetry condition, $\bar{\Gamma} = \Gamma$ with orientation, and

$$S^{-1}(s) = \sigma_\lambda \overline{S(\bar{s})}\sigma_\lambda^{-1}, \quad \sigma_\lambda = \begin{bmatrix} 0 & \lambda^{1/2} \\ -\lambda^{-1/2} & 0 \end{bmatrix}.$$

To see this, one should check that $\tilde{M}(z) = \sigma_\lambda \overline{M(\bar{z})}\sigma_\lambda^{-1}$ is a solution of (3.20) and thus $\tilde{M} = M$. This implies $q_2 = -\lambda \bar{q}_1$. We combine Proposition 3.9 with Lemma 3.8 to obtain the final result of this section.

Theorem 3.10. *Assume $q_0 \in \mathcal{S}_\delta(\mathbb{R})$. Then $S(z)$ is analytic for $|\operatorname{Im} z| < \delta/2$, $e^{i(\diamond x - 2\diamond^2 t)\hat{\sigma}_3}S(\diamond) - I \in H^k(\mathbb{R})$ for all $k > 0$ and $\rho \in \mathcal{S}(\mathbb{R})$. Furthermore, the RH problem in* (3.17) *has a unique solution $\Phi(x,t,z)$ for every x and t. The solution of the NLS equation with $q(x,0) = q_0(x)$ is given by*

$$q(x,t) = \lim_{|z|\to\infty} i[\sigma_3, z\Phi(x,t,z)]_{12} = \lim_{|z|\to\infty} 2iz\Phi_{12}(x,t,z),$$

where the limit is taken in a direction that is not tangential to $\mathbb{R}$.

3.2 ▪ Nonlinear steepest descent

We discuss the Deift and Zhou *method of nonlinear steepest descent* as applied to (3.17). Nonlinear steepest descent has been used to analyze the asymptotic behavior of many RH problems [27, 35, 36, 38, 39]. The full theory for the NLS equation applies to initial conditions lying in appropriate weighted Sobolev spaces [39]. For simplicity and ease of exposition we restrict ourselves to $q_0 \in \mathcal{S}_\delta(\mathbb{R})$.

We begin by introducing factorizations of $G(x,t,z)$. It follows from (3.19) that

$$\sup_{z\in\mathbb{R}} |\rho(z)| < 1,$$

and this implies that G admits both a UL and an LDU factorization for all $x,t,z \in \mathbb{R}$:

$$\begin{aligned}
&G(x,t,z) = M(z)P(z) = L(z)D(z)U(z),\\
&P(z) = \begin{bmatrix} 1 & 0 \\ \rho(z)e^{2i(zx+2z^2t)} & 1 \end{bmatrix}, \quad M(z) = \begin{bmatrix} 1 & \lambda\overline{\rho(\bar{z})}e^{-2i(zx+2z^2t)} \\ 0 & 1 \end{bmatrix},\\
&L(z) = \begin{bmatrix} 1 & 0 \\ \frac{\rho(z)}{\tau(z)}e^{2i(zx+2z^2t)} & 1 \end{bmatrix}, \quad D(z) = \begin{bmatrix} \tau(z) & 0 \\ 0 & 1/\tau(z) \end{bmatrix},\\
&U(z) = \begin{bmatrix} 1 & \frac{\lambda\overline{\rho(\bar{z})}}{\tau(z)}e^{-2i(zx+2z^2t)} \\ 0 & 1 \end{bmatrix}, \quad \tau(z) = 1 + \lambda\rho(z)\overline{\rho(\bar{z})}.
\end{aligned} \tag{3.21}$$

From the results of the previous section, each of these factors is analytic and converges rapidly to the identity in the strip $|\mathrm{Im}\, z| < \delta/2$ as $|z| \to \infty$. The deformations of the RH problem that follow are centered around the stationary phase point for $e^{-2i(zx+2z^2t)}$: $z_0 \triangleq -x/(4t)$ where the derivative of the exponent vanishes. We break the asymptotic calculations up into two cases: $t = 0$ and $t \gg 0$.

3.2.1 ▪ Asymptotics for $t = 0$ as $|x| \to \infty$

These asymptotics are already known, as we have specified the initial condition. It is instructive to see how this behavior can be extracted from the RH problem. For $x < 0$, we use only the LDU factorization and the lensing process; see Section 2.8.2. Define for $0 < \delta' < \delta/2$

$$\Phi_1(z) = \begin{cases} \Phi(z)U^{-1}(z) & \text{if } 0 < \mathrm{Im}\, z < \delta',\\ \Phi(z)L(z) & \text{if } -\delta' < \mathrm{Im}\, z < 0,\\ \Phi(z) & \text{if } |\mathrm{Im}\, z| > \delta', \end{cases}$$

where we have suppressed the x and t dependence for convenience. The function Φ_1 solves the RH problem

$$\Phi_1^+(z) = \Phi_1^+(z)G_1(z), \quad G_1(z) = \begin{cases} U(z) & \text{if } \mathrm{Im}\, z = \delta',\\ L(z) & \text{if } \mathrm{Im}\, z = -\delta',\\ D(z) & \text{if } \mathrm{Im}\, z = 0. \end{cases}$$

The exponential in $U(z)$, the only jump in the upper-half plane, is of the form $\exp(2x\delta' + 2ixs)$ for $s \in \mathbb{R}$. It is clear that $(x < 0)$

$$\|U - I\|_{L^1\cap L^\infty(\mathbb{R}+i\delta')} < Ce^{2x\delta'} \text{ and } \|L - I\|_{L^1\cap L^\infty(\mathbb{R}-i\delta')} < Ce^{2x\delta'}.$$

Hence the same bounds hold for $L^2(\mathbb{R} \pm i\delta)$. Lemma 2.76 implies that u, the solution of $\mathscr{C}[G_1;\Gamma_1]u = G - I$ with $\Gamma_1 = \mathbb{R} \cup (\mathbb{R} + i\delta') \cup (\mathbb{R} - i\delta')$, tends exponentially to the solution of $\mathscr{C}[D;\mathbb{R}]v = D - I$ in the L^2 norm. The solution of this diagonal problem can be found explicitly (see Problem 2.4.1):

$$\Delta(z;z_0) = \begin{bmatrix} \delta(z;z_0) & 0 \\ 0 & \delta^{-1}(z;z_0) \end{bmatrix}, \qquad \delta(z;z_0) = \exp\left(\mathscr{C}_{(-\infty,z_0)} \log \tau(z)\right),$$

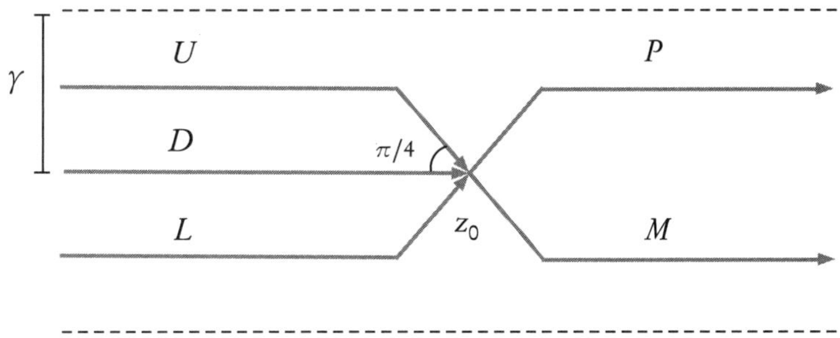

Figure 3.1. *Jump contours after lensing for the RH problem for Φ_2. The distance between the horizontal contours in the upper-lower-half planes is δ'.*

with $z_0 = \infty$. It is important that the $(1,2)$ component of $\Delta(z;z_0)$ is identically zero. From the expression

$$u = \mathscr{C}_{\Gamma_1}^{-} u(G_1 - I) + G_1 - I, \quad \Phi_1 = I + \mathscr{C}_{\Gamma} u,$$

we use that u has a bounded L^2 norm as $x \to -\infty$ and see that the $(1,2)$ component of u satisfies

$$||u_{12}||_{L^1(\Gamma_1)} \leq C||u||_{L^2(\Gamma_1)}||G_1 - I||_{L^2(\Gamma_1)} + ||G_1 - I||_{L^1(\Gamma_1)} \leq C' e^{2x\delta'}$$

for a new constant C'. From Lemma 2.13, $q(x,0)$ does indeed exponentially decay as $x \to -\infty$. Similar arguments show the same exponential behavior as $x \to +\infty$ when the UL factorization is used.

3.2.2 ▪ Asymptotics as $t \to \infty$

We now concern ourselves with the asymptotics of the solution in the more physically relevant region $|z_0| = |-x/(4t)| < M$ for some $M > 0$ as $t \to \infty$. The derivation of these asymptotics requires significantly more machinery. We follow [37]. We lens the RH problem for Φ on $(-\infty, z_0)$ using the LDU factorization and on (z_0, ∞) using the MP factorization. Let Φ_2 denote the new function we seek after lensing. The jump contours and jump matrices for Φ_2 are shown in Figure 3.1.

With the exception of the jump matrix on $(-\infty, z_0)$, all jump matrices associated with Φ_2 tend exponentially to the identity matrix away from the stationary phase point z_0. We remove the jump on $(-\infty, z_0)$. Define $\Phi_3(z) = \Phi_2(z)\Delta^{-1}(z;z_0)$. A simple exercise shows that Φ_3 satisfies the jumps in Figure 3.2.

The exponent present in the jump matrices can be written in the form $\exp(\pm t h(z))$ with $h(z) \triangleq -4iz_0^2 + 4i(z - z_0)^2$. This indicates increased localization as $t \to \infty$. Near the stationary point the jump matrices locally behave like the following approximation:

$$[M](z) = \begin{bmatrix} 1 & \lambda\overline{\rho(\bar{z}_0)} e^{-th(z)} \\ 0 & 1 \end{bmatrix}, \qquad [P](z) = \begin{bmatrix} 1 & 0 \\ \rho(z_0) e^{th(z)} & 1 \end{bmatrix},$$

$$[L](z) = \begin{bmatrix} 1 & 0 \\ \frac{\rho(z_0)}{\tau(z_0)} e^{th(z)} & 1 \end{bmatrix}, \qquad [D](z) = \begin{bmatrix} \tau(z_0) & 0 \\ 0 & 1/\tau(z_0) \end{bmatrix}, \tag{3.22}$$

$$[U](z) = \begin{bmatrix} 1 & \frac{\lambda\overline{\rho(\bar{z}_0)}}{\tau(z_0)} e^{-th(z)} \\ 0 & 1 \end{bmatrix}, \qquad [\Delta](z;z_0) = \Delta_s(z;z_0)\Delta_r(z_0;z_0),$$

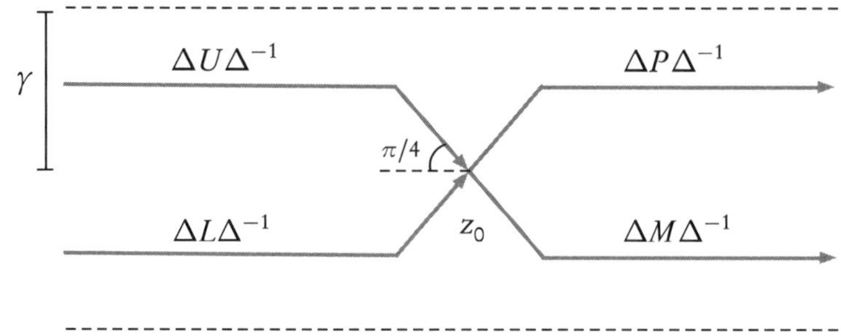

Figure 3.2. *The jump matrices and jump contours for Φ_3. This includes the removal of the jump on $(-\infty, z_0)$.*

with

$$\begin{aligned}\Delta_s(z;z_0) &= \operatorname{diag}((z-z_0)^{2\varphi(z_0)},(z-z_0)^{-2\varphi(z_0)}),\\ \varphi(z) &= \log(\tau(z))/(4\pi \mathrm{i}),\\ \Delta_r(z;z_0) &\triangleq \Delta(z;z_0)\Delta_s^{-1}(z;z_0) \text{ is Hölder continuous at } z=z_0.\end{aligned}$$

In the next section, we solve a problem with the localized jumps explicitly in terms of parabolic cylinder functions.

3.2.3 ▪ Construction of the global parametrix

We define *global parametrix* to be the solution of

$$\Psi^+(\zeta) = \Psi^-(\zeta)\begin{cases} [P]_\Delta(\zeta) & \text{if } \zeta \in \mathrm{e}^{\mathrm{i}\pi/4}(0,\infty),\\ [U]_\Delta(\zeta) & \text{if } \zeta \in \mathrm{e}^{\mathrm{i}3\pi/4}(0,\infty),\\ [L]_\Delta(\zeta) & \text{if } \zeta \in \mathrm{e}^{\mathrm{i}5\pi/4}(0,\infty),\\ [M]_\Delta(\zeta) & \text{if } \zeta \in \mathrm{e}^{\mathrm{i}7\pi/4}(0,\infty), \end{cases} \quad \Psi(\infty)=I, \tag{3.23}$$

where $[P]_\Delta(\zeta) = [\Delta](\zeta/\sqrt{8t}+z_0;z_0)[P](\zeta/\sqrt{8t}+z_0)[\Delta]^{-1}(\zeta/\sqrt{8t}+z_0;z_0)$ and similarly for L, U, and M. Note that the $(1,1)$ component of $\Delta_s(z;z_0)$ can be written as

$$(\Delta_s)_{11}(\zeta/\sqrt{8t}+z_0;z_0) = (8t)^{\varphi(z_0)}\zeta^{2\varphi(z_0)}.$$

The RH problem can be simplified. It is clear that $(8t)^{\varphi(z_0)}$, $\mathrm{e}^{2\mathrm{i}z_0^2 t}$, and $\Delta_r(z_0;z_0)$ have no dependence on ζ (or on z). We define

$$\Psi_1(\zeta) = \Delta_r^{-1}(z_0;z_0)\mathrm{e}^{-2\mathrm{i}z_0^2 t\hat{\sigma}_3}\left[(8t)^{-\varphi(z_0)\hat{\sigma}_3}[\Psi(\zeta)]\right]\Delta_r(z_0;z_0). \tag{3.24}$$

It follows that Ψ_1 solves

$$\Psi_1^+(\zeta) = \Psi_1^-(\zeta)\,\zeta^{2\varphi(z_0)\hat{\sigma}_3}[\mathrm{e}^{-\mathrm{i}\zeta^2/4\hat{\sigma}_3}V_{z_0}],\ \ \zeta \in \mathrm{e}^{\mathrm{i}\pi/4}\cdot(\mathbb{R}\cup \mathrm{i}\mathbb{R}),\ \ \Psi_1(\infty)=I,$$

where V_{z_0} and the orientation of $\mathrm{e}^{\mathrm{i}\pi/4}\cdot(\mathbb{R}\cup\mathrm{i}\mathbb{R})$ are shown in Figure 3.3. This RH problem is now deformed to one on the real axis using a reverse lensing process. The resulting problem is not an L^2 RH problem, but it is exactly solvable. For $\zeta \notin \mathrm{e}^{\mathrm{i}\pi/4}\cdot(\mathbb{R}\cup\mathrm{i}\mathbb{R})$,

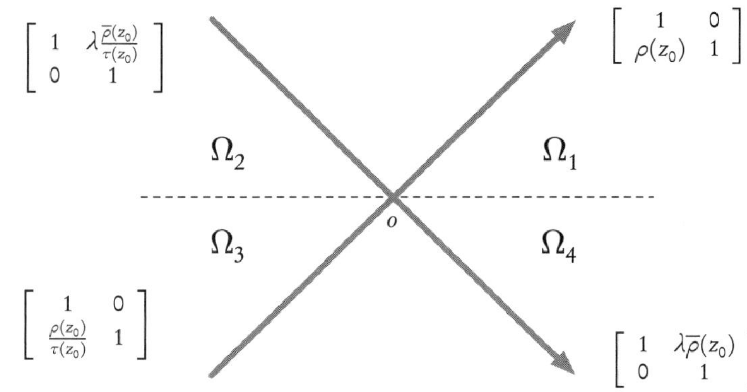

Figure 3.3. *The jump contours for* Φ *with piecewise definition* V_{z_0} *and the regions* Ω_i, $i = 1,\ldots,4$.

define

$$\Psi_2(\zeta) = \Psi_1(\zeta)\begin{cases} \zeta^{2\varphi(z_0)\hat\sigma_3}\left(e^{-i\zeta^2/4\hat\sigma_3}\begin{bmatrix} 1 & 0 \\ \rho(z_0) & 1 \end{bmatrix}\right) & \text{if } \zeta\in\Omega_1, \\ \zeta^{2\varphi(z_0)\hat\sigma_3}\left(e^{-i\zeta^2/4\hat\sigma_3}\begin{bmatrix} 1 & \frac{\lambda\bar\rho(z_0)}{\tau(z_0)} \\ 0 & 1 \end{bmatrix}\right) & \text{if } \zeta\in\Omega_2, \\ \zeta^{-2\varphi(z_0)\hat\sigma_3}\left(e^{-i\zeta^2/4\hat\sigma_3}\begin{bmatrix} 1 & 0 \\ -\frac{\rho(z_0)}{\tau(z_0)} & 1 \end{bmatrix}\right) & \text{if } \zeta\in\Omega_3, \\ \zeta^{-2\varphi(z_0)\hat\sigma_3}\left(e^{-i\zeta^2/4\hat\sigma_3}\begin{bmatrix} 1 & -\lambda\bar\rho(z_0) \\ 0 & 1 \end{bmatrix}\right) & \text{if } \zeta\in\Omega_4. \end{cases}$$

It follows that Ψ_2 satisfies

$$\Psi_2^+(\zeta) = \Psi_2^-(\zeta)e^{-i\zeta^2/4\hat\sigma_3}\left(\zeta_-^{2\varphi(z_0)\sigma_3}G(0,0,z_0)\zeta_+^{-2\varphi(z_0)\sigma_3}\right), \quad \Psi_2(\infty) = I.$$

The branching properties of $\zeta^{2\varphi(z_0)}$, $\zeta_+^{2\varphi(z_0)}\zeta_-^{-2\varphi(z_0)} = \tau(z_0)$ for $\zeta < 0$, need to be incorporated in this calculation. We introduce a final transformation. Define $\Psi_3(\zeta) = \Psi_2(\zeta)\zeta^{2\varphi(z_0)\sigma_3}e^{-i\zeta^2/4\sigma_3}$, which satisfies

$$\Psi_3^+(\zeta) = \Psi_3^-(\zeta)G(0,0,z_0), \quad \zeta\in\mathbb{R}, \tag{3.25}$$

$$\Psi_3(\zeta) = (I + \Psi_{3,1}/\zeta + \cdots)\zeta^{2\varphi(z_0)\sigma_3}e^{-i\zeta^2/4\sigma_3} \text{ as } \zeta\to\infty. \tag{3.26}$$

Note that $\Psi_2(\zeta) = I + \Psi_{3,1}/\zeta + \mathcal{O}(\zeta^{-2})$. Such an expansion is not guaranteed because the jump condition for Ψ_2 has no decay. Because we construct a solution explicitly, this is not an issue.

The jump matrix has no ζ dependence, and it can be seen that $\partial_\zeta\Psi_3(\zeta)\cdot\Psi_3^{-1}(\zeta) = \mathcal{O}(\zeta)$ has no jump on the real axis. Liouville's theorem shows that $\partial_\zeta\Psi_3(\zeta)\cdot\Psi^{-1}(\zeta)$ must be a polynomial. A straightforward calculation shows

$$\partial_\zeta\Psi_3(\zeta)\Psi_3^{-1}(\zeta) = \left(-\frac{i\zeta}{2}\sigma_3 - \frac{i}{2}\Psi_{3,1}\sigma_3 + \mathcal{O}(1/\zeta)\right) \times\left(I - \frac{1}{2}\Psi_{3,1} + \mathcal{O}(1/\zeta^2)\right) = -\frac{i\zeta}{2}\sigma_3 + \frac{i}{2}[\sigma_3,\Psi_{3,1}].$$

We obtain a differential equation for Ψ_3:

$$\partial_\zeta \Psi_3(\zeta) = \begin{bmatrix} -\mathrm{i}\zeta/2 & \kappa \\ \overline{\kappa} & \mathrm{i}\zeta/2 \end{bmatrix} \Psi_3(\zeta), \quad \kappa = \mathrm{i}(\Psi_{3,1})_{12},$$

where the subscripts denote the $(1,2)$ component of $\Psi_{3,1}$. Note that at this point κ is not known and once we obtain it, we have determined the leading asymptotic behavior of the $(1,2)$ component of Ψ_2. Also, the fact that we assume $\overline{\kappa}$ is the quantity in the $(2,1)$ component is justified by completing the construction. Examining the first column and differentiating, we see

$$\begin{aligned}
\partial_\zeta(\Psi_3)_{11}(\zeta) &= -\zeta\frac{\mathrm{i}}{2}(\Psi_3)_{11}(\zeta) + \kappa(\Psi_3)_{21}(\zeta),\\
\partial_\zeta(\Psi_3)_{21}(\zeta) &= \overline{\kappa}(\Psi_3)_{11}(\zeta) + \zeta\frac{\mathrm{i}}{2}(\Psi_3)_{21}(\zeta),\\
\partial_\zeta^2(\Psi_3)_{11}(\zeta) &= -\zeta\frac{\mathrm{i}}{2}\partial_\zeta(\Psi_3)_{11}(\zeta) - \frac{\mathrm{i}}{2}(\Psi_3)_{11}(\zeta) + \kappa\partial_\zeta(\Psi_3)_{21}(\zeta).
\end{aligned}$$

Therefore

$$\partial_\zeta^2(\Psi_3)_{11}(\zeta) = \left(-\frac{\zeta^2}{4} - \frac{\mathrm{i}}{2} + |\kappa|^2\right)(\Psi_3)_{11}(\zeta), \tag{3.27}$$

$$\partial_\zeta^2(\Psi_3)_{21}(\zeta) = \left(-\frac{\zeta^2}{4} + \frac{\mathrm{i}}{2} + |\kappa|^2\right)(\Psi_3)_{21}(\zeta). \tag{3.28}$$

The *parabolic cylinder function* $D_\nu(\zeta)$ solves [91]

$$\frac{\mathrm{d}^2 D_\nu(\zeta)}{\mathrm{d}\zeta^2} + \left(-\frac{\zeta^2}{4} + \frac{1}{2} + \nu\right) D_\nu(\zeta) = 0. \tag{3.29}$$

It follows that $(\Psi_3(\zeta))_{11} = \alpha_{11} D_{\mathrm{i}|\kappa|^2}(\mathrm{e}^{\mathrm{i}\pi/4}\zeta)$ and similarly $(\Psi_3(\zeta))_{21} = \alpha_{21} D_{-1+\mathrm{i}|\kappa|^2}(\mathrm{e}^{\mathrm{i}\pi/4}\zeta)$. Applying this reasoning to the second column of Ψ_3 we find the expression

$$\Psi_3(\zeta) = \begin{bmatrix} \alpha_{11} D_{\mathrm{i}|\kappa|^2}(\mathrm{e}^{\mathrm{i}\pi/4}\zeta) & \alpha_{12} D_{-1-\mathrm{i}|\kappa|^2}(\mathrm{e}^{-\mathrm{i}\pi/4}\zeta) \\ \alpha_{21} D_{-1+\mathrm{i}|\kappa|^2}(\mathrm{e}^{i\pi/4}\zeta) & \alpha_{22} D_{\mathrm{i}|\kappa|^2}(\mathrm{e}^{-\mathrm{i}\pi/4}\zeta) \end{bmatrix} S(\zeta)$$

for some undetermined matrix $S(\zeta)$. We choose $S(\zeta)$ to be piecewise constant so that condition (3.26) is satisfied. It is shown in [29] that to match the asymptotics we should choose

$$\alpha_{11} = \mathrm{e}^{\pi|\kappa|^2/4}, \quad \alpha_{12} = \mathrm{e}^{\pi|\kappa|^2/4 - 3\mathrm{i}\pi/4}\kappa, \quad \alpha_{21} = \overline{\alpha}_{12}, \quad \alpha_{22} = \alpha_{11},$$

and

$$S(\zeta) = \begin{cases} \begin{bmatrix} 1 & 0 \\ \dfrac{2\pi}{\kappa\,\Gamma(-\mathrm{i}|\kappa|^2)}\mathrm{e}^{-\pi|\kappa|^2/2+\mathrm{i}\pi/4} & 1 \end{bmatrix} & \text{if } \operatorname{Im}\zeta > 0, \\[2ex] \begin{bmatrix} 1 & \dfrac{2\pi}{\overline{\kappa}\,\Gamma(\mathrm{i}|\kappa|^2)}\mathrm{e}^{-\pi|\kappa|^2/2-\mathrm{i}\pi/4} \\ 0 & 1 \end{bmatrix} & \text{if } \operatorname{Im}\zeta < 0, \end{cases}$$

where $\Gamma(\zeta)$ is the *Gamma function* [91]. The condition $S^+(\zeta) = S^-(\zeta)G(0,0,z_0)$ from (3.25) dictates the choice of κ:

$$\begin{bmatrix} \tau(z_0) & -\lambda\overline{\rho}(z_0) \\ \rho(z_0) & 1 \end{bmatrix} = \begin{bmatrix} e^{-2\pi|\kappa|^2} & -\frac{\sqrt{2\pi}}{\kappa\,\Gamma(i|\kappa|^2)}e^{-\pi|\kappa|^2/2-i\pi/4} \\ \frac{\sqrt{2\pi}}{\overline{\kappa}\,\Gamma(-i|\kappa|^2)}e^{-\pi|\kappa|^2/2+i\pi/4} & 1 \end{bmatrix},$$

where we used that $\Gamma(\zeta)\Gamma(-\zeta) = -\pi/(\zeta\sin(\pi\zeta))$. Thus

$$|\kappa|^2 = -\frac{1}{2\pi}\log\tau(z_0), \quad \arg\kappa = \frac{\pi}{4} + \arg\Gamma(i|\kappa|^2) - \arg\rho(z_0).$$

This determines κ and hence determines $(\Psi_{3,1})_{12}$. Since $\Psi_1(\zeta) = \Psi_2(\zeta)$ on the positive imaginary axis, we find $-i\kappa = \lim_{\zeta\to\infty}\zeta(\Psi_1(\zeta))_{12}$. From (3.24)

$$\lim_{\zeta\to\infty}\zeta\Psi_{12}(\zeta) = -(8t)^{-2\varphi(z_0)}e^{-2r(z_0)+4itz_0^2}i\kappa,$$

where

$$e^{-r(z_0)} = (\Delta_r(z_0;z_0))_{11}, \quad r(z_0) = \int_{-\infty}^{z_0}\log(s-z_0)\,\mathrm{d}\log(1-\lambda|\rho(s)|^2).$$

This is derived by integrating $\mathscr{C}_{(-\infty,z_0)}\log\tau(z)$ by parts. We have derived the leading-order behavior and if we can show that Ψ is close to Ψ_2, we show that (see Theorem 3.10)

$$q(x,t) = \frac{1}{\sqrt{2t}}(8t)^{-2\varphi(z_0)}e^{-2r(z_0)+4itz_0^2}\kappa + \mathscr{O}(t^{-1}\log t). \tag{3.30}$$

The error bound is established in the next section. The factor of $1/\sqrt{t}$ appears from the transformation back to the z-plane:

$$\lim_{z\to\infty} z(\Psi_2(z))_{12} = \frac{1}{\sqrt{8t}}\lim_{\zeta\to\infty}\zeta(\Psi_2(\zeta/\sqrt{8t}+z_0))_{12}. \tag{3.31}$$

3.2.4 ▪ Error analysis

The error analysis is best performed in the ζ-plane. That is, we must scale and shift the jump contour and jump matrices for Φ_2: define $\hat{\Phi}_2(\zeta) = \Phi_2(\zeta/\sqrt{8t}+z_0)$. The jump contour Γ for Ψ and $\hat{\Gamma}$ for $\hat{\Phi}_2$ are shown in Figure 3.4. Let V be the jump matrix for Ψ and $\hat{V}$ be the jump matrix for $\hat{\Psi}_2$. We extend both jump matrices to $\Gamma\cup\hat{\Gamma}$ by defining them to be the identity matrix outside of their initial domains of definition, which forces both u and $\hat{u}$ to be zero outside their initial domains of definition. We show that

$$\|u-\hat{u}\|_{L^1(\Gamma\cup\hat{\Gamma})} = \mathscr{O}(t^{-1/2}\log t).$$

We claim that it suffices to show that

$$\|V-\hat{V}\|_{L^1\cap L^\infty(\Gamma\cup\hat{\Gamma})} = \mathscr{O}(t^{-1/2}\log t). \tag{3.32}$$

Indeed, in this case Lemma 2.76 shows that

$$\|u-\hat{u}\|_{L^2(\Gamma\cup\hat{\Gamma})} = \mathscr{O}(t^{-1/2}\log t). \tag{3.33}$$

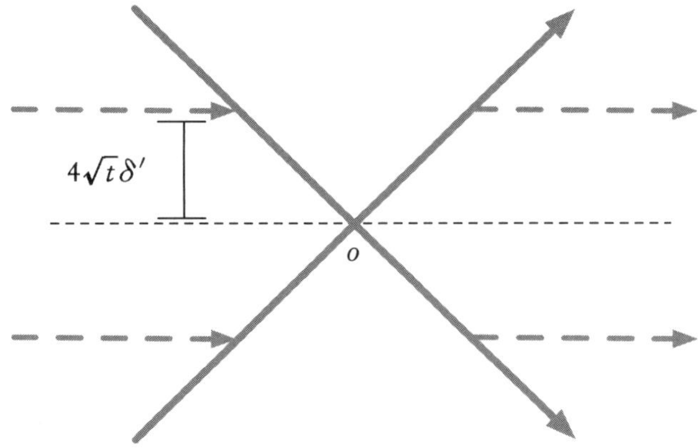

Figure 3.4. *The contours* Γ *(solid) and* $\hat{\Gamma}$ *(dashed) and their overlap region* C_0.

From the integral equations satisfied by u and $\hat{u}$,

$$u-\hat{u}=\mathscr{C}_{\Gamma\cup\hat{\Gamma}}(u-\hat{u})(V-I)-\mathscr{C}_{\Gamma\cup\hat{\Gamma}}u(V-\hat{V})+V-\hat{V}.$$

The Cauchy–Schwarz inequality, along with the triangle inequality, proves the result. Indeed, $||V-I||_{L^2(\Gamma\cup\hat{\Gamma})}$ and $||u||_{L^2(\Gamma\cup\hat{\Gamma})}$ are uniformly bounded[32] and (3.33) along (3.32) shows the sufficiency.

Remark 3.2.1. *The sufficiency of* (3.32) *requires the norm of the Cauchy integral operator to behave well with respect to* t. *The contours in Figure* 3.4 *are just a scaling of a* t*-independent contour. A simple calculation shows that the operator norm for the Cauchy operator for the contour* Γ *is the same for* $\alpha\Gamma$, $\alpha\in\mathbb{C}$. *Therefore, the operator norm is constant.*

It follows that the contour of intersection $C_0=\Gamma\cap\hat{\Gamma}$ is defined by

$$C_0=4t\delta' e^{i\pi/4}([-1,1]\cup[-i,i]),$$

with appropriate orientation. The difference $||V-\hat{V}||_{L^1\cap L^\infty(\hat{\Gamma}\setminus C_0)}$ is exponentially small with respect to t. This follows directly from the arguments given in Section 3.2.1. Next, we consider the difference $||V-\hat{V}||_{L^1\cap L^\infty(\Gamma\setminus C_0)}$. Since $\sup_{\zeta\in\mathbb{C}}|\zeta^{ic}|<\infty$, $c\in\mathbb{R}$, it is clear that the contribution on each connected component of $\Gamma\setminus C_0$ is bounded by a constant times

$$I(t)=\int_{4\sqrt{t}\delta'}^{\infty}e^{-x^2/4}dx.$$

The asymptotics of the error function shows that $I(t)$ decays beyond all orders as $t\to\infty$.

We consider $||V-\hat{V}||_{L^1\cap L^\infty(C_0)}$. We must analyze the difference

$$\Delta(z(\zeta);z_0)P(z(\zeta))\Delta^{-1}(z(\zeta);z_0)-[P](\zeta)$$

for $z(\zeta)=\zeta/\sqrt{8t}+z_0$. We briefly go back to the z-plane. Since ρ can be recovered by a Cauchy integral of its boundary values on the boundary of any strip about the real

[32]This follows because $u\to\hat{u}$ in $L^2(\Gamma\cup\hat{\Gamma})$ and $\hat{u}=\Psi^+-\Psi^-$ has a bounded $L^2(\Gamma\cup\hat{\Gamma})$ norm.

axis with width less than δ, we know that it must be uniformly Lipschitz in the strip $|\operatorname{Im} z| < \delta'$ for $\delta' < \delta/2$. Therefore, with $4t\delta' > 1$ and for some $C > 0$

$$\begin{aligned}
|U(z_0) - U(z)| &\leq C|z - z_0|,\\
|L(z_0) - L(z)| &\leq C|z - z_0|,\\
|P(z_0) - P(z)| &\leq C|z - z_0|,\\
|M(z_0) - M(z)| &\leq C|z - z_0|.
\end{aligned}$$

When setting $z(\zeta) = \zeta/\sqrt{8t} + z_0$ we obtain bounds on the L^∞ norms of these differences that are $\mathcal{O}(t^{-1/2})$. In addition, we must consider

$$\Delta(z;z_0) - \Delta_r(z_0;z_0)\Delta_s(z;z_0) = (\Delta_r(z;z_0) - \Delta_r(z_0;z_0))\Delta_s(z;z_0). \tag{3.34}$$

This can be further simplified. Note that since $0 < \tau(z) < 1$ and $y > 0$ we have

$$1 \geq |\delta(x + \mathrm{i}y; z_0)| \geq \exp\left(\frac{1}{2\pi}\int_{-\infty}^{z_0} \frac{\log \tau(s)}{(s-x)^2 + y^2} y \,\mathrm{d}s\right) \geq (1 - \|\rho\|_{L^\infty(\mathbb{R})})^{1/2}, \tag{3.35}$$

where we have used that the Poisson kernel integrates to unity and (3.19). Similar arguments show for $y < 0$ that $1 \leq |\delta(x + \mathrm{i}y, z_0)| \leq (1 - \|\rho\|_{L^\infty(\mathbb{R})})^{-1/2}$. Thus δ is uniformly bounded in both z and z_0. To analyze the difference (3.34) it suffices to consider

$$I - \Delta_r^{-1}(z;z_0)\Delta_r(z_0;z_0).$$

This is further reduced to the study of

$$\begin{aligned}
I_1(z;z_0) &= \int_{-\infty}^{z_0} \log\left(\frac{s-z}{s-z_0}\right) \mathrm{d}\log\tau(s) = (z_0 - z)\log(z_0 - z)\frac{\tau'(z_0)}{\tau(z_0)}\\
&\quad - (z_0 - z)\int_{-\infty}^{z_0} \log(s - z_0)\mathrm{d}\left(\frac{\tau'(s)}{\tau(s)}\right) + \int_{-\infty}^{z_0} \log(s - z)\log\left(\frac{s-z}{s-z_0}\right)\mathrm{d}\left(\frac{\tau'(s)}{\tau(s)}\right).
\end{aligned}$$

For z in a neighborhood of $z = z_0$, we have $|I_1(z;z_0)| \leq |z_0 - z|\log|z_0 - z|$. Therefore $\|V - \hat{V}\|_{L^\infty(C_0)} = \mathcal{O}(t^{-1/2}\log t)$. To bound the L^1 norm we notice that it involves integrals of the form

$$\begin{gathered}
\int_0^{4\delta'\sqrt{t}} \mathrm{e}^{-x^2/4}\delta^{-2}(\varphi(x);z_0)\left[\rho(\varphi(x)) - \rho(\varphi(0))\mathrm{e}^{I_1(\varphi(x);z_0)/(\pi i)}\right]\mathrm{d}x,\\
\varphi(x) = x\frac{\mathrm{e}^{\mathrm{i}\theta}}{\sqrt{8t}} + z_0 \quad \text{for } \theta = (2n-1)\pi/4.
\end{gathered}$$

Since the L^∞ norm of the term in brackets is $\mathcal{O}(t^{-1/2}\log t)$, we find $\|V - \hat{V}\|_{L^1(C_0)} = \mathcal{O}(t^{-1/2}\log t)$. This shows $\|u - \hat{u}\|_{L^1(\Gamma\cup\hat{\Gamma})} = \mathcal{O}(t^{-1/2}\log t)$. Upon considering (3.31) we obtain the error bound in (3.30).

Part II

Numerical Solution of Riemann–Hilbert Problems

Chapter 4

Approximating Functions

We overview a computational framework for approximating functions on a variety of simple, one-dimensional domains. The key tools are *Fourier series* and their siblings: *Laurent series*, *Taylor series*, and *Chebyshev series*. Our interest is in computational aspects: given a function that we can evaluate pointwise, how do we calculate the series coefficients, in order to approximate functions? The fundamental building block for resolving this question is the discrete Fourier transform. This chapter relies heavily on Appendix B, which includes a background on Fourier, Laurent, and Chebyshev series.

4.1 ▪ The discrete Fourier transform

Our starting point is to numerically calculate the Fourier coefficients of a function, given the constraint that we are only allowed to evaluate the function pointwise. There are two approaches: replace the inner product with a *discrete inner product* via quadrature, or *interpolate* the function by complex exponentials. We see below that these two approaches are equivalent.

4.1.1 ▪ Discrete inner product

The *Fourier coefficients* of a function f are defined by

$$\hat{f}_k \triangleq \left\langle e^{ik\theta}, f \right\rangle = \frac{1}{2\pi} \int_{-\pi}^{\pi} f(\theta) e^{-ik\theta} d\theta,$$

leading to the *Fourier series*

$$f(\theta) \sim \sum_{k=-\infty}^{\infty} \hat{f}_k e^{ik\theta}.$$

We denote the operator that maps integrable functions to their Fourier coefficients by

$$\mathcal{F} f \triangleq \left[\ldots, \hat{f}_{-2}, \hat{f}_{-1}, \hat{f}_0, \hat{f}_1, \hat{f}_2, \ldots \right]^\top.$$

The decay in coefficients of $\mathcal{F} f$ can be inferred from the smoothness of f. An important property for numerical analysis is that $\mathcal{F} f \in \ell^1$ (absolutely summable Fourier coefficients, also known as the *Wiener algebra*; see Appendix B) is sufficient for the Fourier

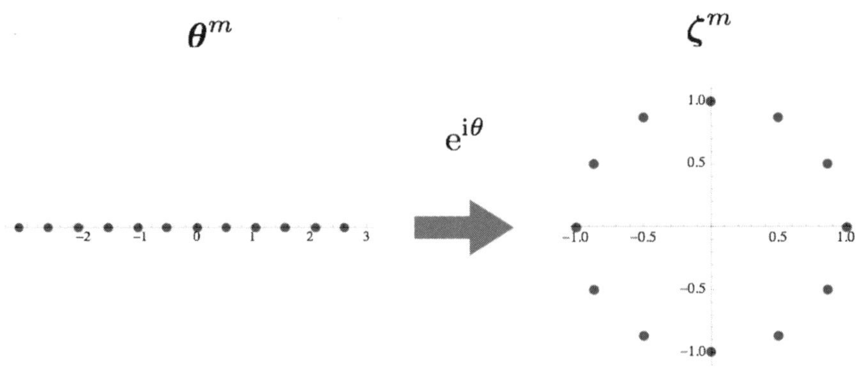

Figure 4.1. *A depiction of m evenly spaced points on the periodic interval, mapped to m evenly spaced points on the unit circle.*

series to converge uniformly (Proposition B.3), and twice differentiability of f is sufficient for $\mathcal{F}f \in \ell^1$ (Proposition B.4).

By discretizing the inner product, we derive an efficient numerical approximation of the Fourier coefficients. We discretize the $L^2(\mathbb{T})$ inner product using the *trapezoidal rule*. First, define m evenly spaced points on the periodic interval:

$$\theta^m \triangleq [\theta_1^m, \ldots, \theta_m^m]^\top \triangleq \pi\left[-1, \frac{2}{m}-1, \frac{4}{m}-1, \ldots, 1-\frac{2}{m}\right]^\top;$$

see Figure 4.1 for a graphical depiction. The trapezoidal rule allows us to approximate an integral by values at θ^m:

$$\int_{-\pi}^{\pi} f(\theta)\mathrm{d}\theta \approx \frac{2\pi}{m}\left[f(\theta_1^m) + \cdots + f(\theta_m^m)\right].$$

Applying the trapezoidal rule to the $L^2(\mathbb{T})$ inner product leads to a *discrete semi-inner product*:

$$\langle f, g\rangle_m \triangleq \frac{1}{m}\sum_{j=1}^{m} \overline{f}(\theta_j^m) g(\theta_j^m) = \frac{1}{m} f(\theta^m)^* g(\theta^m).$$

Then

$$\hat{f}_k^m \triangleq \left\langle e^{ik\theta}, f\right\rangle_m \approx \left\langle e^{ik\theta}, f\right\rangle = \hat{f}_k.$$

We use the discrete inner product to approximate Fourier coefficients, which we use to approximate the Fourier series itself:

$$f_{\alpha,\beta}(\theta) \triangleq \sum_{k=\alpha}^{\beta} \hat{f}_k^m e^{ik\theta} \approx f(\theta),$$

where α and β are two integers satisfying $\beta - \alpha + 1 = m$. The approximation $f_{\alpha,\beta}$ is referred to as the *discrete Fourier series*.

4.1.2 ▪ The DFT as a matrix operator

We can look at the discrete inner product in another way: it is a map from function values at θ^m to approximate coefficients. We refer to this as the *discrete Fourier transform* (DFT), which is represented by the following matrix:

$$\mathrm{F}_{\alpha,\beta}\begin{bmatrix} f(\theta_1^m) \\ \vdots \\ f(\theta_m^m) \end{bmatrix} = \begin{bmatrix} \hat{f}_\alpha^m \\ \vdots \\ \hat{f}_\beta^m \end{bmatrix} \quad \text{for} \quad \mathrm{F}_{\alpha,\beta} = \frac{1}{m}\begin{bmatrix} e^{-i\alpha\theta_1^m} & \cdots & e^{-i\alpha\theta_m^m} \\ \vdots & \cdots & \vdots \\ e^{-i\beta\theta_1^m} & \cdots & e^{-i\beta\theta_m^m} \end{bmatrix}.$$

The standard engineering definition of the DFT — and the one most likely encountered in fast Fourier transform (FFT) routines — is $m\mathrm{F}_{0,m-1}$. However, we can use the following property of $e^{ik\theta_j^m}$ to express $\mathrm{F}_{\alpha,\beta}$ as a simple transformation of $\mathrm{F}_{0,m-1}$.

Proposition 4.1.

$$e^{ik\theta_j^m} = (-)^m e^{i(k+m)\theta_j^m}.$$

Corollary 4.2. *If $\alpha < 0 < \beta$, then*

$$\mathrm{F}_{\alpha,\beta} = \begin{bmatrix} \overbrace{0 \ \cdots \ 0}^{\beta+1 \text{ times}} & \overbrace{(-)^m \quad\quad\quad\quad}^{-\alpha \text{ times}} \\ & \ddots \\ & (-)^m \\ \underline{1} & \\ \ddots & \\ 1 & \end{bmatrix} \mathrm{F}_{0,m-1}.$$

(The underline corresponds to the zero index row and column.)

The question remains how to choose α and β. We choose them to ensure that the discrete Fourier series converges. To understand convergence properties, we have to deduce the error in $\hat{f}_k^m$. This relies on understanding the orthogonality properties of the discretized inner product.

4.1.3 ▪ Orthogonality of the Fourier basis with the discrete inner product

We investigate orthogonality of complex exponentials under the discrete inner product. The following proposition encodes the fact that sums of roots of unity may exhibit cancellation.

Proposition 4.3. *For $m \geq 1$,*

$$\sum_{j=1}^{m} e^{ik\theta_j^m} = (-)^k m \quad \textit{for } k = \ldots, -2m, -m, 0, m, 2m \ldots$$

$$\sum_{j=1}^{m} e^{ik\theta_j^m} = 0 \quad \textit{for all other integers } k.$$

Proof. Note that

$$\zeta_j = -e^{2\pi i(j-1)/m} = -\omega^{j-1}$$

for the mth root of unity $\omega = e^{2\pi i/m}$ (i.e., $\omega^m = 1$). Therefore,

$$\sum_{j=1}^{m} \zeta_j^k = (-)^k \sum_{j=0}^{m-1} \omega^{kj}.$$

If $k = \alpha m$ is a multiple of m, then

$$\sum_{j=0}^{m-1} \omega^{kj} = \sum_{j=0}^{m-1} (\omega^m)^{\alpha j} = \sum_{j=0}^{m-1} 1^{\alpha j} = m.$$

For k not a multiple of m, the geometric sum formula implies

$$\sum_{j=0}^{m-1} \omega^{kj} = \sum_{j=0}^{m-1} (\omega^k)^j = \frac{\omega^{km} - 1}{\omega^k - 1}.$$

Because k is not a multiple of m, the denominator is nonzero. Because $\omega^{km} = (\omega^m)^k = 1^k$, the numerator is zero. □

A consequence of this simple cancellation is that the complex exponentials behave "almost orthogonally" with respect to the discrete inner product.

Corollary 4.4. *For* $m \geq 1$,

$$\left\langle e^{ik\theta}, e^{ij\theta} \right\rangle_m = (-1)^{j-k} \quad \textit{for} \quad j-k = \ldots, -2m, -m, 0, m, 2m, \ldots$$
$$\left\langle e^{ik\theta}, e^{ij\theta} \right\rangle_m = 0 \quad \textit{otherwise}.$$

4.1.4 ▪ Interpolation

Before we discuss convergence, we relate the discrete Fourier series to *interpolation*. Suppose we want to choose $c_\alpha, \ldots, c_\beta$ so that

$$\sum_{k=\alpha}^{\beta} c_k e^{ik\theta_j^m} = f(\theta_j^m),$$

i.e., we interpolate $f(\theta)$ at the evenly spaced points. We rewrite the interpolation criteria in a *Vandermonde system*:

$$V_{\alpha,\beta} \begin{bmatrix} c_\alpha \\ \vdots \\ c_\beta \end{bmatrix} = \begin{bmatrix} f(\theta_1^m) \\ \vdots \\ f(\theta_m^m) \end{bmatrix} \quad \text{for} \quad V_{\alpha,\beta} = \begin{bmatrix} e^{i\alpha\theta_1^m} & \cdots & e^{i\beta\theta_1^m} \\ \vdots & \ddots & \vdots \\ e^{i\alpha\theta_m^m} & \cdots & e^{i\beta\theta_m^m} \end{bmatrix} = m\mathrm{F}^*_{\alpha,\beta}.$$

Then $V_{\alpha,\beta}$ is in fact the *inverse discrete Fourier transform*:

$$\mathrm{F}_{\alpha,\beta} V_{\alpha,\beta} = \begin{bmatrix} \left\langle e^{i\alpha\theta}, e^{i\alpha\theta} \right\rangle_m & \cdots & \left\langle e^{i\alpha\theta}, e^{i\beta\theta} \right\rangle_m \\ \vdots & \ddots & \vdots \\ \left\langle e^{i\beta\theta}, e^{i\alpha\theta} \right\rangle_m & \cdots & \left\langle e^{i\beta\theta}, e^{i\beta\theta} \right\rangle_m \end{bmatrix} = I.$$

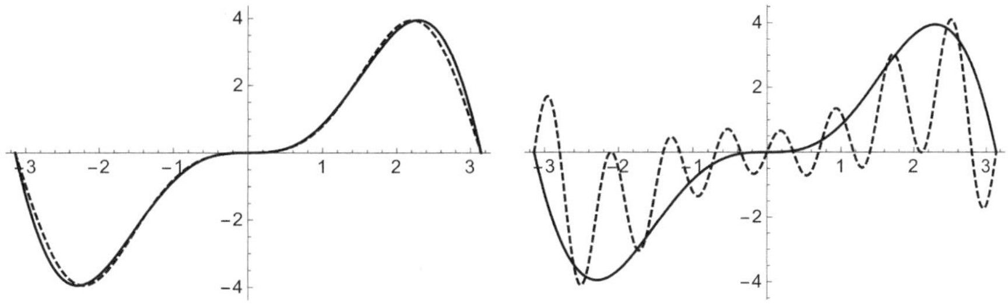

Figure 4.2. *Approximation of $f(\theta) = \theta^2 \sin\theta$ by $f_{-4,4}$ (left) versus $f_{0,8}$ (right). Both interpolate f at the same points.*

In other words, $c_k = \hat{f}_k^m$ and the discrete Fourier transform gives the coefficients of the interpolating complex exponential with

$$\mathrm{F}_{\alpha,\beta}^{-1} = V_{\alpha,\beta} = m\mathrm{F}_{\alpha,\beta}^{*}.$$

While we interpolate for any choice of α and β, the behavior of the interpolants varies greatly, as seen in Figure 4.2. This is the well-known phenomenon of *aliasing*.

4.1.5 ▪ Convergence of discrete Fourier series

We do not want spurious high frequencies: we want the discrete Fourier series to converge to the true function. Thus we must choose α and β carefully. The key to determining α and β is that we can write the numerical coefficients $\hat{f}_k^m$ directly in terms of the true Fourier coefficients $\hat{f}_k$, using the almost-orthogonality property.

Proposition 4.5. *If $\mathcal{F}f \in \ell^1$, then for $m \geq 1$*

$$\hat{f}_k^m = \cdots + \hat{f}_{k-2m} + (-)^m \hat{f}_{k-m} + \hat{f}_k + (-)^m \hat{f}_{k+m} + \hat{f}_{k+2m} - \cdots.$$

Furthermore,

$$\left\|\mathrm{F}_{\alpha,\beta} f(\theta^m)\right\|_{\ell^1} \leq \|\mathcal{F}f\|_{\ell^1}.$$

Proof. The first statement follows from

$$\hat{f}_k^m = \left\langle e^{ik\theta}, f\right\rangle_m = \sum_{j=-\infty}^{\infty} \hat{f}_j \left\langle e^{ik\theta}, e^{ij\theta}\right\rangle_m = \sum_{\tau=-\infty}^{\infty} (-)^{\tau m} \hat{f}_{k+m\tau}.$$

The second statement follows because each mode appears precisely once. □

A consequence of this is convergence of the Fourier coefficients for k fixed as the number of sample points m tends to infinity because $\sum_{\tau\neq 0}(-)^{\tau m}\hat{f}_{k+\tau m} \to 0$ as $m \to \infty$. But α and β must change with m (since $\beta - \alpha + 1 = m$). Thus the convergence of all the coefficients is not quite so simple.

Indeed, consider the naive choice of $\alpha = 0$ and $\beta = m-1$. This choice is inappropriate: how can we expect to represent a function that has both positive and negative Fourier coefficients by an approximation with only positive terms? We can make this statement concrete: we have that

$$\hat{f}^m_{m-1} = \cdots + \hat{f}_{-m-1} + (-)^m \hat{f}_{-1} + \hat{f}_{m-1} + (-)^m \hat{f}_{2m-1} + \hat{f}_{3m-1} + \cdots.$$

Thus $\hat{f}^m_{m-1} \to (-)^m \hat{f}_{-1}$ as $m \to \infty$. In other words, the approximation has a large oscillatory term $(-)^m \hat{f}_{-1} e^{i(m-1)\theta}$ that never decays as $m \to \infty$ (unless, of course, $\hat{f}_{-1} = 0$).

Now consider the choice so that zero is in the middle of α and β:

$$\{\alpha_m, \beta_m\} \triangleq \begin{cases} \left\{-\frac{m-1}{2}, \frac{m-1}{2}\right\} & \text{if } m \text{ is odd,} \\ \left\{-\frac{m}{2}, \frac{m}{2} - 1\right\} & \text{if } m \text{ is even.} \end{cases}$$

The DFT and discrete Fourier series with this choice of α and β is our standard choice, which we denote by

$$\mathrm{F}_m \triangleq \mathrm{F}_{\alpha_m,\beta_m} \qquad \text{and} \qquad f_m(\theta) \triangleq f_{\alpha_m,\beta_m}(\theta).$$

In this case we have (for m even)

$$\hat{f}^m_{\beta_m} = \hat{f}^m_{\frac{m}{2}-1} = \cdots + \hat{f}_{-\frac{3m}{2}} + \hat{f}_{-\frac{m}{2}} + \hat{f}_{\frac{m}{2}-1} + \hat{f}_{\frac{3m}{2}-1} + \cdots.$$

This choice is sufficient to show that the discrete Fourier series f_m tends to f, and that the rate of convergence is dictated by the decay rate of the coefficients, which in turn can be inferred by smoothness of f (Lemma B.7; see also [73]).

Theorem 4.6. *If $\mathcal{F} f \in \ell^1$, then the discrete Fourier series converges uniformly: $\|f - f_m\|_u \to 0$ as $m \to \infty$. If $\mathcal{F} f \in \ell^{(\lambda,R),1}$, then the convergence rate increases with λ and R:*

$$\|f - f_m\|_u \leq 2\|\mathcal{F} f\|_{\ell^{(\lambda,R),1}} \left(\frac{m}{2} + 1\right)^{-\lambda} R^{-\frac{m}{2}}.$$

Proof. We have two sources of errors: the error in truncating the infinite series into a finite series, and the error in replacing the finite Fourier coefficients with those computed from the DFT. The former error is bounded due to Proposition B.9, so we need only worry about the latter. We have (here $\alpha \triangleq \alpha_m$ and $\beta \triangleq \beta_m$ with $|\alpha| \geq |\beta|$)

$$\begin{aligned} \left\| f_m - \sum_{k=\alpha}^{\beta} \hat{f}_k e^{ik\theta} \right\|_u &\leq \sum_{k=\alpha}^{\beta} \left| \hat{f}_k - \hat{f}^m_k \right| = \sum_{k=\alpha}^{\beta} \left| \hat{f}_k - \sum_{\tau=-\infty}^{\infty} (-)^{m\tau} \hat{f}_{k+m\tau} \right| \\ &\leq \sum_{k=\alpha}^{\beta} \sum_{\substack{\tau=-\infty \\ \tau\neq 0}}^{\infty} \left| \hat{f}_{k+m\tau} \right| = \sum_{\substack{k=-\infty \\ k\notin[\alpha,\beta]}}^{\infty} \left| \hat{f}_k \right| \\ &\leq (|\alpha| + 2)^{-\lambda} R^{-|\alpha|-1} \sum_{\substack{k=-\infty \\ k\notin[\alpha,\beta]}}^{\infty} (|k| + 1)^{\lambda} R^{|k|} \left| \hat{f}_k \right| \\ &\leq \|\mathcal{F} f\|_{\ell^{(\lambda,R),1}} (|\alpha| + 2)^{-\lambda} R^{-|\alpha|-1}. \end{aligned}$$

Bounding $|\alpha| \geq \frac{m}{2} - 1$ proves the result. □

4.1.6 ▪ Laurent series

Sufficiently regular functions defined on the unit circle can be expanded in *Laurent series*:

$$f(\zeta) = \sum_{k=-\infty}^{\infty} \hat{f}_k \zeta^k.$$

Under the transformation $\zeta = \mathrm{e}^{\mathrm{i}\theta}$, Laurent series become Fourier series:

$$f(\mathrm{e}^{\mathrm{i}\theta}) = \sum_{k=-\infty}^{\infty} \hat{f}_k \mathrm{e}^{\mathrm{i}k\theta}.$$

All the numerical and analytical study of Fourier series transfers over to Laurent series via this simple map. For example, define the m evenly spaced points on the unit circle

$$\zeta^m \triangleq \exp(\mathrm{i}\theta^m).$$

Then the DFT gives approximation to the Laurent coefficients of f via $\mathrm{F}_m f(\zeta^m)$, which we refer to as the *discrete Laurent series*. Because of this strong link, we also use $\mathscr{F}$ to denote the map from an integrable function defined on the unit circle to its Laurent coefficients. The discrete Laurent series converges *exponentially fast* for functions analytic in a neighborhood of the unit circle, as can be seen by combining the argument of Theorem 4.6 with the decay of the Laurent coefficients from Theorem B.10.

4.2 ▪ Chebyshev series

We turn our attention to approximating nonperiodic functions on $\mathbb{I} = [-1, 1]$, via the *Chebyshev expansion*

$$f(x) = \sum_{k=0}^{\infty} \check{f}_k T_k(x) \quad \text{for} \quad T_k(x) = \cos(k \arccos x).$$

Under the Joukowsky map,

$$J(\zeta) = \frac{1}{2}\left(\zeta + \frac{1}{\zeta}\right),$$

this Chebyshev series becomes a Laurent series with symmetric coefficients,

$$f(J(\zeta)) = \check{f}_0 + \sum_{k=1}^{\infty} \check{f}_k \left(\frac{\zeta^k + \zeta^{-k}}{2}\right),$$

or equivalently the cosine series,

$$f(\cos\theta) = \sum_{k=0}^{\infty} \check{f}_k \cos k\theta,$$

with $\zeta = \mathrm{e}^{\mathrm{i}\theta}$.

We denote the map from an integrable function f to the *Chebyshev coefficients* by $\mathscr{T}$:

$$\mathscr{T} f \triangleq \left[\check{f}_0, \check{f}_1, \check{f}_2, \ldots\right]^{\top}.$$

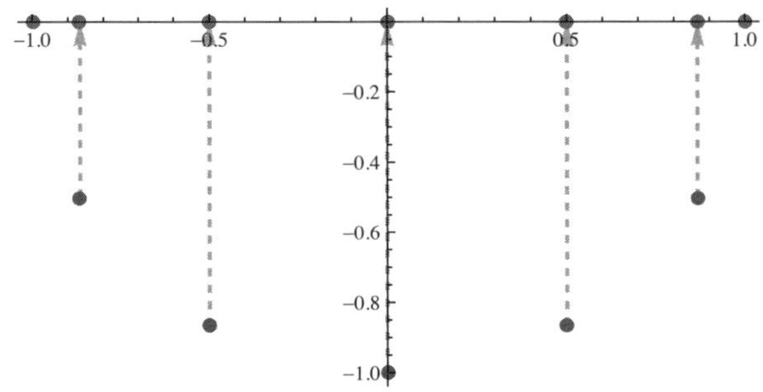

Figure 4.3. *Chebyshev points are the projection of an even number of evenly spaced points on the unit circle to the unit interval.*

Note that Chebyshev coefficients can be related to Laurent coefficients:

$$\left[\check{f}_0, \check{f}_1, \check{f}_2, \ldots\right]^\top = \left[\hat{f}_0, 2\hat{f}_1, 2\hat{f}_2, \ldots\right]^\top,$$

where $\hat{f}_k$ are the Laurent coefficients of $f(J(\zeta))$, which satisfy $\hat{f}_k = \hat{f}_{-k}$. See Appendix B.2 for a more detailed background.

4.2.1 ▪ The discrete cosine transform

Take $m = 2n - 2$, which is even so that both -1 and 1 are in ζ^m. Under the Joukowsky map, the m evenly spaced points on the unit circle get mapped to n Chebyshev points on the unit interval; see Figure 4.3.

Definition 4.7. *The* Chebyshev points *are*

$$\boldsymbol{x}^n \triangleq [x_1^n, \ldots, x_n^n]^\top \triangleq \left[J(\zeta_1^{2n-2}), \ldots, J(\zeta_n^{2n-2})\right]^\top = \left[-1, \cos\theta_2^{2n-2}, \ldots, \cos\theta_{n-1}^{2n-2}, 1\right]^\top.$$

The transform from f evaluated at Chebyshev points to approximate Chebyshev coefficients can be phrased in terms of the DFT applied to $f(J(\zeta))$. We first note that, because we have an even number of points on the unit circle ($m = 2n-2$), our standard choice of discrete Fourier series

$$\hat{f}_{n-1}^m \zeta^{1-n} + \hat{f}_{n-2}^m \zeta^{2-n} + \cdots + \hat{f}_{n-3}^m \zeta^{n-3} + \hat{f}_{n-2}^m \zeta^{n-2}$$

is not symmetric, where $\hat{f}_k^m = \hat{f}_{-k}^m$ follows from Proposition 4.5 and $\hat{f}_k = \hat{f}_{-k}$. However, ζ^{1-n} is indistinguishable from ζ^{n-1} at every interpolation point ζ^m; thus the symmetric function

$$\frac{\hat{f}_{n-1}^m}{2} \zeta^{1-n} + \hat{f}_{n-2}^m \zeta^{2-n} + \cdots + \hat{f}_{n-3}^m \zeta^{n-3} + \hat{f}_{n-2}^m \zeta^{n-2} + \frac{\hat{f}_{n-1}^m}{2} \zeta^{n-1}$$

still interpolates $f(J(\zeta))$ at ζ^m. We can thus construct the *discrete Chebyshev series*

$$f_n(x) = \sum_{k=0}^{n-1} \check{f}_k^n T_k(x) \quad \text{for} \quad \begin{bmatrix} \check{f}_0^n \\ \check{f}_1^n \\ \vdots \\ \check{f}_{n-2}^n \\ \check{f}_{n-1}^n \end{bmatrix} \triangleq \begin{bmatrix} \hat{f}_0^m \\ 2\hat{f}_1^m \\ \vdots \\ 2\hat{f}_{n-2}^m \\ \hat{f}_{n-1}^m \end{bmatrix}$$

which interpolates f at the Chebyshev points $\boldsymbol{x}^n$. We construct the *discrete cosine transform* (DCT):

$$\mathrm{T}_n \triangleq \underbrace{\begin{bmatrix} & & & \underline{1} & 0 & \cdots & 0 \\ & & 2 & & & & \\ & \iddots & & & & & \\ 2 & & & & & & \\ 1 & & & & & & \end{bmatrix}}_{\text{Drop positive coefficients}} \mathrm{F}_{2n-2} \underbrace{\begin{bmatrix} 1 & & & \\ & 1 & & \\ & & \ddots & \\ & & & 1 \\ & & \iddots & \\ & 1 & & \end{bmatrix}}_{\text{Project from } \boldsymbol{x}^n \text{ to } \zeta^{2n-2}}.$$

This matrix maps values at Chebyshev points to approximate Chebyshev coefficients. By using the definition in terms of the Fourier series, we see that the inverse operator is

$$\mathrm{T}_n^{-1} = \underbrace{\begin{bmatrix} 0 & \cdots & 0 & \underline{1} & & \\ & & & & \ddots & \\ & & & & & 1 \end{bmatrix}}_{\text{Drop negative values}} \mathrm{F}_{2n-2}^{-1} \underbrace{\begin{bmatrix} & & & 1 \\ & & \frac{1}{2} & \\ & \iddots & & \\ & \frac{1}{2} & & \\ \underline{1} & & & \\ & \frac{1}{2} & & \\ & & \ddots & \\ & & \frac{1}{2} & \\ & & & 0 \end{bmatrix}}_{\text{Convert to Laurent coefficients}}.$$

Remark 4.2.1. *The transform* T_n *and its inverse can also be applied in* $\mathcal{O}(n \log n)$ *operations, by using the fast cosine transform* [87].

We finally note the following aliasing properties.

Corollary 4.8. *If* $\mathcal{T}f \in \ell^1$*, then*

$$\begin{aligned} \check{f}_0^n &= \check{f}_0 + \check{f}_{2n-2} + \check{f}_{4n-4} + \check{f}_{6n-6} + \cdots, \\ \check{f}_k^n &= \check{f}_k + \check{f}_{2n-2-k} + \check{f}_{2n-2+k} + \check{f}_{4n-4-k} + \check{f}_{4n-4+k} + \cdots \quad \textit{for} \quad k = 1, \ldots, n-2, \\ \check{f}_{n-1}^n &= \check{f}_{n-1} + \check{f}_{3n-3} + \check{f}_{5n-5} + \check{f}_{7n-7} + \cdots. \end{aligned}$$

Therefore,

$$\|\mathrm{T}_n f(\boldsymbol{x}^n)\|_{\ell^1} \leq \|\mathcal{T}f\|_{\ell^1}.$$

Proof. The first part follows from Proposition 4.5 and the fact that $\hat{f}_k = \hat{f}_{-k}$. The second part follows since each mode again appears precisely once. □

4.3 ▪ Mapped series

Given an invertible map M from the unit circle to a contour Γ, we can approximate a function f defined on Γ by projecting to the unit circle and using the Laurent series,

$$f(M(\zeta)) = \sum_{k=-\infty}^{\infty} \hat{f}_k \zeta^k,$$

or equivalently

$$f(s) = \sum_{k=-\infty}^{\infty} \hat{f}_k M^{-1}(s)^k.$$

The regularity of $f(M(\zeta))$ determines the convergence properties of this series. Similarly, if we are given an invertible map from the unit interval to a contour Γ, we can approximate a function f using the Chebyshev series,

$$f(M(x)) = \sum_{k=0}^{\infty} \check{f}_k T_k(x),$$

or equivalently

$$f(s) = \sum_{k=0}^{\infty} \check{f}_k T_k(M^{-1}(s)).$$

Below we specialize this simple approach to several concrete maps.

4.3.1 ▪ Affine and Möbius maps

The simplest maps are affine: for example, $M(x) = \frac{b-a}{2}x + \frac{b+a}{2}$ maps $\mathbb{I}$ to $[a,b]$, which has the simple inverse $M^{-1}(x) = \frac{a+b-2x}{a-b}$. A generalization of affine maps are the *Möbius transformations*:

$$M(x) = \frac{ax+b}{cx+d}$$

provided that $ad - bc \neq 0$. Depending on the choice of parameters, a Möbius transformation maps the unit circle to either a line or circle[33] and the unit interval to either a line segment, circular arc, or ray.

Möbius transformation to the real line: A domain of fundamental importance is the real line. In analysis, the Fourier transform plays a central role. Unfortunately, the Fourier transform over the real line is an inherently continuous object that does not lend itself to practical approximation of functions. In our setting, however, regularity comes to the rescue: we are not interested in arbitrary $L^2(\mathbb{R})$ functions but rather high regularity

[33]Möbius transformations map circles to circles on the Riemann sphere.

functions, e.g., the Schwartz class. Thus, rather than attempting to do function approximation via Fourier transforms, we reduce the problem to Laurent series through a *Möbius transformation* between the real line and the unit circle:

$$M(\zeta) = a - \frac{\mathrm{i}}{L}\frac{\zeta - 1}{1+\zeta} \quad \text{and} \quad M^{-1}(x) = \frac{\mathrm{i} - L(x-a)}{\mathrm{i} + L(x-a)}. \tag{4.1}$$

We can thus approximate by a rational expansion:

$$f(M(\zeta)) = \sum_{k=-\infty}^{\infty} \hat{f}_k \zeta^k.$$

L and a are free parameters: they could be tuned to increase the convergence rate of the expansion, though for our purposes it suffices to take $a = 0$ and $L = 1$.

Note that the convergence properties of this approximation are dictated by the regularity of $f(M(\zeta))$. If $f \in \mathscr{S}(\mathbb{R})$, i.e., is a Schwartz class function, then $f(M(\zeta)) \in C^\infty(\mathbb{U})$, and $\mathscr{F} f \in \ell^{\lambda,p}$ for all λ, demonstrating that the coefficients decay faster than any inverse polynomial (recall Lemma B.7). Schwartz class functions are not the only class of functions such that $f(M(\zeta)) \in C^\infty(\mathbb{U})$. Suppose f is smooth on the real line and has a full asymptotic series at both $\pm\infty$:

$$f(x) \sim \alpha_0 + \frac{\alpha_1}{x} + \frac{\alpha_2}{x^2} + \cdots \quad \text{as} \quad x \to \pm\infty.$$

Then $f(M(\zeta))$ is also smooth on the unit circle: the asymptotic expansion of $f(x)$ as $x \to \infty$ induces differentiability of $f(M(\zeta))$ as $\zeta \to -1$. See [112] for sufficient conditions for such an expansion.

Möbius transformation to a ray: The Möbius transformation

$$M(x) = a + L\mathrm{e}^{\mathrm{i}\varphi}\frac{1+x}{1-x}$$

can be used to map the interval to the ray $[a, \mathrm{e}^{\mathrm{i}\varphi}\infty)$. This is easily verified: Möbius transformations map lines and circles to lines and circles, and we have $M(-1) = a$, $M(0) = a + L\mathrm{e}^{\mathrm{i}\varphi}$, and $M(1) = \infty$. The regularity of $f(M(x))$ dictates the convergence properties of the mapped Chebyshev expansion. In particular, if f is smooth and has a full asymptotic expansion at $+\infty$, then $f(M(x))$ is smooth.

4.3.2 ▪ Differentiation and integration of mapped series

The DCT along with properties of the Chebyshev polynomials allow for effective approximation of derivatives and integrals on general mapped contours. Assume

$$f(M(x)) \approx f_n(x) = \sum_{k=0}^{n-1} \check{f}_k^n T_k(x).$$

If $f(M(x))$ is sufficiently regular, then

$$f'(M(x))M'(x) \approx f_n'(x) \Rightarrow f'(M(x)) \approx \frac{f_n'(x)}{M'(x)}.$$

The formulae in Appendix B.2.6 allow for the Chebyshev coefficients of $f_n'(x)$ to be determined, from which we find

$$f'(M(x)) \approx \sum_{k=0}^{n-1} \check{g}_k T_k(x), \qquad \begin{bmatrix} \check{g}_0 \\ \check{g}_1 \\ \vdots \\ \check{g}_{n-1} \end{bmatrix} = \mathtt{T}_n \frac{f_n'(\boldsymbol{x}^n)}{M'(\boldsymbol{x}^n)}.$$

In the case that $M'(x)$ is constant, the application of $\mathtt{T}_n$ in the last step is not necessary.

We can also integrate functions using

$$\int^s f(t)\mathrm{d}t = \int^{M^{-1}(s)} f(M(\tau))M'(\tau)\mathrm{d}\tau.$$

In a similar way, we have

$$\int f(M(x))M'(x)\mathrm{d}x \approx \sum_{k=0}^{n-1} \check{h}_k \int T_k(x)\mathrm{d}x, \qquad \begin{bmatrix} \check{h}_0 \\ \check{h}_1 \\ \vdots \\ \check{h}_{n-1} \end{bmatrix} = \mathtt{T}_n f_n(\boldsymbol{x}^n)M'(\boldsymbol{x}^n).$$

The integration formulae in Appendix B.2.6 allow for

$$\sum_{k=0}^{n-1} \check{h}_k \int T_k(x)\mathrm{d}x$$

to be expressed as a standard Chebyshev series. Again, if $M'(x)$ is constant, then the application of $\mathtt{T}_n$ is not needed. Finally, we remark that as a special case, indefinite integration allows one to accurately compute contour integrals.

4.4 ▪ Vanishing bases

For RH problems, we need to work in special subspaces that encode vanishing conditions of functions: functions defined on the real line which vanish at ∞, and functions defined on the unit interval which vanish at ± 1 or both. In this section, we introduce coefficient spaces that incorporate these conditions.

Vanishing spaces on the real line: Consider functions on the real line approximated by the mapped Fourier series

$$f(M(\zeta)) = \sum_{k=-\infty}^{\infty} \hat{f}_k \zeta^k.$$

When $f(x)$ decays at infinity, $f \circ M$ vanishes at -1:

$$f(M(-1)) = \sum_{k=-\infty}^{\infty} (-)^k \hat{f}_k = 0,$$

provided that $\mathcal{F} f \in \ell^1$. In other words, $\mathcal{F} f \in \ell^1_{-z}$; recall Definition A.14.

We can incorporate the decay of f directly into the basis. The simplest option is

$$f(M(\zeta)) = \sum_{k=-\infty}^{\infty} \hat{f}_k \zeta^k = \sum_{k=-\infty}^{\infty} \hat{f}_k \zeta^k - \sum_{k=-\infty}^{\infty} (-)^k \hat{f}_k = \sum_{k=-\infty}^{\infty} \hat{f}_k \left[\zeta^k - (-1)^k\right],$$

i.e.,

$$f(x) = \sum_{k=-\infty}^{\infty} \hat{f}_k \left[M^{-1}(x)^k - (-1)^k\right].$$

However, this basis has issues with regularity as k increases: the decay as $x \to \pm\infty$ is not uniform in k. Instead, we use

$$f(M(\zeta)) = \sum_{k=-\infty}^{-1} \hat{f}_k^{-z}(\zeta^k + \zeta^{k+1}) + \hat{f}_0^{-z} + \sum_{k=1}^{\infty} \hat{f}_k^{-z}\left(\zeta^k + \zeta^{k-1}\right),$$

where the discrepancy at $k = 0$ serves the role of relating regularity of f to decay in $\hat{f}_k^{-z}$, with the condition $\hat{f}_0^{-z} = 0$ corresponding to vanishing at -1. It is then clear that

$$\left|M^{-1}(x)^k + M^{-1}(x)^{k+1}\right| \le \left|1 + M^{-1}(x)\right| = \left|1 + \frac{\mathrm{i} - L(x-a)}{\mathrm{i} + L(x-a)}\right| \le 2\frac{L|x-a|+1}{L^2(x-a)^2+1}.$$

Converting from the coefficients of the modified basis to standard Laurent coefficients is straightforward:

$$\mathcal{F} f = \begin{bmatrix} \ddots & & & & & & & \\ \ddots & 1 & & & & & & \\ & 1 & 1 & & & & & \\ & & 1 & 1 & & & & \\ & & & 1 & \underline{1} & 1 & & \\ & & & & & 1 & 1 & \\ & & & & & & 1 & \ddots \\ & & & & & & & \ddots \end{bmatrix} \begin{bmatrix} \vdots \\ \hat{f}_{-2}^{-z} \\ \hat{f}_{-1}^{-z} \\ \hat{f}_0^{-z} \\ \hat{f}_1^{-z} \\ \hat{f}_2^{-z} \\ \vdots \end{bmatrix}.$$

Inverting this operator, we obtain a transform for computing the *vanishing basis coefficients*:

$$\mathcal{F}^{-z} \triangleq \begin{bmatrix} \ddots & & & & & & \\ \ddots & 1 & & & & & \\ \ddots & -1 & 1 & & & & \\ \cdots & 1 & -1 & \underline{1} & -1 & 1 & \cdots \\ & & & & 1 & -1 & \ddots \\ & & & & & 1 & \ddots \\ & & & & & & \ddots \end{bmatrix} \mathcal{F} \quad \text{so that} \quad \mathcal{F}^{-z} f = \begin{bmatrix} \vdots \\ \hat{f}_{-2}^{z} \\ \hat{f}_{-1}^{z} \\ \hat{f}_0^{z} \\ \hat{f}_1^{z} \\ \hat{f}_2^{z} \\ \vdots \end{bmatrix}.$$

From this transformation and the property of norms of operators (Proposition A.22), we bound this operator from $\ell^{2,1}$ to ℓ^1 to obtain the following.

Proposition 4.9. *If $\mathcal{F} f \in \ell^{2,1}$, then $\mathcal{F}^{-z} f \in \ell^1$. If, in addition, $\mathcal{F} f \in \ell^{2,1}_z$, then $\hat{f}_0^{-z} = 0$.*

Vanishing spaces on the interval: For the case of Chebyshev polynomials, we have $T_k(\pm 1) = (\pm 1)^k$. Thus if f vanishes at either ± 1, we expand in the vanishing basis:

$$f(x) \sim \sum_{k=0}^{\infty} \check{f}_k^{\pm z} T_k^{\pm z}(x)$$

for

$$T_0^{\pm z}(x) \triangleq 1 \quad \text{and} \quad T_k^{\pm z}(x) \triangleq T_k(x) \mp T_{k-1}(x), \qquad k = 1, 2, \ldots .$$

The condition $\check{f}_0^{\pm z} = 0$ encodes the vanishing of f at ± 1. In this case, we transform from expansion in the vanishing basis to the standard Chebyshev polynomial basis:

$$\mathcal{T} f = \begin{bmatrix} 1 & \mp 1 & & & \\ & 1 & \mp 1 & & \\ & & 1 & \mp 1 & \\ & & & 1 & \ddots \\ & & & & \ddots \end{bmatrix} \begin{bmatrix} \check{f}_0^{\pm z} \\ \check{f}_1^{\pm z} \\ \check{f}_2^{\pm z} \\ \vdots \end{bmatrix}.$$

By inverting the transformation, we convert from Chebyshev coefficients to vanishing basis coefficients via

$$\mathcal{T}^{\pm z} \triangleq \begin{bmatrix} 1 & \pm 1 & 1 & \pm 1 & 1 & \cdots \\ & 1 & \pm 1 & 1 & \pm 1 & \cdots \\ & & 1 & \pm 1 & 1 & \cdots \\ & & & \ddots & \ddots & \ddots \end{bmatrix} \mathcal{T}.$$

If f vanishes at both ± 1, then we use a basis that vanishes at both endpoints:

$$f(x) \sim \sum_{k=0}^{\infty} \check{f}_k^{z} T_k^{z}(x) \quad \text{for}$$
$$T_0^z(x) \triangleq 1, \quad T_1^z(x) \triangleq T_1(x), \quad \text{and} \quad T_k^z(x) \triangleq T_k(x) - T_{k-2}(x), \qquad k = 2, 3, \ldots,$$

where now the condition $\check{f}_0^z = \check{f}_1^z = 0$ encodes vanishing of f at both ± 1. The transform from vanishing basis coefficients to standard Chebyshev coefficients is

$$\mathcal{T} f = \begin{bmatrix} 1 & 0 & -1 & & \\ & 1 & 0 & -1 & \\ & & 1 & 0 & \ddots \\ & & & 1 & \ddots \\ & & & & \ddots \end{bmatrix} \begin{bmatrix} \check{f}_0^{z} \\ \check{f}_1^{z} \\ \check{f}_2^{z} \\ \vdots \end{bmatrix};$$

thus we can compute the vanishing basis coefficients using

$$\mathscr{T}^{\mathrm{z}} \triangleq \begin{bmatrix} 1 & 0 & 1 & 0 & 1 & \cdots \\ & 1 & 0 & 1 & 0 & \ddots \\ & & 1 & 0 & 1 & \ddots \\ & & & \ddots & \ddots & \ddots \end{bmatrix} \mathscr{T}.$$

Transforming the basis reduces the rate of decay in the coefficients slightly (Proposition A.22).

Proposition 4.10. *If $\mathscr{T} f \in \ell^{2,1}$, then $\mathscr{T}^{\mathrm{z}} f \in \ell^1$. If, in addition, $\mathscr{T} f \in \ell_{\mathrm{z}}^{2,1}$, then $\check{f}_0^{\mathrm{z}} = \check{f}_1^{\mathrm{z}} = 0$.*

We finally note that we can view the difference between a function and its discrete Chebyshev series as living in a vanishing space.

Proposition 4.11. *If $\mathscr{T} f \in \ell^{2,1}$, then*

$$\mathscr{T} f - \mathrm{T}_n f(\boldsymbol{x}^n) \in \ell_{\mathrm{z}}^{2,1}.$$

Furthermore,

$$\|\mathscr{T} f - \mathrm{T}_n f(\boldsymbol{x}^n)\|_{\ell^{2,1}} \to 0 \quad as \quad n \to \infty.$$

Vanishing spaces on rays: For bounded Möbius transformations M, the behaviors of $T_k^{\mathrm{z}} \circ M^{-1}$ and $T_k^{\pm\mathrm{z}} \circ M^{-1}$ are clear. Now consider the Möbius transformation from $\mathbb{I}$ to the ray $[a, \mathrm{e}^{\mathrm{i}\varphi}\infty)$:

$$M(x) = a + L\mathrm{e}^{\mathrm{i}\varphi} \frac{x+1}{1-x} \qquad \text{and} \qquad M^{-1}(y) = \frac{a + L\mathrm{e}^{\mathrm{i}\varphi} - y}{a - L\mathrm{e}^{\mathrm{i}\varphi} - y}.$$

Then $T_k^{+\mathrm{z}} \circ M^{-1}$ decays at ∞, uniformly in k. This follows from

$$T_k^{+\mathrm{z}}(M^{-1}(y)) = \frac{\zeta^{k-1} - \zeta^{-k}}{2}(\zeta - 1)$$

for $\zeta = J_\downarrow^{-1}(M^{-1}(y))$ on the unit circle.

Chapter 5

Numerical Computation of Cauchy Transforms

Consider the numerical calculation of the Cauchy transform

$$\mathscr{C}_\Gamma f(z) = \frac{1}{2\pi \mathrm{i}} \int_\Gamma \frac{f(s)}{s-z} \mathrm{d}s = \int_\Gamma \frac{f(s)}{s-z} \,\mathrm{đ}s.$$

The Cauchy transform of an integrable function is analytic off Γ, and the smoothness of its left and right limits are inherited from the smoothness of f, as established in Section 2.2.2. Our aim is to have a scheme that is uniformly accurate for all z: we want to be able to reliably and accurately approximate the Cauchy transform, including the limits as z approaches Γ. In this chapter we use the term "transform" as opposed to "operator" or "integral" because we concentrate on how $\mathscr{C}_\Gamma$ transforms bases, not its mapping properties on function spaces or its properties as an integral.

Our approach is to determine exact expressions for the Cauchy transforms of the canonical bases introduced in the preceding chapter. This immediately leads to an approximation of the Cauchy transform: after approximating f, defined on Γ, by

$$f(s) \approx f_m(s) = \sum_{k=0}^{m-1} f_k^m \psi_k(s),$$

where ψ_k is a prescribed basis,[34] we obtain an approximation to the Cauchy transform via

$$\mathscr{C}_\Gamma f(z) \approx \mathscr{C}_\Gamma f_m(z) = \sum_{k=0}^{m-1} f_k^m \mathscr{C}_\Gamma \psi_k(z).$$

In the next section, we establish conditions on $\mathscr{C}_\Gamma \psi_k(z)$ to ensure that this approximation of the Cauchy transform converges to the true Cauchy transform and at the same rate as f_m converges to f.

The fundamental tool that we exploit to determine exact formulae for the Cauchy transform is Problem 2.3.1: we look for functions analytic off Γ that satisfy

$$\phi_+(s) - \phi_-(s) = \psi_k(s) \qquad \text{and} \qquad \phi(\infty) = 0.$$

As discussed in Problem 2.3.1, the solution of this problem is unique if we impose local integrability and that the solution decays at infinity, and this solution must be the Cauchy

[34]The Fourier-based expansions in the preceding chapter used both positive and negative indices but can be trivially reordered to the preceding form, e.g., $\psi_0(\zeta), \psi_1(\zeta), \psi_2(\zeta), \psi_3(\zeta), \ldots = 1, \zeta^{-1}, \zeta, \zeta^{-2}, \ldots$.

transform of ψ_k by Lemma 2.7. We refer to this fact as Plemelj's lemma, in addition to Lemma 2.7. Bases based on Laurent series are used for the unit circle (Section 5.2) and the real line (Section 5.2.1). A weighted Chebyshev basis is used for functions on the unit interval with square root singularities (Section 5.4.1). An unweighted Chebyshev basis is used for smooth functions on the unit interval (Section 5.6), Möbius transformations of the unit interval (Section 5.7.3), and polynomial maps of the unit interval (Section 5.7.4).

For smooth functions, Cauchy transforms can have logarithmic singularities at endpoints of the integration contour. These singularities may be canceled due to special properties, such as the zero-sum condition seen in Lemma 2.51. Thus it is important to capture the behavior of the Cauchy transform at the singularity, which we accomplish via a *finite-part Cauchy transform* introduced in Section 5.7.2. We use this to describe the behavior of the Cauchy transform at nonsmooth points of Γ in Section 5.7.5.

Finally, an important companion to the Cauchy transform is the *Hilbert transform*:

$$\mathcal{H}_\Gamma f(s^*) = \frac{1}{\pi} \fint_\Gamma \frac{f(s)}{s-s^*} \mathrm{d}s.$$

However, Plemelj's lemma ensures that it is sufficient to calculate the left and right limits of the Cauchy transform $\mathcal{C}_\Gamma^\pm$:

$$\mathcal{H}_\Gamma f(s) = \mathrm{i}\left[\mathcal{C}_\Gamma^+ f(s) + \mathcal{C}_\Gamma^- f(s)\right].$$

5.1 ▪ Convergence of approximation of Cauchy transforms

Let Γ be a contour, f a function on that contour, and ψ_k a basis such that, for $s \in \Gamma$,

$$f(s) = \sum_{k=0}^{\infty} f_k \psi_k(s) \qquad \text{and} \qquad \sup_{k,s} |\psi_k(s)| < \infty.$$

In this section we investigate the validity of replacing $\mathcal{C}_\Gamma f(z)$ by the Cauchy transform applied to the basis

$$\mathcal{C}_\Gamma f(z) \stackrel{?}{=} \sum_{k=0}^{\infty} f_k \mathcal{C}_\Gamma \psi_k(z).$$

To positively answer this question, including for z near Γ, we require the following:

1. an exact formula for $\mathcal{C}_\Gamma \psi_k(z)$ and
2. a bound on $|\mathcal{C}_\Gamma \psi_k(z)|$ that is uniform in z and k.

5.1.1 ▪ Pointwise Cauchy transform for bounded contours

Suppose Γ is bounded and the coefficients $f = [f_0, f_1, \ldots]^\top$ are in ℓ^1, i.e., absolutely summable. Absolute summability combined with uniform boundedness of the basis ensures that, for z off Γ, we can exchange summation and integration.

Proposition 5.1. *Suppose Γ is bounded, $\|\psi_k\|_{L^\infty(\Gamma)}$ is uniformly bounded with respect to k, and $f \in \ell^1$. Then, for $z \in \mathbb{C} \setminus \Gamma$,*

$$\mathcal{C}_\Gamma f(z) = \sum_{k=0}^{\infty} f_k \mathcal{C}_\Gamma \psi_k(z) \qquad \textit{for} \qquad f(s) = \sum_{k=0}^{\infty} f_k \psi_k(s).$$

Furthermore, it follows that

$$|\mathscr{C}_\Gamma f(z)| \leq \|f\|_{\ell^1} \sup_k |\mathscr{C}_\Gamma \psi_k(z)|.$$

Thus the action of the Cauchy transform on the basis immediately gives the Cauchy transform of a function expanded in the basis. The second part of the proposition allows us to relate the error in approximating coefficients to approximating Cauchy transforms. In the preceding chapter, we used the DFT and DCT to calculate approximate coefficients

$$\boldsymbol{f}_m = \left[f_0^m, \ldots, f_{m-1}^m\right]^\top \approx \left[f_0, \ldots, f_{m-1}\right]^\top$$

so that $\|\boldsymbol{f}_m - \boldsymbol{f}\|_{\ell^1} \to 0$ as $m \to \infty$. The approximation for Cauchy transforms converges at exactly the same rate as approximating the related coefficients:

$$\left|\mathscr{C}_\Gamma f(z) - \sum_{k=0}^{m-1} f_k^m \mathscr{C}_\Gamma \psi_k(z)\right| \leq \|\boldsymbol{f} - \boldsymbol{f}_m\|_{\ell^1} \sup_k |\mathscr{C}_\Gamma \psi_k(z)|.$$

Therefore, approximating Cauchy transforms for z away from Γ requires only an exact expression for $\mathscr{C}_\Gamma \psi_k(z)$ to construct a convergent numerical scheme.

5.1.2 ▪ Left and right limits of the Cauchy transform

A subsequent question is whether we can use the same expansion when taking the left/right limits of the Cauchy transform:

$$\mathscr{C}_\Gamma^\pm f(s) \stackrel{?}{=} \sum_{k=0}^{\infty} f_k \mathscr{C}_\Gamma^\pm \psi_k(s).$$

We cannot appeal directly to Lemma 2.6: this would require a uniform α-Hölder continuous condition that holds for *all* k, a property that simply is not satisfied by our bases. Instead, we use the fact that we calculate $\mathscr{C}_\Gamma \psi_k(z)$ explicitly, and for each basis under consideration we can directly bound the resulting formulae uniformly in k, for z not near an endpoint. Thus continuity ensures that we can pass to the limit $z \to s$.

Proposition 5.2. *Suppose Γ is bounded, $\|\psi_k\|_{L^\infty(\Gamma)}$ is uniformly bounded with respect to k, and $s \in \Gamma$, and let B be a neighborhood of s restricted to the left/right of Γ; more precisely, $B = B(s, \epsilon) \cap \Omega_\pm$ for a completion of Γ (see Definition 2.1). Suppose, further, that $\mathscr{C}_\Gamma \psi_k(z)$ is continuous and uniformly bounded in B by a k-independent constant and $\boldsymbol{f} \in \ell^1$. Then the left/right limit exists for all $s \in B \cap \Gamma$ and satisfies*

$$\mathscr{C}_\Gamma^\pm f(s) = \sum_{k=0}^{\infty} f_k \mathscr{C}_\Gamma^\pm \psi_k(s) \qquad \textit{for} \qquad f(s) = \sum_{k=0}^{\infty} f_k \psi_k(s).$$

Furthermore, it follows that

$$|\mathscr{C}_\Gamma^\pm f(s)| \leq \|\boldsymbol{f}\|_{\ell^1} \sup_k |\mathscr{C}^\pm \psi_k(s)|.$$

5.1.3 ▪ Unbounded contours

When Γ is unbounded, there are technical difficulties with the canonical bases of mapped Chebyshev and Laurent polynomials not decaying at infinity. However, we can exploit the uniform decay of the vanishing bases introduced in Section 4.4, to successfully exchange integration and summations.

Proposition 5.3. *Suppose $|\psi_k(s)| \leq \frac{C}{s}$ holds uniformly in k for all $|s| > s_0$ and $|\psi_k(s)|$ is bounded by a k- and s-independent constant. If $f \in \ell^1$, then, for all $z \in \mathbb{C} \setminus \Gamma$,*

$$\mathcal{C}_\Gamma f(z) = \sum_{k=0}^{\infty} f_k \mathcal{C}_\Gamma \psi_k(z) \quad \textit{for} \quad f(s) = \sum_{k=0}^{\infty} f_k \psi_k(s).$$

Furthermore, it follows that

$$|\mathcal{C}_\Gamma f(z)| \leq \|f\|_{\ell^1} \sup_k |\mathcal{C}_\Gamma \psi_k(z)|.$$

5.2 ▪ The unit circle

The canonical basis on the unit circle is $\{\zeta^k\}$, and to employ our numerical approach to Cauchy transforms we need only to determine $\mathcal{C}[\diamond^k]$, $\mathcal{C} \triangleq \mathcal{C}_{\mathbb{U}}$. We use Plemelj's lemma: we want to find a function ϕ that

1. is analytic off the unit circle,
2. decays at infinity, $\phi(\infty) = 0$, and
3. has continuous limits $\phi_\pm(\zeta)$ that satisfy the jump $\phi_+(\zeta) - \phi_-(\zeta) = \zeta^k$.

In this case the answer follows from residue calculations:

$$\mathcal{C}[\diamond^k](z) = \begin{cases} z^k & \text{if } k \geq 0 \text{ and } |z| < 1, \\ -z^k & \text{if } k < 0 \text{ and } |z| > 1, \\ 0 & \text{otherwise.} \end{cases}$$

We also have $\left|\mathcal{C}[\diamond^k](z)\right| \leq 1$ for all z, including on the unit circle itself, which ensures convergence to the true Cauchy transform uniformly in z.

5.2.1 ▪ The real line

For the real line we are using the map from (4.1),

$$M(\zeta) = a - \frac{\mathrm{i}}{L} \frac{\zeta - 1}{1 + \zeta},$$

to obtain the mapped Laurent expansion

$$f(x) = \sum_{k=-\infty}^{\infty} \hat{f}_k M^{-1}(x)^k.$$

To determine $\mathcal{C}[M^{-1}(\diamond)^k](z)$, we use the following lemma, which states that Cauchy transforms are invariant under Möbius transformations.

Lemma 5.4. *Let $M : \Omega \to \Gamma$ be a Möbius transformation. Provided that f and $f \circ M$ are $H^1(\Omega)$ and $H^1(\Gamma)$, respectively, we have for $z \notin \Gamma$, including boundary values,*

$$\mathcal{C}_\Gamma f(z) = \mathcal{C}_\Omega[f \circ M](M^{-1}(z)) - \mathcal{C}_\Omega[f \circ M](M^{-1}(\infty)).$$

Proof. Define

$$\phi(z) = \mathcal{C}_\Omega[f \circ M](M^{-1}(z)) - \mathcal{C}_\Omega[f \circ M](M^{-1}(\infty)).$$

Then

$$\phi_+(s) - \phi_-(s) = \mathcal{C}_\Omega^+[f \circ M](M^{-1}(s)) - \mathcal{C}_\Omega^-[f \circ M](M^{-1}(s)) = f(M(M^{-1}(s))) = f(s).$$

Also, it decays at infinity. Thus the lemma holds; see the discussion in Section 2.3.1. □

We can employ this lemma on a basis that vanishes at ∞ and the fact that M^{-1} maps the upper-half plane to inside the unit circle and the lower-half plane to outside the unit circle to obtain

$$\begin{aligned}\mathcal{C}_\mathbb{R}[M^{-1}(\diamond)^k - (-1)^k](z) &= \mathcal{C}[\diamond^k - (-1)^k](M^{-1}(z)) - \mathcal{C}[\diamond^k - (-1)^k](-1) \\ &= \begin{cases} M^{-1}(z)^k - (-1)^k & \text{if } \operatorname{Im} z > 0 \text{ and } k > 0, \\ (-1)^k - M^{-1}(z)^k & \text{if } \operatorname{Im} z < 0 \text{ and } k < 0, \\ 0 & \text{otherwise.} \end{cases}\end{aligned}$$

Using this expression, we derive the formula for the vanishing basis:

$$\mathcal{C}_\mathbb{R}[M^{-1}(\diamond)^k + M^{-1}(\diamond)^{k-1}](z) = \begin{cases} M^{-1}(z)^k + M^{-1}(z)^{k-1} & \text{if } \operatorname{Im} z > 0 \text{ and } k > 0, \\ -M^{-1}(z)^{k-1} - M^{-1}(z)^k & \text{if } \operatorname{Im} z < 0 \text{ and } k \le 0, \\ 0 & \text{otherwise.} \end{cases}$$

We bound this uniformly in k and z by

$$\left| \mathcal{C}_\mathbb{R}[M^{-1}(\diamond)^k + M^{-1}(\diamond)^{k-1}](z) \right| \le 2.$$

Thus we successfully fit into the criteria of Proposition 5.3.

In practice, it is more convenient to work with the unmodified basis. If we understand the Cauchy transform as a principle value integral, we have

$$\mathcal{C}_\mathbb{R} 1(z) = ⨍_{-\infty}^{\infty} \frac{1}{x - z} \mathrm{d}x = \lim_{R \to \infty} \int_{-R}^{R} \frac{1}{x - z} \mathrm{d}x = \frac{\operatorname{sgn} \operatorname{Im} z}{2}.$$

Thus we can infer the Cauchy transform of the mapped Laurent basis without decay imposed as

$$\begin{aligned}\mathcal{C}_\mathbb{R}[M^{-1}(\diamond)^k](z) &= \mathcal{C}_\mathbb{R}[M^{-1}(\diamond)^k - (-1)^k](z) + (-1)^k \frac{\operatorname{sgn} \operatorname{Im} z}{2} \\ &= \begin{cases} M^{-1}(z)^k - \frac{(-1)^k}{2} & \text{if } \operatorname{Im} z > 0 \text{ and } k \ge 0, \\ \frac{(-1)^k}{2} & \text{if } \operatorname{Im} z > 0 \text{ and } k < 0, \\ \frac{(-1)^k}{2} - M^{-1}(z)^k & \text{if } \operatorname{Im} z < 0 \text{ and } k < 0, \\ -\frac{(-1)^k}{2} & \text{if } \operatorname{Im} z < 0 \text{ and } k > 0, \\ 0 & \text{otherwise.} \end{cases}\end{aligned}$$

We note that, provided we have $\mathcal{F} f \in \ell^{2,1}_{-z}$, using this expression with mapped Laurent coefficients is guaranteed to give the same result as with the mapped vanishing basis Laurent coefficients. This carries over to the discrete Laurent series setting: if $f(\infty)=0$, the discrete Laurent series also vanishes at infinity.

5.3 ▪ Case study: Computing the error function

In this section we discuss the numerical solution of Problem 1.1.1 discussed in Section 1.1. We use simple Möbius transformations to map the imaginary axis to the unit circle and apply one fast Fourier transform to solve the problem. Let $f(z) = 2e^{z^2}$ and consider the transformation

$$T(\theta) = \mathrm{i}M(\mathrm{e}^{\mathrm{i}\theta}), \quad T : \mathbb{T} \to \mathrm{i}\mathbb{R}.$$

Then define $F = f \circ T : \mathbb{T} \to \mathbb{C}$. Now, we may expand F in a Fourier series with the discrete Fourier transform that converges superalgebraically by Theorem 4.6:

$$F(\theta) \approx \sum_{k=\alpha_m}^{\beta_m} \hat{f}_k^m \mathrm{e}^{\mathrm{i}k\theta}.$$

Thus

$$f(z) \approx \left(\sum_{k=\alpha_m}^{-1} + \sum_{k=1}^{\beta_m} \right) \hat{f}_k^m \left[M^{-1}(-\mathrm{i}z)^k - (-1)^k \right],$$

where we used $\sum_{k=\alpha_m}^{\beta_m} (-)^k \hat{f}_k^m = f(\infty) = 0$.

It is then clear that by the results of Section 5.2.1, with Lemma 5.4 to account for the rotation of $\mathrm{e}^{\mathrm{i}\pi/2}$ which takes the real axis to the imaginary axis, we have

$$\psi(z) = \mathcal{C}_{\mathrm{i}\mathbb{R}} f(z) \approx \psi_m(z) \triangleq \begin{cases} -\displaystyle\sum_{k=1}^{\beta_m} \hat{f}_k^m \left[M^{-1}(-\mathrm{i}z)^k - (-1)^k \right] & \text{if } \operatorname{Re} z < 0, \\ \displaystyle\sum_{k=\alpha_m}^{-1} \hat{f}_k^m \left[M^{-1}(-\mathrm{i}z)^k - (-1)^k \right] & \text{if } \operatorname{Re} z > 0. \end{cases}$$

This leads to an approximation of erfc:

$$\operatorname{erfc} z \approx \begin{cases} 2 + \mathrm{e}^{-z^2} \displaystyle\sum_{k=1}^{\beta_m} \hat{f}_k^m \left[M^{-1}(-\mathrm{i}z)^k - (-1)^k \right] & \text{if } \operatorname{Re} z < 0, \\ -\mathrm{e}^{-z^2} \displaystyle\sum_{k=\alpha_m}^{-1} \hat{f}_k^m \left[M^{-1}(-\mathrm{i}z)^k - (-1)^k \right] & \text{if } \operatorname{Re} z > 0. \end{cases}$$

This approximation converges uniformly in the entire complex plane at the same rate that the approximation of $F(\theta)$ converges uniformly: superalgebraically. Asymptotic approximations of $\psi(z)$ are easily obtained [91]. Let $\tilde{\psi}_n(z)$ be the asymptotic approximation of order n. We examine the difference $\tilde{\psi}_n(z) - \psi_m(z)$ with $n = 8$ and $m = 256$ in Figure 5.1.

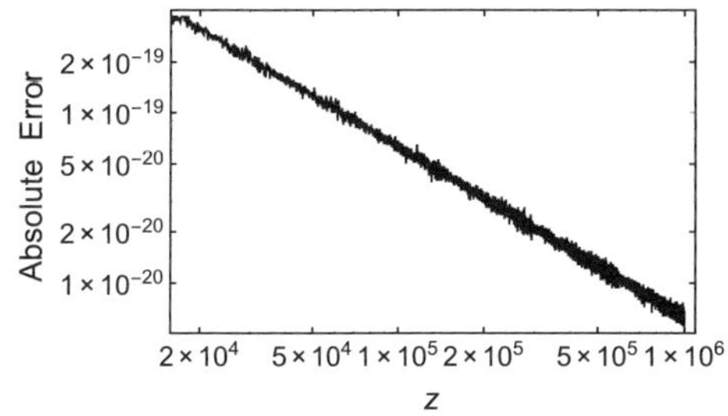

Figure 5.1. *The absolute difference* $|\tilde{\psi}_n(z)-\psi_m(z)|$ *with* $n=8$ *and* $m=256$ *plotted versus* z *for* $z\in\mathbb{R}^+$. *It is clear that the absolute error decreases for large* z.

5.4 ▪ The unit interval and square root singularities

To handle functions on the real line, we mapped to the circle. We follow a similar approach for functions on $\mathbb{I} \triangleq [-1,1]$, but now we use the Joukowsky map, $J(\zeta)=\frac{1}{2}(\zeta+\zeta^{-1})$. To determine the Cauchy transform using this map, we develop an analogue to Lemma 5.4. Unlike Möbius transforms, we have two inverses of our map. Choosing one of the inverses allows us to relate the Cauchy transforms.

Lemma 5.5. *Suppose f is α-Hölder continuous. Then*

$$\mathcal{C}_{\mathbb{I}}f(z)=-\mathcal{C}[f(J(\diamond))\operatorname{sgn}\arg\diamond](J_+^{-1}(z)),$$

where $J_+^{-1}(z)=z-\sqrt{z-1}\sqrt{z+1}$ *is one of the right inverses of the Joukowsky map.*

Proof. Define

$$\Phi(z)=-\mathcal{C}[f(J(\diamond))\operatorname{sgn}\arg\diamond](z).$$

We first want to show that

$$\psi(z)=\frac{\Phi(J_+^{-1}(z))+\Phi(J_-^{-1}(z))}{2}$$

is the Cauchy transform of f, where $J_-^{-1}(z)=z+\sqrt{z-1}\sqrt{z+1}$ is the other right inverse of J. It satisfies the correct jump on $(-1,1)$ (recall the relationship between the four inverses from Proposition B.12):

$$\begin{aligned}\psi_+(x)-\psi_-(x) &= \frac{\Phi_+(J_\downarrow^{-1}(x))+\Phi_-(J_\uparrow^{-1}(x))}{2}-\frac{\Phi_+(J_\uparrow^{-1}(x))+\Phi_-(J_\downarrow^{-1}(x))}{2}\\ &=\frac{f(J(J_\downarrow^{-1}(x)))-f(J(J_\uparrow^{-1}(x)))}{2}=f(x).\end{aligned}$$

Furthermore,

$$\psi(\infty)=\frac{\Phi(0)-\Phi(\infty)}{2}=0,$$

where $\Phi(0)=0$ follows from symmetry:

$$\Phi(0)=-\int_{\mathbb{U}}\frac{f(J(\zeta))\operatorname{sgn}\arg\zeta}{\zeta}\mathrm{d}\zeta=\frac{1}{2\pi}\left[\int_{-\pi}^{0}f(\cos\theta)\mathrm{d}\theta-\int_{0}^{\pi}f(\cos\theta)\mathrm{d}\theta\right]=0.$$

Thus Plemelj's lemma ensures that $\psi(z)$ must be $\mathcal{C}_{\mathbb{I}}f(z)$.

We now need only show that $\Phi(J_-^{-1}(z)) = \Phi(J_+^{-1}(z))$, or equivalently $\Phi(z) = \Phi(1/z)$. This also follows from Plemelj's lemma: $\Phi(1/z)$ decays at infinity and satisfies the jump

$$\begin{aligned}[\Phi(1/\zeta)]^+ - [\Phi(1/\zeta)]^- &= \Phi_-(1/\zeta) - \Phi_+(1/\zeta) = f(J(1/\zeta))\operatorname{sgn}\arg\frac{1}{\zeta}\\ &= -f(J(\zeta))\operatorname{sgn}\arg\zeta. \qquad \square\end{aligned}$$

5.4.1 ▪ Square root singularities

We use this lemma to compute the Cauchy transform in a chosen basis. For smooth functions, the Chebyshev basis is the natural choice; however, this leads to significant complications. Thus we first consider functions with square root singularities:

$$f(x) = \sqrt{1-x^2}\sum_{k=0}^{\infty} u_k U_k(x),$$

where $U_k(x)$ are the Chebyshev U polynomials defined in Appendix B.2.5.

We determine the Cauchy transform of the basis $\sqrt{1-x^2}U_k(x)$ and its bound.

Lemma 5.6. *For $k \geq 0$ and $z \notin \mathbb{I}$,*

$$\mathcal{C}_{\mathbb{I}}[\sqrt{1-\diamond^2}U_k](z) = \frac{i}{2}J_+^{-1}(z)^{k+1},$$

which satisfies

$$\left|\mathcal{C}_{\mathbb{I}}[\sqrt{1-\diamond^2}U_k](z)\right| \leq \frac{1}{2}.$$

Proof. We begin by projecting to the unit circle, recalling the formula for $U_k(J(\zeta))$ from Proposition B.22:

$$\begin{aligned}-\sqrt{1-J(\zeta)^2}U_k(J(\zeta))\operatorname{sgn}\arg\zeta &= -\frac{1}{2}\operatorname{sgn}\arg\zeta\sqrt{-(\zeta-\zeta^{-1})^2}\frac{\zeta^{k+1}-\zeta^{-k-1}}{\zeta-\zeta^{-1}}\\ &= i\frac{\zeta^{k+1}-\zeta^{-k-1}}{2}.\end{aligned}$$

From the knowledge of the Cauchy transform on the unit circle, we immediately find

$$\Phi(z) = -\mathcal{C}[\sqrt{1-J(\diamond)^2}U_k(J(\diamond))\operatorname{sgn}\arg\diamond](z) = \frac{i}{2}\begin{cases} z^{k+1} & \text{if } |z|<1,\\ z^{-k-1} & \text{if } |z|>1.\end{cases}$$

The lemma thus follows from Lemma 5.5, and the uniform bound is trivial. □

We remark that the Hilbert transform has a particularly convenient form, mapping between weighted Chebyshev polynomials of the second and first kinds.

Corollary 5.7. *For $k \geq 0$ and $x \in (-1,1)$,*

$$\mathcal{H}_{\mathbb{I}}[\sqrt{1-\diamond^2}U_k](x) = -T_{k+1}(x).$$

Proof. We have

$$\begin{aligned}\mathcal{H}[\sqrt{1-\diamond^2}U_k](x) &= \mathrm{i}(\mathcal{C}^+ + \mathcal{C}^-)[\sqrt{1-\diamond^2}U_k](x)\\ &= -\frac{[J_+^{-1}(x)^{k+1}]^+ + [J_+^{-1}(x)^{k+1}]^-}{2} = -\frac{J_\downarrow^{-1}(x)^{k+1} + J_\uparrow^{-1}(x)^{k+1}}{2}\\ &= -T_{k+1}(x),\end{aligned}$$

by Definition B.13. □

5.4.2 ▪ Case study: Solving $\phi_+(x)+\phi_-(x)=f(x)$ and $\phi(\infty)=0$

We are now in a position to solve the following simple RH problem with the given tools.

Problem 5.4.1. *Find all $\phi(z)$ that satisfy the following:*

1. *$\phi(z)$ is analytic off $[a,b]$,*
2. *$\phi(z)$ has weaker than pole singularities throughout the complex plane,*
3. *$\phi(\infty)=0$, and*
4. *$\phi(z)$ has the jump*

$$\phi_+(x)+\phi_-(x)=f(x) \quad \textit{for} \quad x\in(a,b).$$

Note that the solution to this problem, as posed, is not unique: the function[35] $\frac{1}{\sqrt{z-a}\sqrt{z-b}}$ satisfies the first three conditions but has a homogeneous jump on $[a,b]$:

$$\frac{1}{\sqrt{x-a}\sqrt{x-b}^+}+\frac{1}{\sqrt{x-a}\sqrt{x-b}^-}=\frac{-\mathrm{i}}{\sqrt{x-a}\sqrt{b-x}}+\frac{\mathrm{i}}{\sqrt{x-a}\sqrt{b-x}}=0.$$

Thus if $\phi(z)$ solves the RH problem, then $\phi(z)+\frac{\alpha}{\sqrt{z-a}\sqrt{z-b}}$ also satisfies the same RH problem for any choice of α.

While this problem is usually reduced to a Cauchy transform of a function with a square root singularity (recall Section 2.3.2), we see that we can, in fact, solve it directly using the basis from Lemma 5.6.

Lemma 5.8. *Suppose $\mathcal{T}f\in\ell^1$. Then every solution to Problem 5.4.1 has the form*

$$\mathcal{P}_\alpha f\left(\frac{a+b}{2}+\frac{b-a}{2}z\right)\triangleq\sum_{k=0}^{\infty}\frac{\check{f}_k}{2}J_+^{-1}(z)^k-\frac{\check{f}_0}{2}\frac{z}{\sqrt{z-1}\sqrt{z+1}}+\frac{\alpha}{\sqrt{z-1}\sqrt{z+1}}.$$

Proof. Without loss of generality, assume that $[a,b]=[-1,1]$. The fact that (directly verifiable by the substitution $x=\cos\theta$)

$$T_k(x)=\frac{J_\downarrow^{-1}(x)^k+J_\downarrow^{-1}(x)^{-k}}{2},$$

[35] We use $\sqrt{z-a}$ to denote the principal branch. It then follows that the product $\sqrt{z-a}\sqrt{a-b}$ has an analytic continuation to $\mathbb{C}\setminus[a,b]$.

where

$$J_{\downarrow}^{-1}(x) = x - \mathrm{i}\sqrt{1-x}\sqrt{1+x} = \lim_{\epsilon \downarrow 0} J_{+}^{-1}(x + \mathrm{i}\epsilon),$$

implies that

$$\phi(z) \triangleq \sum_{k=0}^{\infty} \frac{\check{f}_k}{2} J_{+}^{-1}(z)^k - \frac{\check{f}_0}{2} \frac{z}{\sqrt{z-1}\sqrt{z+1}}$$

satisfies the correct jumps (using absolute convergence of the series to interchange limits).

Now suppose $\tilde{\phi}$ also satisfies

$$\tilde{\phi}_+(x) + \tilde{\phi}_-(x) = f(x) \quad \text{for} \quad x \in (-1,1) \qquad \text{and} \qquad \tilde{\phi}(\infty) = 0$$

with weaker than pole singularities at ± 1. Then $\kappa = \phi - \tilde{\phi}$ satisfies

$$\kappa_+(x) + \kappa_-(x) = 0 \quad \text{for} \quad x \in (-1,1).$$

Let $\alpha(z) = \kappa(z)\sqrt{z-1}\sqrt{z+1}$. Since ϕ and $\tilde{\phi}$ are analytic and decay at ∞, we have that α is bounded at ∞. We also have $\alpha_+(x) = \alpha_-(x)$ on $(-1,1)$, i.e., α is continuous (and thus analytic) on $(-1,1)$. Because κ has weaker than pole singularities at ± 1, α also has weaker than pole singularities at ± 1. Since these singularities are isolated, it follows that α is analytic at ± 1 and hence analytic everywhere: α is constant. This shows that κ is a constant multiple of $(\sqrt{z-1}\sqrt{z+1})^{-1}$, completing the proof. □

From this lemma, we can guarantee uniqueness of the solution when additional boundedness conditions are specified.

Corollary 5.9. *Suppose* $\mathcal{T}f \in \ell^1$. *Then* $\mathcal{P}_{\check{f}_0/2}$ *is the unique solution to Problem* 5.4.1 *which is bounded at* a *and* $\mathcal{P}_{-\check{f}_0/2}$ *is the unique solution which is bounded at* b. *In the special case where* $\check{f}_0 = 0$, $\mathcal{P}_0$ *is the unique solution that is bounded at both* a *and* b.

This result should be compared with the solution of Problem 2.3.3 and Example 2.15.

5.4.3 ▪ Inverse square root singularities

We now consider the case where the function has inverse square root singularities. Using the vanishing basis leads to a simple expression.

Lemma 5.10. *For* $k \geq 2$ *and* $z \notin \mathbb{I}$, *including boundary values,*

$$\begin{aligned}
\mathcal{C}_{\mathbb{I}}\left[\frac{1}{\sqrt{1-\diamond^2}}\right](z) &= \frac{\mathrm{i}}{2\sqrt{z-1}\sqrt{z+1}}, \\
\mathcal{C}_{\mathbb{I}}\left[\frac{\diamond}{\sqrt{1-\diamond^2}}\right](z) &= \frac{\mathrm{i}z}{2\sqrt{z-1}\sqrt{z+1}} - \frac{\mathrm{i}}{2}, \\
\mathcal{C}_{\mathbb{I}}\left[\frac{T_k^{\mathrm{z}}}{\sqrt{1-\diamond^2}}\right](z) &= -\mathrm{i}J_{+}^{-1}(z)^{k-1}.
\end{aligned}$$

Proof. For $k \geq 2$, we have

$$\begin{aligned}\frac{T_k^z(J(\zeta))\mathrm{sgn}\arg\zeta}{\sqrt{1-J(\zeta)^2}} &= \frac{\zeta^k+\zeta^{-k}-\zeta^{k-2}-\zeta^{2-k}}{\sqrt{4-(\zeta+\zeta^{-1})^2}\mathrm{sgn}\arg\zeta} \\ &= \mathrm{i}\frac{\zeta^k+\zeta^{-k}-\zeta^{k-2}-\zeta^{2-k}}{\zeta-\zeta^{-1}} \\ &= \mathrm{i}(\zeta^{k-1}-\zeta^{1-k}).\end{aligned}$$

The lemma then follows from Lemma 5.5, with the first two formulae following in a similar manner. □

We again have a particularly convenient form for the Cauchy transform, which follows from Corollary 5.7.

Corollary 5.11. *For $k \geq 2$ and $x \in (-1,1)$,*

$$\mathcal{H}_{\mathbb{I}}\left[\frac{1}{\sqrt{1-\diamond^2}}\right](x)=0, \quad \mathcal{H}_{\mathbb{I}}\left[\frac{\diamond}{\sqrt{1-\diamond^2}}\right](x)=1, \quad \textit{and} \quad \mathcal{H}_{\mathbb{I}}\left[\frac{T_k^z}{\sqrt{1-\diamond^2}}\right](x)=2T_{k-1}(x).$$

5.4.4 ▪ Indefinite integrals

In the preceding formulae, we can directly compute the indefinite integrals

$$\begin{aligned}\int^z \frac{1}{\sqrt{s-1}\sqrt{s+1}}\mathrm{d}s &= -\log J_+^{-1}(z), \\ \int^z \frac{s}{\sqrt{s-1}\sqrt{s+1}}\mathrm{d}s &= \sqrt{z-1}\sqrt{z+1}, \\ 2\int^z J_+^{-1}(s)\mathrm{d}s &= \frac{J_+^{-1}(z)^2}{2} - \log J_+^{-1}(z), \\ 2\int^z J_+^{-1}(s)^k\mathrm{d}s &= \frac{J_+^{-1}(z)^{k+1}}{k+1} - \frac{J_+^{-1}(z)^{k-1}}{k-1}\end{aligned}$$

for $k \geq 2$.

5.5 ▪ Case study: Computing elliptic integrals

We now turn our attention to calculating the elliptic integral

$$g(z;a) = \int_0^z \psi(s)\mathrm{d}s = \int_0^z \frac{1}{\sqrt{1-s^2}\sqrt{a^2-s^2}}\mathrm{d}s,$$

where we assume that $a > 1$. We take the following approach: (1) re-expand $\psi(z)$ as a Cauchy transform, and (2) indefinitely integrate the resulting formula. We then confirm that this is the only solution to Problem 1.2.1.

On the branch cut $(1,a)$, the integrand $\psi(z)$ satisfies the jump

$$\begin{aligned}\psi_+(x)-\psi_-(x) &= \frac{1}{\sqrt{1+x}\sqrt{a+x}\sqrt{x-1}\sqrt{x-a}^+} - \frac{1}{\sqrt{1+x}\sqrt{a+x}\sqrt{x-1}\sqrt{x-a}^-} \\ &= \frac{2\mathrm{i}}{\sqrt{1+x}\sqrt{a+x}\sqrt{x-1}\sqrt{a-x}}.\end{aligned}$$

On $(-a,-1)$, we have

$$\psi_+(x)-\psi_-(x)=\frac{-2\mathrm{i}}{\sqrt{-1-x}\sqrt{a+x}\sqrt{1-x}\sqrt{a-x}}.$$

These jumps, combined with the decay at infinity, inform us that

$$\begin{aligned}\psi(z) &= -\mathcal{C}_{[-a,-1]}\left[\frac{2\mathrm{i}}{\sqrt{1+\diamond}\sqrt{a+\diamond}\sqrt{\diamond-1}\sqrt{a-\diamond}}\right](z)\\ &\quad -\mathcal{C}_{[a,1]}\left[\frac{2\mathrm{i}}{\sqrt{-1-\diamond}\sqrt{a+\diamond}\sqrt{1-\diamond}\sqrt{a-\diamond}}\right](z),\end{aligned}$$

where the reversion orientation of $(a,1)$ is chosen for symmetry.

We now expand the nonsingular portion of the integrand of the Cauchy transform in terms of the vanishing series

$$\frac{-2\mathrm{i}}{\sqrt{1+M(x)}\sqrt{a+M(x)}}=\sum_{k=0}^{\infty} f_k T_k^{\mathrm{z}}(x),$$

where $M(x)=\frac{1+a+x-ax}{2}$, $M:\mathbb{I}\to[a,1]$. Therefore, we have

$$\begin{aligned}\psi(z) &= -\sum_{k=0}^{\infty} f_k\left\{\mathcal{C}_{[a,1]}\left[\frac{T_k^{\mathrm{z}}(M^{-1}(\diamond))}{\sqrt{\diamond-1}\sqrt{a-\diamond}}\right](z)+\mathcal{C}_{[-a,-1]}\left[\frac{T_k^{\mathrm{z}}(M^{-1}(-\diamond))}{\sqrt{-1-\diamond}\sqrt{a+\diamond}}\right](z)\right\}\\ &= \frac{2}{a-1}\sum_{k=0}^{\infty} f_k\left\{\mathcal{C}_{\mathbb{I}}\left[\frac{T_k^{\mathrm{z}}}{\sqrt{1-\diamond^2}}\right](M^{-1}(z))+\mathcal{C}_{\mathbb{I}}\left[\frac{T_k^{\mathrm{z}}}{\sqrt{1-\diamond^2}}\right](M^{-1}(-z))\right\}\end{aligned}$$

using Lemma 5.4. We integrate this formula explicitly, using Section 5.4.4, giving

$$\begin{aligned}\frac{2}{\mathrm{i}}g(z) &= f_0\left[\log(J_+^{-1}(M(z)))-\log(J_+^{-1}(M(-z)))\right]+f_1\left[J_+^{-1}(M^{-1}(z))-J_+^{-1}(M^{-1}(-z))\right]\\ &+f_2\left[\frac{J_+^{-1}(M^{-1}(z))^2}{2}-\log J_+^{-1}(M^{-1}(z))-\frac{J_+^{-1}(M^{-1}(-z))^2}{2}+\log J_+^{-1}(M^{-1}(-z))\right]\\ &+\sum_{k=3}^{\infty} f_k\left[\frac{J_+^{-1}(M^{-1}(z))^k-J_+^{-1}(M^{-1}(-z))^k}{k}+\frac{J_+^{-1}(M^{-1}(-z))^{k-2}-J_+^{-1}(M^{-1}(z))^{k-2}}{k-2}\right].\end{aligned}$$

This expression is built out of elementary functions (logarithms and square roots); hence the only computational task is calculating the coefficients f_k. This can be accomplished readily via the DCT.

5.6 ▪ Smooth functions on the unit interval

The functions we are most interested in do not have square root singularities: they are smooth on $[-1,1]$. Thus, while the formulae for functions with square root singularities have a simple form, the numerical approximation obtained via the DCT may not converge unless $f(x)(1-x^2)^{1/2}$ is sufficiently smooth. Even if we knew the coefficients of $\mathcal{T}g$ exactly, the convergence rate of the approximation would be extremely slow. To overcome this, we determine the Cauchy transform for our standard canonical basis on the unit interval: $T_k(x)$.

Remark 5.6.1. *An alternative approach is to use Legendre polynomials. The benefit of the Chebyshev series is the fast transform, though a recently developed fast Chebyshev–Legendre transform* [62] *may diminish this benefit.*

5.6.1 ▪ Cauchy transform of Chebyshev polynomials

The building block for our approach to Cauchy transforms is determining a closed form expression for $\mathcal{C}_{\mathbb{I}} T_k$. To tackle this question we again map to the unit circle using Lemma 5.5.

Proposition 5.12. *For* $k \geq 0$,

$$\mathcal{C}_{\mathbb{I}} T_k(z) = -\mathcal{C}[T_k(J(\diamond))\text{sgn arg}\,\diamond](J_+^{-1}(z))$$

and

$$\mathcal{C}[T_k(J(\diamond))\text{sgn arg}\,\diamond](z) = \frac{\mathcal{C}[\diamond^k \text{sgn arg}\,\diamond](z) + \mathcal{C}[\diamond^{-k}\text{sgn arg}\,\diamond](z)}{2}.$$

In other words, we need to calculate $\mathcal{C}[\diamond^k \text{sgn arg}\,\diamond]$ for both positive and negative k. We first investigate the simplest case of $k = 0$, i.e., calculation of $\mathcal{C}[\text{sgn arg}\,\diamond]$.

Proposition 5.13. *For* $z \notin \mathbb{I}$, *including boundary values,*

$$\mathcal{C}_{\mathbb{I}} 1(z) = \frac{\log(z-1) - \log(z+1)}{2\pi \mathrm{i}} = \frac{\log(1-z) - \log(-1-z)}{2\pi \mathrm{i}}$$

and

$$\mathcal{C}[\text{sgn arg}\,\diamond](z) = \frac{2}{\pi \mathrm{i}} \begin{cases} \operatorname{arctanh} z & \text{if } |z| < 1, \\ \operatorname{arctanh} \frac{1}{z} & \text{if } |z| > 1. \end{cases}$$

Proof. The first expression follows from direct computation. The second part of the proposition follows since $\operatorname{arctanh} z - \operatorname{arctanh} \frac{1}{z} = \frac{\pi \mathrm{i}}{2}$. We sketch a proof of this elementary result. Recall that

$$\operatorname{arctanh} z = \frac{\log(z+1) - \log(1-z)}{2} = \frac{1}{2}\int_0^z \left(\frac{1}{z+1} + \frac{1}{1-z}\right) dz = \int_0^z \frac{1}{1-z^2} dz,$$

where the path of integration is taken so that $\operatorname{arctanh} z$ is analytic apart from the branch cuts $(-\infty, -1]$ and $[1, \infty)$. On the branch cuts it satisfies

$$\operatorname{arctanh}^+ x - \operatorname{arctanh}^- x = \begin{cases} \frac{-\log^-(1-x) + \log^+(1-x)}{2} & \text{if } x > 1, \\ \frac{\log^+(x+1) - \log^-(x+1)}{2} & \text{if } x < -1 \end{cases} = \mathrm{i}\pi.$$

Furthermore, the $\sim \frac{1}{z^2}$ decay of the integrand ensures that $\operatorname{arctanh} z$ is bounded and tends to $\frac{\mathrm{i}\pi}{2}\text{sgn arg}\, z$ as $z \to \infty$. On the other hand, $\operatorname{arctanh} \frac{1}{z}$ is analytic apart from a branch on $[-1, 1]$, on which it satisfies

$$\left[\operatorname{arctanh} \frac{1}{x}\right]^+ - \left[\operatorname{arctanh} \frac{1}{x}\right]^- = -\mathrm{i}\pi.$$

Consider

$$\begin{aligned}\kappa(z) &= \operatorname{arctanh} z - \operatorname{arctanh}\frac{1}{z} - \frac{i\pi}{2}\operatorname{sgn}\arg z\\ &= \operatorname{arctanh} z - \operatorname{arctanh}\frac{1}{z} - \frac{i\pi}{2}\begin{cases} 1 & \text{if } \operatorname{Im} z > 0,\\ -1 & \text{if } \operatorname{Im} z < 0,\end{cases}\end{aligned}$$

which is analytic off the real line. From the relevant jumps, we determine that $\kappa_+(x) = \kappa_-(x)$ on the real line; hence κ is entire. (The possible singularities at ± 1 are at most logarithmic.) Furthermore, κ is bounded and tends to zero at ∞. Thus κ is zero, and $\operatorname{arctanh} z - \operatorname{arctanh}\frac{1}{z} = \frac{\pi i}{2}\operatorname{sgn}\arg z$. □

We use the $k = 1$ case to derive the Cauchy transform for other k. Consider a naïve construction $\phi(z) = z\,\mathcal{C}[\operatorname{sgn}\arg\diamond](z)$. This satisfies the correct jump:

$$\phi_+(\zeta) - \phi_-(\zeta) = \zeta\,\mathcal{C}^+[\operatorname{sgn}\arg\diamond](\zeta) - \zeta\,\mathcal{C}^-[\operatorname{sgn}\arg\diamond](\zeta) = \zeta\operatorname{sgn}\arg\zeta.$$

Unfortunately, it does not decay at infinity. Recall that the Taylor series of arctanh is

$$\operatorname{arctanh} z = z + \frac{z^3}{3} + \frac{z^5}{5} + \frac{z^7}{7} + \cdots.$$

Thus

$$z\operatorname{arctanh}\frac{1}{z} \sim 1 \neq 0,$$

but we can modify this definition of ϕ:

$$\phi(z) = z\,\mathcal{C}[\operatorname{sgn}\arg\diamond](z) - \frac{2}{i\pi}.$$

This decays at infinity and still satisfies the correct jump. By subtracting out higher terms in the series we get the following.

Theorem 5.14. *For $k \in \mathbb{Z}$ and $z \notin \mathbb{U}$, including boundary values,*

$$\begin{aligned}\mathcal{C}[\diamond^k\operatorname{sgn}\arg\diamond](z) &= z^k\,\mathcal{C}[\operatorname{sgn}\arg\diamond](z)\\ &\quad - \frac{2}{i\pi}\begin{cases} z^{k-1} + \frac{z^{k-3}}{3} + \cdots + \frac{1}{k} & \text{if } k > 0 \text{ and } k \text{ odd},\\ z^{k-1} + \frac{z^{k-3}}{3} + \cdots + \frac{z}{k-1} & \text{if } k > 0 \text{ and } k \text{ even},\\ z^{k+1} + \frac{z^{k+3}}{3} + \cdots + \frac{z^{-3}}{-k} & \text{if } k < 0 \text{ and } k \text{ odd},\\ z^{k+1} + \frac{z^{k+3}}{3} + \cdots + \frac{z^{-1}}{-k-1} & \text{if } k < 0 \text{ and } k \text{ even}.\end{cases}\end{aligned}$$

Relating this back to T_k, we have the following.

Corollary 5.15. *For $k \geq 0$ and $z \notin \mathbb{I}$, including boundary values,*

$$\mathcal{C}_{\mathbb{I}} T_k(z) = T_k(z)\,\mathcal{C}_{\mathbb{I}}1(z) + \frac{2}{i\pi}\begin{cases} T_{k-1}(z) + \frac{T_{k-3}(z)}{3} + \cdots + \frac{T_2(z)}{k-2} + \frac{1}{2k} & \text{if } k \text{ odd},\\ T_{k-1}(z) + \frac{T_{k-3}(z)}{3} + \cdots + \frac{T_1(z)}{k-1} & \text{if } k \text{ even}.\end{cases}$$

5.6.2 ▪ Formulae in terms of special functions

From a mathematical point of view, we are done: we have found a closed form representation for $\mathcal{C}_{\mathbb{I}} T_k$ in Corollary 5.15. However, this expression is flawed from a numerical standpoint: we know $\mathcal{C}_{\mathbb{I}} T_k$ decays at infinity, but the expression involves subtracting two very large terms. We overcome this issue by returning to the unit circle and the computation of $\mathcal{C}[\diamond^k \operatorname{sgn} \arg \diamond]$, this time reducing it to evaluation of special functions.

To start, we perform steepest descent. The Cauchy transform $\mathcal{C}[\diamond^k \operatorname{sgn} \arg \diamond]$ becomes increasingly oscillatory on the unit circle as k increases. But on the unit interval it is decaying. Thus we want to deform from the unit circle to the unit interval. This is straightforward using Plemelj's lemma.

Proposition 5.16. *For $k \geq 0$ and $z \notin \mathbb{U}$,*

$$\mathcal{C}[\diamond^k \operatorname{sgn} \arg \diamond](z) = -2\mathcal{C}_{\mathbb{I}}[\diamond^k](z) + \begin{cases} z^k & \text{if } |z| < 1 \text{ and } \operatorname{Im} z > 0, \\ -z^k & \text{if } |z| < 1 \text{ and } \operatorname{Im} z < 0, \\ 0 & \text{otherwise.} \end{cases}$$

For $k < 0$ and $z \notin \mathbb{U}$,

$$\mathcal{C}[\diamond^k \operatorname{sgn} \arg \diamond](z) = 2\mathcal{C}_{\mathbb{I}}[\diamond^{-k}](0) - 2\mathcal{C}_{\mathbb{I}}[\diamond^{-k}](1/z) + \begin{cases} -z^k & \text{if } |z| > 1 \text{ and } \operatorname{Im} z > 0, \\ z^k & \text{if } |z| > 1 \text{ and } \operatorname{Im} z < 0, \\ 0 & \text{otherwise.} \end{cases}$$

The standard reasons for deforming an oscillatory integral along its path of steepest descent — reasons which are used extensively later on — are for obtaining *asymptotics* and efficient *numerics*. However, there is a third reason for deforming along the path of steepest descent: reducing an integral to one known in closed form.

So with that in mind, we want to relate

$$\mathcal{C}_{\mathbb{I}}[\diamond^k](z) = \int_{-1}^{1} \frac{x^k}{x - z} \mathrm{d}x$$

to standard special functions, beginning with the *Lerch transcendent*

$$\phi(z, 1, a) \triangleq \int_0^\infty \frac{e^{-at}}{1 - ze^{-t}} \mathrm{d}t = \int_1^\infty \frac{s^{-a}}{s - z} \mathrm{d}s,$$

where $a > 0$. Plemelj's lemma tells us the jump for $x \in [1, \infty)$:

$$\phi_+(x, 1, a) - \phi_-(x, 1, a) = 2\pi i x^{-a}.$$

Furthermore, we have decay at infinity:

$$\phi(z, 1, a) \sim -\frac{1}{z} \int_1^\infty s^{-a} \mathrm{d}s.$$

Now consider $\phi(z^{-2}, 1, a)$, where a is an integer. This has a jump on $(-1, 1)$ where it satisfies

$$[\phi(x^{-2}, 1, a)]^+ - [\phi(x^{-2}, 1, a)]^- = \phi_-(x^{-2}, 1, a) - \phi_+(x^{-2}, 1, a) = -2\pi i x^{2a}$$

for $0 < x < 1$ and

$$[\phi(x^{-2},1,a)]^{+} - [\phi(x^{-2},1,a)]^{-} = \phi_{+}(x^{-2},1,a) - \phi_{-}(x^{-2},1,a) = 2\pi \mathrm{i} x^{2a}$$

for $-1 < x < 0$. Thus we have the following.

Lemma 5.17. *For $k \geq 0$,*

$$\mathscr{C}_{\mathbb{I}}[\diamond^k](z) = \frac{\mathrm{i}}{2\pi} \begin{cases} z^{-1}\phi(z^{-2},1,\frac{k+1}{2}) & \text{if } k \text{ is even,} \\ z^{-2}\phi(z^{-2},1,\frac{k+2}{2}) & \text{if } k \text{ is odd.} \end{cases}$$

While the Lerch transcendent has a particularly simple integral representation, special function routines for the Lerch transcendent are not very well developed. Thus we derive an alternative expression in terms of hypergeometric functions.

Proposition 5.18. *For $z \in \mathbb{C}$,*

$$\phi(z,1,a) = \frac{\Gamma(a)}{1-z} \mathbf{F}\left(\begin{matrix}1,1\\1+a\end{matrix}; \frac{z}{z-1}\right),$$

where $\mathbf{F}$ *is the regularized (Gauss) hypergeometric function.*

Proof. The *(Gauss) hypergeometric function* is defined inside the unit circle by

$$\begin{aligned} {}_2F_1\left(\begin{matrix}1,1\\ \gamma\end{matrix}; z\right) &= \Gamma(\gamma)\mathbf{F}\left(\begin{matrix}1,1\\ \gamma\end{matrix}; z\right), \\ \mathbf{F}\left(\begin{matrix}1,1\\ \gamma\end{matrix}; z\right) &= \sum_{s=0}^{\infty} \frac{s!}{\Gamma(\gamma+s)} z^s, \end{aligned}$$

and by analytic continuation in $\mathbb{C}\backslash[1,\infty)$. On its branch cut $(1,\infty)$ it satisfies the jump [91, equation (15.2.3)]

$$\begin{aligned} \mathbf{F}\left(\begin{matrix}1,1\\ \gamma\end{matrix}; x\right)^{+} - \mathbf{F}\left(\begin{matrix}1,1\\ \gamma\end{matrix}; x\right)^{-} &= 2\pi \mathrm{i}(x-1)^{\gamma-2}\mathbf{F}\left(\begin{matrix}\gamma-1,\gamma-1\\ \gamma-1\end{matrix}; 1-x\right) \\ &= 2\pi \mathrm{i}(x-1)^{\gamma-2}\frac{x^{1-\gamma}}{\Gamma(\gamma-1)}, \end{aligned}$$

where we used [91, equation (15.4.6)] to simplify the jump.

The function

$$\Phi(z) = \frac{1}{1-z}\mathbf{F}\left(\begin{matrix}1,1\\ \gamma\end{matrix}; \frac{z}{z-1}\right)$$

also has a branch cut along $[1,\infty)$, where it satisfies

$$\Phi_{+}(x) - \Phi_{-}(x) = 2\pi \mathrm{i}\frac{(x-1)^{2-\gamma}x^{1-\gamma}(x-1)^{\gamma-1}}{(x-1)\Gamma(\gamma-1)} = 2\pi \mathrm{i}\frac{x^{1-\gamma}}{\Gamma(\gamma-1)}.$$

Thus it satisfies the same jump as ϕ. It also decays as $z \to \infty$ since $\mathbf{F}$ has only a $(z-1)^{\gamma-2}$ singularity at 1. Thus $\Gamma(\gamma-1)\Phi$ with $\gamma = 1+a$ must be the same as ϕ. □

Remark 5.6.2. *There are numerous similarity transformations for the hypergeometric function, some leading to simpler formulae. Our motivation in using the precise form in the preceding proposition is that it seems to be the most efficient for Mathematica to evaluate.*

Remark 5.6.3. *The original derivation [95] of the preceding special function representations was based on series representations, rather than deformation of contours and integral representations.*

5.6.3 ▪ Recurrence relationship

We have derived an exact representation for the Cauchy transform $\mathcal{C}[\diamond^k \operatorname{sgn} \arg \diamond]$ in terms of special functions. The idea of reducing the problem to special functions is that we can use well-developed software (for example, *Mathematica*). However, special function routines have some computational cost attached. Fortunately, we can greatly reduce the number of special function calls by utilizing the following recurrence relationship.

Corollary 5.19. *For $k \in \mathbb{Z}$ and $z \notin \mathbb{U}$,*

$$\mathcal{C}[\diamond^{k+1} \operatorname{sgn} \arg \diamond](z) = z\,\mathcal{C}[\diamond^k \operatorname{sgn} \arg \diamond](z) + \frac{2\mathrm{i}}{\pi} \begin{cases} \frac{1}{k+1} & \text{if } k \text{ even,} \\ 0 & \text{otherwise.} \end{cases}$$

Proof. This follows from

$$\begin{aligned} \mathcal{C}[\diamond^{k+1} \operatorname{sgn} \arg \diamond](z) &= z\,\mathcal{C}[\diamond^k \operatorname{sgn} \arg \diamond](z) - \lim_{z\to\infty} z\,\mathcal{C}[\diamond^k \operatorname{sgn} \arg \diamond](z) \\ &= z\,\mathcal{C}[\diamond^k \operatorname{sgn} \arg \diamond](z) + \oint_{\mathbb{U}} \zeta^k \operatorname{sgn} \arg \zeta \, \mathrm{d}\zeta \\ &= z\,\mathcal{C}[\diamond^k \operatorname{sgn} \arg \diamond](z) + \frac{\mathrm{i}}{\pi} \int_0^\pi \sin(k+1)\theta \, \mathrm{d}\theta. \qquad \square \end{aligned}$$

The question now is whether to apply this relationship forward or backwards, which we determine by the stability of the process. Begin with $k \geq 0$ and $|z| < 1$, where the $|z| > 1$ case follows from symmetry, due to the argument of Lemma 5.5. Our first stage is to rewrite the recurrence in matrix language. The recurrence then takes the form

$$\begin{bmatrix} -z & 1 & & & \\ & -z & 1 & & \\ & & -z & 1 & \\ & & & -z & \ddots \\ & & & & \ddots \end{bmatrix} \begin{bmatrix} \mathcal{C}[\operatorname{sgn} \arg \diamond](z) \\ \mathcal{C}[\diamond \operatorname{sgn} \arg \diamond](z) \\ \mathcal{C}[\diamond^2 \operatorname{sgn} \arg \diamond](z) \\ \mathcal{C}[\diamond^3 \operatorname{sgn} \arg \diamond](z) \\ \vdots \end{bmatrix} = \frac{2\mathrm{i}}{\pi} \begin{bmatrix} 1 \\ 0 \\ 1/3 \\ 0 \\ 1/5 \\ \vdots \end{bmatrix}.$$

This equation as posed does not lend itself to solution, as we cannot truncate the right-hand side without introducing significant error.

Fortunately, we have the additional information of $\mathcal{C}[\operatorname{sgn}\arg\diamond](z)$ in closed form; thus we also have access to forward substitution. This consists of appending an initial condition to the linear system:

$$\mathcal{L}_z \begin{bmatrix} \mathcal{C}[\operatorname{sgn}\arg\diamond](z) \\ \mathcal{C}[\diamond\operatorname{sgn}\arg\diamond](z) \\ \mathcal{C}[\diamond^2\operatorname{sgn}\arg\diamond](z) \\ \vdots \end{bmatrix} = \begin{bmatrix} \mathcal{C}[\operatorname{sgn}\arg\diamond](z) \\ 2\mathrm{i}/\pi \\ 0 \\ 2\mathrm{i}/(3\pi) \\ 0 \\ \vdots \end{bmatrix}$$

for

$$\mathcal{L}_z \triangleq \begin{bmatrix} 1 & & & \\ -z & 1 & & \\ & -z & 1 & \\ & & \ddots & \ddots \end{bmatrix}.$$

Thus we can calculate the first n terms in $\mathcal{O}(n)$ operations using forward substitution.

Numerical error in applying recurrence relationships can have drastic consequences if one is not careful. To control this error, we use the following simplified model of *numerical stability*: we assume that the entries of the right-hand side of the equation $\boldsymbol{r}_z$ may be perturbed by a small relative amount, i.e., a vector $\boldsymbol{\epsilon}_z$ satisfying $\|\boldsymbol{\epsilon}_z\|_{\ell^\infty} \leq \epsilon \|\boldsymbol{r}_z\|_{\ell^\infty}$. (Due to the decay in the coefficients and the fact that the round-off error is relative, we could in fact view the perturbation error as small in ℓ^2, though the bounding of operators in ℓ^2 is more delicate.)

The numerical approximation of the Cauchy transform without round-off error is

$$\mathcal{C}f_n(z) = \frac{1}{2}(\mathrm{T}_n f(\boldsymbol{x}^n))^\top \left(\begin{bmatrix} \mathcal{C}[\operatorname{sgn}\arg\diamond](z) \\ \mathcal{C}[\diamond\operatorname{sgn}\arg\diamond](z) \\ \vdots \\ \mathcal{C}[\diamond^{n-1}\operatorname{sgn}\arg\diamond](z) \end{bmatrix} + \begin{bmatrix} \mathcal{C}[\operatorname{sgn}\arg\diamond](z) \\ \mathcal{C}[\diamond^{-1}\operatorname{sgn}\arg\diamond](z) \\ \vdots \\ \mathcal{C}[\diamond^{1-n}\operatorname{sgn}\arg\diamond](z) \end{bmatrix} \right),$$

where

$$f_n(x) = \sum_{k=0}^{n-1} \check{f}_k^n T_k(x)$$

is the discrete Chebyshev series. The first term of this expression as calculated by the recurrence relationship with round-off error can be modeled as

$$(\mathrm{T}_n f(\boldsymbol{x}^n))^\top \mathcal{L}_z^{-1}(\boldsymbol{r}_z + \boldsymbol{\epsilon}_z) \quad \text{for} \quad \boldsymbol{r}_z = \begin{bmatrix} \mathcal{C}[\operatorname{sgn}\arg\diamond](z) \\ 2\mathrm{i}/\pi \\ 0 \\ 2\mathrm{i}/(3\pi) \\ 0 \\ \vdots \end{bmatrix}.$$

We thus bound the error as

$$\begin{aligned} \left|(\mathrm{T}_n f(\boldsymbol{x}^n))^\top \mathcal{L}_z^{-1}\boldsymbol{\epsilon}_z\right| &\leq \|\mathrm{T}_n f(\boldsymbol{x}^n)\|_{\ell^1} \left\|\mathcal{L}_z^{-1}\boldsymbol{\epsilon}_z\right\|_{\ell^\infty} \leq \epsilon \|\mathrm{T}_n f\|_{\ell^1} \left\|\mathcal{L}_z^{-1}\right\|_{\ell^\infty} \|\boldsymbol{r}_z\|_{\ell^\infty} \\ &\leq \epsilon \|\mathcal{T} f\|_{\ell^1} \left\|\mathcal{L}_z^{-1}\right\|_{\ell^\infty} \max\left\{\frac{2}{\pi}, |\operatorname{arctanh} z|\right\}, \end{aligned}$$

where we used Corollary 4.8. Thus to determine the effect of round-off error in this model, we simply need to bound $\|\mathcal{L}_z^{-1}\|_{\ell^\infty}$. This is easy to calculate directly:

$$\left\|\mathcal{L}_z^{-1}\right\|_{\ell^\infty} = \sum_{k=0}^{\infty} |z|^k = \frac{1}{1-|z|}.$$

For z on the unit circle, this bound explodes. Instead, we utilize the fact that $\mathrm{T}_n f$ has only n nonzero entries: thus we only need the error in the first n terms. Hence we bound

$$\left\|\mathcal{L}_z^{-1}\right\|_{\mathbb{C}^n} \leq n,$$

where $\mathbb{C}^n$ is equipped with the ℓ^∞ norm.

Remark 5.6.4. *Note that $\mathbf{r}_z$ itself is unbounded at ± 1. See Section 5.7.1 for a discussion of computation near the singularity. We omit a discussion on error in applying the recurrence relationship, though this can be controlled with a bit of care by studying the relative error of the approximation.*

Negative k: We now turn to negative k, where we have

$$\begin{bmatrix} \ddots & & & \\ \ddots & 1 & & \\ & -z & 1 & \\ & & -z & 1 \end{bmatrix} \begin{bmatrix} \vdots \\ \mathcal{C}[\diamond^{-2}\operatorname{sgn}\arg\diamond](z) \\ \mathcal{C}[\diamond^{-1}\operatorname{sgn}\arg\diamond](z) \\ \mathcal{C}[\operatorname{sgn}\arg\diamond](z) \end{bmatrix} = \frac{2}{i\pi} \begin{bmatrix} \vdots \\ \frac{1}{3} \\ 0 \\ 1 \\ 0 \end{bmatrix}.$$

If we try to impose the boundary condition on $\mathcal{C}[\operatorname{sgn}\arg\diamond](z)$, we get

$$\begin{bmatrix} \ddots & & & \\ \ddots & 1 & & \\ & -z & 1 & \\ & & -z & 1 \\ & & & -z \end{bmatrix} \begin{bmatrix} \vdots \\ \mathcal{C}[\diamond^{-2}\operatorname{sgn}\arg\diamond](z) \\ \mathcal{C}[\diamond^{-1}\operatorname{sgn}\arg\diamond](z) \\ \mathcal{C}[\operatorname{sgn}\arg\diamond](z) \end{bmatrix} = \begin{bmatrix} \vdots \\ 2/(3i\pi) \\ 0 \\ 2/(i\pi) \\ 0 \\ -z\mathcal{C}[\operatorname{sgn}\arg\diamond](z) \end{bmatrix}.$$

The inverse of the operator is

$$-\begin{bmatrix} \ddots & \ddots & \ddots & \vdots \\ & \frac{1}{z} & \frac{1}{z^2} & \frac{1}{z^3} \\ & & \frac{1}{z} & \frac{1}{z^2} \\ & & & \frac{1}{z} \end{bmatrix}.$$

When z is on the unit circle, the norm of the operator on $\mathbb{C}^n$ with the ℓ^∞ norm attached is n, and back substitution can proceed. For z in the interior of the unit circle, the operator norm grows exponentially with n; hence back substitution is extremely prone to numerical instability.

On the other hand, we can use the exact formulae from Section 5.6.2 to calculate the last entry $\mathcal{C}[\diamond^{1-n}\operatorname{sgn}\arg\diamond](z)$ directly. Then we can impose the initial condition as follows (assuming n is odd):

$$\begin{bmatrix} 1 & & & & \\ -z & 1 & & & \\ & \ddots & \ddots & & \\ & & -z & 1 & \\ & & & -z & 1 \end{bmatrix} \begin{bmatrix} \mathcal{C}[\diamond^{1-n}\operatorname{sgn}\arg\diamond](z) \\ \vdots \\ \mathcal{C}[\diamond^{-2}\operatorname{sgn}\arg\diamond](z) \\ \mathcal{C}[\diamond^{-1}\operatorname{sgn}\arg\diamond](z) \\ \mathcal{C}[\operatorname{sgn}\arg\diamond](z) \end{bmatrix} = \begin{bmatrix} \mathcal{C}[\diamond^{1-n}\operatorname{sgn}\arg\diamond](z) \\ 2/((2n-1)\mathrm{i}\pi) \\ \vdots \\ 2/(3\mathrm{i}\pi) \\ 0 \\ 2/(\mathrm{i}\pi) \\ 0 \end{bmatrix}.$$

This is precisely the same matrix considered for the positive k case, and we know the sense in which forward elimination can be applied stably.

5.7 ▪ Approximation of Cauchy transforms near endpoint singularities

In this section, we consider computation of the Cauchy transform near the endpoint singularities, beginning with the case where f itself vanishes at the endpoints, so that the Cauchy transform is well-defined. We then consider the case where f does not vanish, observing that we can still attach a meaningful value — the *finite-part Cauchy transform* — at the singularity. This proves critical for approximating Cauchy transforms over unbounded contours.

5.7.1 ▪ Bounding the Cauchy transform of Chebyshev polynomials near endpoint singularities

Having determined explicit formulae for the Cauchy transform of Chebyshev polynomials, we now consider the second requirement for robust numerics: bounding the Cauchy transform uniformly in k. For z bounded away from $[-1,1]$, we obtain a uniform bound from the boundedness of the basis itself. For z near $(-1,1)$ but bounded away from ± 1 we can find a uniform bound for $\mathcal{C}[\diamond^k \operatorname{sgn}\arg\diamond]$ by deforming the integral to $[-1,1]$ as in Proposition 5.16 so that uniform boundedness of the integrands with k shows uniform boundedness of the Cauchy transforms. This procedure still breaks down for z near ± 1, which continues to be an endpoint of the deformed Cauchy transform.

For z near ± 1, the Cauchy transforms of Chebyshev polynomials have logarithmic singularities (see Lemma 2.11) and hence are *not* bounded. Thus we instead consider expansion in the vanishing basis introduced in Section 4.4:

$$T_0^z(x) = 1, \quad T_1^z(x) = T_1(x), \quad \text{and} \quad T_k^z(x) = T_k(x) - T_{k-2}(x).$$

The vanishing of the basis (for $k = 2,3,\ldots$) at ± 1 ensures that the Cauchy transforms are bounded for all z. We can take this further and show that they are in fact uniformly bounded with k.

Lemma 5.20. $\left|\mathcal{C}_{\mathbb{I}} T_k^z(z)\right|$ *is uniformly bounded for* $k = 2,3,4,\ldots$ *and* $z \in \mathbb{C}\setminus\mathbb{I}$.

Proof. From Proposition 5.12 we have

$$\mathcal{C}_{\mathbb{I}} T_k^z(J(z)) = \frac{\mathcal{C}[(\diamond^k - \diamond^{k-2})\operatorname{sgn}\arg\diamond](z) + \mathcal{C}[(\diamond^{-k} - \diamond^{2-k})\operatorname{sgn}\arg\diamond](z)}{2}.$$

From Proposition 5.16, the first term can be deformed to the unit interval, reducing the problem to bounding $\mathscr{C}_{\mathbb{I}}[\diamond^k - \diamond^{k-2}]$ for $k > 0$. Embed x in a small interval $[z-\epsilon, z+\epsilon]$ that does not intersect $[-1,1]$ and define

$$w_k(x) = \begin{cases} x^{k-2}(1-x^2) & \text{if } x \in [-1,1], \\ 0 & \text{if } x \in [z-\epsilon, z+\epsilon]. \end{cases}$$

This is in $H_z^1(\Gamma)$ for $\Gamma = [-1,1] \cup [z-\epsilon, z+\epsilon]$, and it has the $H^1(\Gamma)$ norm uniformly bounded in k: $w_k(x) \le 1$ and

$$|w_k'(x)| \le |w_k'(\pm 1)| = 2.$$

For $k = 0, \ldots, 3$ we verify this inequality directly. Otherwise, $w_k'(x) = x^{k-3}[k(1-x^2)-2]$, and by symmetry we only need to show it holds for $0 \le x \le 1$. We reduce the inequality to a simpler form:

$$1 - x \le \frac{x^{3-k}}{k} \Rightarrow 1 - x^2 \le x^{3-k}\frac{1+x}{k} \le \frac{2x^{3-k}}{k} \Rightarrow k(1-x^2) - 2 \le 2x^{3-k} \Rightarrow w_k'(x) \le 2.$$

The first inequality is satisfied for $k \ge 4$. We omit the details, but this result can be shown by finding the (only) saddle point of $\frac{x^{3-k}}{k} + x - 1$ by differentiating with respect to x. Evaluating at this saddle point, the subsequent expression can in turn be differentiated with respect to k to find its (only) saddle point, at which it is positive.

We then have, using Theorem 2.50,

$$|\mathscr{C}_{\mathbb{I}} w_k(z)| = |\mathscr{C}_\Gamma w_k(z)| \le \|\mathscr{C}_\Gamma w_k\|_u \le C \|\mathscr{C}_\Gamma w_k\|_{H_z^1(\Gamma)} \le C \|w_k\|_{H_z^1(\Gamma)} \le \tilde{C}.$$

A similar approach is used to bound $\mathscr{C}[(\diamond^{-k} - \diamond^{-k-2})\operatorname{sgn}\arg\diamond]$. □

It follows from Section 5.1 that if we expand a function in T_k^z with coefficients in ℓ^1, then the numerical approximation of the Cauchy transform converges uniformly in the complex plane. For the case where f does not vanish at ± 1, we state the uniform convergence result utilizing Proposition 4.11.

Corollary 5.21. *Suppose* $\mathscr{T} f \in \ell^{2,1}$. *Then*

$$\mathscr{C}_{\mathbb{I}} f_n(z) - \mathscr{C}_{\mathbb{I}} f(z) \to 0$$

uniformly in $z \in \mathbb{C} \setminus \mathbb{I}$ *as* $n \to \infty$, *where*

$$f_n(x) = \sum_{k=0}^{n-1} \check{f}_k^n T_k(x)$$

is the discrete Chebyshev series.

5.7.2 ▪ Finite-part Cauchy transform

If we remove the blowup of the logarithmic singularity of the Cauchy transform near an endpoint, we define what we call a *finite-part Cauchy transform*. This proves useful when, due to subsequent manipulations, the logarithmic singularities are canceled and one wishes to evaluate the remaining bounded value.

Remark 5.7.1. *The finite-part Cauchy transform is nonclassical and was introduced in* [92, 96].

Definition 5.22. *Let* Γ *be a bounded, smooth arc from* a *to* b*. The finite-part Cauchy transform*[36] $\mathrm{FP}\mathcal{C}_\Gamma$ *is defined as follows: For* $z \neq a, b$*, it is equal to the Cauchy transform*

$$\mathrm{FP}\mathcal{C}_\Gamma f(z) \triangleq \mathcal{C}_\Gamma f(z) \qquad \text{and} \qquad \mathrm{FP}\mathcal{C}_\Gamma^\pm f(z) \triangleq \mathcal{C}_\Gamma^\pm f(z).$$

For $z = a$ *or* b*, we include a parameter* θ *corresponding to the angle of approach,*[37] *and subtract out the logarithmic singularity:*

$$\mathrm{FP}\mathcal{C}_\Gamma f(a + 0\mathrm{e}^{\mathrm{i}\theta}) \triangleq \lim_{\epsilon \downarrow 0}\left[\mathcal{C}_\Gamma f(a + \epsilon \mathrm{e}^{\mathrm{i}\theta}) + \frac{f(a)}{2\pi \mathrm{i}} \log \epsilon\right],$$
$$\mathrm{FP}\mathcal{C}_\Gamma f(b + 0\mathrm{e}^{\mathrm{i}\theta}) \triangleq \lim_{\epsilon \downarrow 0}\left[\mathcal{C}_\Gamma f(b + \epsilon \mathrm{e}^{\mathrm{i}\theta}) - \frac{f(b)}{2\pi \mathrm{i}} \log \epsilon\right].$$

We similarly define for $z = a, b$ *and* θ *the angle that* Γ *leaves* z, $\mathrm{FP}\mathcal{C}_\Gamma^\pm f(z + 0\mathrm{e}^{\mathrm{i}\theta})$ *being the limit of* $\mathrm{FP}\mathcal{C}_\Gamma f(z + 0\mathrm{e}^{\mathrm{i}\varphi})$ *as* φ *approaches* θ *from the left or right of* Γ.

Using Corollary 5.15 and Proposition 5.13 we determine the values of the finite-part Cauchy transform of Chebyshev polynomials directly.

Proposition 5.23. *For* $k \geq 0$,

$$\begin{aligned}
\mathrm{FP}\mathcal{C}_{\mathbb{I}} T_k(-1 + 0\mathrm{e}^{\mathrm{i}\theta}) &= (-)^k \frac{\log 2 - \mathrm{i}\arg(-\mathrm{e}^{\mathrm{i}\theta})}{2\pi \mathrm{i}} \\
&\quad - \frac{2(-)^k}{\pi \mathrm{i}} \begin{cases} 1 + \frac{1}{3} + \cdots + \frac{1}{k-2} + \frac{1}{2k} & \text{if } k \text{ odd}, \\ 1 + \frac{1}{3} + \cdots + \frac{1}{k-3} + \frac{1}{k-1} & \text{if } k \text{ even}, \end{cases} \\
\mathrm{FP}\mathcal{C}_{\mathbb{I}}^\pm T_k(1 + 0\mathrm{e}^{\mathrm{i}\theta}) &= \frac{\mathrm{i}\theta - \log 2}{2\pi \mathrm{i}} + \frac{2}{\pi \mathrm{i}} \begin{cases} 1 + \frac{1}{3} + \cdots + \frac{1}{k-2} + \frac{1}{2k} & \text{if } k \text{ odd}, \\ 1 + \frac{1}{3} + \cdots + \frac{1}{k-3} + \frac{1}{k-1} & \text{if } k \text{ even}. \end{cases}
\end{aligned}$$

The results in Section 5.1 do not address whether

$$\mathrm{FP}\mathcal{C}_{\mathbb{I}} f(\pm 1 + 0\mathrm{e}^{\mathrm{i}\theta}) \stackrel{?}{=} \sum_{k=0}^{\infty} \hat{f}_k \mathrm{FP}\mathcal{C}_{\mathbb{I}} T_k(\pm 1 + 0\mathrm{e}^{\mathrm{i}\theta}).$$

Furthermore, the terms in Proposition 5.23 grow logarithmically with k; hence there exists f satisfying $\mathcal{T} f \in \ell^1$ for which the finite-part Cauchy transform is unbounded. To resolve these issues, we write Cauchy transforms of the vanishing bases T_k^z and $T_k^{\pm z}$ in terms of the finite-part Cauchy transform.

Proposition 5.24. *For the vanishing bases* $T_k^{\pm z}$, $k \geq 0$*, the logarithmic terms cancel and*

$$\mathcal{C}_{\mathbb{I}} T_k^{\pm z}(z) = \mathrm{FP}\mathcal{C}\, T_k(z) \mp \mathrm{FP}\mathcal{C}\, T_{k-1}(z)$$

holds for all $k \geq 2$ *and* $z \neq \mp 1$*. Similarly,*

$$\mathcal{C}_{\mathbb{I}} T_k^{z}(z) = \mathrm{FP}\mathcal{C}\, T_k(z) - \mathrm{FP}\mathcal{C}\, T_{k-2}(z)$$

[36] Not to be confused with the Hadamard finite-part integral.

[37] In other words, we use complex *dual numbers*.

holds for all $k \geq 2$ and z, and in particular

$$\mathscr{C}_{\mathbb{I}} T_k^z(\pm 1) = (\pm)^{k+1} \frac{2}{\mathrm{i}\pi} \begin{cases} \frac{1}{2k} + \frac{1}{2k-4} & \text{if } k \text{ odd}, \\ \frac{1}{k-1} & \text{if } k \text{ even}. \end{cases}$$

It follows immediately that the decay of the coefficients in the modified basis dictates the applicability of the basis expansion for the finite-part Cauchy transform.

Corollary 5.25. *If $\mathscr{T} f \in \ell^{2,1}$, then*

$$\mathrm{FP}\mathscr{C}_{\mathbb{I}} f(\pm 1 + 0\mathrm{e}^{\mathrm{i}\theta}) = \sum_{k=0}^{\infty} \hat{f}_k \mathrm{FP}\mathscr{C}_{\mathbb{I}} T_k(\pm 1 + 0\mathrm{e}^{\mathrm{i}\theta}).$$

5.7.3 ▪ Möbius transformations of the unit interval

Consider a Möbius transformation $M : \mathbb{I} \to \Gamma$ where Γ is bounded (hence, either an arc or a line segment). Using Lemma 5.4, we immediately obtain the Cauchy transform of the basis $T_k(M^{-1}(s))$:

$$\mathscr{C}_\Gamma[T_k \circ M^{-1}](z) = \mathscr{C}_{\mathbb{I}} T_k(M^{-1}(z)) - \mathscr{C}_{\mathbb{I}} T_k(M^{-1}(\infty)).$$

We also determine the finite-part Cauchy transform.

Corollary 5.26. *Suppose M is a Möbius transformation and $M(\mathbb{I})$ is bounded. Then we have, for $a = M(-1)$ and $b = M(1)$,*

$$\mathrm{FP}\mathscr{C}_\Gamma[T_k \circ M^{-1}](a + 0\mathrm{e}^{\mathrm{i}\theta}) = \mathrm{FP}\mathscr{C}_{\mathbb{I}} T_k(-1 + 0\mathrm{e}^{\mathrm{i}(\theta - \varphi_a)}) - \mathscr{C}_{\mathbb{I}} T_k(M^{-1}(\infty)) + \frac{(-)^k \log|M'(-1)|}{2\pi\mathrm{i}},$$
$$\mathrm{FP}\mathscr{C}_\Gamma[T_k \circ M^{-1}](b + 0\mathrm{e}^{\mathrm{i}\theta}) = \mathrm{FP}\mathscr{C}_{\mathbb{I}} T_k(1 + 0\mathrm{e}^{\mathrm{i}(\theta - \varphi_b)}) - \mathscr{C}_{\mathbb{I}} T_k(M^{-1}(\infty)) - \frac{\log|M'(1)|}{2\pi\mathrm{i}}$$

for $k \geq 0$, where $\varphi_a = \arg M'(-1)$ and $\varphi_b = \arg M'(1)$ are the angles that Γ leaves a and approaches b, respectively.

Proof. We have, for the right endpoint $b = M(1)$,

$$\begin{aligned} \mathrm{FP}\mathscr{C}_\Gamma[T_k \circ M^{-1}](b + 0\mathrm{e}^{\mathrm{i}\theta}) &= \lim_{\epsilon \to 0} \mathscr{C}_{\mathbb{I}} T_k(M^{-1}(b + \epsilon \mathrm{e}^{\mathrm{i}\theta})) - \mathscr{C}_{\mathbb{I}} T_k(M^{-1}(\infty)) - \frac{\log \epsilon}{2\pi\mathrm{i}} \\ &= \lim_{\epsilon \to 0} \mathscr{C}_{\mathbb{I}} T_k(1 + M^{-1\prime}(b)\epsilon \mathrm{e}^{\mathrm{i}\theta}) - \frac{\log\left|M^{-1\prime}(b)\right|\epsilon}{2\pi\mathrm{i}} \\ &\quad - \mathscr{C}_{\mathbb{I}} T_k(M^{-1}(\infty)) - \frac{\log \epsilon - \log\left|M^{-1\prime}(b)\right|\epsilon}{2\pi\mathrm{i}} \\ &= \mathrm{FP}\mathscr{C}_{\mathbb{I}} T_k\left(1 + 0\mathrm{e}^{\mathrm{i}\theta + \mathrm{i}\arg M^{-1\prime}(b)}\right) - \mathscr{C}_{\mathbb{I}} T_k(M^{-1}(\infty)) \\ &\quad + \frac{\log\left|M^{-1\prime}(b)\right|}{2\pi\mathrm{i}}. \end{aligned}$$

Inverting the series we find that $M^{-1\prime}(b) = M(1)^{-1}$, which simplifies the expression. The result near a is proved similarly. □

Now consider the case where Γ is not bounded, i.e., a ray $[a, e^{i\varphi}\infty)$. In this case Lemma 5.4 fails: $T_k \circ M^{-1}$ does not decay at infinity and $\mathcal{C}_{\mathbb{I}} T_k(1) = \infty$. We instead construct another definition for the finite-part Cauchy transform, motivated by Lemma 5.28.

Definition 5.27. *Suppose Γ is a contour that is unbounded at one endpoint. Then we define the* finite-part Cauchy transform *by*

$$\mathrm{FP}\mathcal{C}_\Gamma f(z) \triangleq \lim_{r\to\infty} \mathrm{FP}\mathcal{C}_{\Gamma\cap B_r} f(z) - \frac{f(\infty)}{2\pi i}\log r,$$

where B_r is the ball of radius r centered around the origin.

By conformal mapping, we can relate the finite-part Cauchy transform for mapped Chebyshev polynomials to the finite-part Cauchy transform for standard Chebyshev polynomials. The first step is to note the following relationship between the finite-part Cauchy transform and a limiting contour.

Lemma 5.28. *Suppose f is α-Hölder continuous. Then*

$$\mathrm{FP}\mathcal{C}_{\mathbb{I}} f(-1-0) = \lim_{\epsilon\downarrow 0}\left[\mathcal{C}_{[\epsilon-1,1]} f(-1) + \frac{f(-1)}{2\pi i}\log\epsilon\right],$$

$$\mathrm{FP}\mathcal{C}_{\mathbb{I}} f(1+0) = \lim_{\epsilon\downarrow 0}\left[\mathcal{C}_{[-1,1-\epsilon]} f(1) - \frac{f(1)}{2\pi i}\log\epsilon\right].$$

Proof. We have, using the affine transformation $M_\epsilon(x) = \frac{2x+\epsilon}{2-\epsilon}$, $M_\epsilon : [-1, 1-\epsilon] \to \mathbb{I}$,

$$\begin{aligned}\mathcal{C}_{[-1,1-\epsilon]} f(1) &= \int_{-1}^{1-\epsilon} \frac{f(x)-f(1)}{x-1}\,\mathrm{d}x + f(1)\mathcal{C}_{[-1,1-\epsilon]}1(1) \\ &= \int_{-1}^{1-\epsilon} \frac{f(x)-f(1)}{x-1}\,\mathrm{d}x + f(1)\mathcal{C}_{\mathbb{I}}1(M_\epsilon(1)).\end{aligned}$$

Thus we have

$$\begin{aligned}&\lim_{\epsilon\to 0}\left[\mathcal{C}_{[-1,1-\epsilon]} f(1) - \frac{f(1)}{2\pi i}\log\epsilon\right] \\ &= \lim_{\epsilon\to 0}\left[\int_{-1}^{1-\epsilon} \frac{f(x)-f(1)}{x-1}\,\mathrm{d}x + f(1)\mathcal{C}_{[-1,1-\epsilon]}1(1) - \frac{f(1)}{2\pi i}\log\epsilon\right] \\ &= \lim_{\epsilon\to 0}\left[\int_{-1}^{1} \frac{f(x)-f(1)}{x-M_\epsilon(1)}\,\mathrm{d}x + f(1)\mathcal{C}_{\mathbb{I}}1(M_\epsilon(1)) - \frac{f(1)}{2\pi i}\log\epsilon\right] \\ &= \lim_{\epsilon\to 0}\left[\mathcal{C}_{\mathbb{I}} f(M_\epsilon(1)) - \frac{f(1)}{2\pi i}\log\epsilon\right] = \mathrm{FP}\mathcal{C}_{\mathbb{I}} f(1+0),\end{aligned}$$

where we used the fact that $M_\epsilon(1) = 1 + \epsilon + \mathcal{O}(\epsilon^2)$. The proof for evaluating at -1 is similar. □

Using this relationship, along with Lemma 5.4, we can reduce the finite-part Cauchy transform for mapped Chebyshev polynomials over unbounded contours to standard Chebyshev polynomials.

Lemma 5.29. *Suppose $\Gamma = [a, e^{i\varphi}\infty) = M(\mathbb{I})$ is a ray oriented outwards and $k \geq 0$. Then for $z \neq a$, including boundary values,*

$$\mathrm{FP}\mathscr{C}_\Gamma[T_k \circ M^{-1}](z) = \mathscr{C}_\mathbb{I} T_k(M^{-1}(z)) - \mathrm{FP}\mathscr{C}_\mathbb{I} T_k(1+0) - \frac{\log 2 + \log L + i\varphi}{2\pi i},$$

and for $z = a$,

$$\mathrm{FP}\mathscr{C}_\Gamma[T_k \circ M^{-1}](a + 0e^{i\theta}) = \mathrm{FP}\mathscr{C}_\mathbb{I} T_k(-1 + 0e^{i(\theta-\varphi_a)}) - \mathrm{FP}\mathscr{C}_\mathbb{I} T_k(1+0) + \frac{((-1)^k - 1)\log L - ((-1)^k + 1)\log 2 - i\varphi}{2\pi i},$$

where

$$M(x) = a + Le^{i\varphi}\frac{1+x}{1-x}.$$

Proof. Note that

$$M(x) \sim_{x\to 1} \frac{2Le^{i\varphi}}{1-x}.$$

We now have for $\Gamma_\epsilon = M([-1, 1-\epsilon])$

$$\begin{aligned}
\mathrm{FP}\mathscr{C}_\Gamma[T_k \circ M^{-1}](z) &= \lim_{\epsilon\to 0}\left\{\mathscr{C}_{\Gamma_\epsilon}[T_k \circ M^{-1}](z) - \frac{\log M(1-\epsilon)}{2\pi i}\right\} \\
&= \lim_{\epsilon\to 0}\left\{\mathscr{C}_{[-1,1-\epsilon]} T_k(M^{-1}(z)) - \mathscr{C}_{[-1,1-\epsilon]} T_k(1) - \frac{1}{2\pi i}\log M(1-\epsilon)\right\} \\
&= \mathscr{C}_\mathbb{I} T_k(M^{-1}(z)) - \frac{\log 2Le^{i\varphi}}{2\pi i} - \mathrm{FP}\mathscr{C}_\mathbb{I} T_k(1+0),
\end{aligned}$$

where we used Lemma 5.28. This expression is then used to determine the finite-part Cauchy transform near a. □

We can use the finite-part Cauchy transform to determine the true Cauchy transform of functions that vanish at ∞.

Corollary 5.30. *Suppose $\Gamma = [a, e^{i\varphi}\infty) = M(\mathbb{I})$ is a ray oriented outwards and $\mathscr{T} f \in \ell^{2,1}_{+z}$. For $z \notin [a, e^{i\varphi}\infty)$,*

$$\mathscr{C}_\Gamma f(z) = \sum_{k=0}^{\infty} \check{f}_k \mathrm{FP}\mathscr{C}_\Gamma[T_k \circ M^{-1}](z).$$

Proof. The condition $\mathscr{T} f \in \ell^{2,1}_{+z}$ ensures that we can re-expand f in the vanishing basis

$$f(M(x)) = \sum_{k=1}^{\infty} \check{f}_k^{+z} T_k^{+z}(x).$$

It follows from Proposition 5.3 that

$$\mathscr{C}_\Gamma f(z) = \sum_{k=0}^{\infty} \check{f}_k^{+z} \mathscr{C}_\Gamma[T_k^{+z} \circ M^{-1}](z).$$

But for the vanishing basis we have

$$\mathcal{C}_\Gamma[T_k^{+z} \circ M^{-1}](z) = \mathrm{FP}\mathcal{C}_\Gamma[T_k \circ M^{-1}](z) - \mathrm{FP}\mathcal{C}_\Gamma[T_{k-1} \circ M^{-1}](z)$$

since the logarithmic terms cancel in the definition of the finite-part Cauchy transform. The uniform boundedness of $\mathrm{FP}\mathcal{C}_\Gamma[T_k \circ M^{-1}](z)$ means we can rearrange the summation to prove the corollary. □

5.7.4 ▪ Polynomial maps

Consider Cauchy transforms of contours Γ defined by *polynomial* maps, in place of Möbius transformations. Möbius transforms have unique inverses, and we saw in Lemma 5.4 that the Cauchy transform could be expressed in terms of the inverse of the Möbius transform. In the case of a degree d polynomial map, there are d inverses. We show that the Cauchy transform is expressible in terms of a sum over *all* inverses.

The proof again appeals to Plemelj's lemma. The first stage is to show analyticity off Γ, which uses the following general lemma.

Lemma 5.31. *Let $p(z)$ be a dth degree polynomial, g an analytic function everywhere except on a set Σ, and $\lambda_1(z), \ldots, \lambda_d(z)$ an enumeration of the d roots to the equation $p(\lambda) = z$. Then*

$$g(\lambda_1(z)) + \cdots + g(\lambda_d(z))$$

is analytic in z everywhere off $p(\Sigma)$.

Proof. Let $p(y) = \sum_{k=0}^d p_k y^k$. Note that the companion matrix

$$A(z) = \begin{bmatrix} 0 & 1 & & & \\ & & 1 & & \\ & & & \ddots & \\ & & & & 1 \\ \frac{z-p_0}{p_d} & -\frac{p_1}{p_d} & -\frac{p_2}{p_d} & \cdots & -\frac{p_{d-1}}{p_d} \end{bmatrix}$$

of the polynomial $p(y) - z$ has the property that its eigenvalues are precisely $\lambda_1(z), \ldots,$ $\lambda_d(z)$. From the spectral mapping theorem we have

$$g(\lambda_1(z)) + \cdots + g(\lambda_d(z)) = \operatorname{tr} g(A(z)) = \operatorname{tr} \oint g(s)(sI - A(z))^{-1} \text{đ}s,$$

where the integral contour surrounds the spectrum of $A(z)$. This is analytic provided that the contour can avoid Σ, i.e., for all z such that $\lambda_1(z), \ldots, \lambda_d(z)$ are not in Σ. If $\lambda_i(z) \in \Sigma$, then $z = p(\lambda_i(z)) \in p(\Sigma)$. □

We now obtain the following.

Theorem 5.32. *Let $p : \mathbb{I} \to \Gamma$ be a one-to-one dth degree polynomial and $\lambda_1(z), \ldots, \lambda_d(z)$ an enumeration of the roots of $p(\lambda) = z$. Provided that $f \in H^1(\Gamma)$, $z \notin \Gamma$, including boundary values,*

$$\mathcal{C}_\Gamma f(z) = \sum_{i=1}^d \mathcal{C}_\mathbb{I}[f \circ p](\lambda_i(z)).$$

Proof. Let

$$\phi(z)=\sum_{i=1}^{d}\mathscr{C}_{\mathbb{I}}[f\circ p](\lambda_i(z)).$$

From Lemma 5.31 it is analytic everywhere off Γ. Moreover, as $z\to\infty$, $\lambda_i(z)\to\infty$ and $\phi(z)\to 0$.

Now for z on Σ, there is precisely one $\lambda_i(z)$ on $[-1,1]$. Thus we have

$$\phi_+(z)-\phi_-(z)=\phi_+(\lambda_i(z))-\phi_-(\lambda_i(z))=f(p(\lambda_i(z)))=f(z).$$

The theorem thus follows from Plemelj's lemma. □

In our applications, we can always deform such a Γ to be a straight segment, avoiding the need for polynomial maps and hence avoiding the calculation of roots of a polynomial. This depends on analyticity of the integrand, however, and future applications may not allow for deformation.

5.7.5 ▪ Multiple contours

We finally consider the computation of Cauchy transforms on contours composed of multiple Möbius transformations of the unit interval. This proceeds by applying the Cauchy evaluation formula to f restricted to each subinterval. The uniform convergence of this approximation away from nonsmooth points of Γ is an immediate consequence. We define a space that encapsulates the zero-sum condition and absolute convergence.

Definition 5.33. *Suppose* $\Gamma=\Gamma_1\cup\cdots\cup\Gamma_L$, *where each component is defined by a Möbius transformation* $M_j:\mathbb{I}\to\Gamma_j$. *We say* $f\in Z(\Gamma)$ *if* $\mathscr{T}f|_{\Gamma_j}\in\ell^{2,1}$ *for* $j=1,\ldots,L$ *and the zero-sum condition holds at all sets of nonsmooth points* γ_0 *of* Γ. *We equip* $Z(\Gamma)$ *with the norm*

$$\|f\|_{Z(\Gamma)}\triangleq\sum_{j=1}^{L}\left\|\mathscr{T}f|_{\Gamma_j}\right\|_{\ell^{2,1}}.$$

We can now determine exactly the behavior of the Cauchy transform near self-intersection points.

Lemma 5.34. *Suppose* $f\in Z(\Gamma)$. *Then we can express the limits of the Cauchy transform at* $a\in\gamma_0$ *in terms of the finite-part Cauchy transform:*

$$\mathscr{C}_\Gamma f(a+0\mathrm{e}^{\mathrm{i}\theta})=\sum_{j=1}^{L}\sum_{k=0}^{\infty}\check{f}_{kj}\,\mathrm{FP}\mathscr{C}_{\Gamma_j}[T_k\circ M_j^{-1}](a+0\mathrm{e}^{\mathrm{i}\theta}),$$

where

$$\mathscr{C}_\Gamma f(a+0\mathrm{e}^{\mathrm{i}\theta})\triangleq\lim_{\epsilon\downarrow 0}\mathscr{C}_\Gamma f(a+\epsilon\mathrm{e}^{\mathrm{i}\theta})$$

and $\check{f}_{kj}$ *are the mapped Chebyshev coefficients of* $f|_{\Gamma_j}$:

$$f(M_j(x))=\sum_{k=0}^{\infty}\check{f}_{kj}\,T_k(x).$$

Proof. Let $\Gamma_{j_1},\ldots,\Gamma_{j_\ell} \Subset \Gamma$ be an enumeration of the subcomponents of Γ that contain a as an endpoint, which we assume are oriented outwards from a for simplicity. For all $j \neq j_1,\ldots,j_\ell$, a is bounded away from Γ_j and the finite-part Cauchy transforms are equivalent to the true Cauchy transforms; thus we need to only consider the contours Γ_{j_i}. The regularity of f on each subcontour means that we can expand in the vanishing basis

$$f(M_{j_i}(x)) = f_i + \sum_{k=1}^{\infty} \check{f}_{kj_i}^{-z} T_k^{-z}(x), \tag{5.1}$$

where the coefficients are in ℓ^1 due to Proposition 4.10 and $f_i = f|_{\Gamma_{j_i}}(a)$. The finite-part Cauchy transform is also equivalent to the standard Cauchy transform at a for the $T_k^{-z} \circ M_j^{-1}$ terms. The zero-sum condition imposes that $\sum_{i=1}^{\ell} f_i = 0$, causing the logarithmic singularities to cancel:

$$\begin{aligned}\lim_{\epsilon\to 0}\sum_{i=1}^{\ell} f_i \mathcal{C}_{\Gamma_{j_i}} 1(a+\epsilon e^{i\theta}) &= \sum_{i=1}^{\ell} f_i \mathrm{FP}\mathcal{C}_{\Gamma_{j_i}} 1(a+0e^{i\theta}) + \lim_{\epsilon\to 0}\sum_{i=1}^{\ell} f_i \frac{\log\epsilon}{2\pi i} \\ &= \sum_{\ell=1}^{L} f_i \mathrm{FP}\mathcal{C}_{\Gamma_{j_i}} 1(a+0e^{i\theta}). \qquad \square\end{aligned}$$

We represent solutions to RH problems as Cauchy transforms of numerically calculated functions. This lemma states that to show convergence of the approximation, we need to only show convergence in $Z(\Gamma)$.

Corollary 5.35. *If $f, f_n \in Z(\Gamma)$, then*

$$|\mathcal{C}f(z) - \mathcal{C}f_n(z)| \leq C\,\|f - f_n\|_{Z(\Gamma)}$$

holds uniformly in the complex plane, including the left and right limits on Γ and along any angle as z approaches a junction point.

Remark 5.7.2. *Note that $Z(\Gamma) \subset H^2(\Gamma) \cap H_z^1(\Gamma)$.*

Argument change of finite-part Cauchy transforms: Define the finite-part Cauchy transform when Γ is a union of contours by linearity.

Definition 5.36. *If $\Gamma = \Gamma_1 \cup \cdots \cup \Gamma_L$, then*

$$\mathrm{FP}\mathcal{C}_\Gamma f(z) \triangleq \sum_{j=1}^{L} \mathrm{FP}\mathcal{C}_{\Gamma_j} f(z).$$

We finally observe a property of the finite-part Cauchy transform for multiple contours that will prove necessary in the next chapter. We know that when the zero-sum condition is satisfied, the Cauchy transform is continuous in each sector near a junction point between two contours. When the zero-sum condition is not satisfied, we lose this property, which the following corollary encapsulates.

Corollary 5.37. *Let $\Gamma = \Gamma_1 \cup \cdots \cup \Gamma_L$, where each component is a Möbius transformation of the unit interval. Assume that $a \in \gamma_0$ and let $\Gamma_{j_1},\ldots,\Gamma_{j_\ell}$ denote the subcomponents of Γ which* *contain a as an endpoint, ordered by increasing argument. Suppose f satisfies $\mathcal{T} f|_{\Gamma_j} \in \ell^{2,1}$ for $j = 1,\ldots,L$. Then*

$$\begin{aligned}
\mathrm{FP}\mathcal{C}_\Gamma^- f(a+0\mathrm{e}^{\mathrm{i}\varphi_{j+1}}) &= \mathrm{FP}\mathcal{C}_\Gamma^+ f(a+0\mathrm{e}^{\mathrm{i}\varphi_j}) + (\varphi_{j+1}-\varphi_j)S,\\
\mathrm{FP}\mathcal{C}_\Gamma^- f(a+0\mathrm{e}^{\mathrm{i}\varphi_1}) &= \mathrm{FP}\mathcal{C}_\Gamma^+ f(a+0\mathrm{e}^{\mathrm{i}\varphi_L}) + (\varphi_1+2\pi-\varphi_L)S,
\end{aligned}$$

where φ_i denotes the angle that Γ_{j_i} leaves/approaches a and

$$S = -\sum_{i=1}^{\ell} \sigma_i \frac{f_{j_i}(a)}{2\pi},$$

where $\sigma_i = +1$ if Γ_{j_i} is oriented outwards and -1 if Γ_{j_i} is oriented inwards. In particular, we have continuity in each sector whenever $f \in Z(\Gamma)$:

$$\begin{aligned}
\mathrm{FP}\mathcal{C}^- f(a+0\mathrm{e}^{\mathrm{i}\varphi_{j+1}}) &= \mathrm{FP}\mathcal{C}^+ f(a+0\mathrm{e}^{\mathrm{i}\varphi_j}),\\
\mathrm{FP}\mathcal{C}^- f(a+0\mathrm{e}^{\mathrm{i}\varphi_1}) &= \mathrm{FP}\mathcal{C}^+ f(a+0\mathrm{e}^{\mathrm{i}\varphi_L}).
\end{aligned}$$

Proof. In this proof, we assume each Γ_{j_i} is oriented so that a is the left endpoint, and hence $\varphi_i = \arg M'(-1)$ and $\sigma_i = 1$. The regularity in f allows us to expand in the zero-sum basis as in (5.1). When we take the finite-part Cauchy transform of this basis, the only terms that vary with the argument are the constant terms. Combining Corollary 5.26 with Proposition 5.23, we see that the dependence on θ is simple, giving us

$$\mathrm{FP}\mathcal{C}_{\Gamma_{j_i}} 1(a+0\mathrm{e}^{\mathrm{i}\theta}) = \frac{\arg\left[-\mathrm{e}^{\mathrm{i}(\theta-\varphi_i)}\right]}{2\pi} + C,$$

where C is some constant independent of θ. Thus, provided $\theta \le \phi < \pi$, we have

$$\mathrm{FP}\mathcal{C}_{\Gamma_{j_i}} 1(a+0\mathrm{e}^{\mathrm{i}\phi}) - \mathrm{FP}\mathcal{C}_{\Gamma_{j_i}} 1(a+0\mathrm{e}^{\mathrm{i}\theta}) = \frac{\theta-\phi}{2\pi}.$$

This shows the first equation, and the second follows because arg differs by a factor of 2π once $\phi > \pi$. $\square$

Chapter 6
The Numerical Solution of Riemann–Hilbert Problems

Before we begin this chapter, we remark that other approaches to the numerical solution of Riemann–Hilbert problems have appeared in the literature (see, for example, [88, 40, 65]). These methods are typically based on discretizing the singular integral, or regularized singular integral, with a quadrature rule, e.g., the trapezoidal rule. Significant care must be employed in evaluating such discretized integrals near the contour of integration, and successful methodologies exist to deal with this issue [70]. Our method relies on a basis expansion and an explicit application of the Cauchy transform. Our evaluations and solutions are valid uniformly up to the contour of integration without the need for additional schemes.

With this in mind we consider the numerical solution of general RH problems, posed on complicated contours $\Gamma = \Gamma_1 \cup \cdots \cup \Gamma_L$, where Γ_j are Möbius transformations of the unit interval $M_j : \mathbb{I} \to \Gamma_j$. Our approach is based on solving the SIE (2.43):

$$\begin{aligned} \mathscr{C}[G;\Gamma]u &= G - I \quad \text{for} \\ \mathscr{C}[G;\Gamma]u &\triangleq \mathscr{C}_\Gamma^+ u - \mathscr{C}_\Gamma^- u G = u - \mathscr{C}_\Gamma^- u(G-I). \end{aligned}$$

Note that each row of the system is independent; therefore it is possible to reduce the problem to several vector-valued RH problems:

$$\mathscr{C}[G;\Gamma](e_j^\top u) = e_j^\top G - e_j^\top.$$

In the presentation below, we primarily use the full matrix version, though in some cases we specialize to the 2×2 case where we solve for the rows separately.

To numerically approximate solutions to this SIE, we replace the infinite-dimensional operator with finite-dimensional matrices, reducing the problem to finite-dimensional linear algebra. In Section 6.1, we introduce the family of *projection methods* for solving general linear equations. Convergence for general projection methods is discussed in Section 6.1.1, subject to an assumption restricting the growth of the norm of the inverse of the underlying linear system.

To construct the required matrices, we exploit the fact that we can readily evaluate the Cauchy transform applied to a mapped Chebyshev basis *pointwise* to construct a *collocation method*, as described in Section 6.1.2. This gives us Chebyshev coefficients such that

$$u(s) \approx \sum_{k=0}^{n_j-1} u_{kj}^n T_k(M_j^{-1}(s)) \qquad \text{for} \qquad s \in \Gamma_j,$$

where the unknown coefficients u_{kj}^n are determined by imposing that the SIE holds at a sequence of points. This leads to an approximation to the solution of the RH problem

$$\Phi(z) \approx \Phi_n(z) \triangleq I + \sum_{j=1}^{L} \sum_{k=0}^{n_j-1} u_{kj}^n \mathcal{C}_{\Gamma_j}[T_k \circ M_j^{-1}](z),$$

where the resulting Cauchy transforms can be evaluated via Chapter 5. Collocation methods are a type of projection method and hence fit directly into the general framework of Section 6.1. For an approximation resulting from a projection method converging to the true solution, it is sufficient to show that the approximation to u converges in $Z(\Gamma)$; see Corollary 5.35.

6.1 ▪ Projection methods

Suppose we are given an infinite-dimensional linear equation

$$\mathcal{A} u = f,$$

where $\mathcal{A} : X \to Y$ is an invertible linear map and $f \in Y$ for two vector spaces X and Y. The goal of projection methods is to use *projection operators* to convert the infinite-dimensional equation to a finite-dimensional linear system.

Assume we are given a projection operator $\mathcal{I}_n : Y \to Y_n$ where $Y_n \subset Y$ is finite-dimensional, and suppose X_n is a finite-dimensional subspace of X. The projection method considers the finite-dimensional operator $\mathcal{A}_n : X_n \to Y_n$ defined by

$$\mathcal{A}_n \triangleq \mathcal{I}_n \mathcal{A}|_{X_n}.$$

An approximation to $u \approx u^n \in X_n$ is obtained by solving the finite-dimensional equation

$$\mathcal{A}_n u^n = \mathcal{I}_n f.$$

Example 6.1. The canonical concrete example of a projection method is the *finite-section method*. In this case $X = Y = \ell^2$, $X_n = Y_n = \mathbb{C}^n$, and $\mathcal{I}_n$ is the *truncation operator*:

$$\mathcal{I}_n \begin{bmatrix} u_1 \\ u_2 \\ u_3 \\ \vdots \end{bmatrix} = \begin{bmatrix} u_1 \\ u_2 \\ \vdots \\ u_n \end{bmatrix}.$$

If we write $\mathcal{A}$ in matrix form by its action on the canonical basis $\boldsymbol{e}_k$,

$$\mathcal{A} = \begin{bmatrix} A_{11} & A_{12} & A_{13} & \cdots \\ A_{21} & A_{22} & A_{23} & \cdots \\ A_{31} & A_{32} & A_{33} & \ddots \\ \vdots & \vdots & \ddots & \ddots \end{bmatrix},$$

then the finite-section method consists of solving the finite-dimensional linear system

$$\mathcal{A}_n u^n = f_n \quad \text{for} \quad \mathcal{A}_n = \mathcal{I}_n \mathcal{A}|_{\mathbb{C}^n} = \begin{bmatrix} A_{11} & \cdots & A_{1n} \\ \vdots & \ddots & \vdots \\ A_{n1} & \cdots & A_{nn} \end{bmatrix}. \quad \blacksquare$$

6.1.1 ▪ Convergence of projection methods

We now turn our attention to bounding the error of projection methods. To facilitate this, it helps to have another projection operator $\mathscr{P}_n : X \to X_n$. Note that there can be many such projection operators $\mathscr{P}_n$, the choice of which can impact the bound.

Lemma 6.2. *Suppose $\mathscr{P}_n : X \to X_n$ and $\mathscr{I}_n : Y \to Y_n$ are projection operators, where $X_n \subset X$ and $Y_n \subset Y$. Then we have*

$$||u - u^n||_X \leq \left(1 + ||\mathscr{A}_n^{-1}||_{\mathscr{L}(Y_n, X_n)} ||\mathscr{I}_n||_{\mathscr{L}(Y, Y_n)} ||\mathscr{A}||_{\mathscr{L}(X,Y)}\right) ||u - \mathscr{P}_n u||_X. \tag{6.1}$$

Proof. We have

$$u^n \triangleq \mathscr{A}_n^{-1} \mathscr{I}_n f = \mathscr{A}_n^{-1} \mathscr{I}_n \mathscr{A} u.$$

Thus we have

$$\begin{aligned}
u - u^n &= u - \mathscr{P}_n u + \mathscr{P}_n u - \mathscr{A}_n^{-1} \mathscr{I}_n \mathscr{A} u \\
&= u - \mathscr{P}_n u + \mathscr{A}_n^{-1} \mathscr{A}_n \mathscr{P}_n u - \mathscr{A}_n^{-1} \mathscr{I}_n \mathscr{A} u \\
&= u - \mathscr{P}_n u + \mathscr{A}_n^{-1} (\mathscr{A}_n \mathscr{P}_n u - \mathscr{I}_n \mathscr{A} u) \\
&= u - \mathscr{P}_n u + \mathscr{A}_n^{-1} \mathscr{I}_n \mathscr{A} (\mathscr{P}_n u - u) \\
&= \left(I - \mathscr{A}_n^{-1} \mathscr{I}_n \mathscr{A}\right)(u - \mathscr{P}_n u),
\end{aligned}$$

where I is taken to mean the identity operator on X. □

Provided the projection $\mathscr{P}_n u$ converges to u faster than the other operator norms grow, the method converges. The power of this lemma is that the rate of decay of $||\mathscr{P}_n u - u||_X$ can be extremely fast due to high regularity properties of the solution u, which implies fast convergence of u^n to u, even though $\mathscr{A}_n$ is constructed in low regularity spaces and no information about high regularity is used in the numerical scheme.

Remark 6.1.1. *This is in contrast to many other numerical methods — e.g., the finite-element method — where the convergence rate is limited by the regularity properties used in the construction of the numerical scheme.*

In the context of the SIEs associated with RH problems, we want to take $X = Z(\Gamma)$ as introduced in Definition 5.33 to ensure uniform convergence, due to Corollary 5.35. In this context, it is clear that the finite-section method is not appropriate: the truncation operator does not preserve the zero-sum condition encoded in $Z(\Gamma)$. We therefore must use an alternative.

6.1.2 ▪ Collocation methods

In practice, we use a *collocation method* for solving RH problems, which will prove convenient for enforcing the zero-sum condition. We first present collocation methods, before seeing that they are, in fact, a special case of projection methods.

Assume that $\mathscr{A} : X \to Y$, where $Y \subset C^0(\Gamma)$ (i.e., piecewise continuous functions; see (A.1)), so that evaluation is a well-defined operation. Then we can apply the operator $\mathscr{A}$ to a basis $\{\psi_k\} \subset X$ pointwise, i.e., we can evaluate $\mathscr{A}\psi_k(s)$ exactly for $s \in \Gamma$. Given n

collocation points $s^n = [s_1, \ldots, s_n]$, $s_j \in \Gamma$, we can construct an $n \times n$ linear system that consists of imposing that the linear equation holds pointwise:

$$\sum_{k=0}^{n-1} u_k^n \mathcal{A} \psi_k(s_j) = f(s_j) \quad \text{for} \quad j = 1, \ldots, L.$$

In other words, we calculate u_k^n by solving the linear system

$$A_n \begin{bmatrix} u_0^n \\ \vdots \\ u_{n-1}^n \end{bmatrix} = \begin{bmatrix} f(s_1) \\ \vdots \\ f(s_n) \end{bmatrix} \quad \text{for} \quad A_n = \begin{bmatrix} \mathcal{A}\psi_0(s_1) & \cdots & \mathcal{A}\psi_{n-1}(s_1) \\ \vdots & \ddots & \vdots \\ \mathcal{A}\psi_0(s_n) & \cdots & \mathcal{A}\psi_{n-1}(s_n) \end{bmatrix}.$$

We then arrive at the approximation

$$u^n = \sum_{k=0}^{n-1} u_k^n \psi_k. \tag{6.2}$$

Convergence bounds: The same argument as in projection methods allows us to bound the error.

Corollary 6.3. *Define the evaluation operator* $\mathcal{E}_n : C^0(\Gamma) \to \mathbb{C}^n$ *by*

$$\mathcal{E}_n f = [f(s_1), \ldots, f(s_n)]^\top,$$

the expansion operator $\mathcal{B}_n : \mathbb{C}^n \to X_n$ *by*

$$\mathcal{B}_n \begin{bmatrix} u_0 \\ \vdots \\ u_{n-1} \end{bmatrix} = \sum_{k=0}^{n-1} u_k \psi_k,$$

and the domain space by $X_n = \operatorname{span}\{\psi_0, \ldots, \psi_{n-1}\}$. *For any projection operator* $\mathcal{P}_n : X \to X_n$, *and any choice of norm on* $\mathbb{C}^n$, *we have the bound*

$$\|u - u^n\|_X \le \left(1 + \|\mathcal{B}_n\|_{\mathcal{L}(\mathbb{C}^n, X)} \|A_n^{-1}\|_{\mathcal{L}(\mathbb{C}^n)} \|\mathcal{E}_n\|_{\mathcal{L}(Y, \mathbb{C}^n)} \|\mathcal{A}\|_{\mathcal{L}(X,Y)}\right) \|u - \mathcal{P}_n u\|_X,$$

where u^n *is* (6.2).

Proof. Similar to the argument of Lemma 6.2, we have

$$u - u^n = (I - \mathcal{B}_n A_n^{-1} \mathcal{E}_n \mathcal{A})(u - \mathcal{P}_n u). \qquad \square$$

In the setting of RH problems, X has special structure that may not be satisfied by the basis: for example, the basis may be piecewise mapped Chebyshev polynomials but X imposes the zero-sum condition. Thus we consider the case where there is a larger space Z that contains both X and the basis: $\psi_k \in Z$ and $X \subset Z$ but $\psi_k \notin X$. This raises an issue: we can no longer define the collocation matrix by applying the operator to the basis, as the operator is only defined on X. Instead, we assume we are given a matrix A_n (in our case, built from the finite-part Cauchy transform) that is equivalent to $\mathcal{A}$ for all expansions in the basis that are also in X. The argument of Corollary 6.3 remains valid, under the condition that the collocation method maps to the correct space.

Corollary 6.4. *Define the domain space* $X_n = \operatorname{span}\{\psi_0, \dots, \psi_{n-1}\} \cap X$. *Suppose we are given an invertible matrix* A_n *satisfying*

$$A_n \begin{bmatrix} u_0 \\ \vdots \\ u_{n-1} \end{bmatrix} = \mathscr{E}_n \mathscr{A} \sum_{k=0}^{n-1} u_k \psi_k$$

whenever $\sum_{k=0}^{n-1} u_k \psi_k \in X_n$, *which we use in the definition of* u^n *via (6.2). For any projection operator* $\mathscr{P}_n : X \to X_n$, *and any choice of norm on*

$$V_n \triangleq \mathscr{E}_n \mathscr{A} X_n \subset \mathbb{C}^n, \tag{6.3}$$

we have the bound

$$\|u - u^n\|_X \le \left(1 + \|\mathscr{B}_n A_n^{-1}\|_{\mathscr{L}(V_n, X)} \|\mathscr{E}_n\|_{\mathscr{L}(Y, V_n)} \|\mathscr{A}\|_{\mathscr{L}(X,Y)}\right) \|u - \mathscr{P}_n u\|_X \tag{6.4}$$

provided that

$$\mathscr{B}_n A_n^{-1} : V_n \to X.$$

Furthermore, for any choice of norm on $\mathbb{C}^n$

$$\|u - u^n\|_X \le \left(1 + \|\mathscr{B}_n\|_{\mathscr{L}(\mathbb{C}^n, Z)} \|A_n^{-1}\|_{\mathscr{L}(\mathbb{C}^n)} \|\mathscr{E}_n\|_{\mathscr{L}(Y, \mathbb{C}^n)} \|\mathscr{A}\|_{\mathscr{L}(X,Y)}\right) \|u - \mathscr{P}_n u\|_X.$$

Proof. We have the identity

$$u - u^n = (I - \mathscr{B}_n A_n^{-1} \mathscr{E}_n \mathscr{A})(u - \mathscr{P}_n u)$$

from before, where the right-hand side must be in X from the assumptions. Taking the X-norm of both sides gives (6.4). We also have

$$\|u - u^n\|_X \le \left(1 + \|\mathscr{B}_n\|_{\mathscr{L}(S_n, X)} \|A_n^{-1}\|_{\mathscr{L}(V_n, S_n)} \|\mathscr{E}_n\|_{\mathscr{L}(Y, V_n)} \|\mathscr{A}\|_{\mathscr{L}(X,Y)}\right) \|u - \mathscr{P}_n u\|_X,$$

where $S_n = A_n^{-1} V_n \subset \mathbb{C}^n$. We use $\{\|v\|_X = 1\}$ to denote the set of all vectors in X with unit norm. Then if V_n and S_n are equipped with the same norm as $\mathbb{C}^n$,

$$\|\mathscr{B}_n\|_{\mathscr{L}(S_n, X)} = \sup_{\|v\|_{S_n}=1} \|\mathscr{B}_n v\|_X = \sup_{\|v\|_{S_n}=1} \|\mathscr{B}_n v\|_Z \le \sup_{\|v\|_{\mathbb{C}^n}=1} \|\mathscr{B}_n v\|_Z = \|\mathscr{B}_n\|_{\mathscr{L}(\mathbb{C}^n, Z)},$$

$$\|A_n^{-1}\|_{\mathscr{L}(V_n, S_n)} = \sup_{\|v\|_{V_n}=1} \|A_n^{-1} v\|_{S_n} = \sup_{\|v\|_{V_n}=1} \|A_n^{-1} v\|_{\mathbb{C}^n} \le \sup_{\|v\|_{\mathbb{C}^n}=1} \|A_n^{-1} v\|_{\mathbb{C}^n} = \|A_n^{-1}\|_{\mathscr{L}(\mathbb{C}^n)},$$

$$\|\mathscr{E}_n\|_{\mathscr{L}(Y, V_n)} = \sup_{\|v\|_Y=1} \|\mathscr{E}_n v\|_{V_n} = \sup_{\|v\|_Y=1} \|\mathscr{E}_n v\|_{\mathbb{C}^n} = \|\mathscr{E}_n\|_{\mathscr{L}(Y, \mathbb{C}^n)},$$

and the corollary follows. □

Collocation methods as projection methods: We remark that collocation methods can also be viewed as a special family of projection methods, via an interpolation operator that takes V_n to Y.

Corollary 6.5. *Assume that there is an interpolation operator* $\mathscr{T}_n : V_n \to Y \subset C^0(\Gamma)$ *defined on* V_n *from (6.3) so that* $\mathscr{T}_n v(s_k) = e_k^\top v$ *for all* $v \in V_n$. *Then collocation methods are equivalent to a projection method with the projection operator* $\mathscr{I}_n = \mathscr{T}_n \mathscr{E}_n$, *the domain space* $X_n = \operatorname{span}\{\psi_0, \dots, \psi_{n-1}\} \cap X$, *and* $\mathscr{A}_n = \mathscr{T}_n A_n \mathscr{B}_n^{-1}$.

This interpretation will prove useful in the next chapter.

Chebyshev collocation methods: We will use the Chebyshev bases introduced in Chapter 4 for the construction of collocation methods for RH problems. In the case where Γ has a single component and $M : \mathbb{I} \to \Gamma$ is a map from the unit interval, we use mapped Chebyshev polynomials as the basis $\psi_k(s) = T_k(M^{-1}(s))$ and mapped Chebyshev points as the collocation points $s^n = M(x^n)$, where x^n was defined in Definition 4.7. When $\Gamma = \Gamma_1 \cup \cdots \cup \Gamma_L$ and $M_j : \mathbb{I} \to \Gamma_j$, we take $n = n_1 + \cdots + n_L$ and use the basis

$$\begin{cases} T_k(M_j^{-1}(s)) & \text{if } s \in \Gamma_j, \\ 0 & \text{otherwise} \end{cases} \quad \text{for} \quad k = 0, \ldots, n_j - 1 \quad \text{and} \quad j = 1, \ldots, L. \tag{6.5}$$

(We omit the relationship between the basis defined above and the precise ordering of ψ_k, as it is immaterial.) We use the union of mapped Chebyshev points as our collocation points:

$$s^n = M_1(x^{n_1}) \cup \cdots \cup M_L(x^{n_L}).$$

Note that this choice repeats nonsmooth points for every contour that contains them. We treat these as different points attached with the direction of approach:

$$M_j(-1) + 0\mathrm{e}^{\mathrm{i}\arg M_j'(-1)} \qquad \text{and} \qquad M_j(1) - 0\mathrm{e}^{\mathrm{i}\arg M_j'(1)},$$

recalling that $\arg M_j'(\pm 1)$ gives the angle that Γ_j leaves $M_j(-1)$ and approaches $M_j(1)$.

Finally, when the spaces are two-component row vectors, we take $n = 2(n_1 + \cdots + n_L)$ and use the bases

$$\begin{cases} \left[T_k(M_j^{-1}(s)),\ 0\right] & \text{if } s \in \Gamma_j, \\ [0, 0] & \text{otherwise} \end{cases} \quad \text{and} \quad \begin{cases} \left[0,\ T_k(M_j^{-1}(s))\right] & \text{if } s \in \Gamma_j, \\ [0, 0] & \text{otherwise} \end{cases}$$

for $k = 0, \ldots, n_j - 1$ and $j = 1, \ldots, L$.

Zero-sum spaces: For our setting, we want to work within the space $Z(\Gamma)$: in particular, we choose $X_n = Z_n(\Gamma) \triangleq Z(\Gamma) \cap \operatorname{span}\{\psi_0(s), \ldots, \psi_{n-1}(s)\}$, where the basis is constructed using mapped Chebyshev polynomials. Applying a truncation operator such as $\mathcal{I}_n$ to each component of Γ does not take $Z(\Gamma)$ to $Z_n(\Gamma)$; hence we cannot use it as $\mathcal{P}_n$. Instead, we use the projection consisting of evaluation and interpolation:

$$\mathcal{P}_n = \mathcal{B}_n \mathrm{T}_n \mathcal{E}_n,$$

which returns the piecewise-mapped Chebyshev polynomial that interpolates at Chebyshev points, including all self-intersection points. We have a guarantee that $\|\mathcal{P}_n u - u\|_{Z(\Gamma)}$ tends to zero by the theory of Chapter 4, with the speed of convergence dictated by the regularity of u.

6.2 ▪ Collocation method for RH problems

We now turn to the problem of constructing a collocation method to solve the SIE

$$\mathcal{C}[G;\Gamma]u = G - I.$$

In order to successfully construct a collocation method, our goal is to choose the basis and collocation points so that (1) we can evaluate the singular integral operator applied to the basis pointwise, (2) the collocation method converges, and (3) the collocation method converges *fast*. The last property is critical, not only because we aim to have high accuracy numerics, but also for the practical reason that the linear collocation system is dense, and dense linear algebra quickly breaks down as the dimension of the system increases.

6.2.1 ▪ Collocation method for a scalar RH problem on the unit interval

The singularities of the Cauchy transform at the endpoints of Γ present several difficulties for the construction of a collocation method. We first consider a simple example that avoids these issues: an RH problem for $\Gamma = \mathbb{I}$, where G is a smooth scalar function satisfying $G(\pm 1) = 1$.

We use the Chebyshev basis $\psi_k(x) = T_k(x)$ with the Chebyshev points. We thus need to evaluate

$$\mathscr{C}[G;\mathbb{I}]T_k(x) = T_k(x) + (1 - G(x))\mathscr{C}_{\mathbb{I}}^- T_k(x)$$

at the Chebyshev points. While $\mathscr{C}_{\mathbb{I}}^- T_k(x)$ has a logarithmic singularity at ± 1, this is canceled out by the decay of $1 - G(x)$, and hence we have $\mathscr{C}[G;\mathbb{I}]T_k(\pm 1) = T_k(\pm 1)$. We can therefore construct the *collocation matrix*

$$C_n[G;\mathbb{I}] \triangleq \begin{bmatrix} T_0(-1) & \cdots & T_{n-1}(-1) \\ \mathscr{C}[G;\mathbb{I}]T_0(x_2^n) & \cdots & \mathscr{C}[G;\mathbb{I}]T_{n-1}(x_2^n) \\ \vdots & \ddots & \vdots \\ \mathscr{C}[G;\mathbb{I}]T_0(x_{n-1}^n) & \cdots & \mathscr{C}[G;\mathbb{I}]T_{n-1}(x_{n-1}^n) \\ T_0(1) & \cdots & T_{n-1}(1) \end{bmatrix}.$$

Provided it is nonsingular, we solve the linear system

$$C_n[G;\mathbb{I}] \begin{bmatrix} u_0^n \\ \vdots \\ u_{n-1}^n \end{bmatrix} = \begin{bmatrix} G(x_1) - 1 \\ \vdots \\ G(x_n) - 1 \end{bmatrix},$$

giving the approximation

$$u(x) \approx u^n(x) \triangleq \sum_{k=0}^{n-1} u_k^n T_k(x).$$

In other words, we approximate the solution to the RH problem as

$$\Phi(z) \approx \Phi_n(z) \triangleq 1 + \mathscr{C}_{\mathbb{I}} u^n(z) = 1 + \sum_{k=0}^{n-1} u_k^n \mathscr{C}_{\mathbb{I}} T_k(z).$$

The terms $\mathscr{C}_{\mathbb{I}} T_k(z)$ can be calculated via the formulae in Section 5.6.

Convergence: The construction of a numerical approximation to the solution of an RH problem thus proceeds without difficulty. Convergence of the approximation is a more delicate question. If we assume a simple bound on the growth of the inverse of the collocation method, we can guarantee spectral convergence.

We recall Definition 2.45 and prove the following.

Theorem 6.6. *Assume that the solution $u \in H^{\lambda+3}(\mathbb{I})$, $u(\pm 1) = 0$, and*

$$\|C_n[G;\mathbb{I}]^{-1}\|_{\mathscr{L}(\ell^\infty)} = O(n^{\beta-3})$$

for some $\beta \geq 3$. Then

$$|\Phi(z) - \Phi_n(z)| = O(n^{\beta-\lambda}) \quad \text{as} \quad n \to \infty$$

holds uniformly in z.

Proof. It is sufficient by Section 5.7.1 to prove that

$$\|u^n - u\|_{Z(\mathbb{I})} = O(n^{\beta-\lambda}),$$

i.e., we need $u^n \in Z(\mathbb{I})$ and the coefficients of u^n to converge to the coefficients of u at the desired rate in the $\ell^{2,1}$ norm. We will bound each operator in Corollary 6.4.

Note that, for some constant D_1,

$$\|f\|_{H_z^1(\mathbb{I})} \leq D_1 \|f\|_{Z(\mathbb{I})}$$

since every function in $Z(\mathbb{I})$ has a uniformly converging derivative and vanishes to first order at ± 1. Therefore, we have that

$$\|\mathcal{C}[G;\mathbb{I}]\|_{\mathcal{L}(Z(\mathbb{I}),H_z^1)} \leq D_1 \|\mathcal{C}[G;\mathbb{I}]\|_{\mathcal{L}(H_z^1)}$$

is bounded, by Theorem 2.50. The evaluation operator is also bounded $H^1(\mathbb{I}) \to \mathbb{C}^n$ by Sobolev embedding, using the supremum norm attached to $\mathbb{C}^n$: there exists a constant D_2 such that for every u

$$\|\mathcal{E}_n u\|_{\mathbb{C}^n} \leq \|u\|_{L^\infty(\mathbb{I})} \leq D_2 \|u\|_{H^1(\mathbb{I})}.$$

Finally, the expansion operator satisfies

$$\|\mathcal{B}_n\|_{\mathcal{L}(\ell^\infty,\ell^{2,1})} \leq n^3,$$

using the trivial estimate

$$\|\boldsymbol{u}\|_{\ell^{2,1}} = \sum_{k=0}^{n-1} |u_k|(k+1)^2 \leq n^3 \|\boldsymbol{u}\|_{\ell^\infty}.$$

We thus have one last criterion, that

$$\mathcal{B}_n C_n[G;\mathbb{I}]^{-1} : V_n \to X_n,$$

where $V_n = \operatorname{ran} \mathcal{E}_n \mathcal{C}[G;\mathbb{I}] Z(\mathbb{I}) \subset \{\boldsymbol{u} : \boldsymbol{e}_1^\top \boldsymbol{u} = \boldsymbol{e}_n^\top \boldsymbol{u} = 0\}$, i.e., the space of vectors whose first and last entries are zero. This follows since the first and last rows of $C_n[G;\mathbb{I}]$ correspond to

$$\sum_{k=0}^{n-1} u_k^n T_k(\pm 1);$$

hence if the right-hand side has zero first and last rows, the approximate solution vanishes at ± 1. As a consequence of Proposition B.9, we have $\|\mathcal{P}_n u - u\|_{Z(\mathbb{I})} = O(n^{-\lambda})$. Putting everything together, we have

$$\begin{aligned}
&\|u^n - u\|_{Z(\mathbb{I})} \\
&\quad \leq \left(1 + \|\mathcal{B}_n\|_{\mathcal{L}(\ell^\infty,Z(\mathbb{I}))} \|C_n[G;\mathbb{I}]^{-1}\|_{\mathcal{L}(\ell^\infty)} \|\mathcal{E}_n\|_{\mathcal{L}(H^1(\mathbb{I}),\ell^\infty)} \|\mathcal{A}\|_{H_z^1(\mathbb{I})}\right) \|\mathcal{P}_n u - u\|_{Z(\mathbb{I})} \\
&\quad = \left[1 + n^3 \mathcal{O}(n^{\beta-3})\right] \mathcal{O}(n^{-\lambda}) \|\mathcal{P}_n u - u\|_{Z(\mathbb{I})} = \mathcal{O}(n^{\beta-\lambda}) \|\mathcal{P}_n u - u\|_{Z(\mathbb{I})}.
\end{aligned}$$

□

6.2.2 ▪ Construction of a collocation method for matrix RH problems on multiple contours

We now consider the construction of a collocation method for our general family of RH problems. In this case, we use the piecewise Chebyshev basis of (6.5), with the mapped Chebyshev points with the argument information included in the self-intersection points. Then we will construct a collocation matrix $\mathrm{FP}\, C_n[G;\Gamma]$ using the finite-part Cauchy transform.

We now construct the collocation matrix in the scalar case. (The construction in the vector and matrix case follows in a straightforward manner.) Begin with a single contour Γ, where $M : \mathbb{I} \to \Gamma$ is a Möbius transformation, so that Γ is a line segment, ray, or circular arc. We obtain a discretization of the Cauchy operator on Γ alone as

$$\mathrm{FP}\, C^{\pm}_{n\Gamma} \triangleq \begin{bmatrix} \mathrm{FP}\mathscr{C}^{\pm}_{\Gamma}[T_0 \circ M^{-1}](M(x^n_1)) & \cdots & \mathrm{FP}\mathscr{C}^{\pm}_{\Gamma}[T_{n-1} \circ M^{-1}](M(x^n_1)) \\ \mathscr{C}^{\pm}_{\Gamma}[T_0 \circ M^{-1}](M(x^n_2)) & \cdots & \mathscr{C}^{\pm}_{\Gamma}[T_{n-1} \circ M^{-1}](M(x^n_2)) \\ \vdots & \ddots & \vdots \\ \mathscr{C}^{\pm}_{\Gamma}[T_0 \circ M^{-1}](M(x^n_{n-1})) & \cdots & \mathscr{C}^{\pm}_{\Gamma}[T_{n-1} \circ M^{-1}](M(x^n_{n-1})) \\ \mathrm{FP}\mathscr{C}^{\pm}_{\Gamma}[T_0 \circ M^{-1}](M(x^n_n)) & \cdots & \mathrm{FP}\mathscr{C}^{\pm}_{\Gamma}[T_{n-1} \circ M^{-1}](M(x^n_n)) \end{bmatrix},$$

recalling that $M(x^n_1)$ and $M(x^n_n)$ are endowed with the argument that Γ leaves the respective point. This gives us a discretization of $\mathscr{C}[G;\Gamma]$ as

$$\mathrm{FP}\, C_n[G;\Gamma] \triangleq I - \boldsymbol{G} \cdot \mathrm{FP}\, C^{-}_{n\Gamma},$$

where $\boldsymbol{G} = \operatorname{diag}\left\{G(M(x^n_1)) - 1, \ldots, G(M(x^n_n)) - 1\right\}$. (In the matrix and vector case, this becomes a matrix corresponding to pointwise multiplication on the right.)

We can similarly construct a discretization of the Cauchy operator on Γ evaluated on another contour Ω given by a Möbius transformation $N : \mathbb{I} \to \Omega$:

$$\mathrm{FP}\, C_{n\Gamma}|_{m\Omega} \triangleq \begin{bmatrix} \mathrm{FP}\mathscr{C}_{\Gamma}[T_0 \circ M^{-1}](N(x^m_1)) & \cdots & \mathrm{FP}\mathscr{C}_{\Gamma}[T_{n-1} \circ M^{-1}](N(x^m_1)) \\ \mathscr{C}_{\Gamma}[T_0 \circ M^{-1}](N(x^m_2)) & \cdots & \mathscr{C}_{\Gamma}[T_{n-1} \circ M^{-1}](N(x^m_2)) \\ \vdots & \ddots & \vdots \\ \mathscr{C}_{\Gamma}[T_0 \circ M^{-1}](N(x^m_{m-1})) & \cdots & \mathscr{C}_{\Gamma}[T_{n-1} \circ M^{-1}](N(x^m_{m-1})) \\ \mathrm{FP}\mathscr{C}_{\Gamma}[T_0 \circ M^{-1}](N(x^m_m)) & \cdots & \mathrm{FP}\mathscr{C}_{\Gamma}[T_{n-1} \circ M^{-1}](N(x^m_m)) \end{bmatrix},$$

noting that the finite-part Cauchy transform becomes a standard Cauchy transform when Γ and Ω are disjoint. We now decompose the full collocation matrix into its action on each subcomponent Γ:

$$\mathrm{FP}\, C_n[G;\Gamma] \triangleq \begin{pmatrix} \mathrm{FP}\, C_{n_1}[G_1;\Gamma_1] & -\boldsymbol{G}_1 \cdot \mathrm{FP}\, C_{n_2\Gamma_2}|_{n_1\Gamma_1} & \cdots & -\boldsymbol{G}_1 \cdot \mathrm{FP}\, C_{n_L\Gamma_L}|_{n_1\Gamma_1} \\ -\boldsymbol{G}_2 \cdot \mathrm{FP}\, C_{n_1\Gamma_1}|_{n_2\Gamma_2} & \mathrm{FP}\, C_{n_2}[G_2,\Gamma_2] & \cdots & -\boldsymbol{G}_2 \cdot \mathrm{FP}\, C_{n_L\Gamma_L}|_{n_2\Gamma_2} \\ \vdots & \vdots & \ddots & \vdots \\ -\boldsymbol{G}_L \cdot \mathrm{FP}\, C_{n_1\Gamma_1}|_{n_L\Gamma_L} & -\boldsymbol{G}_L \cdot \mathrm{FP}\, C_{n_2\Gamma_2}|_{n_L\Gamma_L} & \cdots & \mathrm{FP}\, C_{n_L}[G_L,\Gamma_L] \end{pmatrix},$$

where $\boldsymbol{G}_j = \operatorname{diag}\left\{G(M_j(x^{n_j}_1)), \ldots, G(M_j(x^{n_j}_{n_j}))\right\}$.

Remark 6.2.1. *More explicit formulae for the collocation matrices in every special case are given in* [96] *and are implemented in* [93].

Convergence: To establish convergence, we need to show that $C_n[G;\Gamma]^{-1}$ maps the correct spaces so that we can appeal to Corollary 6.4. Before continuing, we use the following proposition, which is in the same vein as Theorem 2.77 and compiled from the arguments that preceded it. This result is used repeatedly to simplify the arguments of the proofs by facilitating the reversal of orientations.

Proposition 6.7. *Suppose* $\Gamma = \Gamma_1 \cup \cdots \cup \Gamma_L$, *where* $M_j : \mathbb{I} \to \Gamma_j$. *Consider* Γ'', *where we reverse the orientation of one contour:* $\Gamma'' = M_1(-\mathbb{I}) \cup \Gamma_2 \cup \cdots \cup \Gamma_L$. *Then the RH problem* $[G,\Gamma]$ *has the same solution as* $[\tilde{G};\Gamma'']$, *where* $\tilde{G}(s) = G^{-1}(s)$ *for* $s \in \Gamma_1$ *and* $\tilde{G}(s) = G(s)$ *otherwise. Furthermore, for any* $u \in L^2(\Gamma)$, $\mathscr{C}[G;\Gamma]u|_{\Gamma_1} = -\mathscr{C}[\tilde{G};\Gamma'']u|_{\Gamma_1}$ *and* $\mathscr{C}[G;\Gamma]u|_{\Gamma_j} = \mathscr{C}[\tilde{G};\Gamma'']u|_{\Gamma_j}$ *for* $j = 2,\ldots,L$. *Specifically, if* u *solves* $\mathscr{C}[G;\Gamma]u = G - I$, *then* $\mathscr{C}[\tilde{G};\Gamma'']\tilde{u} = \tilde{G} - I$, *where* $\tilde{u}|_{\Gamma_1} = -u|_{\Gamma_1}$ *and* $\tilde{u}|_{\Gamma_j} = u|_{\Gamma_j}$ *for* $j = 2,\ldots,L$. *These properties carry over to the collocation discretization.*

We now describe precisely the space in which the right-hand side of the collocation lives.

Remark 6.2.2. *Unlike the analytic treatment (recall Definition 2.54) we do not use a decomposition of the jump matrix* $G(s) = X_-^{-1}(s)X_+(s)$ *in what follows. Part of the reasoning for this is that we only need to account for the first-order zero-sum/product condition in our discretization.*

Definition 6.8. *A function* f *defined on* Γ *satisfies* the cyclic junction condition with respect to G *if, for every* $a \in \gamma_0$, *we have*

$$\sigma_1 f_1 G_2^{\sigma_2} \cdots G_\ell^{\sigma_\ell} + \sigma_2 f_2 G_3^{\sigma_3} \cdots G_\ell^{\sigma_\ell} + \cdots + \sigma_{\ell-1} f_{\ell-1} G_\ell^{\sigma_\ell} + \sigma_\ell f_\ell = 0,$$

where $\Gamma_{j_1},\ldots,\Gamma_{j_\ell}$ *are the components of* Γ *that contain* a *as an endpoint,* $G_i = G|_{\Gamma_{j_i}}(a)$ *and* $f_i = f|_{\Gamma_{j_i}}(a)$, *and* $\sigma_i = +1$ *if* Γ_{j_i} *is oriented outwards and* -1 *if* Γ_{j_i} *is oriented inwards from* a.

Lemma 6.9. *If* G *satisfies the product condition and* $u \in Z(\Gamma)$, *then* $\mathscr{C}[G;\Gamma]u$ *satisfies the cyclic junction condition with respect to* G.

Proof. Define $f = \mathscr{C}[G;\Gamma]u$, let a be a nonsmooth point of Γ, and let $\Gamma_{i_1},\ldots,\Gamma_{i_\ell}$ be the contours with a as an endpoint, which we assume are oriented outwards (using Proposition 6.7). Because $u \in Z(\Gamma)$, we know $\mathscr{C}u(z)$ is continuous in each sector of the complex plane off Γ in a neighborhood of a, and therefore $\mathscr{C}_\Gamma^- u(a + 0e^{i\varphi_{i+1}}) = \mathscr{C}_\Gamma^+ u(a + 0e^{i\varphi_i})$, where $\varphi_i = \arg M'_{j_i}(-1)$ is the angle that Γ_{j_i} leaves a. Also, from Plemelj's lemma, we have $\mathscr{C}_\Gamma^+ u(a + 0e^{i\varphi_i}) = u_i + \mathscr{C}_\Gamma^- u(a + 0e^{i\varphi_i})$. Putting these together, we have for $i = 2,\ldots,\ell$

$$\begin{aligned}
f_i &= u_i + \mathscr{C}_\Gamma^- u(a + 0e^{i\varphi_i})(I - G_i) = u_i + \mathscr{C}_\Gamma^+ u(a + 0e^{i\varphi_{i-1}})(I - G_i)\\
&= u_i + u_{i-1} - u_{i-1}G_i + \mathscr{C}_\Gamma^- u(a + 0e^{i\varphi_{i-1}})(I - G_i)\\
&= u_i + u_{i-1} - u_{i-1}G_i + \mathscr{C}_\Gamma^+ u(a + 0e^{i\varphi_{i-2}})(I - G_i)\\
&= u_i + u_{i-1} + u_{i-2} - (u_{i-1} + u_{i-2})G_i + \mathscr{C}_\Gamma^- u(a + 0e^{i\varphi_{i-2}})(I - G_i)\\
&\;\;\vdots\\
&= u_i + \cdots + u_1 - (u_{i-1} + \cdots + u_1)G_i + \mathscr{C}_\Gamma^- u(a + 0e^{i\varphi_1})(I - G_i).
\end{aligned}$$

Now consider plugging this expression into

$$f_1 G_2 \cdots G_\ell + f_2 G_3 \cdots G_\ell + \cdots + f_{\ell-1} G_\ell + f_\ell,$$

to show that it vanishes. Due to cancellation between terms, we have

$$(I-G_1)G_2 \cdots G_\ell + (I-G_2)G_3 \cdots G_\ell + \cdots + (I-G_{\ell-1})G_\ell + (I-G_\ell) = I - G_1 \cdots G_\ell = 0,$$

where the final equality follows from the product condition, which shows that the sum over the $\mathscr{C}_\Gamma^- u(a+0\mathrm{e}^{\mathrm{i}\varphi_1})(I-G_i)G_{i+1}\cdots G_\ell$ terms are canceled. Similar cancellation implies that

$$\begin{aligned} u_1 G_2 \cdots G_\ell + [(u_2+u_1) - u_1 G_2] G_3 \cdots G_\ell + \cdots + \left[(u_\ell + \cdots + u_1) - (u_{\ell-1} + \cdots + u_1) G_\ell\right] \\ = u_\ell + \cdots + u_1 = 0, \end{aligned}$$

where the final equality follows from the zero-sum condition. Putting these two cancellations together shows that f satisfies the required condition. □

We now show that, since the right-hand side satisfies the cyclic jump condition, the solution to the collocation system satisfies the zero-sum condition. There is a technical requirement on G for this to be true.

Definition 6.10. *The nonsingular junction condition is satisfied if, for all $a \in \gamma_0$,*

$$(\varphi_1 + 2\pi - \varphi_\ell)I + \sum_{j=2}^{L} (\varphi_j - \varphi_{j-1}) G_j \cdots G_\ell$$

is nonsingular (in other words, its determinant is nonzero), where we use the notation of Definition 6.8.

Lemma 6.11. *If $C_n[G;\Gamma]$ is nonsingular, G satisfies the product condition and nonsingular junction condition, and f satisfies the cyclic junction condition, then $\mathscr{B}_n C_n[G;\Gamma]^{-1}\mathscr{E}_n f \in Z_n(\Gamma)$.*

Proof. Let $u^n \triangleq \mathscr{B}_n C_n[G;\Gamma]^{-1}\mathscr{E}_n f$ be the solution to the collocation system, which we want to show is in $Z(\Gamma)$. Using the same notation as in the preceding proof, let $a \in \gamma_0$ and again assume the subcontours which contain a as an endpoint are oriented outwards. Define

$$C_i^\pm = \mathrm{FP}\,\mathscr{C}_\Gamma^\pm u^n(a + 0\mathrm{e}^{\mathrm{i}\varphi_i}).$$

By construction, the collocation system imposes that

$$C_i^+ - C_i^- G_i = f_i,$$

where f_i is $f|_{\Gamma_{j_i}}(a)$. From Corollary 5.37 we have

$$C_{i+1}^- = C_i^+ + (\varphi_{i+1} - \varphi_i)S,$$

where $S = -\frac{1}{2\pi}\sum_{i=1}^{\ell} u^n(a+0e^{i\varphi_i})$ (which we want to show is zero). We get

$$\begin{aligned} C_\ell^+ &= C_\ell^- G_\ell + f_i = C_{\ell-1}^+ G_\ell + S(\varphi_\ell - \varphi_{\ell-1})G_\ell + f_\ell \\ &= C_{\ell-1}^- G_{\ell-1} G_\ell + S(\varphi_\ell - \varphi_{\ell-1})G_\ell + f_{\ell-1}G_\ell + f_\ell \\ &= C_{\ell-2}^- G_{\ell-2}G_{\ell-1}G_\ell + S\left[(\varphi_{\ell-2} - \varphi_{\ell-1})G_{\ell-1}G_\ell + (\varphi_\ell - \varphi_{\ell-1})G_\ell\right] \\ &\qquad + f_{\ell-2}G_{\ell-1}G_{\ell-2} + f_{\ell-1}G_\ell + f_\ell \\ &\ \vdots \\ &= C_1^- G_1 \cdots G_\ell + S\sum_{i=2}^{\ell}(\varphi_i - \varphi_{i-1})G_i \cdots G_\ell + \sum_{i=1}^{\ell} f_i G_{i+1}\cdots G_\ell \\ &= C_\ell^+ + S\left[(\varphi_1 + 2\pi - \varphi_\ell)I + \sum_{i=2}^{\ell}(\varphi_i - \varphi_{i-1})G_i \cdots G_\ell\right], \end{aligned}$$

where we used the cyclic junction condition to remove the sums involving f_i. The nonsingular junction condition means that the equality holds only if $S = 0$. □

With this property out of the way, the proof of convergence of the collocation method follows almost identically to Theorem 6.6.

Corollary 6.12. *Assume that* Γ *is bounded, the solution* $u \in H^{\lambda+3}(\Gamma)$*, and*

$$\|C_n[G;\Gamma]^{-1}\|_{\mathcal{L}(\ell^\infty)} = \mathcal{O}(n^{\beta-3}).$$

Then

$$|\Phi(z) - \Phi_n(z)| = \mathcal{O}(n^{\beta-\lambda}) \quad \textit{as} \quad n \to \infty$$

holds uniformly in z.

Failure of the nonsingular junction condition: Now suppose the nonsingular junction condition is not satisfied at a single point $a \in \gamma_0$. Consider the modified collocation system with the condition

$$C_\ell^+ - C_\ell^- G_\ell = f_\ell$$

replaced with the condition that $S = \sum_{i=1}^{\ell} u^n(a+0e^{i\varphi_i}) = 0$, ensuring the solution is in the correct space. Assuming that the resulting system is nonsingular, we have an approximation that satisfies the zero-sum condition, and now we will show that $C_\ell^+ - C_\ell^- G_\ell = f_\ell$ is still satisfied. We have continuity of the Cauchy transform, and hence

$$C_{j+1}^- = C_j^+ \quad \text{and} \quad C_1^- = C_\ell^+.$$

Thus, using the cyclic jump condition and the product condition,

$$\begin{aligned} C_\ell^- G_\ell + f_\ell &= C_{\ell-1}^+ G_{\ell-1} G_\ell + f_\ell = \cdots = C_1^- G_1 \cdots G_{\ell-1} G_\ell + \sum_{j=1}^{\ell} f_j G_{j+1} \cdots G_\ell \\ &= C_\ell^+, \end{aligned}$$

and the removed condition is still satisfied.

Remark 6.2.3. *The implementation in* [93] *does not take into account this possible failure or the remedy, as it does not arise generically.*

6.3 ▪ Case study: Airy equation

In this section we numerically solve the RH problem for the Airy equation introduced in Section 1.3. This demonstrates the deformation procedure that is required to obtain an RH problem that is sufficiently regular and has smooth solutions. Initially, one may want to apply the numerical method directly to the RH problem in Figure 1.5, but a simple computation shows that it does not satisfy the product condition (Definition 2.55) that is required for smooth solutions which is, in turn, required for convergence. The issue arises from singularities of $z^{-1/4}$ and $z^{3/2}$ at $z = 0$. So, we demonstrate an important technique for moving contours away from such singularities.

Deformations: When deforming, it is beneficial to consider modifying $\boldsymbol{y}$ first because its jump matrices have no branch cuts. Define

$$J_0 = \begin{bmatrix} 0 & -\mathrm{i} \\ -\mathrm{i} & 0 \end{bmatrix}, \quad J_1 = \begin{bmatrix} 1 & 0 \\ -\mathrm{i} & 1 \end{bmatrix}, \quad J_2 = \begin{bmatrix} 1 & -\mathrm{i} \\ 0 & 1 \end{bmatrix},$$

and the jump conditions for $\boldsymbol{y}$ are shown in Figure 6.1. While $\boldsymbol{y}$ does not have nice asymptotic behavior at infinity, it is apparent that this RH problem satisfies the product condition of any order. We define $\boldsymbol{y}_1$ in terms of $\boldsymbol{y}$ within an open neighborhood, say $|z| < 1$, of the origin in Figure 6.2. Note that $\boldsymbol{y}_1 = \boldsymbol{y}$ for $|z| > 1$. The jumps satisfied by $\boldsymbol{y}_1$ are shown in the right panel of Figure 6.2. The next task is to deal with the large z behavior of $\boldsymbol{y}_1$ so that the resulting function tends to $[1,\ 1]$ at infinity. While this was accomplished in some sense in Section 1.3, the jump matrix should tend to the identity matrix at infinity for the RH problem to be k-regular for any k. As an intermediate step define

$$\boldsymbol{\phi}_1(z) = \boldsymbol{y}_1(z) \begin{cases} 2\sqrt{\pi} z^{1/4} \begin{bmatrix} e^{\frac{2}{3}z^{3/2}} & 0 \\ 0 & e^{-\frac{2}{3}z^{3/2}} \end{bmatrix} & \text{if } |z| > 1, \\ I & \text{if } |z| < 1. \end{cases}$$

Then, the same computations as in Section 1.3 show that $\boldsymbol{\phi}_1^+(z) = \boldsymbol{\phi}_1^-(z) J_0$ for $z \in \mathbb{R}^- \setminus \{|z| \le 1\}$ and $\boldsymbol{\phi}_1 \sim [1,\ 1]$ at infinity. Next, we modify $\boldsymbol{\phi}_1$ so that the resulting function has jumps that all tend to the identity. We look to find a 2×2 matrix solution of

$$H^+(z) = H^-(z) J_0,$$

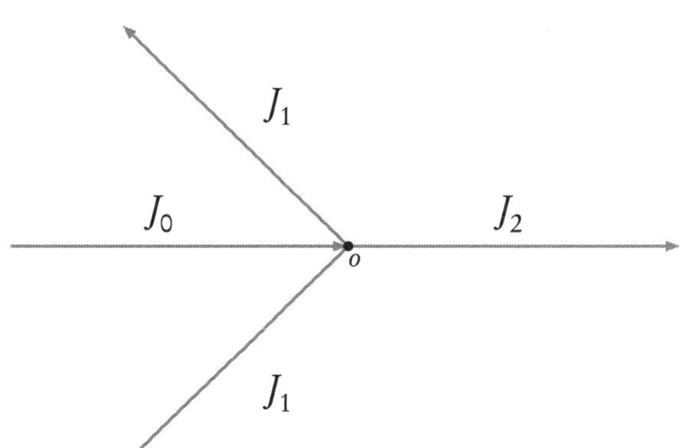

Figure 6.1. *The jumps satisfied by $\boldsymbol{y}$.*

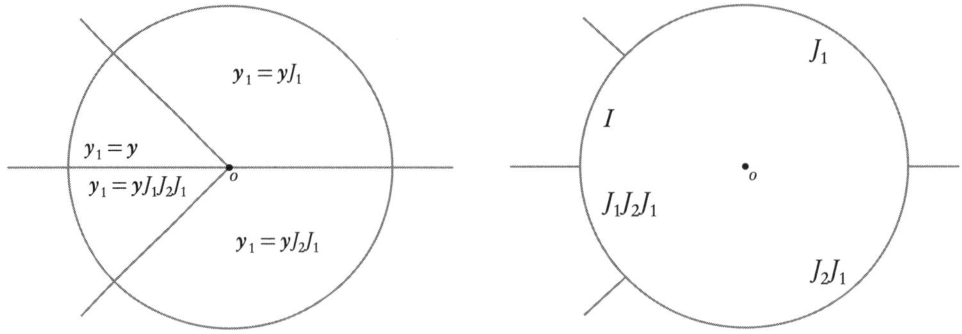

Figure 6.2. *The definition of* $\mathbf{y}_1$ *in B (left). The jumps satisfied by* $\mathbf{y}_1$ *(right). Circular contours have counterclockwise orientation.*

under the condition that $([1,\ 1]+\mathcal{O}(z^{-1}))H(z)=[1,\ 1]+\mathcal{O}(z^{-1})$. We multiply the function ϕ_1 by H^{-1} for $|z|>1$ to effectively remove the jump of J_0 on $\{z<0\}$. When we do this, we want to preserve the $[1,\ 1]$ condition at infinity. A suitable choice is

$$H(z)=\frac{1}{2}\begin{bmatrix} 1 & 1-\dfrac{1}{\sqrt{z}} \\ 1 & 1+\dfrac{1}{\sqrt{z}} \end{bmatrix}, \tag{6.6}$$

and a straightforward calculation verifies the required properties when $\sqrt{z}$ is given its canonical branch cut.

Finally, define

$$\phi_2(z)=\mathbf{y}_1(z)\begin{cases} R(z) & \text{if } |z|>1, \\ I & \text{if } |z|<1, \end{cases} \quad R(z)=2\sqrt{\pi}z^{1/4}\begin{bmatrix} e^{\frac{2}{3}z^{3/2}} & 0 \\ 0 & e^{-\frac{2}{3}z^{3/2}} \end{bmatrix}H(z).$$

So then $\phi_2(z)=[1,1]+\mathcal{O}(z^{-1})$ and it satisfies the jumps in Figure 6.3. In practice, infinite contours can be truncated following Proposition 2.78 when, say, $|z|\geq 15$ so that $e^{-2/3|z|^{3/2}}\approx 10^{-17}$.

Numerical results: To examine accuracy for large z an asymptotic expansion $\tilde{\phi}_m(z)$ of $\phi(z)$ to order m is readily computed [91]. We compare our numerical approximation $\phi^{6n}(z)$ of ϕ, with n collocation points per contour, to $\tilde{\phi}_m(z)$ in Figure 6.4. Accuracy is maintained for large values z.

6.4 ▪ Case study: Monodromy of an ODE with three singular points

In this section we demonstrate the statements made in Section 1.4 numerically. This involves first solving the ODE

$$Y'(z)=\sum_{k=1}^{3}\frac{A_k}{z-k}Y(z), \tag{6.7}$$

to compute the monodromy matrices M_1 and M_2. With the monodromy matrices in hand, we must solve Problem 1.4.1.

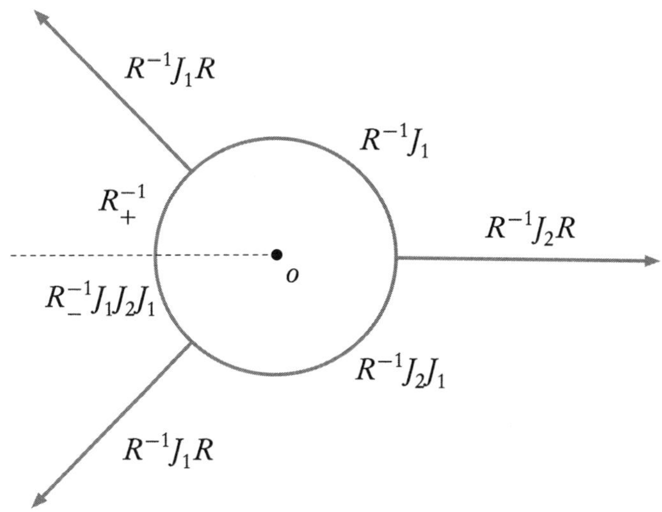

Figure 6.3. *The jump contours and jump conditions for ϕ_3. This RH problem can be solved numerically. Circular contours have counterclockwise orientation.*

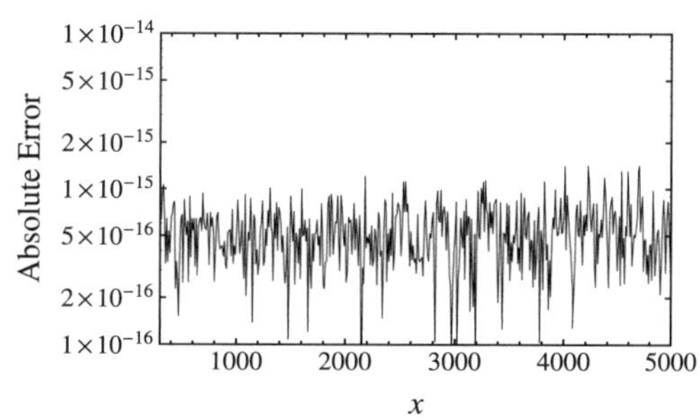

Figure 6.4. *A plot of $\|\tilde{\phi}_6^+(x)-\hat{\phi}_-^{360}(x)\|_2$ versus x. High accuracy is maintained as a function of x.*

Computing the monodromy matrices: We integrate (6.7) along paths in the complex plane. This is accomplished using complex "time-stepping." For example, the first-order Euler scheme becomes

$$Y(z+\mathrm{e}^{\mathrm{i}\theta}\Delta|z|)\approx Y(z)+\Delta|z|\mathrm{e}^{\mathrm{i}\theta}\sum_{k=1}^{3}\frac{A_k}{z-k}Y(z),\quad A_3=-A_1-A_2,$$

where $\Delta|z|$ is the time step. In this example, we use Runge–Kutta 4 instead of the Euler scheme; see [76]. We integrate this along the six contours in Figure 6.5. The contours that are close to one another can be taken to overlap and thus reduce the overall computation. Let $\tilde{Y}$ be the approximation of Y found using this method. We define approximations of the monodromy matrices

$$\tilde{M}_1=(\tilde{Y}^-(1.5))^{-1}\tilde{Y}^+(1.5),$$
$$\tilde{M}_2=(\tilde{Y}^-(2.5))^{-1}\tilde{Y}^+(2.5),$$
$$\tilde{M}_3=(\tilde{Y}^-(3.5))^{-1}\tilde{Y}^+(3.5).$$

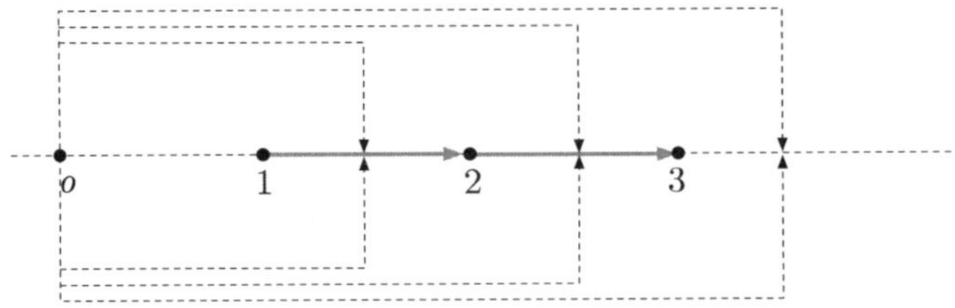

Figure 6.5. *Integration paths for the ODE* (6.7) *to compute* M_1, M_2, *and* M_3.

Because $A_1 + A_2 + A_3 = 0$, ∞ is a regular point and $\tilde{M}_3 \approx I$. It is clear that these can be obtained to any accuracy desired by reducing $\Delta|z|$.

Solving the RH problem: We proceed as if $\tilde{M}_k = M_k$ and leave the discussion of propagation of errors to a remark. Because $M_1 \neq I$ and generically $M_2 \neq M_1$ the RH problem in Problem 1.4.1 is not k-regular for any k. For the numerical method discussed here to have success we must convert it to a problem with smooth solutions.

We use local parametrices near $z = 1,2,3$ to remove singularities using the methodology of Problem 2.4.2. Let P_1 be a solution of $P_1^+(s) = P_1^-(s)M_1$ for $s \in (1,2)$, P_2 be a solution of $P_2^+(s) = P_2^-(s)M_2$ for $s \in (2,3)$, and P_{mid} be a solution of $P_{\text{mid}}^+(s) = P_{\text{mid}}^-(s)M_2^{-1}M_1$ for $s \in (1,2)$.

For our choice of A_k, we find positive eigenvalues for the monodromy matrices so that there is only one locally integrable solution of these problems and our parametrices are uniquely defined (up to multiplication on the left by a constant matrix).

The deformation process is described in Figure 6.6. As in Problem 1.4.1, Y is the solution of the undeformed problem. First, Y_1 is defined by a local redefinition of Y near $z = 2$ inside a circle of radius $0 < r < 1/2$; see Figure 6.6(a). This modifies the jumps to those given in Figure 6.6(b). This is essentially a lensing process, as described in Section 2.8.2. Using the local parametrices as shown in Figure 6.6(c) we find the jumps in Figure 6.6(d).

One can now easily check that all jump matrices are C^∞ smooth and satisfy the product condition to all orders. Furthermore, for our choice below, $\operatorname{tr} A_k = 0$ and therefore $\det Y(z) = 1$, and the determinant of all the jumps is constant by Liouville's formula. From here it follows that the determinants of all the jumps in Figure 6.6(d) are constant, and hence the index of the problem is zero; see Theorem 2.69. Therefore, the problem might be uniquely solvable. Indeed, because we have an invertible solution $Y(z)$, Lemma 2.67 shows that the associated singular integral operator is invertible.

Therefore, since the RH problem is uniquely solvable, the solution of the corresponding SIE $\mathscr{C}[G;\Gamma]u = G - I$ must have infinitely smooth solutions. The methodology of this chapter is readily applied to solve this RH problem with the normalization $Y_2(\infty) = I$. The circular contours in Figure 6.6 can also be replaced with piecewise linear ones.

To approximate $Y(z)$ and compute A_1, A_2, and A_3 we first calculate an approximation $Y_2^n(z) \approx Y_2(z)$ using collocation. We then renormalize at the origin via $Y(z) \approx Y_2^n(0)^{-1}Y_2^n(z) \triangleq Y_n(z)$ for z outside the circles in Figure 6.6(d). The first derivative $Y_n'(z)$ is also computed; see Section 4.3.2. Finally, the integrals

$$A_k \approx A_k^n \triangleq \int_{\partial B(k,2r)} Y_n'(z)Y_n(z)^{-1}\text{đ}z$$

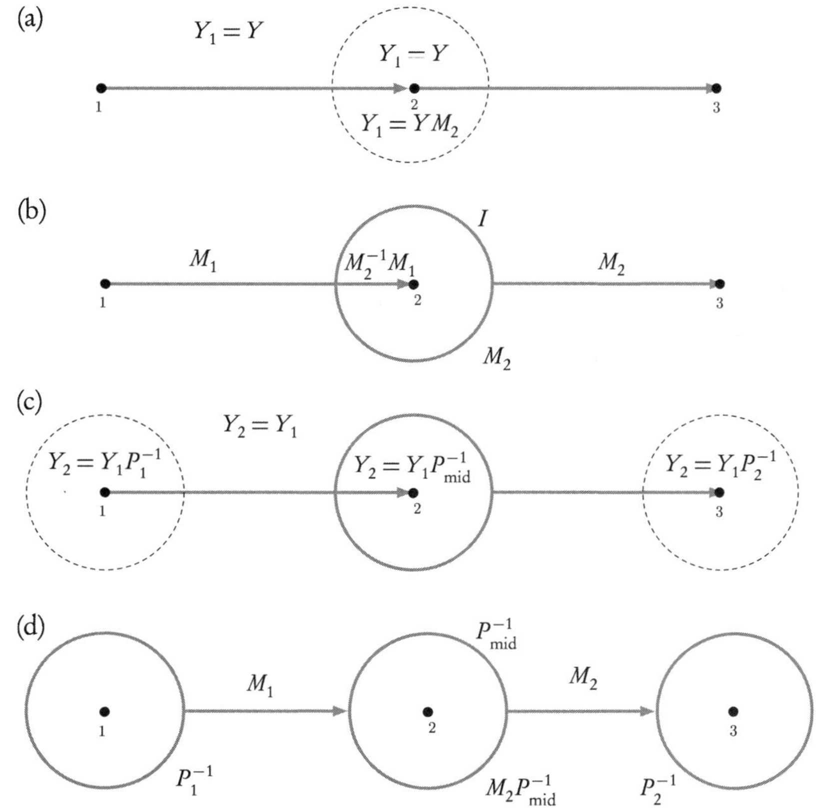

Figure 6.6. *The deformation of the monodromy RH problem for numerical purposes. All circular contours have radius* $0 < r < 1/2$ *with counterclockwise orientation.* (a) *A local lensing near* $z = 2$ *and the definition of* Y_1. (b) *The resulting jumps for* Y_1. (c) *The use of all three local parametrices to define* Y_2. (d) *The resulting jumps for* Y_2. *This RH problem has smooth solutions and is tractable numerically.*

for $k = 1,2,3$ are readily computed with the trapezoidal rule. Note that the integration contour has radius $2r$ so that the deformation regions are avoided.

Specifically, we consider the case

$$A_1 = \begin{bmatrix} 0 & 0.1 \\ -0.1 & 0 \end{bmatrix}, \quad A_2 = \begin{bmatrix} 0 & 0.2 \\ -0.6 & 0 \end{bmatrix}.$$

We find

$$\tilde{M}_1 = \begin{bmatrix} 1.20397 - 0.0707178\mathrm{i} & -0.650764\mathrm{i} \\ 0.698487\mathrm{i} & 1.20397 + 0.0707178\mathrm{i} \end{bmatrix},$$
$$\tilde{M}_2 = \begin{bmatrix} 8.92914 - 0.663479\mathrm{i} & -6.16164\mathrm{i} \\ 12.8488\mathrm{i} & 8.92914 + 0.663479\mathrm{i} \end{bmatrix},$$
$$\tilde{M}_3 \approx I,$$

with the step size $\Delta|z| = 0.0005$. Then, using 40 collocation points per contour to solve the RH problem for Y_2 (and 40 points for the trapezoidal rule) we find that $||A_k - A_k^n|| \approx 3 \times 10^{-14}$.

Remark 6.4.1. *Assume M_1 and M_2 have distinct eigenvalues and pick a normalization for the eigenvectors. It follows from the construction in Problem* 2.4.2 *that if* $||\tilde{M}_1 - M_1|| = \mathcal{O}(\epsilon) = ||\tilde{M}_2 - M_2||$, *then any derivative of the jumps in Figure* 6.6(d) *will differ from those with M_i replaced by $\tilde{M}_i$ by $\mathcal{O}(\epsilon)$ because the eigenvalues and eigenvectors must also converge. Lemma* 2.76 *demonstrates that the difference between any derivatives of the solutions of the two RH problems (one with M_i and the other with $\tilde{M}_i$) is $\mathcal{O}(\epsilon)$, measured appropriately.*

Chapter 7

Uniform Approximation Theory for Riemann–Hilbert Problems

In the previous chapter we discussed a numerical method for RH problems based around discretizing and solving the SIE $\mathscr{C}[G;\Gamma]u = G - I$. Thus far the desired output for our case studies in Sections 5.3, 5.5, 6.3, and 6.4 has been the solution of one RH problem evaluated accurately anywhere in the complex plane. The spectral problems in Sections 1.5 and 1.6 and the IST RH problem (3.17) are of a different nature because there is an infinite family of RH problems that need to be solved, and in the end only the asymptotics of the solution as $z \to \infty$ is used. In the case of Jacobi operators, the parameter is discrete ($n \in \mathbb{N}$) and in the case of the Schrödinger operator and the IST it is a continuous variable ($x \in \mathbb{R}$ and $t \geq 0$). To effectively solve these parameter-dependent RH problems we need to be able to accurately and efficiently solve them for any parameter value. In this chapter we discuss sufficient conditions for the accuracy of methods for RH problems to be uniform with respect to parameters. We ignore uniform convergence in the complex variable z.

The preconditioning of the associated SIE via deformation and lensing is critical to ensure accurate numerics. To more precisely understand the need for preconditioning, consider the jump matrix in (3.17). We must consider the solution of

$$u - \mathscr{C}_\Gamma^- u \cdot (G - I) = G - I, \quad G(x,t,z) = \begin{bmatrix} 1 - \lambda\overline{\rho(\overline{z})}\rho(z) & -\overline{\rho(\overline{z})}e^{-2izx-4iz^2t} \\ \rho(z)e^{2izx+4iz^2t} & 1 \end{bmatrix}. \quad (7.1)$$

Rearranging, we find that $u = (I + \mathscr{C}_\Gamma^- u)(G - I)$. The matrix $G(x,t,z)$ has *rapid oscillations* for x and t nonzero, and unless there is some highly unlikely cancellation, u will also contain rapid oscillations. The conditioning and convergence of a numerical method is tightly connected to the magnitude of the derivatives of the solution. We expect to lose accuracy for large x and t, and in practice, due to finite-precision arithmetic, one can only solve (7.1) accurately in a small neighborhood of the origin in the (x,t)-plane without preconditioning. This is analogous to the problem of evaluating oscillatory integrals, where an increasing number of quadrature points are required to resolve oscillations. This issue can be mitigated for oscillatory integrals by first deforming the integral along the path of steepest descent and then applying quadrature. Similarly, RH problems can be preconditioned using deformations motivated by the method of nonlinear steepest descent. This fact is demonstrated below in Sections 8.3, 9.6, 10.2.2, and 11.8 using the theory developed in this chapter. We remark that progress has been made on automatically deforming RH problems, see [119]. A full-scale implementation of these ideas would alleviate the laborious deformations performed in Part III. The framework of this chapter was originally published in [98], and a modified version appeared in [111].

The method of nonlinear steepest descent [34] as discussed in Section 3.2 allows the transformation of an oscillatory Riemann–Hilbert problem to a Riemann–Hilbert problem localized near the associated stationary points of an oscillatory factor in the jump matrix. In lieu of local parametrices, which are not always available and are delicate to construct, we scale the isolated Riemann–Hilbert problems to capture the local behavior. The analysis of this technique is more difficult, and the focus of this chapter is to derive sufficient conditions for which we can prove that the approach of solving RH problems numerically on scaled contours is guaranteed to be accurate to a given tolerance inside and outside of asymptotic regimes (i.e., for arbitrarily large values of parameters) with bounded computational cost. We refer to this type of behavior as *uniform approximation*.

We use the following rule of thumb for the scaling rates for contours, which is motivated by the method of steepest descent for integrals.

Assumption 7.0.1. *If the jump matrix G has a factor $e^{\xi\theta}$ and β_j corresponds to a qth-order stationary point (i.e., $\theta(z) \backsim C_1 + C_2(z-\beta_j)^q$), then the scaling of a contour near β_j which achieves asymptotic stability is $\alpha_j(\xi) = |\xi|^{-1/q}$.*

The heuristic reason for this choice is as follows. If $z = |\xi|^{-1/q}\zeta + \beta_j$, then for ζ in a bounded set we have

$$\xi\theta(z) = C_1\xi + C_2 e^{i\arg\xi}\zeta^q + o(1) \text{ as } |\xi| \to \infty.$$

It turns out in many cases that $C_1\xi$ is purely imaginary. This has ideal boundedness properties once exponentiated. In lieu of a rigorous guarantee, we demonstrate the validity of this assumption for the deformations below on a case-by-case basis.

In addition, we show the deep connection between the success of the numerical method and the success of the method of nonlinear steepest descent. A notable conclusion is that one can expect that whenever the method of nonlinear steepest descent produces an asymptotic formula, the numerical method can be made asymptotically stable. Achieving this requires varying amounts of preconditioning of the RH problem. This can vary from not deforming the RH problem at all, all the way to using the full deformation needed by the analytical method. An important question is, "When can we stop deforming to have a reliable numerical method?" Our main results are in Section 7.3, and these results provide an answer to this question. In short, although we do not require the knowledge of local and global parametrices[38] to construct the numerical method, their existence ensures that the numerical method remains accurate, and their explicit knowledge allows us to analyze the error of the approximation directly.

This chapter is structured as follows. We use an abstract framework for the numerical solution of RH problems which allows us to address asymptotic accuracy in a more concise way (Section 7.1). Additionally, other numerical methods (besides the one described in Chapter 6) may fit within the framework. The results would apply equally well to a Galerkin or finite-section method as they do to the method in Chapter 6 since we no longer require uniform convergence of the approximation to the solution of the RH problem itself, only convergence of the residue at infinity. We prove our main results which provide sufficient conditions for uniform approximation (Section 7.3). The numerical approach of Chapter 6 is placed within the general framework, along with necessary assumptions which allow a realization of uniform approximation (Section 7.4).

[38]Local and global parametrices are solutions of "model" RH problems that are used in the asymptotic solution of these RH problems. The function Ψ in (3.23) qualifies as both a local and global parametrix because it approximately solves the full RH problem locally and globally.

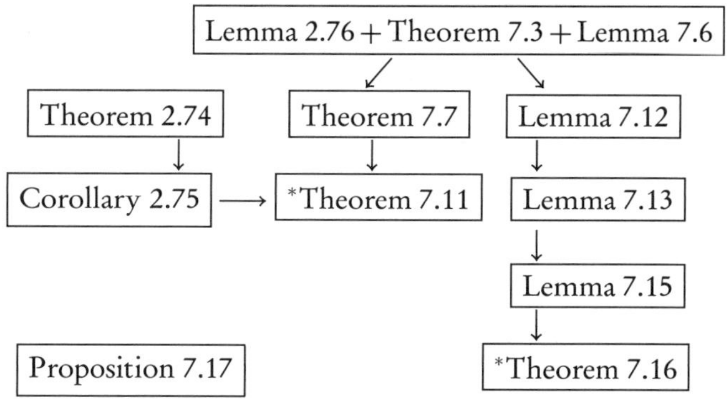

Figure 7.1. *The interdependency of results related to the development of the framework in this chapter. The fundamental results of this chapter are marked with an asterisk. Reprinted with permission from John Wiley and Sons, Inc.* [98].

The fundamental results of this chapter are necessarily encoded in the notation, definitions, and intermediate results that follow. Here we provide a road map to guide the reader through this chapter; see Figure 7.1. Theorems 7.11 and 7.16 represent the fundamental results of this chapter. Both present a detailed asymptotic analysis of Algorithm 7.1. To enhance readability we present a summary of each theorem and proposition.

- **Theorem 7.3:** Sufficient conditions for the convergence of projection-based numerical methods for operator equations are restated within the framework.

- **Algorithm 7.1:** The approach of scaling contours and solving a sequence of RH problems is encoded.

- **Theorem 7.7:** General conditions for the convergence of Algorithm 7.1 are established.

- **Theorem 7.11:** One set of sufficient conditions for the uniform accuracy of Algorithm 7.1 is established.

- **Theorem 7.16:** A set of relaxed conditions, with weakened results, for the uniform accuracy of numerical methods is provided. This result is derived using a *numerical parametrix*; see Definition 7.14.

- **Proposition 7.17:** Checking the conditions of Theorem 7.11 or the requirements of Definition 7.14 can be difficult. This proposition assists in that process.

7.1 ▪ A numerical Riemann–Hilbert framework

The goal of this section is to introduce additional tools to understand approximate solutions of the SIE

$$\mathscr{C}[G;\Gamma]u = G - I, \tag{7.2}$$

as parameters present in G vary. We start with the finite-dimensional approximation of $\mathscr{C}[G;\Gamma]$ via a general projection method (recall Section 6.1):

$$\mathscr{C}_n[G;\Gamma] = \mathscr{I}_n \mathrm{FP}\mathscr{C}[G;\Gamma]\mathscr{P}_n,$$
$$\mathrm{FP}\mathscr{C}[G;\Gamma]u(z) \triangleq u(z) - \mathrm{FP}\mathscr{C}_\Gamma u(z) \cdot (G(z) - I),$$
$$\mathscr{C}_n[G;\Gamma] : X_n \to Y_n.$$

As in Chapter 6, n will refer to the number of degrees of freedom in the discretization. The projections $\mathscr{P}_n$ and $\mathscr{I}_n$ are both defined on $B(\Gamma)$, the space of functions f such that $|f(x)| < \infty$ for each $x \in \Gamma$. The finite-dimensional ranges can be described by

$$X_n \triangleq \mathscr{P}_n B(\Gamma) \quad \text{and} \quad Y_n \triangleq \mathscr{I}_n B(\Gamma). \tag{7.3}$$

For concreteness we equip X_n and Y_n with the $L^2(\Gamma)$ norm. This plays a critical role in simplifying the results below. It is important that if $u \in H^1_z(\Gamma)$, then $\mathrm{FP}\mathscr{C}[G;\Gamma]u = \mathscr{C}[G;\Gamma]u$. The pair $(\mathscr{I}_n, \mathscr{P}_n)$ is used to refer to this numerical method. To simplify notation, define $\mathscr{H}[G;\Gamma]u = \mathrm{FP}\mathscr{C}^-_\Gamma u(G-I)$, so that $\mathrm{FP}\mathscr{C}[G;\Gamma] = I - \mathscr{H}[G;\Gamma]$, and $\mathscr{H}_n[G;\Gamma] = \mathscr{I}_n \mathscr{H}[G;\Gamma]\mathscr{P}_n$.

Definition 7.1. *The approximation $\mathscr{C}_n[G;\Gamma]$ of $\mathscr{C}[G;\Gamma]$ is said to be of type (α, β, γ) if, whenever $[G;\Gamma]$ is 1-regular and $\mathscr{C}[G;\Gamma]$ is invertible, there exist $N > 0$ and an invertible operator $\mathscr{T}_n : \mathbb{C}^n \to Y_n$ such that for $n > N$, $\mathscr{C}_n[G;\Gamma] : X_n \to Y_n$ is invertible and*

- $\|\mathscr{T}_n^{-1}\mathscr{I}_n\|_{\mathscr{L}(C^0(\Gamma),\mathbb{C}^n)} \leq C_1 n^\alpha$,
- $\|\mathscr{C}_n[G;\Gamma]^{-1}\mathscr{T}_n\|_{\mathscr{L}(\mathbb{C}^n,X_n)} \leq C_2 n^\beta \|\mathscr{C}[G;\Gamma]^{-1}\|_{\mathscr{L}(L^2(\Gamma))}$, *and*
- $\|\mathscr{T}_n^{-1}\mathscr{H}_n[G;\Gamma]\|_{\mathscr{L}(X_n,\mathbb{C}^n)} \leq C_3 n^\gamma \|G - I\|_{L^\infty(\Gamma)}$.

The constants N, C_1, C_2, and C_3 are allowed to depend on Γ.

Remark 7.1.1. *We defer the discussion of whether the collocation method constructed in the preceding chapter satisfies these conditions to Section 7.4.*

The first and second conditions in Definition 7.1 are necessary for the convergence of the numerical method. This will be made more precise below. The first and third conditions are needed to control operator norms as G varies. One should think of $\mathscr{T}_n v$ as being an interpolant of the data v, as in the previous chapter.

Definition 7.2. *The pair $(\mathscr{I}_n, \mathscr{P}_n)$ is said to produce an admissible numerical method if the following hold:*

- *The method is of type (α, β, γ).*
- *For each $m > 0$ there exists $k > 0$ such that if $u \in H^k(\Gamma)$, then $\|\mathscr{I}_n u - u\|_{H^1(\Gamma)}$ and $\|\mathscr{P}_n u - u\|_{H^1(\Gamma)}$ tend to zero faster than n^{-m} as $n \to \infty$.*

Remark 7.1.2. *We assume spectral convergence — i.e., faster than any algebraic power with n — of the projections. This assumption can be relaxed, but one has to spend considerable effort to ensure α, β, and γ are sufficiently small. The absence of an infinite number of bounded derivatives does not mean the method will not converge but that a proof of uniform approximation is more technically involved.*

Next, we state the generalized convergence theorem with this notation (compare with Lemma 6.2).

Theorem 7.3. *Assume that $(\mathcal{I}_n, \mathcal{P}_n)$ produces an admissible numerical method. If $[G;\Gamma]$ is 1-regular and $\mathcal{C}[G;\Gamma]$ is invertible on $L^2(\Gamma)$, we have*

$$\begin{aligned} \|u-u^n\|_{L^2(\Gamma)} &\leq (1+cn^{\alpha+\beta})\|\mathcal{P}_n u - u\|_{H^1(\Gamma)} \quad \textit{with} \\ c &= C\|\mathcal{C}[G;\Gamma]^{-1}\|_{\mathcal{L}(L^2(\Gamma))}(1+\|G-I\|_{L^\infty(\Gamma)}), \\ u &= \mathcal{C}[G;\Gamma]^{-1}(G-I), \quad u^n \triangleq \mathcal{C}_n[G;\Gamma]^{-1}\mathcal{I}_n(G-I). \end{aligned} \tag{7.4}$$

Proof. It is straightforward to check the identity from Lemma 6.2:

$$u-u^n = u-\mathcal{P}_n u + \mathcal{C}_n[G;\Gamma]^{-1}\mathcal{T}_n\mathcal{T}_n^{-1}\mathcal{I}_n\mathcal{C}[G;\Gamma](\mathcal{P}_n u - u).$$

Taking the $L^2(\Gamma)$ norm of both sides, applying Sobolev embedding of $H^1((a,b))$ into $C^0([a,b])$, and using that the method is admissible produces the result. □

In this setting, we use a weaker notation of convergence and do not require the approximate solution of the RH problem to converge uniformly for all of $\mathbb{C}$.

Corollary 7.4. *Under the assumptions of Theorem 7.3 and assuming that $[G;\Gamma]$ is k-regular for large k (large is determined by Definition 7.2), we have that $\Phi_n = I + \mathcal{C}_\Gamma u^n$ is an approximation of Φ, the solution of $[G;\Gamma]$, in the following sense:*

- $\Phi_n^\pm - \Phi^\pm \to 0$ *in $L^2(\Gamma)$, and*
- $\|\Phi_n - \Phi\|_{W^{j,\infty}(S)} \to 0$ *for all $j \geq 0$ whenever S is bounded away from Γ.*

Proof. The first claim follows from the boundedness of the Cauchy operator on $H^1(\Gamma)$, and, as before, the Cauchy–Schwarz inequality gives the second. □

Below, we always assume the numerical method considered is admissible.

7.2 ▪ Solving an RH problem on disjoint contours

In specific cases the contour Γ consists of disjoint components. Instead of solving the RH problem all at once, we decompose the RH problem into a sequence of RH problems defined on each disjoint component of Γ.

Example 7.5. Consider the RH problem $[G;\Gamma]$ with $\Gamma = \Gamma_1 \cup \Gamma_2$, where Γ_1 and Γ_2 are disjoint. To solve the full RH problem, we first solve for Φ_1 — the solution of $[G|_{\Gamma_1};\Gamma_1]$ — assuming that this subproblem has a unique solution. The jump on Γ_2 is modified through conjugation by Φ_1. Define

$$\tilde{G}_2 = \Phi_1 G|_{\Gamma_2}\Phi_1^{-1}.$$

Next, the solution Φ_2 of $[\tilde{G}_2;\Gamma_2]$ is found. A simple calculation shows that $\Phi = \Phi_2\Phi_1$ solves the original RH problem $[G;\Gamma]$. This process parallels the method used in Theorem 2.63. ∎

This idea allows us to treat each disjoint contour separately, solving in an iterative way. When using this algorithm numerically, the dimension of the linear system solved at each step is a fraction of that of the full discretized problem. This produces significant computational savings. We now generalize these ideas.

Consider an RH problem $[G;\Gamma]$ where $\Gamma = \Gamma_1 \cup \cdots \cup \Gamma_L$. Here each Γ_i is disjoint and $\Gamma_i = \alpha_i \Lambda_i + \beta_i$ for some contour Λ_i. We emphasize for this discussion that each Λ_i need not be an interval because this is a *different* decomposition of Γ than the decomposition into its smooth components. We define $G_i = G|_{\Gamma_i}$ and $H_i(z) = G_i(\alpha_i z + \beta_i)$. As a notational remark, in this chapter we always associate H_i and G in this way.

Remark 7.2.1. *The motivation for introducing the representation of the contours in this fashion is made clear below. Mainly, this formulation is important when α_i and/or β_i depend on a parameter but Λ_i does not.*

We now describe the general iterative solver.

***Algorithm* 7.1.**
(Scaled and shifted RH solver)

1. Solve the RH problem $[H_1;\Lambda_1]$ to obtain $\tilde{\Phi}_1$. We denote the solution of the associated SIE as U_1 with domain Λ_1. Define $\Phi_1(z) = \tilde{\Phi}_1\left(\frac{z-\beta_1}{\alpha_1}\right)$.

2. For each $j = 2,\ldots,L$ define $\Phi_{i,j}(z) = \Phi_i(\alpha_j z + \beta_j)$ and solve the RH problem $[\tilde{H}_j;\Lambda_j]$ with
$$\tilde{H}_j = \Phi_{j-1,j} \cdots \Phi_{1,j} H_j \Phi_{1,j}^{-1} \cdots \Phi_{j-1,j}^{-1},$$
to obtain $\tilde{\Phi}_j$. Again, the solution of the integral equation is denoted by U_j with domain Λ_j. Define $\Phi_j(z) = \tilde{\Phi}_j\left(\frac{z-\beta_j}{\alpha_j}\right)$.

3. Construct $\Phi = \Phi_L \cdots \Phi_1$, which satisfies the original problem.

When this algorithm is implemented numerically, the jump matrix corresponding to $\tilde{H}_j$ is not exact. It depends on the approximations of each of the Φ_i for $i < j$, and more specifically, it depends on the order of approximation of the RH problem on Λ_i for $i < j$. We use the notation $\boldsymbol{n}_i = (n_1,\ldots,n_i)$, where each n_i is the order of approximation on Λ_i. Further, we use $\boldsymbol{n} > \boldsymbol{m}$ whenever the vectors are of the same length and $n_j > m_j$ for all j. The statement $\boldsymbol{n} \to \infty$ means that each component of $\boldsymbol{n}$ tends to ∞. Let $\Phi_{i,j,\boldsymbol{n}_i}$ be the approximation of $\Phi_{i,j}$ and define
$$\tilde{H}_{j,\boldsymbol{n}_j} = \Phi_{j-1,j,\boldsymbol{n}_{j-1}} \cdots \Phi_{1,j,\boldsymbol{n}_1} H_j \Phi_{1,j,\boldsymbol{n}_1}^{-1} \cdots \Phi_{j-1,j,\boldsymbol{n}_{j-1}}^{-1}.$$

If the method converges, then $\tilde{H}_{j,\boldsymbol{n}_j} \to \tilde{H}_j$ uniformly as $\boldsymbol{n}_j \to \infty$.

A significant question remains: "How do we know solutions exist at each stage of this algorithm?" In general, this is not the case. $\mathscr{C}[G;\Gamma]$ can be expressed in the form $\mathscr{K} - \mathscr{H}$, where $\mathscr{K}$ is the block-diagonal operator with blocks $\mathscr{C}[G_i;\Gamma_i]$ and $\mathscr{H}$ is a compact operator. Here $\mathscr{H}$ represents the effect of one contour on another, and if the operator norm of $\mathscr{H}$ is sufficiently small, solutions exist at each iteration of Algorithm 7.1, due to a Neumann series argument. This is true if the arc length of each Γ_i is sufficiently small; see Example 2.80. An implicit assumption in our numerical framework is that such equations are uniquely solvable, which follows in our examples where the arclength tends to zero.

Additionally, if each of the scale factors $\alpha_i = \alpha_i(\xi)$ is parameter dependent such that $\alpha_i(\xi) \to 0$ as $\xi \to 0$, then the norms of the inverse operators are related. When each of the $\alpha_i(\xi)$ are sufficiently small, there exists $C > 1$ such that

$$\frac{1}{C}\|\mathscr{C}[G;\Gamma]^{-1}\|_{\mathscr{L}(L^2(\Gamma))} \leq \max_i \|\mathscr{C}[G_i;\Gamma_i]^{-1}\|_{\mathscr{L}(L^2(\Gamma_i))} \leq C\|\mathscr{C}[G;\Gamma]^{-1}\|_{\mathscr{L}(L^2(\Gamma))}. \tag{7.5}$$

Due to the simplicity of the allowed scalings, the norms of the operators $\mathscr{C}[G_i;\Gamma_i]$ and $\mathscr{C}[G_i;\Gamma_i]^{-1}$ are equal to those of their scaled counterparts $\mathscr{C}[H_i;\Lambda_i]$ and $\mathscr{C}[H_i;\Lambda_i]^{-1}$.

The final question is one of convergence. For a single fixed contour we know that if $(\mathscr{I}_n, \mathscr{P}_n)$ produces an admissible numerical method and the RH problem is sufficiently regular, the numerical method converges. This means that the solution of this RH problem converges uniformly, away from the contour on which it is defined. This is the basis for proving that Algorithm 7.1 converges. Lemma 2.76 aids us when considering the infinite-dimensional operator for which the jump matrix is uniformly close, but we need an additional result for the finite-dimensional case.

Lemma 7.6. *Consider a family of RH problems* $([G_\xi;\Gamma])_{\xi\geq 0}$ *on the fixed contour* Γ *which are k-regular. Assume* $G_\xi \to G$ *in* $L^\infty(\Gamma)\cap L^2(\Gamma)$ *as* $\xi \to \infty$ *and* $[G;\Gamma]$ *is k-regular. Then the following hold:*

- *If* $\mathscr{C}_n[G;\Gamma]$ *is invertible, then there exists* $T(n) > 0$ *such that* $\mathscr{C}_n[G_\xi;\Gamma]$ *is also invertible for* $\xi > T(n)$.
- *If* $\Phi_{n,\xi}$ *is the approximate solution of* $[G_\xi;\Gamma]$ *and* Φ_n *is the approximate solution of* $[G;\Gamma]$, *then* $\Phi_{n,\xi} - \Phi_n \to 0$ *in* $L^2(\Gamma)$ *as* $\xi \to \infty$ *for fixed n.*
- $\|\Phi_{n,\xi} - \Phi_n\|_{W^{j,\infty}(S)} \to 0$, *as* $\xi \to \infty$, *for all* $j \geq 1$ *whenever S is bounded away from* Γ *for fixed n.*

Proof. We consider the two equations

$$\mathscr{C}_n[G_\xi;\Gamma]u_\xi^n = \mathscr{I}_n(G_\xi - I),$$
$$\mathscr{C}_n[G;\Gamma]u^n = \mathscr{I}_n(G - I),$$

with solutions given by

$$u_\xi^n = \mathscr{C}_n[G_\xi;\Gamma]^{-1}\mathscr{T}_n\mathscr{T}_n^{-1}\mathscr{I}_n(G_\xi - I),$$
$$u^n = \mathscr{C}_n[G;\Gamma]^{-1}\mathscr{T}_n\mathscr{T}_n^{-1}\mathscr{I}_n(G - I).$$

Since the method is of type (α,β,γ), we have (see Definition 7.1)

$$\|\mathscr{T}_n^{-1}(\mathscr{C}_n[G_\xi;\Gamma] - \mathscr{C}_n[G;\Gamma])\|_{\mathscr{L}(X_n,\mathbb{C}^n)} \leq C_3 n^\gamma \|G_\xi - G\|_{L^\infty(\Gamma)} = E(\xi)n^\gamma.$$

For fixed n, by increasing ξ, we can make $E(\xi)$ small so that

$$\begin{aligned}\|\mathscr{T}_n^{-1}(\mathscr{C}_n[G_\xi;\Gamma] - \mathscr{C}_n[G;\Gamma])\|_{\mathscr{L}(X_n,\mathbb{C}^n)} &\leq \frac{1}{C_2}\frac{1}{\|\mathscr{C}[G;\Gamma]^{-1}\|_{\mathscr{L}(L^2(\Gamma))}} n^{-\beta} \\ &\leq \frac{1}{\|\mathscr{C}_n[G;\Gamma]^{-1}\|_{\mathscr{L}(Y_n,X_n)}}.\end{aligned}$$

Specifically, we choose ξ small enough so that

$$E(\xi) \leq \frac{1}{2}\frac{1}{C_2 C_3}\frac{1}{\|\mathscr{C}[G;\Gamma]^{-1}\|_{\mathscr{L}(L^2(\Gamma))}} n^{-\gamma-\beta}.$$

Using Theorem A.20, $\mathscr{C}_n[G_\xi;\Gamma]$ is invertible, and we bound

$$\|(\mathscr{C}_n[G_\xi;\Gamma]^{-1} - \mathscr{C}_n[G;\Gamma]^{-1})\mathscr{T}_n\|_{\mathscr{L}(\mathbb{C}^n, X_n)} \leq 2C_2 n^{2\beta+\gamma}\|\mathscr{C}[G;\Gamma]^{-1}\|^2_{\mathscr{L}(L^2(\Gamma))}E(\xi). \quad (7.6)$$

Importantly, the quantity on the left tends to zero as $\xi \to \infty$. We use the triangle inequality:

$$\begin{aligned}\|u^n - u^n_\xi\|_{L^2(\Gamma)} &\leq \|(\mathscr{C}_n[G_\xi;\Gamma]^{-1} - \mathscr{C}_n[G;\Gamma]^{-1})\mathscr{T}_n\mathscr{T}_n^{-1}\mathscr{I}_n(G-I)\|_{L^2(\Gamma)} \\ &+ \|\mathscr{C}_n[G;\Gamma]^{-1}\mathscr{T}_n\mathscr{T}_n^{-1}\mathscr{I}_n(G-G_\xi)\|_{L^2(\Gamma)}.\end{aligned}$$

Since we have assumed that Γ is bounded and that the norm of $\mathscr{T}_n^{-1}\mathscr{I}_n : C^0(\Gamma) \to \mathbb{C}^n$ is slowly growing, we obtain L^2 convergence of u^n to u^n_ξ as $\xi \to \infty$:

$$\begin{aligned}\|u^n - u^n_\xi\|_{L^2(\Gamma)} &\leq C_3 n^{2\beta+\gamma+\alpha}E(\xi)\|G-I\|_{L^\infty(\Gamma)} + C_4 n^{\beta+\alpha}\|G-G_\xi\|_{L^\infty(\Gamma)} \quad (7.7) \\ &\leq C_5 n^{2\beta+\gamma+\alpha}E(\xi).\end{aligned}$$

This proves the three required properties. □

Remark 7.2.2. *A good way to interpret this result is to see $E(\xi)$ as the difference in norm between the associated infinite-dimensional operators, which is proportional to the uniform difference in the jump matrices. Then (7.6) gives the resulting error between the finite-dimensional operators. It is worthwhile to note that if $\alpha = \beta = \gamma = 0$, then T can be chosen independently of n.*

Now we have the tools needed to address the convergence of the solver. We introduce some notation to simplify matters. At stage j in the solver we solve an SIE on Λ_j. On this domain we need to compare two RH problems:

$$[\tilde{H}_j;\Lambda_j] \text{ and } [\tilde{H}_{j,\boldsymbol{n}_j};\Lambda_j].$$

Let u_j be the exact solution of the SIE which is obtained from $[\tilde{H}_j;\Lambda_j]$. As an intermediate step we need to consider the numerical solution of $[\tilde{H}_j;\Lambda_j]$, which differs from the numerical solution of $[\tilde{H}_{j,\boldsymbol{n}_j};\Lambda_j]$ because the jump matrix is exact. We use $u_j^{n_j}$ to denote the numerical approximation of u_j of order n_j. Also, $u_j^{\boldsymbol{n}_j}$ is used to denote the numerical approximation of the solution of the SIE associated with $[\tilde{H}_{j,\boldsymbol{n}_j};\Lambda_j]$.

Theorem 7.7. *Assume that each problem in Algorithm 7.1 is solvable and k-regular for sufficiently large k. Then the algorithm converges to the true solution of the RH problem. More precisely, there exist $\boldsymbol{N}_i$ and constants $C_i > 0$ such that for $\boldsymbol{n}_i > \boldsymbol{N}_i$ we have*

$$\|u_i^{\boldsymbol{n}_i} - u_i\|_{L^2(\Lambda_i)} \leq C_i(\max \boldsymbol{n}_i)^{i(2\beta+\alpha+\gamma)} \max_{j\leq i}\|\mathscr{P}_{n,j}u_j - u_j\|_{H^1(\Lambda_j)},$$

where $\mathscr{P}_{n,j}$ is taken to be the appropriate projection for Λ_j.

Proof. We prove this by induction. Since $u_1^{\boldsymbol{n}_1} = u_1^{n_1}$, the claim follows from Theorem 7.3 for $i = 1$. Now assume the claim is true for all $j < i$. We use Lemma 7.6 to show it is true for i. Using the triangle inequality we have

$$||u_i^{\boldsymbol{n}_i} - u_i||_{L^2(\Lambda_i)} \leq ||u_i^{\boldsymbol{n}_i} - u_i^{n_i}||_{L^2(\Lambda_i)} + ||u_i - u_i^{n_i}||_{L^2(\Lambda_i)}.$$

Using Theorem 7.3, we bound the second term:

$$||u_i - u_i^{n_i}||_{L^2(\Lambda_i)} \leq C n_i^{\alpha+\beta} ||\mathscr{P}_{n,i} u_i - u_i||_{H^1(\Lambda_i)}.$$

To bound the first term we use (7.7):

$$||u_i^{\boldsymbol{n}_i} - u_i^{n_i}||_{L^2(\Lambda_i)} \leq C n_i^{2\beta+\gamma+\alpha} E(\boldsymbol{n}_{i-1}). \tag{7.8}$$

$E(\boldsymbol{n}_{i-1})$ is proportional to the uniform difference of $\tilde{H}_i$, and its approximation is obtained through the numerical method $\tilde{H}_{i,\boldsymbol{n}_{i-1}}$. By the induction hypothesis, if k is sufficiently large, Lemma 7.6 tells us that this difference tends to zero as $\boldsymbol{n}_{i-1} \to \infty$, and the use of (7.7) is justified. More precisely, consider

$$\begin{aligned}\tilde{H}_{i,\boldsymbol{n}_i} - \tilde{H}_i = {} & (\Phi_{i-1,i,\boldsymbol{n}_i} - \Phi_{i-1,i})\Phi_{i-2,i,\boldsymbol{n}_{i-2}} \cdots \Phi_{1,i,\boldsymbol{n}_1} H_i \Phi_{1,i,\boldsymbol{n}_1}^{-1} \cdots \Phi_{i-1,i,\boldsymbol{n}_i}^{-1} \\ & \Phi_{i-1,i}(\Phi_{i-2,i,\boldsymbol{n}_{i-2}} \cdots \Phi_{1,i,\boldsymbol{n}_1} H_i \Phi_{1,i,\boldsymbol{n}_1}^{-1} \cdots \Phi_{i-1,i,\boldsymbol{n}_i}^{-1} \\ & - \Phi_{i-2,i} \cdots \Phi_{1,i} H_i \Phi_{1,i}^{-1} \cdots \Phi_{i-1,i}^{-1}).\end{aligned}$$

Then

$$\begin{aligned}||\tilde{H}_{i,\boldsymbol{n}_i} - \tilde{H}_i||_{L^\infty(\Lambda_i)} \leq {} & C_1 ||u_{i-1}^{\boldsymbol{n}_{i-1}} - u_{i-1}||_{L^2(\Lambda_{i-1})} \\ & + C_2 ||\Phi_{i-2,i,\boldsymbol{n}_{i-2}} \cdots \Phi_{1,i,\boldsymbol{n}_1} H_i \Phi_{1,i,\boldsymbol{n}_1}^{-1} \cdots \Phi_{i-1,i,\boldsymbol{n}_i}^{-1} \\ & - \Phi_{i-2,i} \cdots \Phi_{1,i} H_i \Phi_{1,i}^{-1} \cdots \Phi_{i-1,i}^{-1}||_{L^\infty(\Lambda_i)}.\end{aligned}$$

Here C_1 and C_2 are constants replacing the uniformly bounded (in $\boldsymbol{n}_i$) functions:

$$C_1 = \sup_{\boldsymbol{n}_i} ||\Phi_{i-2,i,\boldsymbol{n}_{i-2}} \cdots \Phi_{1,i,\boldsymbol{n}_1} H_i \Phi_{1,i,\boldsymbol{n}_1}^{-1} \cdots \Phi_{i-1,i,\boldsymbol{n}_i}^{-1}||_{L^\infty(\Lambda_i)}, \quad C_2 = ||\Phi_{i-1,i}||_{L^\infty(\Lambda_i)}.$$

This process can be continued with Theorem A.20 being used to deal with inverses. Thus, repeated triangle inequalities result in

$$||\tilde{H}_i - \tilde{H}_{i,\boldsymbol{n}_{i-1}}||_{L^\infty(\Lambda_i)} \leq C \sum_{j=1}^{i-1} ||u_j - u_j^{\boldsymbol{n}_j}||_{L^2(\Lambda_j)}. \tag{7.9}$$

Combining (7.8) and (7.9) we complete the proof. □

Remark 7.2.3. *The requirement that k be large can be made more precise using Definition 7.2 with $m = L(2\beta + \gamma + \alpha)$, where L is the number of disjoint contours Γ_i that make up the full contour Γ. There is little restriction if $(\alpha, \beta, \gamma) = (0,0,0)$.*

7.3 ▪ Uniform approximation

We use the ideas presented above to explain how numerics can be used to provide asymptotic approximations. Before we proceed, we define two types of uniform approximation. Let $(u_n^\xi)_{\xi\geq 0}$ be a sequence, depending on the parameter ξ, in a Banach space such that for each ξ, $\|u_n^\xi - u^\xi\| \to 0$ as $n \to \infty$ for some u^ξ.

Definition 7.8. *We say the sequence $(u_n^\xi)_{\xi\geq 0}$ is weakly uniform if for every $\epsilon > 0$ there exists a function $N(\xi) : \mathbb{R}^+ \to \mathbb{N}$ taking finitely many values such that*

$$\|u_{N(\xi)}^\xi - u^\xi\| < \epsilon.$$

Definition 7.9. *We say the sequence $(u_n^\xi)_{\xi\geq 0}$ is strongly uniform (or just* uniform*) if for every $\epsilon > 0$ there exists $N \in \mathbb{N}$ such that for $n \geq N$, $\xi \geq 0$*

$$\|u_n^\xi - u^\xi\| < \epsilon.$$

The necessity for the definition of a weakly uniform sequence is mostly a technical detail, as we do not see it arise in practice. To illustrate how it can arise we give an example.

Example 7.10. Consider the sequence

$$(u_n^\xi)_{n,\xi\geq 0} = \left(\sin\xi + e^{-n^2} + e^{-(\xi-n)^2}\right)_{n,\xi\geq 0}.$$

For fixed ξ, $u_n^\xi \to \sin\xi$. We want, for $\epsilon > 0$, while keeping n bounded,

$$|u_n^\xi - \sin\xi| = |e^{-n^2} + e^{-(\xi-n)^2}| < \epsilon.$$

We choose $n > \xi$, or if ξ is large enough, we choose $0 < n < \xi$. To maintain error that is uniformly less than ϵ we cannot choose a fixed n; it must vary with respect to ξ. When relating to RH problems the switch from $n > \xi$ to $0 < n < \xi$ is related to transitioning into the asymptotic regime. ∎

7.3.1 ▪ Direct estimates

As before, we are assuming we have an RH problem $[G^\xi; \Gamma^\xi]$ that depends on a parameter ξ and Γ^ξ is bounded. Here we use a priori bounds on the solution of the associated SIE which are uniform in ξ to prove the uniform approximation. In general, when this is possible, it is the simplest way to proceed.

Our main tool is Corollary 2.75. We can easily estimate the regularity of the solution of each problem $[H_i; \Lambda_i]$ provided we have some information about $\|\mathscr{C}[G_i^\xi; \Gamma_i^\xi]^{-1}\|_{\mathscr{L}(L^2(\Gamma_i))}$ or equivalently $\|\mathscr{C}[H_i^\xi; \Lambda_i]^{-1}\|_{\mathscr{L}(L^2(\Lambda_i))}$. We address how to estimate this later in this section. First, we need a statement about how regularity is preserved throughout Algorithm 7.1. Specifically, we use information from the scaled jumps H_i and the local inverses $\mathscr{C}[H_i; \Lambda_i]^{-1}$ to estimate global regularity. The following theorem uses this to prove strong uniformity of the numerical method.

Theorem 7.11. *Assume*

- $([G^\xi;\Gamma^\xi])_{\xi\geq 0}$ *is a sequence of k-regular RH problems,*
- *the norm of* $\mathscr{C}[H_i^\xi;\Lambda_i]^{-1}$ *is uniformly bounded in* ξ,
- $\|H_i^\xi\|_{W^{k,\infty}(\Lambda_i)} \leq C$, *and*
- $\alpha_i(\xi) \to 0$ *as* $\xi \to \infty$ *if* $L > 1$.[39]

Then, if k and ξ *are sufficiently large,*

- *Algorithm* 7.1 *applied to* $([G^\xi;\Gamma^\xi])_{\xi\geq 0}$ *has solutions at each stage,*
- $\|u_j^\xi\|_{H^k(\Lambda_i)} \leq P_k$, *where* P_k *depends on* $\|H_i^\xi\|_{H^k(\Lambda_i)\cap W^{k,\infty}(\Lambda_i)}$, $\|\mathscr{C}[H_i^\xi;\Lambda_i]^{-1}\|_{\mathscr{L}(L^2(\Lambda_i))}$, *and* $\|u_j^\xi\|_{L^2(\Lambda_j)}$ *for* $j < i$, *and*
- *the approximation* $u_i^{n_i,\xi}$ *of* u_i^ξ *(the solution of the SIE) at each step in Algorithm* 7.1 *converges in* L^2 *uniformly in* ξ *as* $\mathbf{n}_i \to \infty$, *i.e., the convergence is strongly uniform.*

Proof. First, we note that since $\alpha_i(\xi) \to 0$, (2.60) shows that jump matrix $\tilde{H}_i^\xi$ for the RH problem solved at stage i in Algorithm 7.1 tends uniformly to H_i^ξ. This implies the solvability of the RH problems at each stage in Algorithm 7.1, as well as the bound

$$\|\mathscr{C}[\tilde{H}_i^\xi;\Lambda_i]^{-1}\|_{\mathscr{L}(L^2(\Gamma))} \leq C\|\mathscr{C}[H_i^\xi;\Lambda_i]^{-1}\|_{\mathscr{L}(L^2(\Gamma))}$$

for sufficiently large ξ. As before, C can be taken to be independent of ξ. We claim that $\|u_i^\xi\|_{H^k(\Lambda_i)}$ is uniformly bounded. We prove this by induction. When $i = 1$, $u_1^\xi = \mathscr{C}[H_i^\xi;\Lambda_i]^{-1}(H_i^\xi - I)$ and the claim follows from Corollary 2.75. Now assume the claim is true for $j < i$. All derivatives of the jump matrix $\tilde{H}_i^\xi$ depend on the Cauchy integral of u_j^ξ evaluated away from Λ_j and H_i^ξ. The former is bounded by the induction hypothesis and the latter is bounded by assumption. Again, using Corollary 2.75 we obtain the uniform boundedness of $\|u_i^\xi\|_{H^k(\Lambda_i)}$. Theorem 7.7 implies that convergence is uniform in ξ. □

The most difficult part about verifying the hypotheses of this theorem is establishing an estimate of $\|\mathscr{C}[H_i^\xi;\Lambda_i^\xi]^{-1}\|_{\mathscr{L}(L^2(\Lambda_i))}$ as a function of ξ. A very useful fact is that once the solution Ψ^ξ of the RH problem $[G^\xi;\Gamma^\xi]$ is known then the inverse of the operator is also known (see Lemma 2.67). When Ψ^ξ is known approximately, i.e., when a parametrix is known, then estimates on the boundedness of the inverse can be reduced to studying the L^∞ norm of the parametrix. Then (7.5) can be used to relate this to each $\mathscr{C}[G_i^\xi;\Gamma_i^\xi]^{-1}$, which gives an estimate on the norm of $\mathscr{C}[H_i^\xi;\Lambda_i]^{-1}$. We study this further in the chapters that follow (see Sections 9.6 and 10.2.2).

[39] Recall that L is the number of disjoint components of the contour Γ^ξ.

7.3.2 ▪ Failure of direct estimates*

We study a toy RH problem to motivate where the direct estimates can fail. Let $\phi(x)$ be a smooth function with compact support in $(-1,1)$ satisfying $\max_{[-1,1]}|\phi(x)| = 1/2$. Consider the following scalar RH problem for a function μ:

$$\mu^+(x) = \mu^-(x)(1+\phi(x)(1+\xi^{-1/2}e^{i\xi x})), \quad \mu(\infty) = 1, \quad x \in (-1,1), \quad \xi > 0. \tag{7.10}$$

This problem can be solved explicitly, but we study it from the linear operator perspective instead. From the boundedness assumption on ϕ, a Neumann series argument gives the invertibility of the singular integral operator and uniform boundedness of the L^2 inverse in ξ. Using the estimates in Corollary 2.75 we obtain useless bounds that all grow with ξ. Intuitively, the solution to (7.10) is close in L^2 to the solution to

$$v^+(x) = v^-(x)(1+\phi(x)), \quad v(\infty) = 1, \quad x \in (-1,1), \quad \xi > 0, \tag{7.11}$$

which trivially has uniform bounds on its Sobolev norms. In the next section we introduce the idea of a numerical parametrix which resolves this complication.

7.3.3 ▪ Extension to indirect estimates*

We assume minimal hypotheses for dependence of the sequence $([G^\xi;\Gamma^\xi])_{\xi\geq 0}$ on ξ. Specifically we require only that the map $\xi \mapsto H_i^\xi$ be continuous from $\mathbb{R}^+$ to $L^\infty(\Lambda_i)$ for each i. We do not want to hypothesize further, as that would alter the connection to the method of nonlinear steepest descent, which only requires uniform convergence of the jump matrix on bounded contours. The results in this section can be extended to the case where $\xi \mapsto H_i^\xi$ is continuous from $\mathbb{R}^+$ to $W^{k,\infty}(\Lambda_i)$, which would allow the estimation of H^k norms as opposed to the L^2 norms.

The fundamental result we need to prove a uniform approximation theorem is the continuity of Algorithm 7.1 with respect to uniform perturbations in the jump matrix. With the jump matrix G we associated H_j, the scaled restriction of G to Γ_j. With G we also associated u_j, the solution of the SIE obtained from $[\tilde{H}_j;\Lambda_j]$, where $\tilde{H}_j$ is defined in Algorithm 7.1. In what follows, if we have another jump matrix J and analogously we use K_j to denote the scaled restriction of J to Γ_j, $\tilde{K}_j$ is analogous to $\tilde{H}_j$ with G replaced with J and p_j is used to denote the solution of the SIE obtained from $[\tilde{K}_j;\Lambda_j]$. So the notation we use is to associate $G, H_j, \tilde{H}_j$, and u_j and $J, K_j, \tilde{K}_j$, and p_j.

Lemma 7.12. *Assume $([G^\xi;\Gamma^\xi])_{\xi\geq 0}$ is a sequence of 1-regular RH problems such that $\xi \mapsto H_i^\xi$ is continuous from $\mathbb{R}^+$ to $L^\infty(\Lambda_i)$ for each i. Then for sufficiently large but fixed $\boldsymbol{n}_i$, the map $\xi \mapsto u_i^{\boldsymbol{n}_i,\xi}$ is continuous from $\mathbb{R}^+$ to $L^2(\Lambda_i)$ for each i.*

Proof. We prove this by induction on i. For $i = 1$ the claim follows from Lemma 7.6. Now assume the claim is true for $j < i$. We prove it holds for i. We show the map is continuous at η for $\eta \geq 0$. First, from Lemma 7.6,

$$\|u_i^{\boldsymbol{n}_i,\eta} - u_i^{\boldsymbol{n}_i,\xi}\|_{L^2(\Lambda)} \leq C n_i^{2\alpha+\gamma} E(\xi,\eta),$$

where $E(\xi,\eta)$ is proportional to $\|\tilde{H}^{\eta}_{i,\boldsymbol{n}_{i-1}} - \tilde{H}^{\xi}_{i,\boldsymbol{n}_{i-1}}\|_{L^\infty(\Lambda_i)}$. An argument similar to that in Theorem 7.7 gives

$$\|\tilde{H}^{\eta}_{i,\boldsymbol{n}_{i-1}} - \tilde{H}^{\xi}_{i,\boldsymbol{n}_{i-1}}\|_{L^\infty(\Lambda_i)} \leq C(\eta,\boldsymbol{n}_i)\sum_{j=1}^{i-1}\|u_j^{\boldsymbol{n}_j,\eta} - u_j^{\boldsymbol{n}_j,\xi}\|_{L^2(\Lambda_j)}$$

for $|\xi-\eta|$ sufficiently small. By assumption, the right-hand side tends to zero as $\xi \to \eta$, which proves the lemma. □

It is worthwhile noting that the arguments in Lemma 7.12 show the same continuity for the infinite-dimensional, nondiscretized problem. Now we show weak uniform convergence of the numerical scheme on compact sets.

Lemma 7.13. *Assume* $([G^\xi;\Gamma^\xi])_{\xi\geq 0}$ *is a sequence of k-regular RH problems such that all the operators in Algorithm 7.1 are invertible for every* ξ. *Assume that k is sufficiently large so that the approximations from Algorithm 7.1 converge for every* $\xi \geq 0$. *Then there exists a vector-valued function* $N(i,\xi)$ *that takes finitely many values such that*

$$\|u_i^{N(i,\xi),\xi} - u_i^\xi\|_{L^2(\Lambda_i)} < \epsilon.$$

Moreover, if the numerical method is of type $(0,0,0)$, *then convergence is strongly uniform.*

Proof. Let $S \subset \mathbb{R}^+$ be compact. It follows from Lemma 7.12 that the function $E(\xi,\boldsymbol{n},i) = \|u_i^{\boldsymbol{n},\xi} - u_i^\xi\|_{L^2(\Gamma_i)}$ is a continuous function of ξ for fixed $\boldsymbol{n}$. For $\epsilon > 0$ find $\boldsymbol{n}_\xi$ such that $E(\xi,\boldsymbol{n}_\xi,i) < \epsilon/2$. By continuity, define $\delta_\xi(\boldsymbol{n}_\xi) > 0$ so that $E(s,\boldsymbol{n}_\xi,i) < \epsilon$ for $|s-\xi| < \delta_\xi$. The open sets $(B(\xi,\delta_\xi))_{\xi\in S}$ cover S, and we can select a finite subcover $(B(\xi_j,\delta_{\xi_j}))_{j=1}^N$. We have $E(s,\boldsymbol{n}_{\xi_j},i) < \epsilon$ whenever $s \in B(\xi_j,\delta_{\xi_j})$. To prove the claim for a method of type $(0,0,0)$, we use the fact that δ_ξ can be taken independently of $\boldsymbol{n}_\xi$ and that $E(s,\boldsymbol{n},i) < \epsilon$ for every $\boldsymbol{n} > \boldsymbol{n}_\xi$. □

Definition 7.14. *Given a sequence of k-regular RH problems* $([G^\xi;\Gamma^\xi])_{\xi\geq 0}$ *such that*

- $\Gamma^\xi = \Gamma_1^\xi \cup \cdots \cup \Gamma_L^\xi$ *and*
- $\Gamma_i^\xi = \alpha_i(\xi)\Lambda_i + \beta_i$,

another sequence of k-regular RH problems $([J^\xi,\Sigma^\xi])_{\xi\geq 0}$ *is said to be a numerical parametrix if*

- $\Sigma^\xi = \Sigma_1^\xi \cup \cdots \cup \Sigma_L^\xi$,
- $\Sigma_i^\xi = \gamma_i(\xi)\Lambda_i + \sigma_i$,
- *for all i*

$$J^\xi(\gamma_i(\xi)z + \sigma_i) - G^\xi(\alpha_i(\xi)z + \beta_i) \to 0, \tag{7.12}$$

  uniformly on Λ_i *as* $\xi \to \infty$,

- *the norms of the operators and inverse operators at each step in Algorithm 7.1 are uniformly bounded in* ξ, *implying uniform boundedness of* J^ξ *in* ξ, *and*

- *the approximation* $p^{\xi}_{i,\boldsymbol{n}_i}$ *of* p^{ξ}_i *(the solution of the SIE) at each step in Algorithm* 7.1 *converges uniformly as* $\min \boldsymbol{n}_i \to \infty$.

This definition hypothesizes desirable conditions on a nearby limit problem for the sequence $([G^{\xi};\Gamma^{\xi}])_{\xi\geq 0}$. Under the assumption of this nearby limit problem we are able to obtain a uniform approximation for the solution of the original RH problem.

Lemma 7.15. *Assume there exists a numerical parametrix* $(J_{\xi},\Sigma_{\xi})_{\xi\geq 0}$ *for a sequence of RH problems* $([G_{\xi};\Gamma_{\xi}])_{\xi\geq 0}$. *Then for every* $\epsilon > 0$ *there exists* $\boldsymbol{N}_i$ *and* $T > 0$ *such that, at each stage in Algorithm* 7.1,

$$\|u_i^{\boldsymbol{N}_i,\xi} - u_i^{\xi}\|_{L^2(\Lambda_i)} < \epsilon \text{ for } \xi > T. \tag{7.13}$$

Furthermore, if the numerical method is of type $(0,0,0)$, *then* (7.13) *is true with* $\boldsymbol{N}_i$ *replaced by any* $\boldsymbol{M}_i > \boldsymbol{N}_i$.

Proof. At each stage in Algorithm 7.1 we have

$$\|u_i^{\boldsymbol{n}_i,\xi} - u_i^{\xi}\|_{L^2(\Lambda_i)} \leq \|u_i^{\boldsymbol{n}_i,\xi} - p_i^{\boldsymbol{n}_i,\xi}\|_{L^2(\Lambda_i)} + \|p_i^{\boldsymbol{n}_i,\xi} - p_i^{\xi}\|_{L^2(\Lambda_i)} + \|p_i^{\xi} - u_i^{\xi}\|_{L^2(\Lambda)}. \tag{7.14}$$

Since $p_i^{\boldsymbol{n}_i,\xi}$ originates from a numerical parametrix, we know that $\|p_i^{\boldsymbol{n}_i,\xi} - p_i^{\xi}\|_{L^2(\Lambda_i)} \to 0$ uniformly in ξ as $\boldsymbol{n}_i$ is increased. Furthermore, $\|p_i^{\xi} - u_i^{\xi}\|_{L^2(\Lambda)}$ depends only on ξ and tends to zero as $\xi \to \infty$. The main complication comes from the fact that a bound on $\|u_i^{\boldsymbol{n}_i,\xi} - p_i^{\boldsymbol{n}_i,\xi}\|_{L^2(\Lambda_i)}$ from (7.6) depends on both $\boldsymbol{n}_{i-1}$ and ξ if the method is not of type $(0,0,0)$. The same arguments as in Lemma 7.12 show this tends to zero. Therefore we choose $\boldsymbol{n}_i$ large enough so that the second term in (7.14) is less than $\epsilon/3$. Next, we choose ξ large enough so that the sum of the remaining terms is less than $2/3\epsilon$. If the method is of type $(0,0,0)$, then this sum remains less than ϵ when $\boldsymbol{n}_i$ is replaced with $\boldsymbol{n}$ for $\boldsymbol{n} > \boldsymbol{n}_i$. This proves the claims. ☐

Now we prove the uniform approximation theorem.

Theorem 7.16. *Assume* $([G^{\xi};\Gamma^{\xi}])_{\xi\geq 0}$ *is a sequence of k-regular RH problems for k sufficiently large so that Algorithm* 7.1 *converges for each* ξ. *Assume there exists a numerical parametrix as* $\xi \to \infty$. *Then Algorithm* 7.1 *produces a weakly uniform approximation to the solution of* $([G^{\xi};\Gamma^{\xi}])_{\xi\geq 0}$. *Moreover, convergence is strongly uniform if the method is of type* $(0,0,0)$.

Proof. Lemma 7.15 provides an $M > 0$ and $N_1(i)$ such that if $\xi > M$, then

$$\|u_i^{N_1(i),\xi} - u_i^{\xi}\|_{L^2(\Lambda_i)} < \epsilon \text{ for every } i.$$

According to Lemma 7.13 there is $N_2(\xi,i)$ such that

$$\|u_i^{N_2(\xi,i),\xi} - u_i^{\xi}\|_{L^2(\Lambda_i)} < \epsilon \text{ for every } i.$$

The function

$$N(\xi,i) = \begin{cases} N_1(i) & \text{if } \xi > M, \\ N_2(\xi,i) & \text{if } \xi \leq M \end{cases}$$

satisfies the required properties for weak uniformity. Strong uniformity follows in a similar way from Lemmas 7.15 and 7.13. □

Remark 7.3.1. *This proves weak uniform convergence of the numerical method for the toy problem introduced in Section 7.3.2: we can take the RH problem for v as a numerical parametrix.*

The seemingly odd restrictions for the general theorem are a consequence of poorer operator convergence rates when n is large. A well-conditioned numerical method does not suffer from this issue. It is worth noting that using direct estimates is equivalent to requiring that the original sequence of RH problems themselves satisfy the properties of a numerical parametrix.

In what follows, we want to show a given sequence of RH problems is a numerical parametrix. The reasoning for the following result is two-fold. First, we hypothesize only conditions which are easily checked in practice. Second, we want to connect the stability of numerical approximation with the use of local model problems in nonlinear steepest descent.

Proposition 7.17. *Assume*

- $([J^\xi,\Sigma^\xi])_{\xi\geq 0}$ *is a sequence of k-regular RH problems,*
- *the norm of* $\mathscr{C}[K_i^\xi;\Lambda_i]^{-1}$ *is uniformly bounded in* ξ,
- $\|K_i^\xi\|_{W^{k,\infty}(\Lambda_i)} \leq C$, *and*
- $\gamma_i(\xi) \to 0$ *as* $\xi \to \infty$.

Then, if k and ξ are sufficiently large,

- *Algorithm 7.1 applied to* $([J^\xi,\Sigma^\xi])_{\xi\geq 0}$ *has solutions at each stage and*
- $([J^\xi,\Sigma^\xi])_{\xi\geq 0}$ *satisfies the last two properties of a numerical parametrix (Definition 7.14).*

Proof. The proof is essentially the same as that of Theorem 7.11. □

Remark 7.3.2. *Due to the decay of* γ_i, *the invertibility of each* $\mathscr{C}[K_i^\xi;\Lambda_i]$ *is equivalent to that of* $\mathscr{C}[G^\xi;\Gamma^\xi]$.

This proposition states that a numerical parametrix only needs to be locally reliable; we can consider each shrinking contour as a separate RH problem as far as the analysis is concerned.

7.4 ▪ A collocation method realization

We address the properties required in Definition 7.2 for the collocation method described in Chapter 6. First, the method is given by the pair $(\mathscr{I}_n,\mathscr{P}_n=\mathscr{I}_n)$, where $\mathscr{I}_n=\mathscr{B}_n\mathrm{T}_n^{-1}\mathscr{E}_n$ maps a function to its Chebyshev interpolant; compare with Corollary 6.5. Of course, the operators are expanded to act on each smooth component Γ_i of a contour $\Gamma = \Gamma_1 \cup \cdots \cup \Gamma_L$. We proceed as if all SIEs are scalar: the extension to matrix equations is accomplished by simply replacing the absolute value with a matrix/vector norm.

We need to specify the operator $\mathscr{T}_n$ in Definition 7.1. $\mathscr{T}_n$ is the operator that constructs Chebyshev interpolants on each component Γ_i of Γ from data at the mapped Chebyshev points $\boldsymbol{s}^n = (M_1(\boldsymbol{x}^{n_1}), \ldots, M_L(\boldsymbol{x}^{n_L}))$ and $n = c(n_1 + \cdots + n_L)$, where c is the number of entries in the solution of the RH problem under consideration.

Lemma 7.18. *When $G \in W^{1,\infty}(\Gamma)$, the numerical method of Chapter 6 satisfies*

- $\|\mathscr{T}_n^{-1} \mathscr{I}_n\|_{\mathscr{L}(C^0(\Gamma),\mathbb{C}^n)} \leq C_1$,
- $\|\mathscr{T}_n^{-1} \mathscr{H}_n[G;\Gamma]\|_{\mathscr{L}(X_n,\mathbb{C}^n)} \leq C_2 n^2 \log n \|G - I\|_{L^\infty(\Gamma)}$,
- $\|\mathscr{I}_n u - u\|_{H^1(\Gamma)} \leq C_s n^{2-s} \|u\|_{H^s(\Gamma)}$.

Recall (7.3) *for the definition of X_n in terms of $\mathscr{P}_n$.*

Proof. The first statement follows directly because both $C^0(\Gamma)$ and $\mathbb{C}^n$ are equipped with the supremum norm and $\mathscr{T}_n^{-1}$ maps a function back to its values at interpolation points. For the second statement consider

$$\|\mathscr{T}_n^{-1} \mathscr{H}_n[G;\Gamma] \mathscr{B}_n^{-1} \mathscr{B}_n\|_{\mathscr{L}(X_n,\mathbb{C}^n)} \leq \|I - C_n[G;\Gamma]\|_{\mathscr{L}(\mathbb{C}^n)} \|\mathscr{B}_n\|_{\mathscr{L}(X_n,\mathbb{C}^n)}.$$

Now, the norm $\|I - C_n[G;\Gamma]\|_{\mathscr{L}(\mathbb{C}^n)}$ is bounded by the maximum absolute row sum of the matrix. The matrix representation of $\mathscr{C}_\Gamma^-$ acting on $\mathbb{C}^n$, the matrix that maps from Chebyshev coefficients to the value of the finite-part Cauchy transform at Chebyshev points (see Corollary 5.15), has entries that consist of $\mathscr{C}_{\mathbb{I}} T_k$ evaluated at a distance $\mathscr{O}(1/n)$ from ± 1 multiplied by $G(x) - I$. We use the following.

Lemma 7.19. *For $x \in (-1,1)$*

$$|\mathscr{C}_{\mathbb{I}}^{\pm} T_k(x)| \leq \frac{1}{2\pi} \left| \log \frac{x-1}{x+1} \right| + \frac{2}{\pi}(1 + \log n).$$

For $z = \pm 1 + r e^{i\theta}$, $r < 1/2$,

$$|\mathscr{C}_{\mathbb{I}} T_k(z)| \leq \frac{\log r^{-1}}{2\pi} + C_{\pm,\theta},$$

where $C_{+,\theta}$ and $C_{-,\theta}$ are unbounded as $\theta \to \pi, 0$, respectively.

Proof. The first claim follows directly from Corollary 5.15, $\|T_k\|_u = 1$, and an estimate on the harmonic series. The second claim follows from a close analysis of

$$\int_{-1}^{1} \frac{\mathrm{d}x}{|x - z|} = \log \left(\frac{\operatorname{Re}(z+1) + |z+1|}{\operatorname{Re}(z-1) + |z-1|} \right). \qquad \square$$

Therefore, each entry of $I - C_n[G;\Gamma]$ is bounded by $C \log n$, $n > 1$, and

$$\|I - C_n[G;\Gamma]\|_{\mathscr{L}(\mathbb{C}^n)} \leq C \|G - I\|_{L^\infty(\Gamma)} n \log n.^{40}$$

It remains to estimate $\|\mathscr{B}_n\|_{\mathscr{L}(X_n,\mathbb{C}^n)}$.

[40] One can check numerically that this norm grows like $\log n$, but this simple estimate suffices.

We must understand the norm of $\mathscr{B}_n$ as it maps from X_n, the space of (n_i-1)th-order polynomials on each component $\Gamma_i = M_i(\mathbb{I})$ of $\Gamma = \Gamma_1 \cup \cdots \cup \Gamma_L$, with the $L^2(\Gamma)$ norm to the corresponding mapped Chebyshev coefficients of these polynomials with the uniform norm. The simplest way to do this is to relate the Chebyshev polynomials to the normalized *Legendre polynomials* L_k:

$$L_k(x) = a_{kk}x^{k-1} + \cdots + a_{k1}x + a_{k0},$$

$$\int_{-1}^{1} L_k(x)L_j(x)\mathrm{d}x = \delta_{jk}.$$

Thus for $u_i \triangleq u|_{\Gamma_i} \circ M_i$

$$u_i(x) = \sum_{k=0}^{n_i-1} a_k L_k(x) \Rightarrow ||\mathbf{a}||_{\ell^2} \le C_i ||u||_{L^2(\Gamma_i)},$$

where the constant C_i arises from the change of variables M_i. Furthermore, it can be shown [103] that $||\mathbf{u}||_{\ell^2} \le n||\mathbf{a}||_{\ell^2}$, where $\mathbf{u}$ represents the vector of Chebyshev coefficients of u_i. Then, it is clear that $||\mathbf{u}||_{\ell^\infty} \le ||\mathbf{u}||_{\ell^2}$ and $||\mathscr{B}_n||_{\mathscr{L}(X_n,\mathbb{C}^n)} \le Cn$. This establishes the second claim. The final claim follows from Theorem 4.6. □

The final property we need to obtain an admissible numerical method, the boundedness of the inverse, is a very difficult problem. We can easily verify, a posteriori, that the norm of the inverse does not grow too much. In general, for this method, we see at most logarithmic growth.

Assumption 7.4.1. *For some $\beta \ge 2$ assume*

$$||C_n[G;\Gamma]^{-1}||_{\mathscr{L}(\mathbb{C}^n)} \le C||\mathscr{C}[G;\Gamma]^{-1}||_{\mathscr{L}(L^2(\Gamma))} n^{\beta-2},$$

where C is allowed to depend on Γ.

Theorem 7.20. *The numerical method of Chapter 6 is of type $(0,\beta,n^2+\epsilon)$ for every $\epsilon > 0$ and is admissible, provided Assumption 7.4.1 holds.*

Proof. We have $\mathscr{C}_n[G;\Gamma]^{-1} = \mathscr{T}_n^{-1} C_n[G;\Gamma]^{-1}\mathscr{B}_n$ so that

$$||\mathscr{C}_n[G;\Gamma]^{-1}||_{\mathscr{L}(Y_n,X_n)} \le ||\mathscr{T}_n^{-1}||_{\mathscr{L}(X_n,\mathbb{C}^n)} ||C_n[G;\Gamma]^{-1}||_{\mathscr{L}(\mathbb{C}^n)} ||\mathscr{B}_n||_{\mathscr{L}(Y_n,\mathbb{C}^n)}.$$

From the proof of the previous lemma, $||\mathscr{B}_n||_{\mathscr{L}(Y_n,\mathbb{C}^n)} \le Cn$. In light of Assumption 7.4.1 it remains to get a bound on $||\mathscr{T}_n^{-1}||_{\mathscr{L}(X_n,\mathbb{C}^n)}$. For a function $u \in X_n$, again consider the expansion of $u_i \triangleq u|_{\Gamma_i} \circ M_i$ in normalized Legendre polynomials L_k:

$$u_i(x) = \sum_{k=0}^{n_i-1} a_k L_k(x) \Rightarrow ||u_i||_{X_n} = ||u_i||_{L^2(\mathbb{I})} = ||\mathbf{a}||_{\ell^2}.$$

Now, $|u_i(x)| \le \max_{0\le k\le n-1} ||L_k||_u ||\mathbf{a}||_{\ell^1} \le \max_{0\le k\le n-1} ||L_k||_u \sqrt{n}||\mathbf{a}||_{\ell^2}$. It is well known that[41] $||L_k||_u = \sqrt{k+1/2}$ so that $||\mathscr{T}_n^{-1}||_{\mathscr{L}(X_n,\mathbb{C}^n)} \le Cn$, and this proves the result. □

[41]In the usual notation, $|P_k(x)| \le |P_k(1)| = 1$, and $||P_k||^2_{L^2(\mathbb{I})} = k + 1/2$ [91].

Remark 7.4.1. *The estimates in Theorem 7.20 and Lemma 7.18 are surely not sharp, but due to the uncertainty coming from Assumption 7.4.1 it is not currently useful to sharpen the estimates. Furthermore, our transformation operators $\mathscr{B}_n$ and $\mathscr{T}_n$ allow us to specify our assumptions directly in terms of something computable: the matrix $C_n[G;\Gamma]$.*

7.4.1 ▪ The integral of the solution

From Corollary 6.4, we realize L^2 convergence of our approximations to the function u (uniform convergence to Φ by Corollary 6.12). Since $\Phi = I + \mathscr{C}_\Gamma u$, we find that (see Lemma 2.12)

$$\Phi(z) \sim I + \frac{\phi_1}{z} + \mathcal{O}(z^{-2}), \quad \phi_1 = -\int_\Gamma u\, \text{đ}z \approx -\int_\Gamma u^n \text{đ}z, \tag{7.15}$$

where u^n is the numerical solution to the RH problem. Here L^2 convergence implies L^1 convergence since Γ is bounded. An approximation of ϕ_1 is critical in inverse scattering; see (1.13). To efficiently and accurately compute $\int_\Gamma u^n \text{đ}z$ one may use the methods described in Section 4.3.2.

Note that if Algorithm 7.1 is used, then an approximation of u such that $\Phi = I + \mathscr{C}_\Gamma u$ is not directly returned from the algorithm since SIEs are solved on each disjoint contour at each stage, not on all of Γ. What we have is $\Phi = \Phi_L \cdots \Phi_1$ with $\Phi_i = I + \mathscr{C}_{\Gamma_i} u_i$. This can be written as

$$\Phi(z) = \prod_{i=1}^{L} \left(I - \frac{1}{z} \int_{\Gamma_i} u_i(s)\text{đ}s + \mathcal{O}(z^{-2}) \right) = I + \frac{1}{z}\left(-\sum_{i=1}^{L} \int_{\Gamma_i} u_i(s)\text{đ}s \right) + \mathcal{O}(z^{-2}).$$

From this we can conclude that

$$\int_\Gamma u(s)\text{đ}s = \sum_{i=1}^{L} \int_{\Gamma_i} u_i(s)\text{đ}s.$$

But since an approximation $u_i^{n_i}$ to the solution u_i is known we find

$$\phi_1 = -\sum_{i=1}^{L} \int_{\Gamma_i} u_i(s)\text{đ}s \approx -\sum_{i=1}^{L} \int_{\Gamma_i} u_i^{n_i}(s)\text{đ}s. \tag{7.16}$$

Part III

The Computation of Nonlinear Special Functions and Solutions of Nonlinear PDEs

Chapter 8

The Korteweg–de Vries and Modified Korteweg–de Vries Equations

In this chapter we consider the numerical solution of the initial-value problem on the whole line for the *Korteweg–de Vries equation* (KdV)

$$\begin{aligned} q_t + 6qq_x + q_{xxx} &= 0, \\ q(\diamond, 0) &= q_0 \in \mathscr{S}_\delta(\mathbb{R}). \end{aligned} \tag{8.1}$$

We also consider the *(defocusing) modified KdV equation*, given by

$$\begin{aligned} q_t - 6q^2 q_x + q_{xxx} &= 0, \\ q(\diamond, 0) &= q_0 \in \mathscr{S}_\delta(\mathbb{R}). \end{aligned} \tag{8.2}$$

Many of the results of this section were originally published in [118]. The KdV equation describes the propagation of long waves in dispersive media, e.g., long surface water waves [72]. Historically, the KdV equation is the first known case of a PDE that is solvable by the IST [57]. The KdV equation and the modified KdV equation can also be thought of as dispersive regularizations of the Burgers and modified Burgers equations, respectively.

Broadly, in this chapter we present the product of the mathematical structure of the KdV equation. We will see that because of the reformulation of (8.1) in terms of an RH problem, asymptotics are readily available for a large class of initial data. Furthermore, the RH problem reformulation allows for accurate computation for all x and t. In this sense, solutions of (8.1) can be considered nonlinear special functions.

The computation of solutions of (8.1) is not without its difficulties. The presence of dispersion makes the quantitative approximation of solutions of the KdV equation and the modified KdV equation through numerical methods especially delicate; see Appendix E.1 for a detailed discussion. Appendix E.1 demonstrates that while the oscillatory nature of the solution is reproduced in many numerical methods, a high degree of accuracy for moderate times is elusive. To see this heuristically, consider solutions with the initial condition $q(x,0) = A\,\mathrm{sech}^2(x)$. With $A = 3$ the solution is a two-soliton solution without any dispersive tail [42]. A *soliton* is a solitary (traveling and decaying) wave that interacts elastically with other solitons in the sense that the individual interacting wave profiles are unchanged after a collision. In Figure 8.1 we approximate the solution of the KdV equation with $q(x,0) = A\,\mathrm{sech}^2(x)$, where $A = 3.2$ using the numerical scheme presented in this chapter. Notice that a significant dispersive tail forms even though the solution is close to a soliton solution. The issue becomes worse when we consider solutions that are further from a soliton solution; see Figure 8.2.

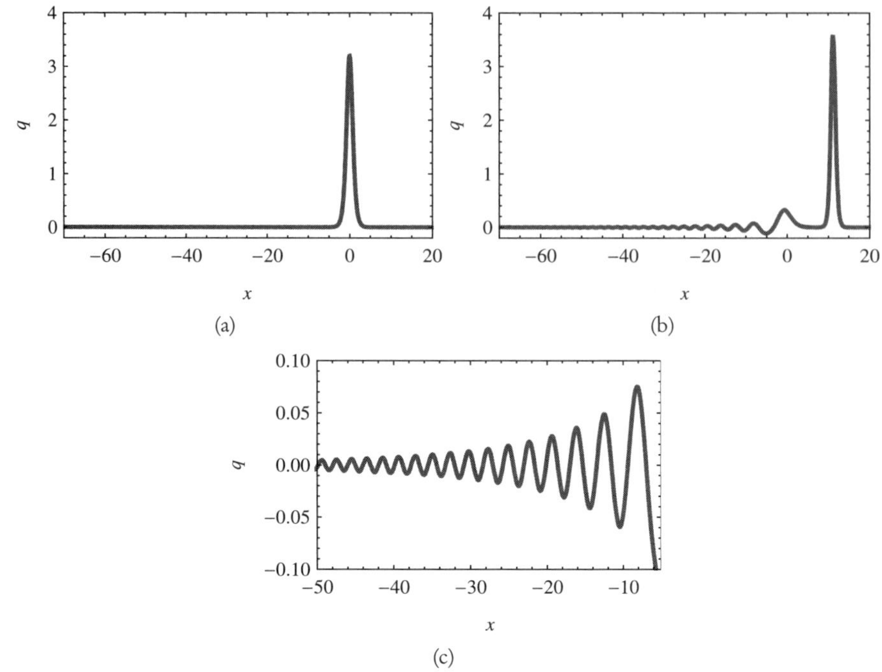

Figure 8.1. *Numerical solution of the KdV equation with initial data that is close to a two-soliton solution.* (a) *Initial condition.* (b) *Solution at $t = 1.5$. The two largest peaks each correspond to a soliton.* (c) *Dispersive tail at $t = 1.5$. Reprinted with permission from Elsevier* [118].

To address this dispersive issue, we exploit the integrability of the KdV equation and the modified KdV equation and evaluate the IST numerically. Computing the IST involves developing techniques to compute the forward transform (direct scattering) and the inverse transform (inverse scattering). The approach to direct scattering employs collocation methods for ODEs (see, for example, the methodology in [9]) and existing spectrum approximation techniques [24]. An in-depth discussion of these methods is not presented here.

For inverse scattering we use the numerical method for RH problems presented in Chapter 6. After deforming the RH problem in the spirit of Chapter 3, the numerical method becomes *uniformly accurate*: the work required to compute the solution at a point to a desired accuracy is seen to be bounded for all x and t. In this method, the roles of x and t are reduced to that of parameters and there is no time stepping to speak of.

The RH problem for the modified KdV equation has a simple form, and the deformations are straightforward. All RH problems stemming from the modified KdV equation have unique matrix solutions. This is readily demonstrated using Theorems 2.69 and 2.73.

Next, the KdV equation is considered. Now one has to deal with the addition of solitons to the problem. After deformation, the RH problem for the KdV equation has a singularity and this requires two additional deformations. Importantly, this singularity gives rise to new asymptotic regimes with distinctly different behaviors. To handle this, we introduce a new transition region corresponding to the use of a new deformation. This transition region allows for uniformly accurate asymptotic computation of the solution in a region where the classical deformations break down numerically and for asymptotic analysis. Numerical results for the KdV equation are presented. For some parameter values, one must consider a vector problem tending to the vector [1, 1] at infinity. This

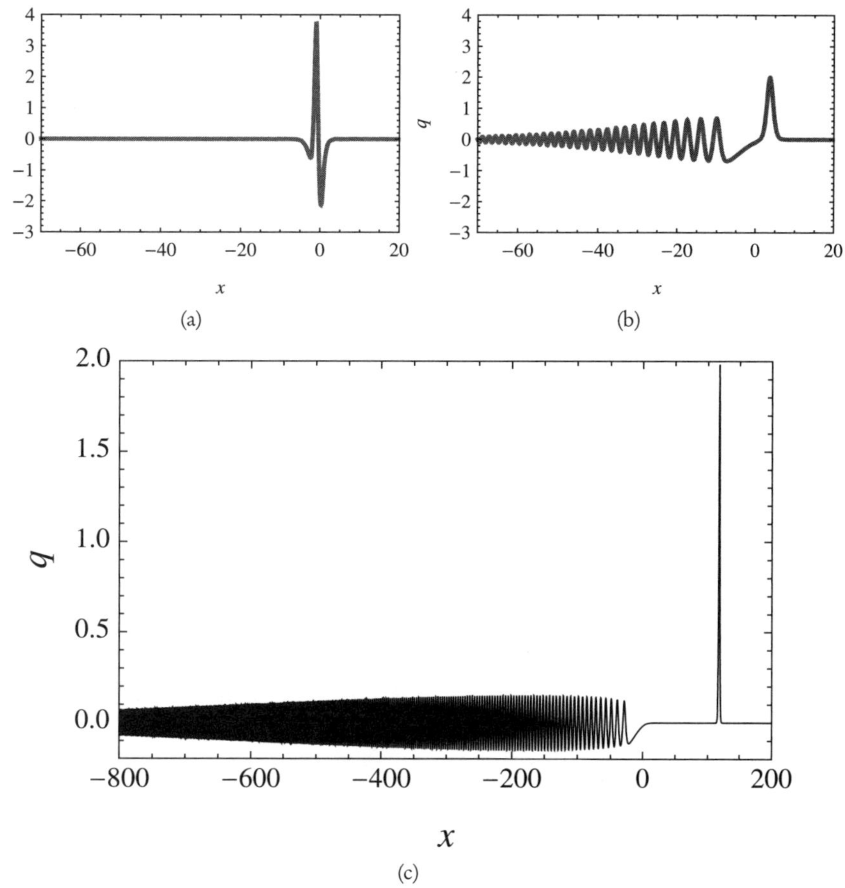

Figure 8.2. *Numerical solution of the KdV equation for an initial condition which is far from a pure soliton initial condition.* (a) *Initial condition obtained by adding a soliton to the RH problem associated with* $q(x,0) = -2.3\,\mathrm{sech}^2(x)$. (b) *Solution at* $t = 1.5$. (c) *Solution at* $t = 30$. *It is not practical to use conventional methods to capture this solution quantitatively for longer times. Reprinted with permission from Elsevier* [118].

is essentially the RH problem that is discussed in Problem 1.6.1. This technical point is discussed further in Section 8.2.4, and it does not affect proceeding naively with computations. Finally, the numerical solutions of the modified KdV equation and the KdV equation are compared using the Miura transformation, which maps solutions of the modified KdV equation to solutions of the KdV equation.

Through the comparison of our results with existing asymptotic expressions we can verify the accuracy of the method. Furthermore, the theory of Chapter 7 can be applied in many situations to put this claim on a firmer theoretical foundation. Finally, as a notational note, when we display circular contours for an RH problem we always assume counterclockwise orientation unless otherwise specified.

8.0.1 ▪ Integrability and Lax pairs

The modified KdV equation and the KdV equation are both *completely integrable*. We take this to mean that for each equation there exist two linear systems of ODEs depending fundamentally on a parameter z,

$$\mu_x = L(z,q)\mu, \quad \mu_t = M(z,q)\mu,$$

such that $\mu_{xt} = \mu_{tx}$ if and only if q satisfies the PDE in question. As has been discussed previously, systems of this form are called Lax pairs. Compare this with Sections 2.5.3 and 3.1. The Lax pair is also known as the *scattering problem* for the PDE. We introduce the *modified Zakharov–Shabat scattering problem*[42] given by

$$\mu_x = \begin{bmatrix} -\mathrm{i}z & q \\ r & \mathrm{i}z \end{bmatrix}\mu, \quad \mu_t = \begin{bmatrix} A & B \\ C & D \end{bmatrix}\mu,$$

where r, A, B, C, and D are scalar functions to be determined [5, p. 11]. If we make the choice

$$\begin{aligned} A &= -4\mathrm{i}z^3 + 2\mathrm{i}zqr - (r_x q - q_x r), \\ B &= 4qz^2 + 2\mathrm{i}zq_x - 2q^2 r - q_{xx}, \\ C &= 4rz^2 - 2\mathrm{i}zr_x + 2qr^2 - r_{xx}, \\ D &= -A, \end{aligned} \tag{8.3}$$

we can obtain Lax pairs for both the modified KdV equation and the KdV equation.

Remark 8.0.1. *These Lax pairs can be expressed as a differential form like* (3.4). *We pursue a more classical approach in this chapter and the next because from a numerical analysis point of view, differential equations are more tractable than the oscillatory integral equations that result from the differential form (see* (3.6)*).*

The modified Korteweg–de Vries equation

To obtain a Lax pair for the (defocusing) modified KdV equation (8.2), let $r = q$ so that the x equation of the Lax pair takes the form

$$\mu_x = \begin{bmatrix} -\mathrm{i}z & q \\ q & \mathrm{i}z \end{bmatrix}\mu. \tag{8.4}$$

In what follows we do not need the explicit form of the equation for μ_t, but it is, of course, used in determining the associated RH problem that we give below.

Remark 8.0.2. *As above, we perform scattering in a more restricted space of functions. We assume* $q(\diamond, 0) \in \mathcal{S}_\delta(\mathbb{R})$*. This simplifies some technical details, as noted below. This assumption is relaxed on a case-by-case basis. The decay rate is needed for analyticity properties, and the smoothness is needed to numerically compute the scattering data defined below.*

1. *Definition of the scattering data.* Consider the problem (8.4). We define the spectral data which is the output of the forward spectral problem. Assuming $q \in \mathcal{S}_\delta(\mathbb{R})$, it follows that there are two matrix-valued eigenfunctions

$$\phi(x;z) \sim \begin{bmatrix} e^{-\mathrm{i}zx} & 0 \\ 0 & -e^{\mathrm{i}zx} \end{bmatrix} \text{ as } x \to -\infty, \quad \psi(x;z) \sim \begin{bmatrix} e^{-\mathrm{i}zx} & 0 \\ 0 & e^{\mathrm{i}zx} \end{bmatrix} \text{ as } x \to \infty. \tag{8.5}$$

[42] One can consider the "scattering problem" as being the spectral theory of the x-part of the Lax pair.

From Liouville's formula, the determinants of these solutions are constant in x; evaluating at $\pm\infty$ we see that the columns do indeed form a linearly independent solution set and hence span the solution space. There exists a scattering (or transition) matrix

$$S(z) = \begin{bmatrix} a(z) & \mathscr{B}(z) \\ b(z) & \mathscr{A}(z) \end{bmatrix}$$

such that

$$\phi(x;z) = \psi(x;z)S(z).$$

Define $\rho(z) = b(z)/a(z)$ to be the reflection coefficient. For the defocusing modified KdV equation we define the scattering data to be only the reflection coefficient. Based on the definition of $\mathscr{S}_\delta(\mathbb{R})$ one can show that the reflection coefficient is analytic in a strip $S_\gamma = \{z : |\operatorname{Im} z| < \gamma\}$ with $\gamma = \delta/2$ by considering Volterra integral equations as in Lemma 3.4.

2. *The inverse problem.* We phrase the inverse problem (or inverse spectral problem) in terms of an RH problem. We seek a 2×2 matrix-valued function Φ that satisfies

$$\Phi^+(s) = \Phi^-(s)G(s), \quad s \in \mathbb{R}, \quad \Phi(\infty) = I,$$
$$G(z) = \begin{bmatrix} 1-\rho(z)\rho(-z) & -\rho(-z)e^{-\theta(z)} \\ \rho(z)e^{\theta(z)} & 1 \end{bmatrix}, \quad \theta(z) = 2izx + 8iz^3t.$$

The solution to the modified KdV equation is given by

$$q(x,t) = -2i \lim_{z\to\infty} z\Phi(z)_{21}, \tag{8.6}$$

where the subscript denotes the $(2,1)$ component [35]. We suppress the x and t dependence for notational simplicity.

The Korteweg–de Vries equation

To obtain the KdV equation (8.1) from (8.3) we set $r = -1$ and the x portion of the Lax pair takes the form

$$\mu_x = \begin{bmatrix} -iz & q \\ -1 & iz \end{bmatrix} \mu.$$

This can be simplified to the time-independent Schrödinger equation

$$-\mu_{xx} - (q + z^2)\mu = 0. \tag{8.7}$$

As before, we do not need the explicit form of the equation for μ_t.

1. *Definition of the scattering data.* We consider the problem (8.7) and assume $q \in \mathscr{S}_\delta(\mathbb{R})$. We define the spectral data which is the output of the forward spectral problem in this context. A detailed description of the scattering data is given in Section 1.6. To be precise we define the set

$$\{\rho, \{z_j\}_{j=1}^n, \{c_j\}_{j=1}^n\} \tag{8.8}$$

to be the scattering data for the KdV equation.

2. *The inverse problem.* We can pose the meromorphic RH problem for the solution of the KdV equation which is the inverse spectral problem. This is the same problem as in Section 1.6 but now with time dependence. We seek a function $\Phi : \mathbb{R} \to \mathbb{C}^{1\times 2}$ that is meromorphic off $\mathbb{R}$ with simple poles at $\pm z_j$ with residue conditions

$$\begin{aligned}
&\Phi^+(z) = \Phi^-(z)G(z), \quad z \in \mathbb{R}, \quad \Phi(\infty) = [1,\ 1],\\
&\operatorname{Res}_{z=z_j} \Phi(z) = \lim_{z\to z_j} \begin{bmatrix} 0 & 0 \\ c_j e^{\theta(z_j)} & 0 \end{bmatrix} \Phi(z),\\
&\operatorname{Res}_{z=-z_j} \Phi(z) = \lim_{z\to -z_j} \begin{bmatrix} 0 & -c_j e^{\theta(z_j)} \\ 0 & 0 \end{bmatrix} \Phi(z).
\end{aligned} \tag{8.9}$$

The solution to the KdV equation is given by the reconstruction formula (compare with Problem 1.6.1)

$$q(x,t) = 2\mathrm{i} \lim_{z\to\infty} z\partial_x \Phi(z)_1.$$

Remark 8.0.3. *This meromorphic problem can be turned into an analytic problem by introducing small circles around each pole and using the appropriate jump on this new contour* [60]. *Fix* $0 < \epsilon < \min_{z\neq j} |z_j - z_k|/2$, *with* $\epsilon < \min_j |z_j|$. *This* ϵ *is chosen so that the circles* $A_j^{\pm} = \{z \in \mathbb{C} : |z - \pm z_j| < \epsilon\}$ *do not intersect each other or the real axis. We define* $\hat{\Phi}$ *by*

$$\hat{\Phi}(z) = \begin{cases} \Phi(z) \begin{bmatrix} 1 & 0 \\ -c_j e^{\theta(z_j)}/(z - z_j) & 1 \end{bmatrix} & \text{if } |z - z_j| < \epsilon,\ j = 1,\ldots,n, \\[2ex] \Phi(z) \begin{bmatrix} 1 & 0 \\ c_j e^{\theta(z_j)}/(z + z_j) & 1 \end{bmatrix} & \text{if } |z + z_j| < \epsilon,\ j = 1,\ldots,n, \\[2ex] \Phi(z) & \text{otherwise.} \end{cases}$$

It is straightforward to show that $\hat{\Phi}$ *solves the RH problem*

$$\hat{\Phi}^+(z) = \begin{cases} \hat{\Phi}^-(z)G(z) & \text{if } z \in \mathbb{R}, \\[2ex] \hat{\Phi}^-(z) \begin{bmatrix} 1 & 0 \\ -c_j e^{\theta(z_j)}/(z - z_j) & 1 \end{bmatrix} & \text{if } z \in A_j^+, \\[2ex] \hat{\Phi}^-(z) \begin{bmatrix} 1 & -c_j e^{\theta(z_j)}/(z + z_j) \\ 0 & 1 \end{bmatrix} & \text{if } z \in A_j^-, \end{cases}$$

$$\hat{\Phi}(\infty) = [1,\ 1],$$

where $A_j^-(A_j^+)$ *has (counter-)clockwise orientation.*

8.0.2 ▪ Asymptotic regions

In this section we present classical results on the long-time asymptotics of the solution of the modified KdV equation and the KdV equation. We introduce constants, C_i, to divide regions. While any valid choice of these will work, the numerical method can be improved by adjusting them on a case-by-case basis.

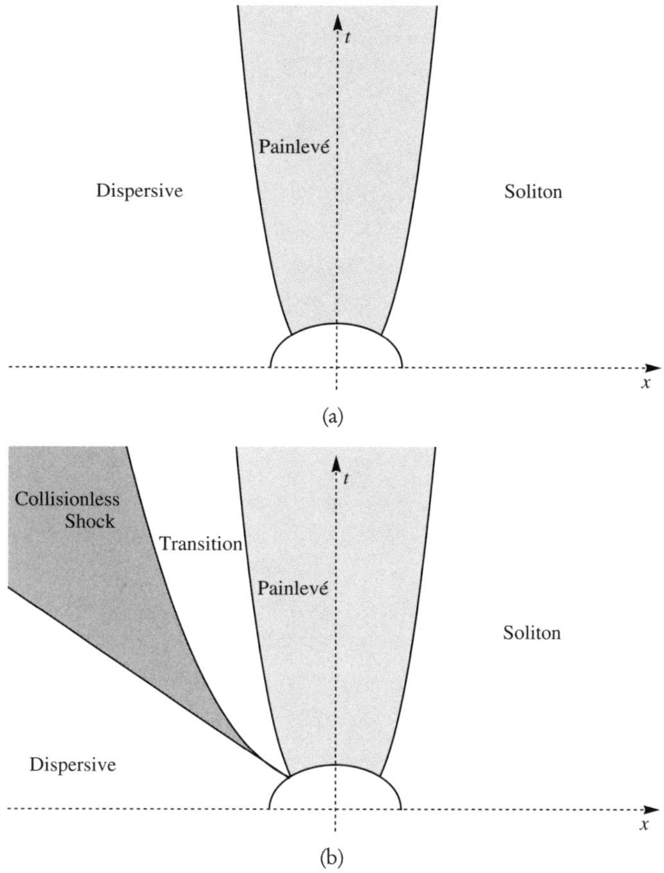

Figure 8.3. (a) *Regions for the asymptotic analysis for the modified KdV equation.* (b) *Regions for the asymptotic analysis for the KdV equation. Reprinted with permission from Elsevier* [118].

The modified Korteweg–de Vries equation

The results presented here are found in [35]. In the (x,t)-plane, the long-time evolution of the modified KdV equation is described in three fundamentally different ways. For a diagram of these regions see Figure 8.3(a).

1. *The soliton region.* This region is defined for $x \geq C_1 t^{1/3}$ for any fixed $C_1 > 0$. The name "soliton region" is a misnomer because there are no solitons present in the defocusing modified KdV equation, but for the sake of uniformity with the KdV equation we retain the name. Here the solution $q(x,t)$ decays beyond all orders, i.e.,

$$q(x,t) = \mathcal{O}((x+t)^{-j}) \text{ as } x,t \to \infty \text{ for all } j > 0. \tag{8.10}$$

2. *The Painlevé region.* This region is defined for $|x| \leq C_1 t^{1/3}$. More general results can be found in [35]. Along a trajectory $x = -Ct^{1/3}$, $C > 0$, the solution satisfies

$$q(x,t) - U(x,t) = \mathcal{O}(t^{-2/3}), \tag{8.11}$$

where

$$U(x,t) = (3t)^{-1/3} v(x/(3t)^{1/3}), \tag{8.12}$$

and $v(x) \triangleq P_{\mathrm{II}}(-\mathrm{i}\rho(0), 0, \mathrm{i}\rho(0); x)$ is the Ablowitz–Segur solution to Painlevé II with Stokes constants $\{s_1, s_2, s_3\} = \{-\mathrm{i}\rho(0), 0, \mathrm{i}\rho(0)\}$. See Chapter 10 for the definition of P_{II} and a numerical method to compute this solution.

3. *The dispersive region.* Historically, this region is defined for $-x > C_2 t > 0$ for $C_2 > 0$. For our purposes, we use $-x > C_1 t^{1/3}$ for the definition of this region. The reasoning for this will become clear below. Along a trajectory $-x = Ct$, $C > 0$, the solution satisfies

$$q(x,t) - R(x,t) = \mathcal{O}(\log(t) t^{-1}), \tag{8.13}$$

where

$$R(x,t) = \sqrt{\frac{\nu(z_0)}{3t z_0}} \cos\left(16t z_0^3 - \nu(z_0)\log(192 t z_0^3) + \delta(z_0)\right),$$

and

$$z_0 = \sqrt{-x/(12t)}, \quad \nu(z_0) = -\frac{1}{2\pi}\log(1 - \rho(z_0)\bar{\rho}(z_0)),$$

$$\delta(z_0) = \frac{\pi}{4} - \arg(\rho(z_0)) + \arg(\Gamma(\mathrm{i}\nu(z_0))) - \frac{1}{\pi}\int_{-z_0}^{z_0} \log\left(\frac{1-\rho(\eta)\bar{\rho}(\eta)}{1-\rho(z_0)\bar{\rho}(z_0)}\right)\frac{1}{\eta - z_0}\,d\eta.$$

The Korteweg–de Vries equation

The results presented here are found in [32, 60]. See Figure 8.3(b) for a diagram of these regions.

1. *The soliton region.* This region is defined for $x \geq C_1 t^{1/3}$ for any fixed $C_1 > 0$. For $x > Ct$, $C > 0$, the solution of the KdV equation in this region satisfies

$$q(x,t) - S(x,t) = \mathcal{O}((x+t)^{-m}) \quad \text{for all } m > 0,$$

where

$$S(x,t) = \sum_{j=1}^{n} 2z_j^2 \operatorname{sech}^2(z_j x - 4z_j^3 t - p_j), \quad p_j = \frac{1}{2}\log\left(\frac{c_j^2}{2z_j}\prod_{l=j+1}^{n}\left(\frac{z_l - z_j}{z_l + z_j}\right)^2\right).$$

The constants z_j and c_j are those in (8.8).

2. *The Painlevé region.* This region is defined for $|x| < C_2 t^{1/3}$ for any fixed $C_2 > 0$. Along a trajectory $x = \pm C t^{1/3}$, $C > 0$, the solution to the KdV equation satisfies

$$q(x,t) - U(x,t) = \mathcal{O}(t^{-1}), \tag{8.14}$$

where

$$U(x,t) = \frac{1}{(3t)^{2/3}}\left(v^2\left(\frac{x}{(3t)^{1/3}}\right) + v'\left(\frac{x}{(3t)^{1/3}}\right)\right),$$

and v is the Hastings–McLeod solution to Painlevé II with Stokes constants $\{s_1, s_2, s_3\} = \{\mathrm{i}, 0, -\mathrm{i}\}$ [64] (see also Chapter 10). The error bound is not present in [32], but we infer it from (8.11) through the Miura transformation in Section 8.2.5.

3. *Transition region.* This region is, to our knowledge, not present in the literature. It is defined by the relation $-C_3 t^{1/3}(\log t)^{2/3} \leq x \leq -C_4 t^{1/3}$, C_3, $C_4 > 0$. While leading-order asymptotics in this region are known [4], they have not been verified rigorously.

4. *The collisionless shock region.* This region is defined by

$$-C_5 t \leq x \leq -C_6 t^{1/3}(\log t)^{2/3},$$

$0 < C_5 \leq 12$, and $C_6 > 0$. The asymptotic formula in [32] is given with the constraint $1/C \leq -x/(t^{1/3}(\log t)^{2/3}) \leq C$ for $C > 1$. With this constraint the RH problem limits to an RH problem on $(-b(s), b(s))$ of the form [32]

$$\zeta^+(z) = \begin{cases} \zeta^-(z)\begin{bmatrix} 0 & \mathrm{e}^{-24\mathrm{i}\tau\int_0^{a(s)} f(p)\mathrm{d}p} \\ -\mathrm{e}^{24\mathrm{i}\tau\int_0^{a(s)} f(p)\mathrm{d}p} & 0 \end{bmatrix} \\ \text{when } a(s) < z < b(s), \\ \zeta^-(z)\begin{bmatrix} 2\nu z^2 & 0 \\ 0 & (2\nu z^2)^{-1} \end{bmatrix} \\ \text{when } -a(s) < z < a(s), \\ \zeta^-(z)\begin{bmatrix} 0 & \mathrm{e}^{-24\mathrm{i}\tau\int_0^{-a(s)} f(p)\mathrm{d}p} \\ -\mathrm{e}^{24\mathrm{i}\tau\int_0^{-a(s)} f(p)\mathrm{d}p} & 0 \end{bmatrix} \\ \text{when } -b(s) < z < -a(s), \end{cases} \tag{8.15}$$

$$\zeta(\infty) = \begin{bmatrix} 1 & 1 \end{bmatrix}, \quad f(p) = \sqrt{(a^2 - p^2)(b^2 - p^2)}.$$

The definitions of a, b, s, and τ can be found in Appendix E.2. See Section 8.2.2 for the definition of ν. Note that the only x and t dependence enters through a, b, and τ. The approximation W of the solution of the KdV equation is obtained by

$$W(x,t) = 2\mathrm{i}\sqrt{-x/(12t)}\lim_{z\to\infty} z\partial_x \zeta(z).$$

Remark 8.0.4. *By adjusting C_2 and C_6, the collisionless shock region can be made to overlap with the Painlevé region up to a finite time. In the absence of the transition region, this will always leave a gap in the (x,t)-plane that is not contained in any region. From a numerical point of view, we introduce the transition region precisely to compute the solution of the KdV equation in this gap.*

5. *The dispersive region.* This region is defined by $-x > C_7 t > 0$, $C_7 > 0$. Along a trajectory $x = -Ct$, $C > 0$, the solution to the KdV equation satisfies

$$q(x,t) - R(x,t) = \mathcal{O}(t^{-1}), \tag{8.16}$$

where

$$R(x,t) = -\sqrt{\frac{4\nu(z_0)z_0}{3t}}\sin\left(16tz_0^3 - \nu(z_0)\log(192tz_0^3) + \delta(z_0)\right),$$

and

$$z_0 = \sqrt{-x/(12t)}, \quad \nu(z_0) = -\frac{1}{2\pi}\log(1-\rho(z_0)\bar{\rho}(z_0)),$$
$$\delta(z_0) = \frac{\pi}{4} - \arg(\rho(z_0)) + \arg(\Gamma(\mathrm{i}\nu(z_0))) + \sum_{j=1}^{n}\arctan\left(\frac{z_j}{z_0}\right)$$
$$-\frac{1}{\pi}\int_{-z_0}^{z_0}\log\left(\frac{1-\rho(\eta)\bar{\rho}(\eta)}{1-\rho(z_0)\bar{\rho}(z_0)}\right)\frac{1}{\eta - z_0}\mathrm{d}\eta.$$

8.1 ▪ The modified Korteweg–de Vries equation

8.1.1 ▪ Numerical computation of the scattering data

We look for solutions of the form (8.5) to (8.4). Define

$$\sigma_3 = \begin{bmatrix} 1 & 0 \\ 0 & -1 \end{bmatrix}, \quad \sigma_1 = \begin{bmatrix} 0 & 1 \\ 1 & 0 \end{bmatrix},$$

and two new functions

$$J(z) = \phi(z)\sigma_3 e^{\mathrm{i}zx\sigma_3} - I, \quad K(z) = \psi(z)e^{\mathrm{i}zx\sigma_3} - I. \tag{8.17}$$

Therefore $J \to 0$ as $x \to -\infty$ and $K \to 0$ as $x \to \infty$. Rewriting (8.5),

$$\mu_x = q\sigma_1\mu - \mathrm{i}z\sigma_3\mu,$$

and we find that K and J both solve

$$M_x - \mathrm{i}z[M, \sigma_3] - q\sigma_1 M = q\sigma_1.$$

For each z, this can be solved with a Chebyshev collocation method on $(-L, 0]$ for J and on $[0, L)$ for K using vanishing boundary conditions at $\pm L$. See [9] for further details. If we use n collocation points, this gives two approximate solutions J_n and K_n for J and K, respectively. From J_n and K_n we obtain ϕ_n and ψ_n, approximations of ϕ and ψ, respectively, by inverting (8.17). Furthermore, ϕ_n and ψ_n share the point $x = 0$ in their domain of definition. Define

$$S_n(z) = \psi_n^{-1}(0; z)\phi_n(0; z).$$

This is an approximation of the scattering matrix $S(z)$, from which we extract an approximation of the reflection coefficient.

8.1.2 ▪ Numerical solution of the inverse problem

The RH problems considered here have the key feature that the jump matrices are highly oscillatory. Deift and Zhou adapted ideas from the asymptotic evaluation of integrals to this problem to obtain asymptotic formulae with rigorous error bounds [34, 35, 32]. To summarize, the main idea of this method is to deform the contours of the RH problem so that it limits (in some sense) to a simple problem that can be solved explicitly. In general, these same ideas translate to the numerics. The exponential decay that is sought in the analytic method also enables fast convergence of the numerical approximation, as the smoothness of the resulting asymptotic expansions ensures that the solution to the RH problem can be well represented by mapped Chebyshev polynomials. In what follows we deform the RH problem for the modified KdV equation. The deformations are guided by the desire to remove oscillations from the jump contours. This is generally accomplished by factoring the jump matrix and deforming the contours by lensing so that each factor is isolated near *stationary points*, away from which they approach the identity exponentially fast.

To remove oscillations from the jump matrix, we need to examine the exponential that appears in these expressions, which we represent as $\exp\theta(z)$, where $\theta(z) = 2\mathrm{i}zx + 8\mathrm{i}z^3 t$. For $x < 0$, in analogy with the method of steepest descent for integrals, we deform the RH problem through the stationary points of θ. We find that $\theta'(z) = 2\mathrm{i}x + 24\mathrm{i}z^2 t$, and solving for $\theta'(z) = 0$ gives the stationary points $z = \pm z_0$, with $z_0 = \sqrt{-x/(12t)}$. The directions of steepest descent, at $\pm z_0$, along which the oscillations of the jump matrix become exponential decay, are given by

$$\theta_s^+ = 3\pi/4 \pm \pi/2, \quad \theta_s^- = \pi/4 \pm \pi/2.$$

Note that the steepest descent angles for $\exp(-\theta(z))$ differ from these by π.

The dispersive region: We present the full deformation from the initial RH problem on the real axis. We introduce two factorizations of the original jump matrix $G(z)$:

$$\begin{aligned}
&G(z) = M(z)P(z),\\
&M(z) = \begin{bmatrix} 1 & -\rho(-z)\mathrm{e}^{-\theta(z)} \\ 0 & 1 \end{bmatrix}, \quad P(z) = \begin{bmatrix} 1 & 0 \\ \rho(z)\mathrm{e}^{\theta(z)} & 1 \end{bmatrix},\\
&G(z) = L(z)D(z)U(z), \quad L(z) = \begin{bmatrix} 1 & 0 \\ \rho(z)\mathrm{e}^{\theta(z)}/(1-\rho(z)\rho(-z)) & 1 \end{bmatrix},\\
&D(z) = \begin{bmatrix} 1-\rho(z)\rho(-z) & 0 \\ 0 & 1/(1-\rho(z)\rho(-z)) \end{bmatrix},\\
&U(z) = \begin{bmatrix} 1 & -\rho(-z)\mathrm{e}^{-\theta(z)}/(1-\rho(z)\rho(-z)) \\ 0 & 1 \end{bmatrix}.
\end{aligned}$$

In what follows, we often suppress x and t dependence for notational simplicity. The factorizations are suggestively defined. M (for "minus") will be deformed into the lower-half plane and P (for "plus") will be deformed into the upper-half plane. L is lower triangular and will be deformed into the lower-half plane, D is diagonal and will not be deformed. Finally, U is upper triangular and will be deformed into the upper-half plane. Throughout our deformations we use the notation $\Phi_{n,\alpha}$ for the solution of the deformed problem. The number n indicates how many deformations have been performed, with $n = 1$ being the original RH problem. The characters α are used to denote the region (e.g., $\alpha = \mathrm{cs}$ for the collisionless shock region).

Since $q \in \mathscr{S}_\delta(\mathbb{R})$ for some $\delta > 0$, ρ has an analytic continuation off the real axis into a symmetric strip of width δ that contains the real axis so that all the deformations are justified. These factorizations are used so that only one of $\exp\theta(z)$ or $\exp(-\theta(z))$ is present in each matrix. This makes it possible to deform the contours to new contours which have angles $\theta_s^\pm$ with the real axis, along which the jump matrices approach the identity exponentially fast. The "ghost" contours introduced in Figure 8.4(a) all satisfy this desired property, and hence we define a new matrix function $\Phi_{2,\mathrm{d}}$ based on these regions. Notice that the new definitions still satisfy the condition at infinity. We compute the jumps that $\Phi_{2,\mathrm{d}}$ satisfies to phrase an RH problem for $\Phi_{2,\mathrm{d}}$; see Figure 8.4(b).

In order to achieve uniform approximation in the sense of Section 7.3, we need the jump matrix to approach the identity away from $\pm z_0$, i.e., we need to remove the contour on $(-z_0, z_0)$. Indeed, numerical results show that the solution on this contour is increasingly oscillatory as $|x|+|t|$ becomes large. We introduce the unique 2×2 matrix-valued function Δ that satisfies the diagonal RH problem

$$\Delta^+(z;z_0) = \Delta^-(z;z_0)D(z), \quad z \in (-z_0, z_0), \quad \Delta(\infty; z_0) = I. \tag{8.18}$$

See Section 2.4.1 for the exact form of Δ. Notice that in general Δ has singularities at $\pm z_0$. To combat this issue we introduce circles around both $\pm z_0$; see Figure 8.4(c). We define $\Phi_{3,\mathrm{d}}$ by the definitions in Figure 8.5(a) where $\Phi_{3,\mathrm{d}} = \Phi_{2,\mathrm{d}}$ when no definition is specified. Computing the jumps we see that $\Phi_{3,\mathrm{d}}$ satisfies the RH problem in Figure 8.5(b). We apply the same procedure at $-z_0$ and obtain the problem shown graphically in Figure 8.6(a). Finally, we define $\Phi_{4,\mathrm{d}} = \Phi_{3,\mathrm{d}}\Delta^{-1}$ and $\Phi_{4,\mathrm{d}}$ satisfies the RH problem shown in Figure 8.6(b). We solve this resulting RH problem numerically.

Remark 8.1.1. *To obtain an RH problem valid for $t = 0$ and $x < 0$ one can take the limit of the above RH problem as $t \to 0^+$. In this limit $z_0 \to \infty$ and Δ has a jump on all of $\mathbb{R}$.*

The Painlevé region: For $x > 0$, this region intersects with the soliton region defined below, and we use that deformation. For $x < 0$, the stationary points are coalescing, and this allows for a new deformation. In this region we reduce the number of contours present, in order to reduce the overall computational cost. Indeed, consider the interval between the two stationary points $[-z_0, z_0]$, where

$$|z| \leq \sqrt{\frac{C_2}{12}t^{-1/3}} \Rightarrow |2zx + 8z^3t| \leq 2C_2 t^{1/3}|z| + 8|z|^3 t \leq \frac{2}{\sqrt{12}}C_2 + \frac{8}{12\sqrt{12}}C_2^{3/2}.$$

This implies that the oscillations are controlled between the two stationary points and the LDU factorization is not needed. See Figure 8.7(a) for the RH problem in this region.

Remark 8.1.2. *The deformations for the dispersive region and the Painlevé regions are valid in overlapping regions of the (x,t)-plane. As $x \to 0, x < 0$, the deformation for the dispersive region can be used until the Painlevé region is reached. Using these deformations in tandem allows the method to retain accuracy in the region $x < 0$, $t \geq 0$ for $|x|$ and t large. Note that for the deformation for the dispersive region to be valid as $z_0 \to 0$ it is necessary that $\|\rho\|_{L^\infty(\mathbb{R})} < 1$ because of the form of D.*

The soliton region: Choose a function $\alpha(x,t)$ so that $0 \leq \alpha(x,t) < \gamma$; then the deformation used in this region is given in Figure 8.7(b). Note that the angle of the contours is chosen so that $\operatorname{Re}\theta(z) \leq 0$ on all contours with $\operatorname{Im} z > 0$, whereas $\operatorname{Re}\theta(z) > 0$ on all contours with $\operatorname{Im} z \leq 0$.

Remark 8.1.3. *There is a lot of freedom in choosing α. For simplicity, we assume the reflection coefficient is analytic and decays in the strip $\{s + it : s \in \mathbb{R},\ t \in (-T, T),\ T < \gamma\}$, and therefore we use $\alpha(x,t) = \min\{\gamma/2, |z_0|\}$.*

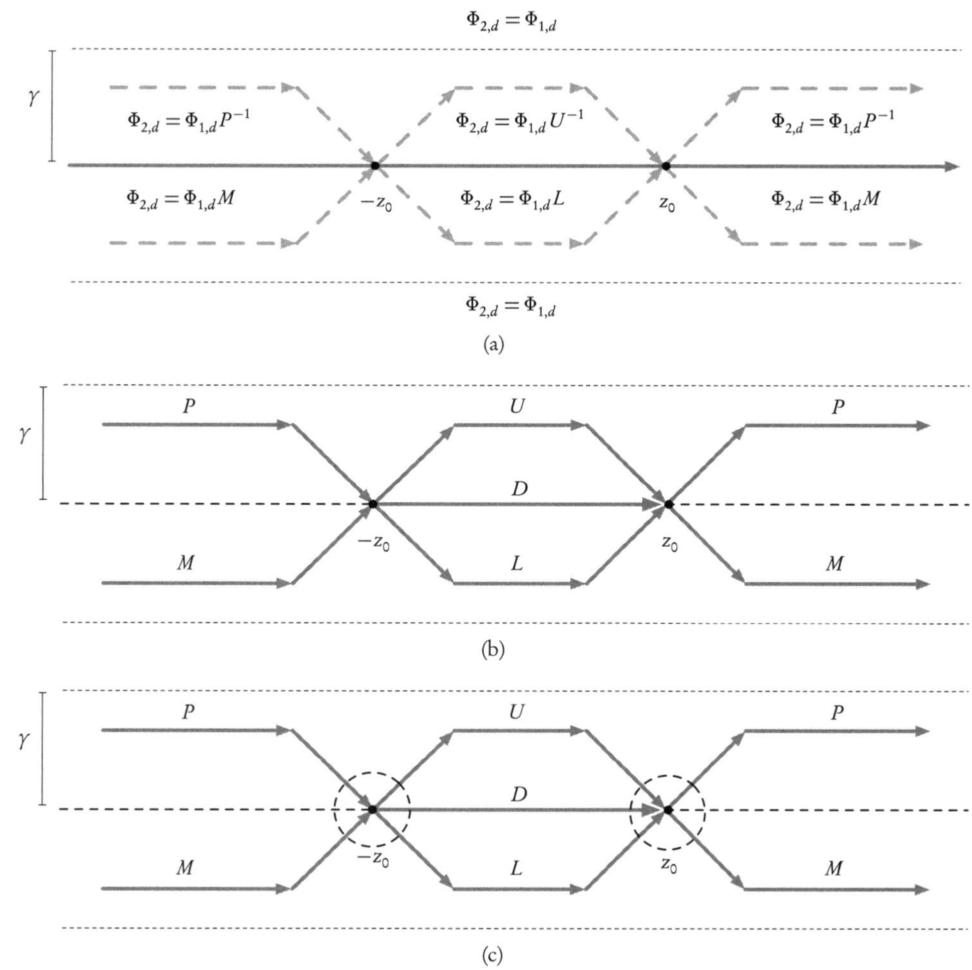

Figure 8.4. (a) *The jump contours and matrices of the initial RH problem with "ghost" contours.* (b) *The jump contours and matrices of the RH problem satisfied by* $\Phi_{2,\mathrm{d}}$. (c) *Ghost circles in preparation for the singularities of* Δ.

8.1.3 ▪ Numerical results

There are additional issues that have to be addressed before these RH problems can be efficiently solved numerically. First, in Section 8.1.2 we opened up circles around two singularities at $\pm z_0$. This deformation is valid provided the radius of the circles is sufficiently small. In addition, we need to shrink the radius of these circles as $|x|$ or t become large, and this is done following Assumption 7.0.1. Second, we truncate contours when the jump matrices are to machine precision, the identity matrix which is justified using Proposition 2.78. This allows us to have only finite contours present in the problem. Furthermore, it allows all the contours to shrink as x and t increase because of the more drastic exponential decay (to the identity matrix) in the jumps. The scaling on these contours is the same as for the circles around the stationary points. Note that if all jump contours are decaying to the identity as x and t becomes large, it is possible for us to truncate all contours and approximate the solution by zero.

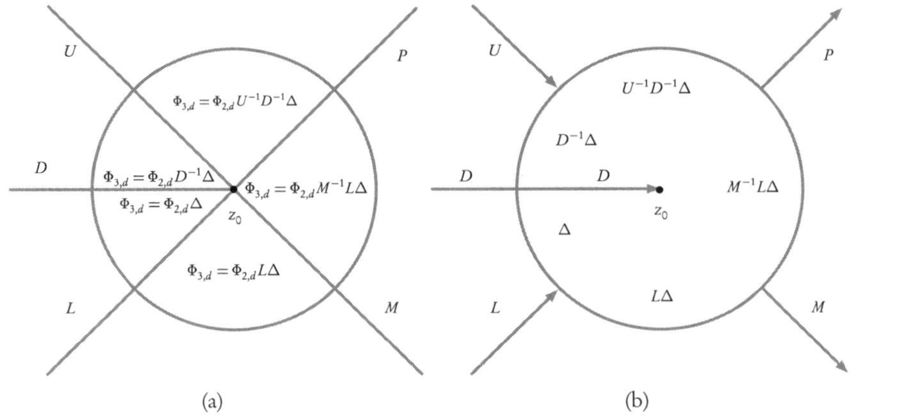

Figure 8.5. (a) *Definition of* $\Phi_{3,\mathrm{d}}$ *near* z_0. (b) *The jump contours and matrices of the RH problem satisfied by* $\Phi_{3,\mathrm{d}}$ *near* z_0.

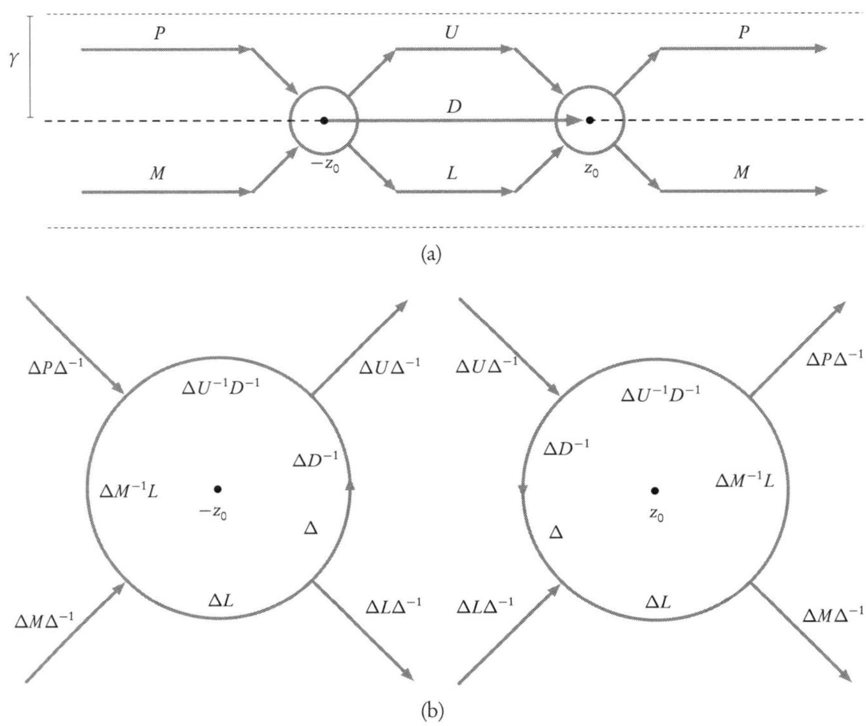

Figure 8.6. (a) *The jump contours and matrices of the RH problem satisfied by* $\Phi_{3,\mathrm{d}}$. (b) *The jump contours and matrices of the RH problem satisfied by* $\Phi_{4,\mathrm{d}}$. *Note that the contours with jumps* $\Delta U\Delta^{-1}$ *and* $\Delta L\Delta^{-1}$ *connect.*

Finally, we define $q_n(x,t)$ as the approximation to the solution of the modified KdV equation with n collocation points on each contour where the initial condition is implied from context. For implementational specifics, we refer the reader to the package [110].

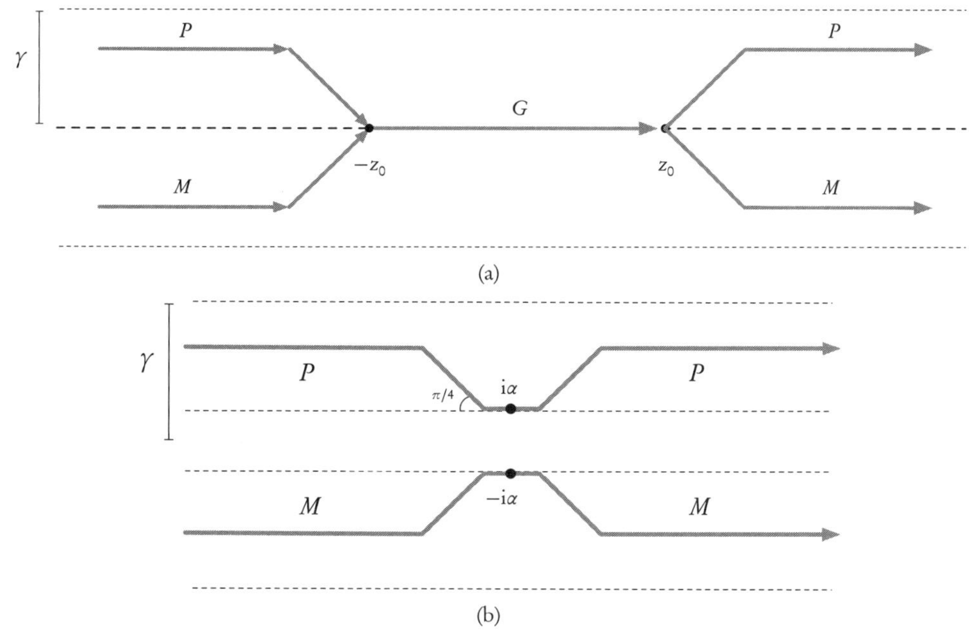

Figure 8.7. (a) *The jump contours and matrices of the RH problem for the modified KdV equation in the Painlevé region with* $x < 0$. (b) *The jump contours and matrices of the RH problem for the modified KdV equation in the soliton region.*

Direct scattering: For an initial condition where the reflection coefficient is not known explicitly we can verify our direct, and in the process inverse, scattering computations by evaluating the solution to the inverse problem at $t = 0$. As an example we start with the initial condition $q(x,0) = -1.3\,\mathrm{sech}^2(x)$. In Figure 8.8(a) we plot the error, $|q(x,0) - q_{80}(x,0)|$, while varying the number of collocation points. Define $\rho^{(m)}(z)$ to be the approximation of the reflection coefficient obtained using m collocation points. In Figure 8.8(b) we show spectral convergence of the computation of the reflection coefficient when $z = 1$.

Inverse scattering: Throughout this section we proceed as if the reflection coefficient is obtained to machine precision. This is often not the case since we do not have an explicit formula for the reflection coefficient. This limits the accuracy obtained in the plots below.

1. *Convergence.* To analyze the error, we introduce some notation. Define

$$Q_n^m(x,t) = |q_n(x,t) - q_m(x,t)|.$$

   Using this notation, Figure 8.23 demonstrates the spectral (Cauchy) convergence with each of the deformations.

2. *Uniform accuracy.* For the method to be uniformly accurate we require that, for a given n and m, $Q_n^m(x,t)$ remain bounded (and small) as $|x| + |t|$ becomes large. In fact, what we numerically demonstrate is that $Q_n^m(x,t)$ tends to zero in all regions. See Figure 8.10 for a demonstration of this. Note that we expect $Q_n^m(x,t)$ to approach zero only when the solution approaches zero as well.

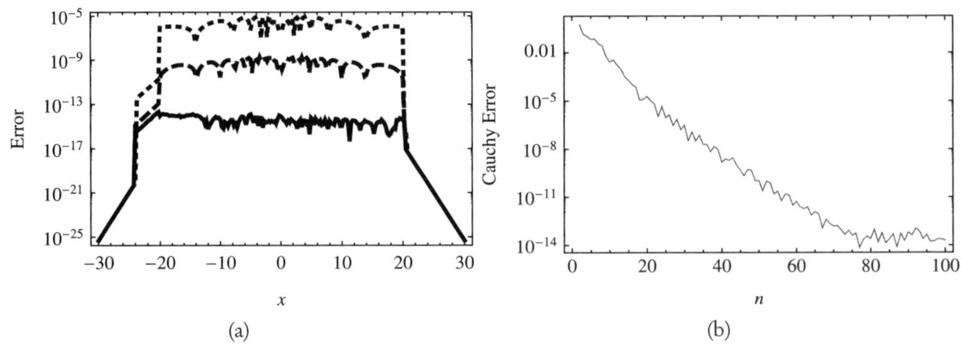

Figure 8.8. (a) *Error in performing the full inverse scattering transformation at $t = 0$ while varying the number of collocation points m for the direct scattering ($m = 20$: dotted line, $m = 40$: dashed line, $m = 80$: solid line). Note that for moderate $|x|$ we approximate $q(x,0)$ by zero after the truncating contours and obtain very small absolute error.* (b) *The Cauchy error, $|\rho^{(200)}(1) - \rho^{(n)}(1)|$, plotted for $n = 2$ to $n = 100$ on a log scale to show the spectral convergence of the reflection coefficient. Reprinted with permission from Elsevier* [118].

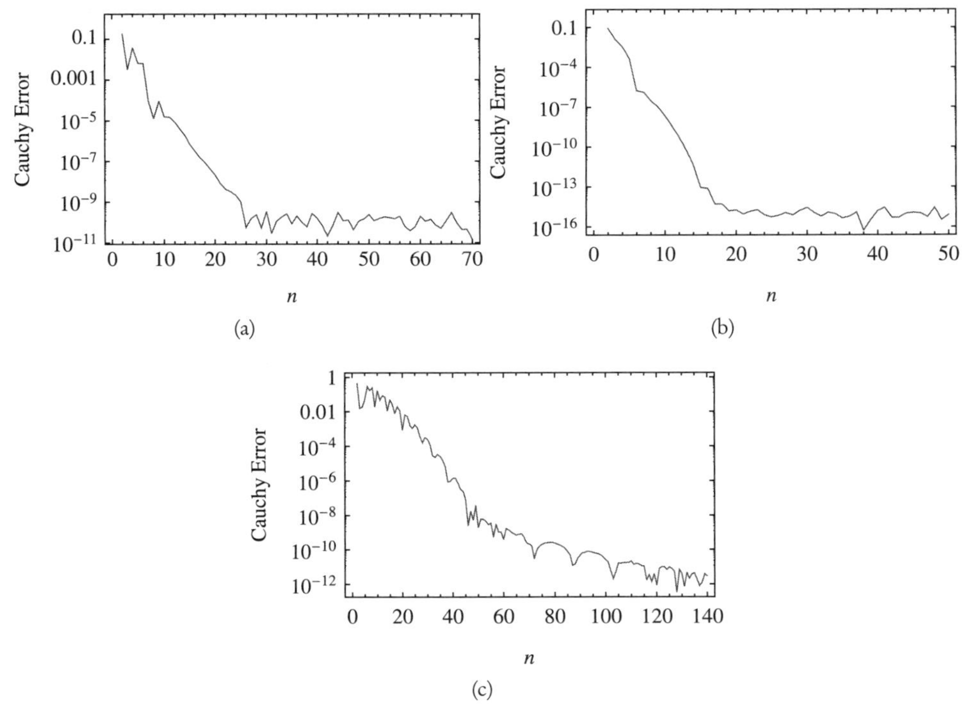

Figure 8.9. *Demonstration of spectral convergence for the modified KdV equation with $q(x,0) = -1.3\ \mathrm{sech}^2(x)$. All plots have $Q^n_{2m}(x,t)$ plotted as a function of n as n ranges from 2 to m.* (a) The dispersive region*: $m = 70$ at the point $(x,t) = (-8.8, 0.6)$.* (b) The Painlevé region*: $m = 50$ at the point $(x,t) = (-0.8, 0.6)$.* (c) *The* soliton/Painlevé region*: $m = 140$ at the point $(x,t) = (0.2, 2.6)$. This deformation requires more collocation points because it only has four contours, so that each contour contains more information about the solution. Machine precision is not achieved since some errors are present in the computation of the reflection coefficient. Reprinted with permission from Elsevier* [118].

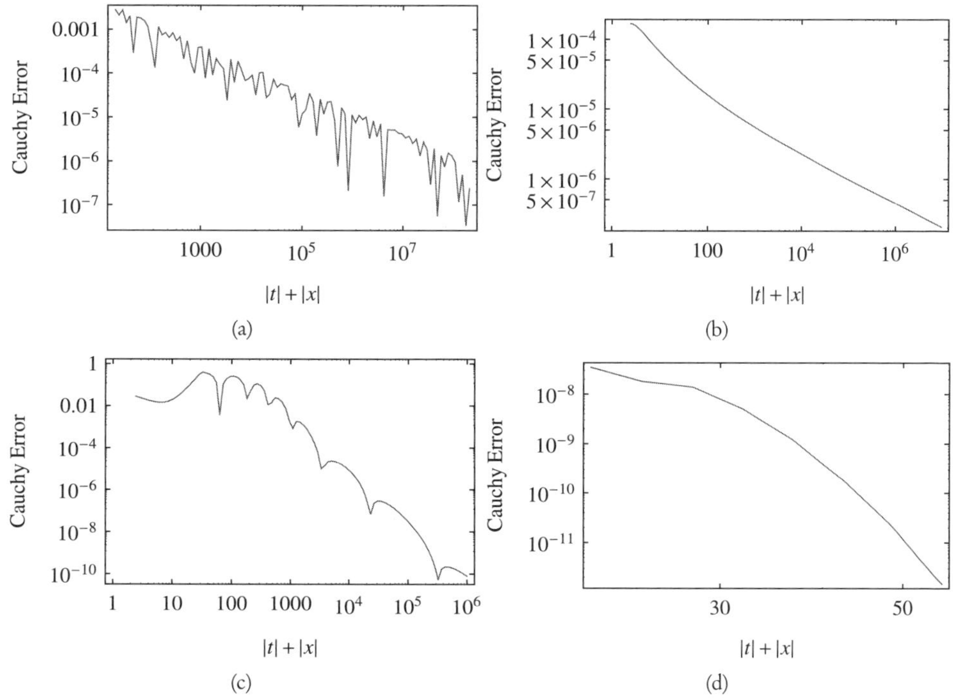

Figure 8.10. *Demonstration of uniform approximation for the modified KdV equation with* $q(x,0)=-1.3\,\mathrm{sech}^2(x)$. *All plots have* $Q_m^n(x,t)$ *plotted as a function of* $|t|+|x|$. (a) The dispersive region*:* $m=10$, $n=5$ *along the trajectory* $x=-20t$. (b) The Painlevé region*:* $m=10$, $n=5$ *along the trajectory* $x=-(3t)^{1/3}$. (c) The Painlevé region*:* $m=20$, $n=10$ *along the trajectory* $x=(3t)^{1/3}$. (d) The soliton region*:* $m=10$, $n=5$ *along the trajectory* $x=20t$. *Reprinted with permission from Elsevier* [118].

Comparison with asymptotic formulae: In Section 8.0.2 asymptotic formulae in various regions for the modified KdV equation were presented. In this section we compare numerical results with these formulae. We skip the soliton region because the asymptotic formula approximates the solution by zero, which is not interesting. Taking into account the verifiable convergence and the fact that convergence of the numerical method has no long-time requirements, it seems reasonable to assume that the computed solutions in the plots below approximate the true solution better than the asymptotic formulae.

1. *The dispersive region.* In Figure 8.11(a, b) we present a numerical verification of the error bound (8.13) along with a plot of both approximations in the dispersive region.

2. *The Painlevé region.* In Figure 8.11(c, d) we present a numerical verification of the error bound (8.11) along with a plot of both approximations in the Painlevé region.

8.2 ▪ The Korteweg–de Vries equation

We discuss numerical inverse scattering for the KdV equation. We can adjust the constants C_2 and C_7 in Section 8.0.2 to make the dispersive region overlap with the Painlevé region up to some finite t. This essentially allows one to use only the deformations needed for the modified KdV equation for small time, eliminating the collisionless shock and transition

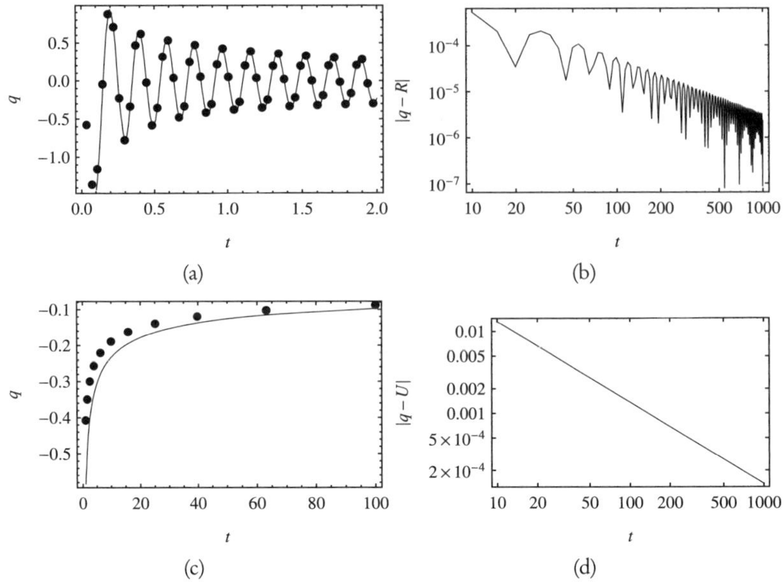

Figure 8.11. *Comparison of numerical results with the asymptotic formulae in the dispersive and Painlevé regions for the modified KdV equation.* (a) The dispersive region: $q_{10}(x,t)$ *and* $R(x,t)$ *plotted as a function of* t *with* $x=-20t$. *The computed solution is shown by the solid line and the asymptotic formula by the dots.* (b) The dispersive region: $|q_{10}(x,t)-R(x,t)|$ *plotted as a function of* t *with* $x=-20t$. *A least-squares fit gives* $|q_{10}(x,t)-R(x,t)|=\mathcal{O}(t^{-1.2})$, *in agreement with the error formula.* (c) The Painlevé region: $q_{10}(x,t)$ *and* $U(x,t)$ *plotted as a function of* t *with* $x=-t^{1/3}$. *The computed solution is shown by the solid line and the asymptotic formula by dots.* (d) The Painlevé region: $|q_{10}(x,t)-U(x,t)|$ *plotted as a function of* t *with* $x=-t^{1/3}$. *A least-squares fit gives* $|q_{10}(x,t)-U(x,t)|=\mathcal{O}(t^{-0.65})$, *in agreement with the error formula. Reprinted with permission from Elsevier* [118].

regions. For practical purposes this is sufficient. However, there always exists a time at which the regions no longer overlap, and since we are interested in the development of a uniformly accurate method, we need to construct the deformations in the collisionless shock and transition regions. These deformations are more complicated.

The RH problem for the KdV equation is generally a meromorphic problem which alters the deformations for $x>0$. Additionally, $\rho(0)=-1$, generically, which complicates the deformations for $x<0$ [32]. The deformation for the dispersive region is only uniformly accurate in its original region of definition, $-x>\alpha t$, $\alpha>0$; it cannot be extended into the Painlevé region for large t. For concreteness we use $-x>12t>0$.

Deift, Venakides, and Zhou used a new deformation of the RH problem for the collisionless shock region [32] (see [4] for the first appearance of this region and formal asymptotics). This deformation is valid into the dispersive region but does not extend to the Painlevé region. Below we present the deformations for the RH problem associated with the KdV equation in these four classical regions. To fill the final gap we introduce a new deformation to transition from the collisionless shock region into the Painlevé region.

8.2.1 ▪ Numerical computation of the scattering data for the KdV equation

Calculating the scattering data numerically relies on two spectral methods: a Chebyshev collocation method for ODEs and *Hill's method* [24] for computing the spectrum of a linear operator.

- Computing ρ.

  For $z \in \mathbb{R}$ we are looking for solutions of $-\mu_{xx} - q_0(x)\mu = z^2\mu$ which behave like $\exp(\pm izx)$ as $x \to \pm\infty$. If $q_0(x) \in \mathcal{S}_\gamma(\mathbb{R})$, the eigenfunctions limit to this asymptotic behavior exponentially fast. For illustration purposes we concentrate on the eigenfunctions at $-\infty$. We set $u(x) = \mu(x)e^{\pm izx} - 1$, where the $\pm$ is chosen when $\mu \sim e^{\mp izx}$. Then $u(x)$ satisfies the ODE

  $$u_{xx} \mp 2izu_x + q_0 u = -q_0 u, \quad u(\pm\infty) = u'(\pm\infty) = 0.$$

  A Chebyshev collocation method (see, for example, [9]) is used to solve this equation on $(-L, 0]$ for each choice of $\pm$ and L sufficiently large. The same ideas apply to the eigenfunctions whose behavior is specified at $+\infty$. We solve for these on $[0, L)$. We enforce the boundary condition at $\pm L$. As in the case of the modified KdV equation, matching the solutions at the origin produces an approximation of the reflection coefficient (see (1.7)).

- Computing $\{z_1, \ldots, z_n\}$.

  As discussed in Section 1.6, calculating these values is equivalent to calculating the $L^2(\mathbb{R})$ eigenvalues of the operator $-d^2/dx^2 - q_0(x)$ and each eigenvalue will correspond to a soliton in the solution of the KdV equation. Through the transformation $x = 2\tan(y/2)$ we map the original ODE to the interval $[-\pi, \pi]$. This is well-defined because of the decay of q_0. If $m(y) = \mu(2\tan(y/2))$ and $Q(y) = q_0(2\tan(y/2))$, then m satisfies the problem

  $$-\cos^2(y/2)\left(\cos^2(y/2)m_y\right)_y - Q(y)m = \lambda m, \quad \lambda = z^2, \quad m(x) = m(x + 2\pi). \quad (8.19)$$

Define $C_p^k([a,b]) = \{f \in C^k([a,b]) : f^{(j)}(a) = f^{(j)}(b), \ 0 \le j \le k\}$. To show the equivalence of this problem with solving the original scattering problem we have the following lemma.

Lemma 8.1. *Assume $q_0(x) \in \mathcal{S}(\mathbb{R})$ and $m \in C_p^2([-\pi, \pi])$ solves (8.19) with $\lambda > 0$. Then $\mu(x) = m(2\arctan(x/2))$ is an L^2 eigenfunction of $-d^2/dx^2 - q_0(x)$. Furthermore, all L^2 eigenfunctions for $-d^2/dx^2 - q_0(x)$ can be found this way.*

Proof. The behavior of the coefficients of (8.19) at $\pm\pi$ forces $m(\pm\pi) = 0$. Also, m is Lipschitz with constant $C = \sup_{y \in [-\pi,\pi]} |m'(y)|$. Therefore

$$|m(y) - m(\pm\pi)| \le C|y \mp \pi| \Rightarrow |m(y)| \le C|y \mp \pi|.$$

Using the asymptotic expansion of $2\arctan(x/2)$ we see that

$$|\mu(x)| \le \min\{C|2\arctan(x/2) - \pi|, C|2\arctan(x/2) + \pi|\} \le C'/(1 + |x|)$$

for a new constant C'. This shows μ is an L^2 eigenfunction. Now assume that μ is an L^2 eigenfunction of the operator $-d^2/dx^2 - q_0(x)$. We know that $\lambda < 0$ (see Section 1.6) and $\mu \sim \exp(-\sqrt{\lambda}|x|)$ as $|x| \to \infty$. Since q is smooth, μ must be smooth, and $\mu(2\tan(y/2))$ is a $C_p^2([-\pi, \pi])$ solution of (8.19). Therefore these eigenvalues and eigenfunctions are in direct correspondence. □

Applying the spectrum approximation techniques in [24] to (8.19) allows us to obtain $\{z_1, \ldots, z_n\}$ with spectral accuracy.

- Computing $\{c_1,\ldots,c_n\}$.

  The norming constants c_j are given by

$$c_j = \frac{b(z_j)}{a'(z_j)},$$

where $b(z_j)$ is the proportionality constant as defined in Section 1.6, not the analytic continuation (if it even exists) of b as defined on $\mathbb{R}$.

Since the above method for calculating $\rho(z)$ gives a method for computing $b(z_j)$, we reduce the problem to computing $a'(z_j)$. We use the important relationship [2, p. 77]

$$a'(z_j) = \frac{1}{\mathrm{i}b(z_j)} \int_{\mathbb{R}} \mu(x;z_j)^2 \mathrm{d}x,$$

where μ is the eigenfunction of the operator $-\mathrm{d}^2/\mathrm{d}x^2 - q_0(x)$ with eigenvalue $\lambda = z_j^2$ such that $\mu \sim \exp(-|\lambda x|)$ as $|x| \to \infty$. This is evaluated using Clenshaw–Curtis quadrature; see Section 4.3.2.

8.2.2 ▪ Numerical solution of the inverse problem

The dispersive region: We proceed as in the case of the modified KdV equation. Assume we performed the deformation in Remark 8.0.3 to introduce small circles around each pole. Examining the exponent, $\exp(2\mathrm{i}z_j x + 8\mathrm{i}z_j^3 t)$, and further recalling that $z_j \in \mathrm{i}\mathbb{R}^+$, we see that the exponent is unbounded in this region. Define the index set

$$K_{x,t} = \{j : |c_j e^{2\mathrm{i}z_j x + 8\mathrm{i}z_j^3 t}| > 1\}. \tag{8.20}$$

Following the approach in [60] we consider for $j \in K_{x,t}$

$$\Phi_{1,\mathrm{d}}(z) = \begin{cases} \Phi(z)\begin{bmatrix} 1 & -(z-z_j)/(c_j e^{\theta(z_j)}) \\ c_j e^{\theta(z_j)}/(z-z_j) & 0 \end{bmatrix} V(z) & \text{if } |z-z_j| < \epsilon, \\ \Phi(z)\begin{bmatrix} 0 & -c_j e^{\theta(z_j)}/(z+z_j) \\ (z+z_j)/(c_j e^{\theta(z_j)}) & 1 \end{bmatrix} V(z) & \text{if } |z+z_j| < \epsilon, \\ \Phi(z)V(z) & \text{otherwise} \end{cases}$$

for

$$V(z) = \begin{bmatrix} \prod_{j\in K_{x,t}} (z-z_j)/(z+z_j) & 0 \\ 0 & \prod_{j\in K_{x,t}} (z+z_j)/(z-z_j) \end{bmatrix}. \tag{8.21}$$

Note that the matrix

$$\begin{bmatrix} 1 & -(z-z_j)/(c_j e^{\theta(z_j)}) \\ c_j e^{\theta(z_j)}/(z-z_j) & 0 \end{bmatrix} V(z)$$

has a removable pole at z_j and $\Phi_{1,\mathrm{d}}$ still tends to the identity at infinity. Recall $A_j^{\pm} = \{z \in \mathbb{C} : |z \mp z_j| = \epsilon\}$, where A_j^+ has counterclockwise orientation and A_j^- clockwise. Further,

ϵ is chosen small enough so that the $A_j^{\pm}$ do not intersect any other contour. We compute the jumps of $\Phi_{1,\text{d}}$:

$$\Phi_{1,\text{d}}^{+}(z)=\begin{cases}\Phi_{1,\text{d}}^{-}(z)V^{-1}(z)G(z)V(z) & \text{if } z\in\mathbb{R},\\ \Phi_{1,\text{d}}^{-}(z)V^{-1}(z)\begin{bmatrix}1 & -(z-z_j)/(c_j e^{\theta(z_j)})\\ 0 & 1\end{bmatrix}V(z) & \text{if } z\in A_j^{+},\\ \Phi_{1,\text{d}}^{-}(z)V^{-1}(z)\begin{bmatrix}1 & 0\\ -(z+z_j)/(c_j e^{\theta(z_j)}) & 1\end{bmatrix}V(z) & \text{if } z\in A_j^{-},\end{cases}$$

$$\Phi_{1,\text{d}}(\infty)=[1,\ 1].$$

This effectively inverts the exponent and turns exponential blowup into decay to the identity. This demonstrates that the solitons exhibit exponential decay. To simplify the notation, define

$$T_{j,+}(z;x,t)=\begin{bmatrix}1 & 0\\ -c_j e^{\theta(z_j)}/(z-z_j) & 1\end{bmatrix},\quad T_{j,-}(z;x,t)=\begin{bmatrix}1 & -c_j e^{\theta(z_j)}/(z+z_j)\\ 0 & 1\end{bmatrix},$$

$$S_{j,+}(z;x,t)=\begin{bmatrix}1 & -(z-z_j)/(c_j e^{\theta(z_j)})\\ 0 & 1\end{bmatrix},\quad S_{j,-}(z;x,t)=\begin{bmatrix}1 & 0\\ -(z+z_j)/(c_j e^{\theta(z_j)}) & 1\end{bmatrix}.$$

As before, the "ghost" contours introduced in Figure 8.12(a) pass along the directions of steepest descent. We define a new matrix function $\Phi_{2,\text{d}}$ based on these regions. Notice that the new definitions still satisfy the normalization condition at infinity. When $j\in K_{x,t}$ we use the jump $S_{j,\pm}$ on $A_j^{\pm}$, otherwise $T_{j,\pm}$ is used. Importantly, for sufficiently large x, $K_{x,t}=\{1,2,\ldots,n\}$ and the exponential decay of solitons in the KdV equation is inferred from the exponential decay of the $S_{j,\pm}$ to the identity matrix. This can be established rigorously with the method of nonlinear steepest descent. We compute the jumps that $\Phi_{2,\text{d}}$ satisfies to phrase an RH problem for $\Phi_{2,\text{d}}$; see Figure 8.12(b). Throughout the figures in this section, the dot inside the circles with jumps $T_{\pm}$ or $S_{\pm}$ represents $\pm z_j$.

We decompose G into its LDU and MP factorizations and deform the jump contour off $\mathbb{R}$ as we did in Section 8.1.2. However, there is a significant difference: if we examine the matrix D, we see that there is a singularity at the origin since generically $\rho(0)=-1$ [4]. We need to remove this singularity in order to represent the solution by Chebyshev polynomials. Additionally, we need to remove the contour on $(-z_0,z_0)$ to attain uniform approximation as mentioned in Section 8.1.2 using Δ in Section 2.4.1. We proceed in the same way and arrive at the RH problem in Figure 8.13, noting that the circles (corresponding to solitons) and the presence of the matrix V are the only aspects that are different.

Remark 8.2.1. *We assumed that $\rho(0)=-1$. If it happens that $|\rho(0)|<1$, then the deformations reduce to those done for the modified KdV equation but now in the (possible) presence of solitons. Numerical results show that this can happen when an initial condition for the KdV equation is obtained through the Miura transformation; see Section* 8.2.5. *In this case, the deformations for the dispersive, Painlevé, and soliton regions cover the (x,t)-plane.*

The Painlevé region: As in the case of the modified KdV equation, for $x>0$ we have an intersection with the soliton region defined below. We use that deformation. The final deformation for the KdV equation when $x<0$ is nearly the same as in the case of the modified KdV equation; see Figure 8.14.

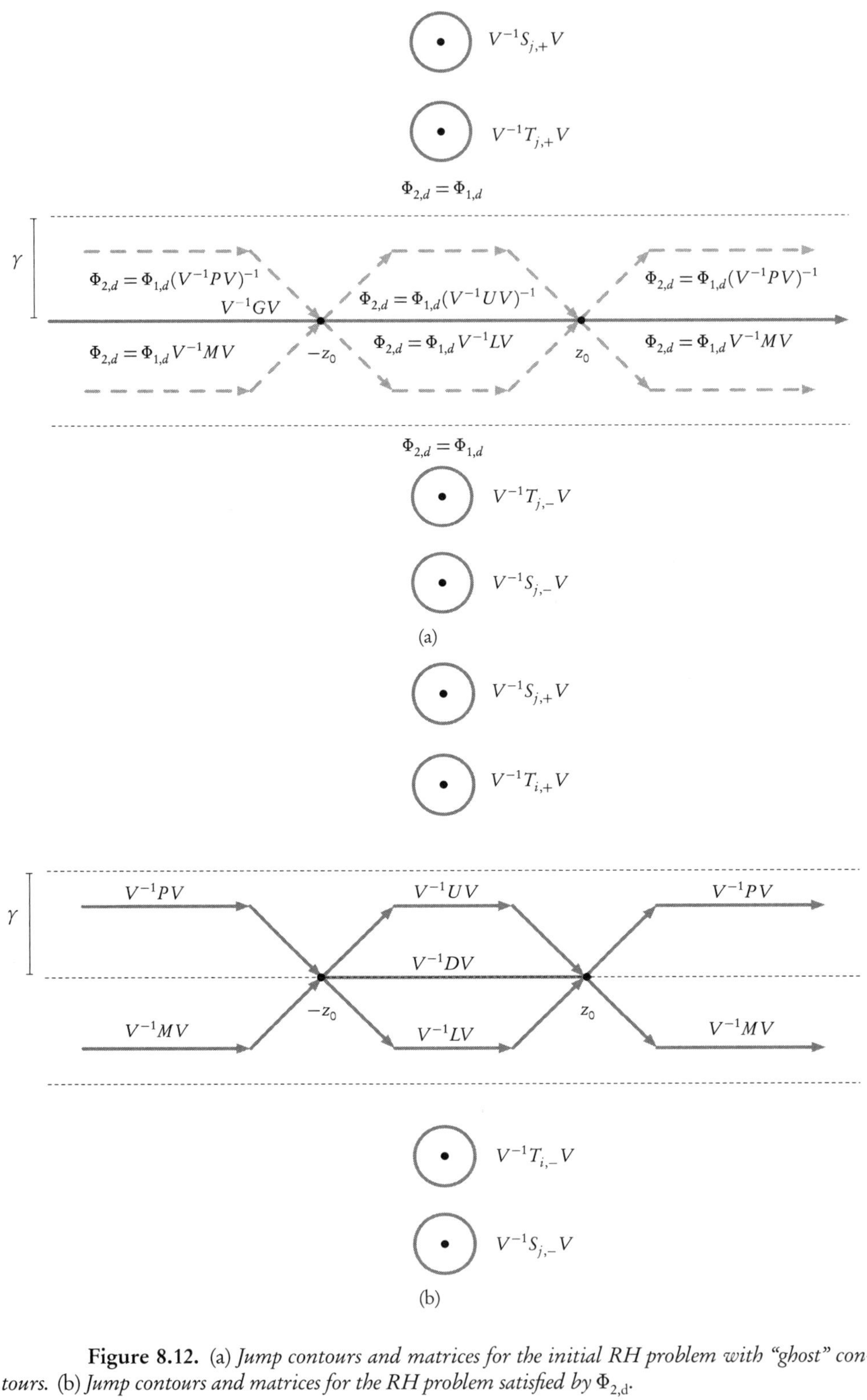

Figure 8.12. (a) *Jump contours and matrices for the initial RH problem with "ghost" contours.* (b) *Jump contours and matrices for the RH problem satisfied by* $\Phi_{2,d}$.

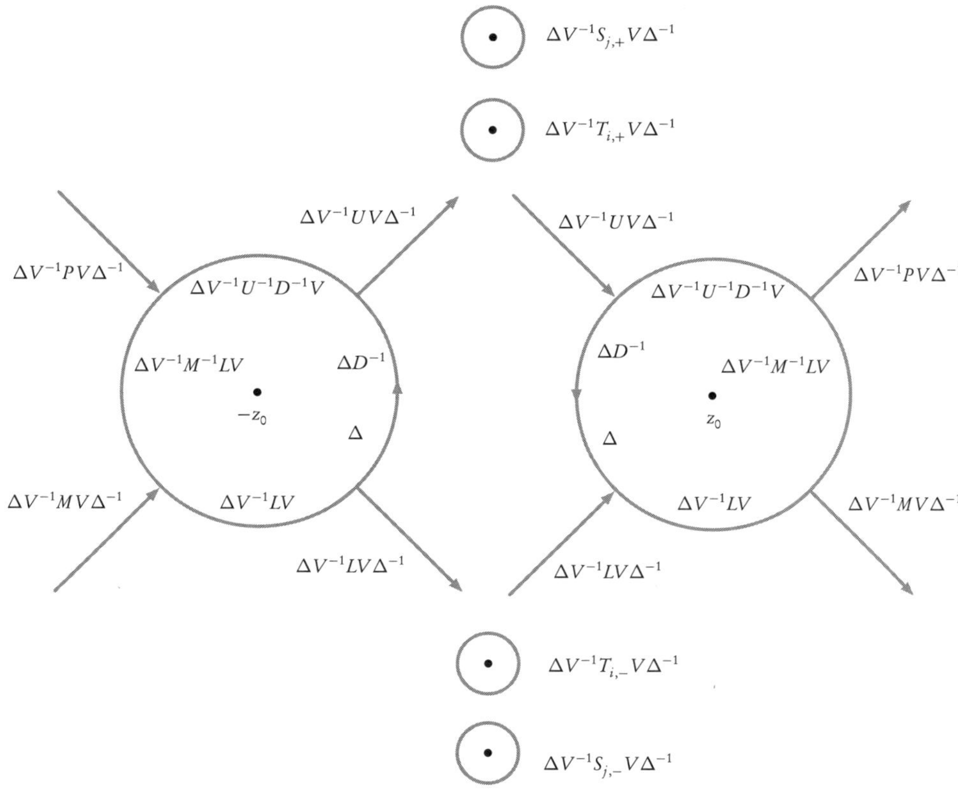

Figure 8.13. *A zoomed view of the jump contours and matrices for the RH problem in the dispersive region of the KdV equation. Note that the contours with jumps* $\Delta V^{-1}UV\Delta^{1}$ *and* $\Delta V^{-1}LV\Delta^{1}$ *connect.*

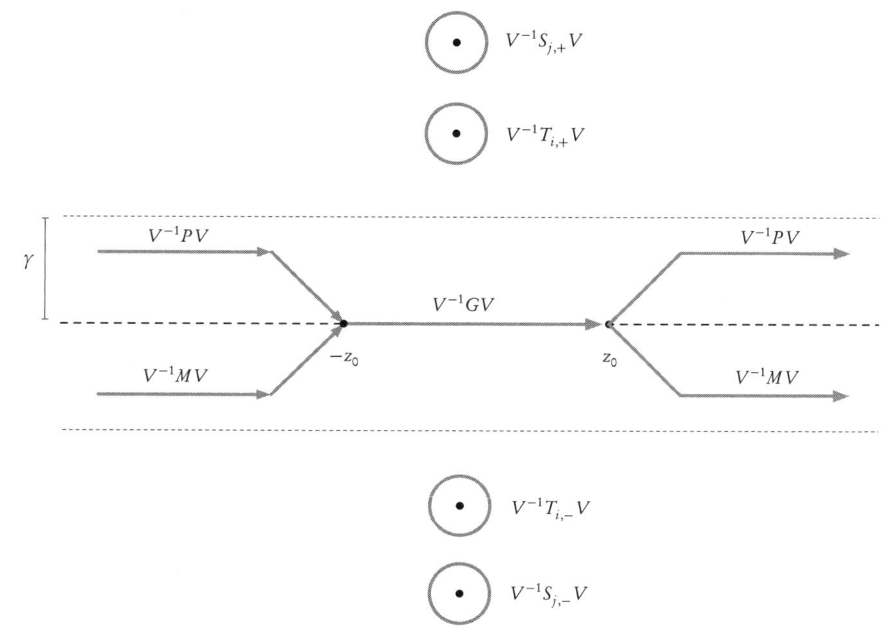

Figure 8.14. *The jump contours and matrices for the RH problem in the Painlevé region with* $x < 0$.

The collisionless shock region: The singularity at $z=0$ in the matrix $D(z)$ destroys the boundedness of $\Delta(z)$, which poses problems that do not occur for the modified KdV equation. As $z \to 0$ the matrices $\Delta V^{-1} P V \Delta^{-1}$ and $\Delta V^{-1} M V \Delta^{-1}$ are unbounded and we cannot link up the dispersive region with the Painlevé region, as we did for the modified KdV equation. We need to introduce additional deformations to bridge the dispersive and Painlevé regions. The first region we address is the collisionless shock region. Ablowitz and Segur [4] introduced this region, and Deift, Venakides, and Zhou derived the needed deformations [32].

The results presented below for this region are from [32]. As x increases in the dispersive region, the stationary points of $\exp\theta$, $\pm z_0$, approach the singularity ($z=0$) of Δ. To prevent this, we replace θ by a so-called g-function [33], whose stationary points, after scaling, do not approach a singularity. For $b>a>0$, we determine constants D_1, D_2 so that there exists a function $g(z)$ which is bounded in the finite plane and satisfies the following properties:

$$
\begin{aligned}
&g^+(z)+g^-(z)=\begin{cases} D_1 & \text{if } z\in(-b,-a),\\ -D_1 & \text{if } z\in(a,b),\end{cases}\\
&g^+(z)-g^-(z)=D_2,\quad z\in(-a,a),\\
&\quad g(z) \text{ is analytic in } z \text{ off } [-b,b],\\
&\quad g(z) \text{ has stationary points at } \pm a, b,\\
&\quad g(z)\sim 4z^3-12z \text{ as } z\to\infty.
\end{aligned}
$$

The constants D_1 and D_2 depend on a and b and have the desired properties to scale away singularities. These will be determined below. Also, once all these constants are fixed, g is uniquely determined.

Remark 8.2.2. *For the KdV equation, g can be determined explicitly (Appendix E.2), but it is more instructive to introduce it as above. It is more convenient to compute it numerically from this formulation since the method in Section 5.5 (see also [94]) is easily adapted to ensure spectral accuracy.*

Define the function $\mathfrak{g}(z)=-i\tau[4z^3-12z-g(z)]$, $\tau=tz_0^3$ and construct

$$
\phi(z)=\begin{bmatrix} e^{\mathfrak{g}(z)} & 0\\ 0 & e^{-\mathfrak{g}(z)}\end{bmatrix}\to I \text{ as } z\to\infty.
$$

It is advantageous to introduce a scaling operator, $\tilde{\ }$, defined by $\tilde f(\diamond;x,t)=f(z_0\diamond;x,t)$ and solve for $\tilde\Phi(z)$. For $z\in\mathbb{R}$ the jump satisfied by $\tilde\Phi(z)\phi(z)$ is $\phi_-^{-1}(z)\tilde G(z)\phi^+(z)$. This assumes the absence of solitons; otherwise we replace G by $V^{-1}GV$. Explicitly

$$
\phi_-^{-1}(z)\tilde G(z)\phi^+(z)=\begin{bmatrix} [1-\rho(z_0z)\rho(-z_0z)]e^{\mathfrak{g}^+(z)-\mathfrak{g}^-(z)} & -\rho(-z_0z)e^{-\theta(z_0z)-\mathfrak{g}^+(z)-\mathfrak{g}^-(z)}\\ \rho(z_0z)e^{\theta(z_0z)+\mathfrak{g}^+(z)+\mathfrak{g}^-(z)} & e^{-\mathfrak{g}^+(z)+\mathfrak{g}^-(z)}\end{bmatrix}.
$$

Note that $\theta(z_0z)=2iz_0zx+8iz_0^3z^3t=2i\tau(-12z+4z^3)$, and $\mathfrak{g}$ satisfies

$$
\begin{aligned}
&\mathfrak{g}^+(z)-\mathfrak{g}^-(z)=i\tau(g^+(z)-g^-(z))=0 \text{ for } z\notin[-b,b],\\
&\mathfrak{g}^+(z)+\mathfrak{g}^-(z)=i\tau(g^+(z)+g^-(z))-\theta(z_0z)\to 0 \text{ as } z\to\infty.
\end{aligned}
$$

We write

$$\phi_-^{-1}(z)\tilde{G}(z)\phi(z)_+ = \begin{cases} \begin{bmatrix} 1-\rho(z_0 z)\rho(-z_0 z) & -\rho(-z_0 z)\mathrm{e}^{-2\mathrm{i}\tau g(z)} \\ \rho(z_0 z)\mathrm{e}^{2\mathrm{i}\tau g(z)} & 1 \end{bmatrix}, \\ \text{when } z\in(-\infty,-b), \\ \begin{bmatrix} [1-\rho(z_0 z)\rho(-z_0 z)]\mathrm{e}^{\mathrm{i}\tau(g^+(z)-g^-(z))} & -\rho(-z_0 z)\mathrm{e}^{-L_1} \\ \rho(z_0 z)\mathrm{e}^{L_1} & \mathrm{e}^{\mathrm{i}\tau(-g^+(z)+g^-(z))} \end{bmatrix}, \\ \text{when } z\in(-b,-a), \\ \begin{bmatrix} [1-\rho(z_0 z)\rho(-z_0 z)]\mathrm{e}^{L_2} & -\rho(-z_0 z)\mathrm{e}^{\mathrm{i}\tau(-g^+(z)-g^-(z))} \\ \rho(z_0 z)\mathrm{e}^{\mathrm{i}\tau(g^+(z)+g^-(z))} & \mathrm{e}^{-L_2} \end{bmatrix}, \\ \text{when } z\in[-a,a], \\ \begin{bmatrix} [1-\rho(z_0 z)\rho(-z_0 z)]\mathrm{e}^{\mathrm{i}\tau(g^+(z)-g^-(z))} & -\rho(-z_0 z)\mathrm{e}^{-L_1} \\ \rho(z_0 z)\mathrm{e}^{L_1} & \mathrm{e}^{\mathrm{i}\tau(-g^+(z)+g^-(z))} \end{bmatrix}, \\ \text{when } z\in(a,b), \\ \begin{bmatrix} 1-\rho(z_0 z)\rho(-z_0 z) & -\rho(-z_0 z)\mathrm{e}^{-2\mathrm{i}\tau g(z)} \\ \rho(z_0 z)\mathrm{e}^{2\mathrm{i}\tau g(z)} & 1 \end{bmatrix}, \\ \text{when } z\in[b,\infty), \end{cases} \tag{8.22}$$

where $L_1/\mathrm{i}\tau = g^+(z)+g^-(z)$ for $z\in[a,b]$ and $L_2/\mathrm{i}\tau = g^+(z)-g^-(z)$ for $z\in[-a,a]$. This successfully removes θ from the problem. As in the dispersive region, we proceed to factor $\tilde{G}=\tilde{L}\tilde{D}\tilde{U}$ on $[-a,a]$. Again, $\tilde{D}$ has a singularity at the origin that we must remove. Before we remove this singularity let us analyze the system in the limit as $z_0\to 0$, as this will guide the choice of the parametrix and the constants L_1 and L_2. On the interval $[-a,a]$ we have

$$\phi_-^{-1}(z)\tilde{D}(z)\phi_+(z) = \begin{bmatrix} [1-\rho(z_0 z)\rho(-z_0 z)]\mathrm{e}^{L_2} & 0 \\ 0 & \left([1-\rho(z_0 z)\rho(-z_0 z)]\mathrm{e}^{L_2}\right)^{-1} \end{bmatrix}.$$

Using $\rho(0)=-1$ and the analyticity of $\rho(z)$ in a neighborhood of the origin, we obtain that $1-\rho(z_0 z)\rho(-z_0 z) = 2\nu z^2 z_0^2 + \mathcal{O}((z z_0)^4)$ near $z=0$ for some constant ν. We left $b>a>0$ mostly arbitrary above. It follows (Appendix E.2) that the boundedness condition along with the prescribed asymptotic behavior requires $a^2+b^2=2$, leaving a single degree of freedom. We use this degree of freedom to enforce $z_0^2\exp(L_2)=1$ so that $(1-\rho(z_0 z)\rho(-z_0 z))\exp(L_2)\sim 2\nu z^2 + \mathcal{O}(z_0^2 z^4)$, removing, to second order, the dependence on z_0. To see that there does exist an a that satisfies this condition, we refer the reader to the explicit construction of g in Appendix E.2. As $z,z_0\to 0$ there is a constant $C>1$ so that

$$\frac{1}{C} \le \frac{1-\rho(z_0 z)\rho(-z_0 z)}{z^2}\mathrm{e}^{L_2} \le C \quad \text{for } z\in[-a,a].$$

Thus, to obtain a global parametrix, we should solve the RH problem

$$\psi_+(z) = \psi_-(z)\phi_-^{-1}(z)\tilde{D}(z)\phi^+(z), \quad z\in(-a,a), \quad \psi(\infty)=I.$$

This diagonal RH problem can be solved explicitly using the method from Problem 2.4.1. We conjugate the problem by ψ in the same way as was done with Δ in Section 8.2.2.

The full deformation for this region now follows. We lens the scaled problem into the form shown in Figure 8.15(a). Near a,b the jumps on the contours are also given there.

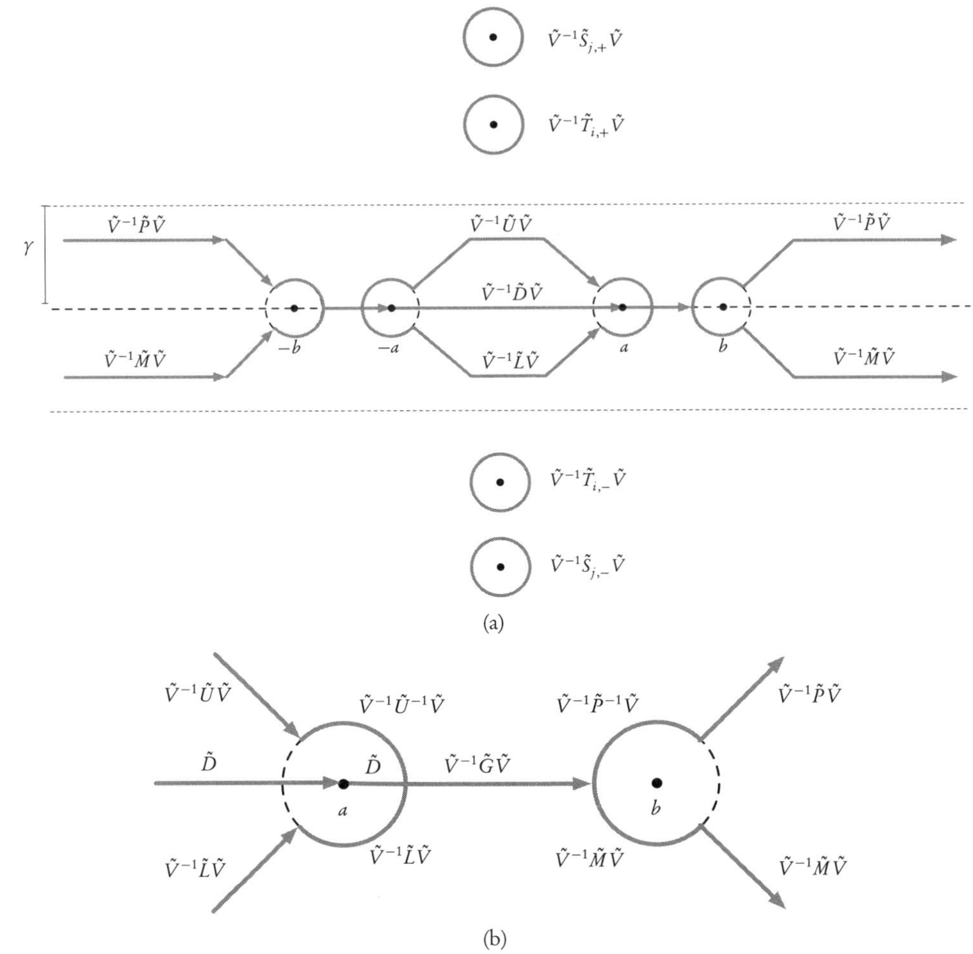

Figure 8.15. (a) *The initial deformation of the RH problem in the collisionless shock region for a function* $\Phi_{1,\mathrm{cs}}$. (b) *The initial jump contours and matrices near* a, b.

Define $\Phi_{2,\mathrm{cs}} = \Phi_{1,\mathrm{cs}}\phi$, where $\Phi_{1,\mathrm{cs}}$ has the jumps shown in Figure 8.15(a). Near a, b, $\Phi_{2,\mathrm{cs}}$ satisfies the problem shown in Figure 8.16. We conjugate by the local parametrix, defining $\Phi_{3,\mathrm{cs}} = \Phi_{2,\mathrm{cs}}\psi^{-1}$. See Figure 8.16 for the RH problem near a, b for $\Phi_{3,\mathrm{cs}}$. By symmetry, what happens at $-a, -b$ is clear. More work is necessary to avoid singularities near a and b. Define the two functions β_m and β_p via diagonal RH problems

$$\beta_m^+(z) = \beta_m^-(z)(\phi_-^{-1}\tilde{D}\phi_+)^{-1}, \quad z \in (-b, -a), \quad \beta_m(\infty) = I,$$
$$\beta_p^+(z) = \beta_p^-(z)(\phi_-^{-1}\tilde{D}\phi_+)^{-1}, \quad z \in (a, b), \quad \beta_m(\infty) = I.$$

For the final deformation define

$$\Phi_{4,\mathrm{cs}} = \begin{cases} \Phi_{3,\mathrm{cs}}\phi^{-1} & \text{inside the circle centered at } -b, \\ \Phi_{3,\mathrm{cs}}\beta_m & \text{inside the circle centered at } -a, \\ \Phi_{3,\mathrm{cs}}\beta_p & \text{inside the circle centered at } a, \\ \Phi_{3,\mathrm{cs}}\phi^{-1} & \text{inside the circle centered at } b, \\ \Phi_{3,\mathrm{cs}} & \text{otherwise.} \end{cases}$$

It follows that $\Phi_{4,\mathrm{cs}}$ solves the RH problem shown in Figure 8.17.

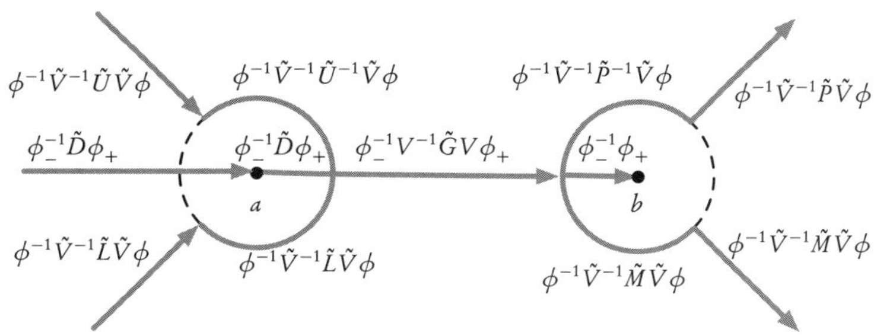

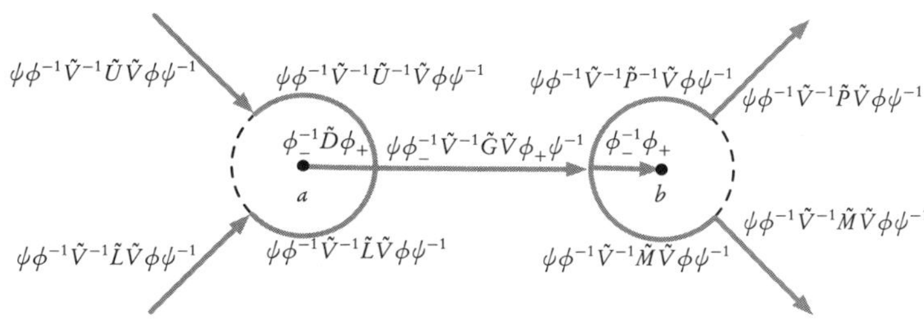

Figure 8.16. (a) *The jump contours and matrices for the RH problem for* $\Phi_{2,\mathrm{cs}}$ *near* a, b. (b) *The jump contours and matrices for the RH problem for* $\Phi_{3,\mathrm{cs}}$ *near* a, b.

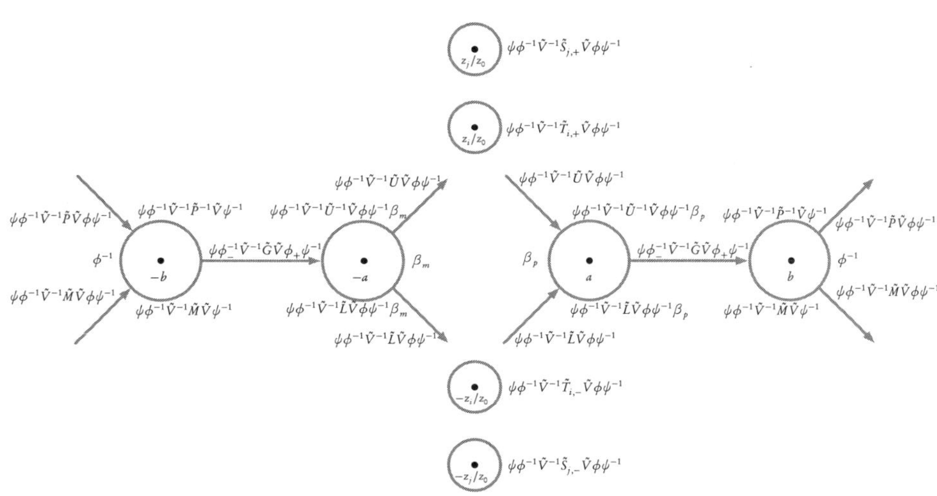

Figure 8.17. *A zoomed view of the jump contours and matrices of the final deformation of the RH problem in the collisionless shock region.*

Remark 8.2.3. *Note that* $s = 0$ *when* $z_0 = 1$ *or* $x = -12t$ *and we switch to the dispersive region. This switch is continuous in the sense that* $s = 0 \Rightarrow a = b = 1$ *and* ϕ *is the identity. The deformation automatically reduces to the deformation in the dispersive region. On the other side of the region, the curve defined by* $8^{3/2} = -\log z_0^2/\tau$ *lies to the right of the curve defined by* $x = -(3t)^{1/3}\log(t)^{2/3}$. *In the next section we address what happens as the curve defined by* $8^{3/2} = -\log z_0^2/\tau$ *is approached.*

The transition region: While the collisionless shock region has extended the values of (x,t) for which there exists a well-behaved RH problem past that of the dispersive region, it is not asymptotically reliable as we approach the Painlevé region: as $|x|$ decreases, a approaches the singularity of the local parametrix at zero. To avoid this issue, we collapse the lensing. To maintain numerical accuracy, we choose a to ensure that the oscillations are controlled on $[-b,b]$. For simplicity let $x=-t^{1/3}R(t)$, where

$$\lim_{t\to\infty}\frac{R(t)}{\log(t)^{2/3}}=0 \text{ and } \lim_{t\to\infty}R(t)=\infty.$$

Given a positive bounded function $f(x,t)$, we choose a so that

$$i\tau(g^+(z)+g^-(z))=if(x,t),\quad z\in[a,b], \tag{8.23}$$

which implies

$$i\tau(g^+(z)+g^-(z))=-if(x,t),\quad z\in[-b,-a].$$

In light of (E.4) this is equivalent to solving

$$f(x,t)/\tau=24\int_0^a\sqrt{(a^2-p^2)(b^2-p^2)}\mathrm{d}p \tag{8.24}$$

for a and b. By adjusting f this can be solved since the right-hand side is a monotone function of a, under the constraint $a^2+b^2=2$, which increases from 0 to 16 as a increases from 0 to 1. Furthermore, $\tau\to\infty$ in this region.

The RH problem, after conjugation by ϕ, is of the form (8.22), and we claim that all entries of the matrices in (8.22) are bounded and the oscillations are controlled. In choosing (8.23) we have that $|i\tau(g^+(z)+g^-(z))|\le f(x,t)$ on $[-b,b]$, which implies that the $(1,2)$ and $(2,1)$ components of the matrix have controlled oscillations and are bounded. Next, consider $i\tau(g^+(z)-g^-(z))$. The choice (8.23) implies

$$h(x,t)/\tau=24\int_a^b\sqrt{(p^2-a^2)(b^2-p^2)}\mathrm{d}p \tag{8.25}$$

for a positive function h such that $1/C<h(x,t)/\tau+f(x,t)/\tau<C$, $C>1$. This comes from the fact that both (8.25) and (8.24) cannot vanish simultaneously. Since f is chosen to be bounded, $h=\mathcal{O}(\tau)$ and $\tau=\mathcal{O}(R(t)^{3/2})$. Using these facts along with $z_0^2=\mathcal{O}(R(t)/t^{2/3})$ we obtain

$$\lim_{t\to\infty}e^{-h(x,t)}=0,\quad \lim_{t\to\infty}z_0^2e^{h(x,t)}\to 0.$$

This shows that the $(1,1)$ and $(2,2)$ components of the matrices in (8.22) are bounded. These matrices produce uniformly accurate numerics without any lensing on $[-b,b]$. After lensing on $(-\infty,-b)\cup(b,\infty)$ we obtain an RH problem for $\Phi_{1,t}$; see Figure 8.18(a). Define $\Phi_{2,t}=\Phi_{1,t}\phi$ and refer to Figure 8.19 for the jump contours and jump matrices of the RH problem for $\Phi_{2,t}$ near a,b. Finally, define

$$\Phi_{3,t}=\begin{cases}\Phi_{2,t}\phi^{-1} & \text{inside the circles centered at } \pm b,\pm a,\\ \Phi_{2,t} & \text{otherwise.}\end{cases}$$

Refer to Figure 8.20 for the jump contours and jump matrices of the final RH problem in the transition region.

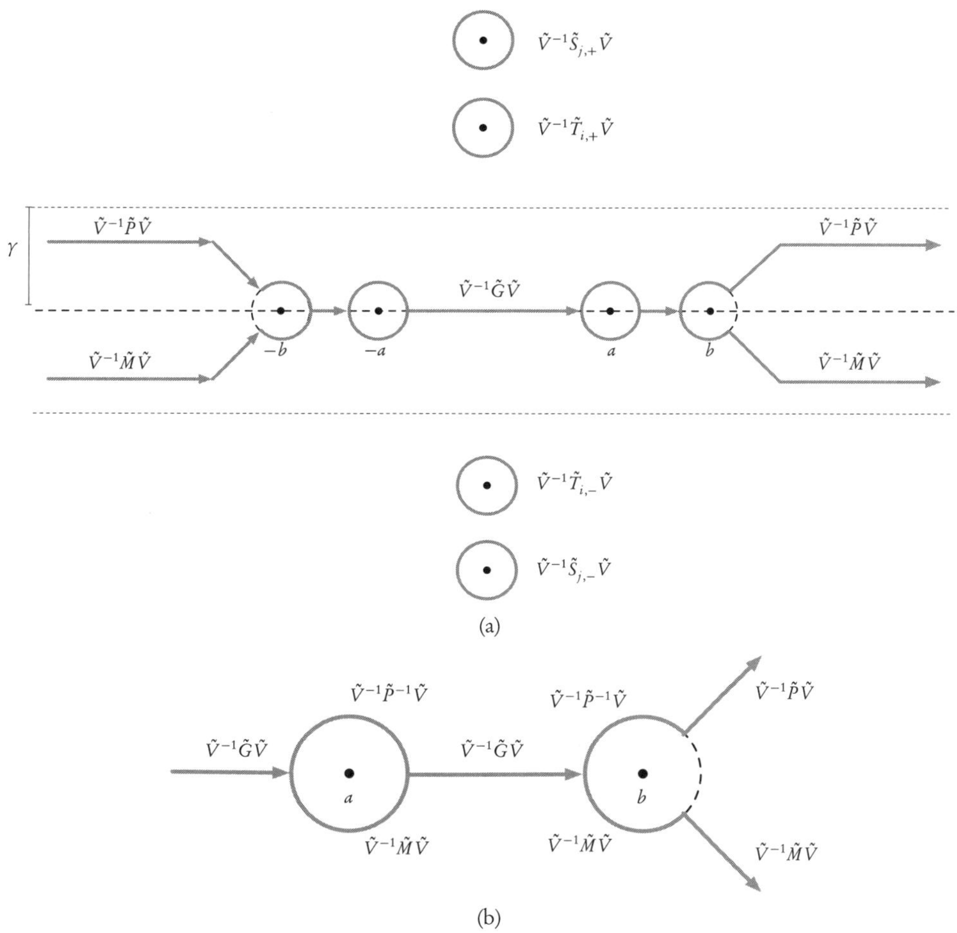

Figure 8.18. (a) *The jump contours and matrices of the RH problem for* $\Phi_{1,t}$. (b) *The jump contours and matrices of the RH problem for* $\Phi_{1,t}$ *near* a, b.

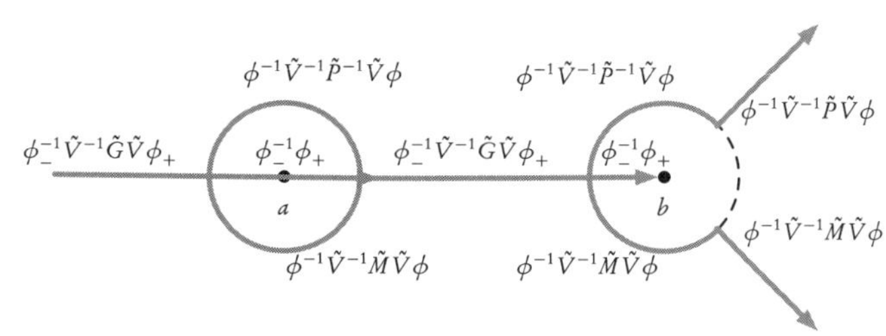

Figure 8.19. *The jump contours and matrices of the RH problem for* $\Phi_{2,t}$ *near* a, b.

The soliton region: This is the region where $x > 0, x = \mathcal{O}(t)$. We present a deformation that is very similar to that used for the modified KdV equation. We use the $G = MP$ factorization, and the only complication arises from dealing with the jumps on $A_j^{\pm}$, but the definition of V and the transition from $T_{j,\pm}$ to $S_{j,\pm}$ as described above mollifies this issue. Again we use a function $0 \le \alpha(x,t) \le |z_0|$. The reader is referred to Figure 8.21 for the final deformation.

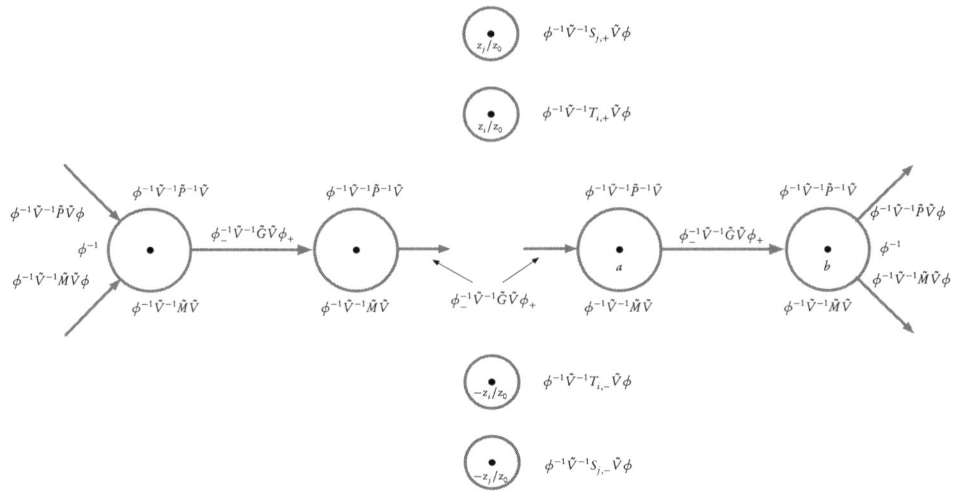

Figure 8.20. *A zoomed view of the jump contours and matrices of the final deformation of the RH problem in the transition region.*

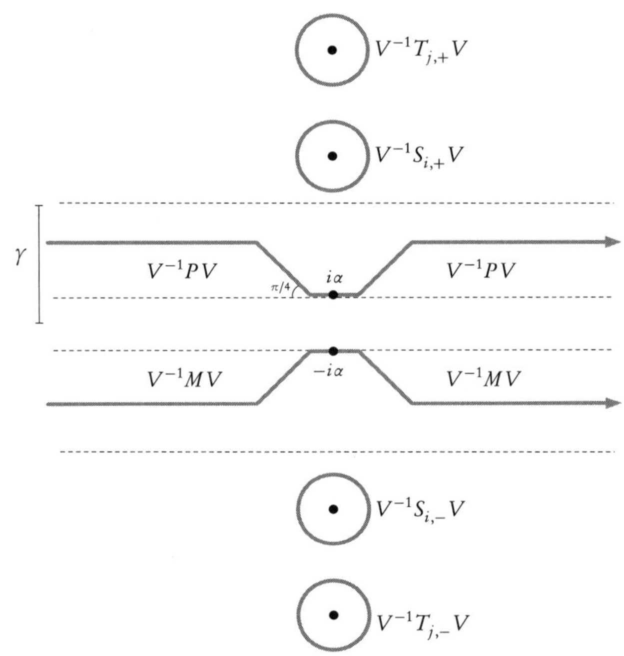

Figure 8.21. *The final deformation of the RH problem in the soliton region for the KdV equation. The solution to this problem contains two solitons to illustrate when $S_{j,\pm}$ needs to be replaced with $T_{j,\pm}$ and vice versa.*

8.2.3 ▪ Numerical results

As in Section 8.1.3, we scale and truncate the contours appropriately and $q_n(x,t)$ is defined to be the solution obtained with n collocation points on each contour. Again, for implementational specifics, we refer the reader to the code in [110].

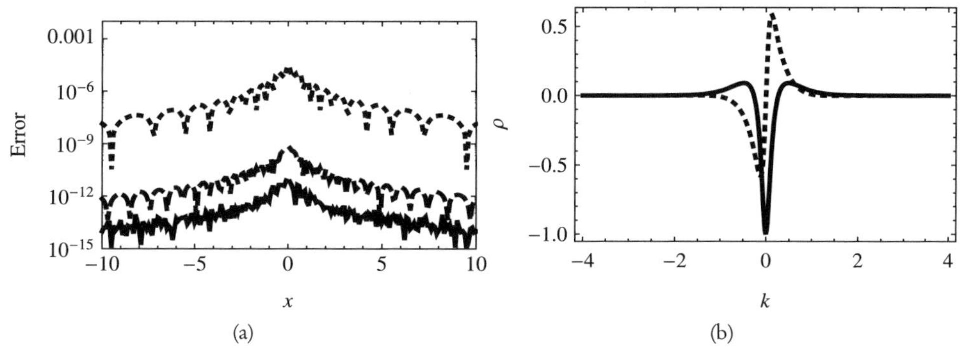

Figure 8.22. *Numerical computation of the reflection coefficient* $\rho(z)$ *with* $q(x,0) = 2.4\,\mathrm{sech}^2(x)$. (a) *Absolute error between computed and actual reflection coefficient plotted vs.* z *when the number of collocation points is* 25 *(dotted),* 50 *(dashed), and* 100 *(solid).* (b) *Plot of the computed reflection coefficient with* 100 *collocation points. The real part is shown as a curve and the imaginary part as a dashed graph. Reprinted with permission from Elsevier* [118].

Direct scattering: As a test case to verify the computed reflection coefficient we use an exact form given in [42]. If $q(x,0) = A\,\mathrm{sech}^2(x)$, then

$$\rho(z) = \frac{a(z)\Gamma(\tilde{c}(z))\Gamma(\tilde{c}(z)-\tilde{a}(z)-\tilde{b}(z))}{\Gamma(\tilde{c}(z)-\tilde{a}(z))\Gamma(\tilde{c}(z)-\tilde{b}(z))}, \quad a(z) = \frac{\Gamma(\tilde{a}(z))\Gamma(\tilde{b}(z))}{\Gamma(\tilde{c}(z))\Gamma(\tilde{a}(z)+\tilde{b}(z)-\tilde{c}(z))},$$
$$\tilde{a}(z) = 1/2 - \mathrm{i}z + (A+1/4)^{1/2}, \quad \tilde{b}(z) = 1/2 - \mathrm{i}z - (A+1/4)^{1/2}, \quad \tilde{c}(z) = 1 - \mathrm{i}z,$$

where Γ is the Gamma function [91]. If $A > 0$, the set of poles is not empty. The poles are located at

$$z_j = \mathrm{i}((A+1/4)^{1/2} - (j+1/4)), \quad j = 1, \ldots, \quad \text{while } ((A+1/4)^{1/2} - (j+1/4)) > 0,$$

and the corresponding norming constants c_j are computed from the expression for ρ. Figure 8.22(a) shows the error between this relation and the computed reflection coefficient when $A = 2.4$ for a varying number of collocation points.

Inverse scattering: As before, throughout this section we assume the reflection coefficient is obtained to machine precision.

1. *Convergence.* To analyze error we again use

   $$Q_n^m(x,t) = |q_n(x,t) - q_m(x,t)|.$$

   See Figure 8.23 for a demonstration of spectral convergence for the new deformations.

2. *Uniform accuracy.* As mentioned before, for the method to be uniformly accurate we need that for a given n and m, $Q_n^m(x,t)$ should remain bounded (and small) as $|x|+|t|$ becomes large. Again, what we numerically demonstrate is that $Q_n^m(x,t)$ tends to zero in all regions. See Figure 8.24 for the demonstration of this for the new deformations. As mentioned before, we expect $Q_n^m(x,t)$ to approach zero only when the solution approaches zero.

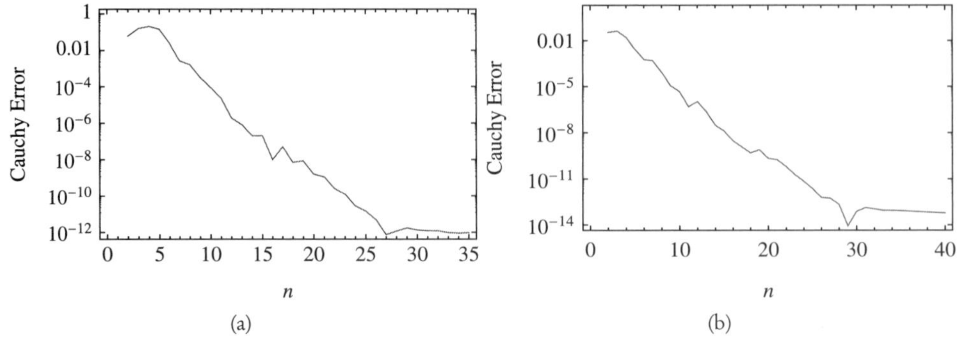

Figure 8.23. *Demonstration of spectral convergence for the KdV equation with* $q(x,0)$ *as shown in Figure* 8.2(a). *Both plots have* $Q^n_{2m}(x,t)$ *plotted as a function of* n *as* n *ranges from 2 to* m. (a) The collisionless shock region*:* $m=30$ *at the point* $(-9.86, 2.8)$. (b) The transition region*:* $m=40$ *at the point* $(-3.12, 7.)$. *The smallest errors achieved are greater than in the other plots for the modified KdV equation due to errors accumulating from the larger number of functions computed to set up the corresponding RH problem. Reprinted with permission from Elsevier* [118].

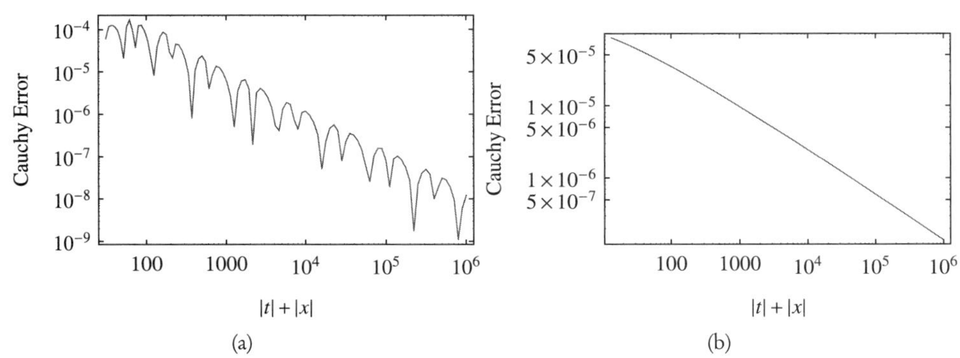

Figure 8.24. *Demonstration of uniform approximation for the KdV equation with* $q(x,0)$ *as shown in Figure* 8.2(a). *Both plots have* $Q^n_m(x,t)$ *plotted as a function of* $|x|+|t|$. (a) The collisionless shock region*:* $m=20$, $n=20$ *along the trajectory* $x=-4(3t)^{1/3}\log(t)^{2/3}$. (b) The transition region*:* $m=16$, $n=8$ *along the trajectory* $x=-(3t)^{1/3}\log(t)^{1/6}$. *Reprinted with permission from Elsevier* [118].

Comparison with asymptotic formulae: In this section we compare our numerics with the asymptotic formulae for the KdV equation in select regions. As before, we emphasize that in view of the verified convergence, the numerical results are believed to be more accurate than the asymptotic results.

1. *The dispersive region.* Numerical results are compared with the asymptotic formula (8.16) in Figure 8.25. The difference between the numerical approximation and the asymptotic approximation is of the correct order.

2. *The Painlevé region.* Numerical results are compared with the asymptotic formula in (8.14) in Figure 8.25. As before, we use the Riemann–Hilbert-based techniques in Chapter 10 to compute v.

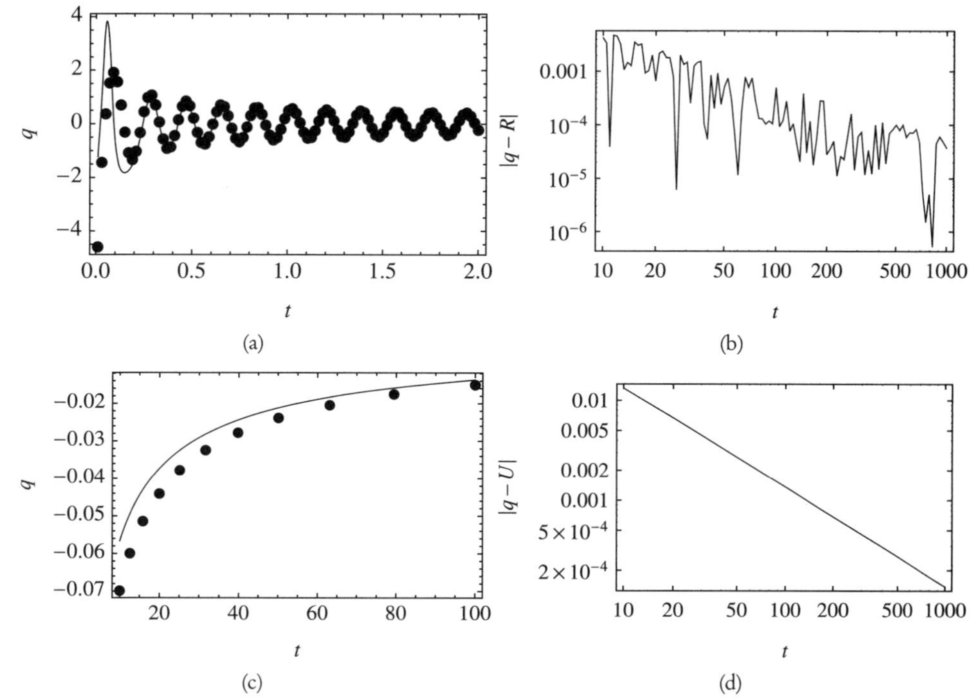

Figure 8.25. *Numerical computations for long time in the dispersive and Painlevé regions for the KdV equation.* (a) The dispersive region*:* $q_{10}(x,t)$ *and* $R(x,t)$ *plotted as a function of t with* $x=-20t$. $R(x,t)$ *is defined in* (8.16)*. Solid: Computed solution. Dots: Asymptotic formula.* (b) The dispersive region*:* $|q_{10}(x,t)-R(x,t)|$ *plotted as a function of t with* $x=-20t$*. A least-squares fit gives* $|q_{10}(x,t)-R(x,t)|=\mathcal{O}(t^{-1.1})$. (c) The Painlevé region*:* $q_{10}(x,t)$ *and* $U(x,t)$ *plotted as a function of t with* $x=-t^{1/3}$. $U(x,t)$ *is defined in* (8.14)*. Solid: Computed solution. Dots: Asymptotic formula.* (d) The Painlevé region*:* $|q_{10}(x,t)-U(x,t)|$ *plotted as a function of t with* $x=-t^{1/3}$*. A least-squares fit gives* $|q_{10}(x,t)-U(x,t)|=\mathcal{O}(t^{-0.99})$ *which is in agreement with the error bound. Reprinted with permission from Elsevier* [118].

3. *The collisionless shock region.* Numerical results are compared with the W from (8.15) in Figure 8.26. In [118] it is demonstrated numerically that

$$q(x,t)-W(x,t)=\mathcal{O}\left(\frac{|x|}{t}(\log t)^{-2/3}\right).$$

8.2.4 ▪ Noninvertibility of the singular integral operator

To demonstrate the importance of symmetries in RH problems we consider the spectrum of the singular integral operator $\mathscr{C}[G;\Gamma]$ from the solid contours in Figure 8.12(a), i.e., the undeformed problem. It is known that the vector RH problem for the KdV equation ($\Phi(\infty)=[1,\ 1]$) has a unique solution once the symmetry condition,

$$\Phi(-z)=\Phi(z)\begin{bmatrix} 0 & 1 \\ 1 & 0 \end{bmatrix},$$

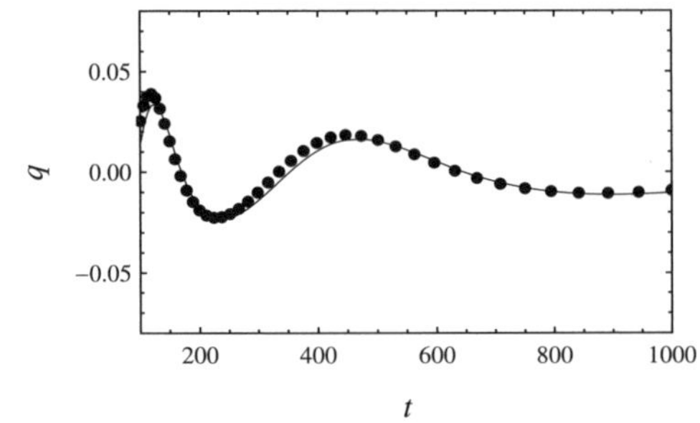

Figure 8.26. *Numerical computations for long time in the collisionless shock region for the KdV equation.* $q_{10}(x,t)$ *and* $W(x,t)$ *are plotted as a function of* t *with* $x = 4(3t)^{1/3}(\log t)^{2/3}$. *Solid: Computed solution. Dots: Computed solution to* (8.15). *Reprinted with permission from Elsevier* [118].

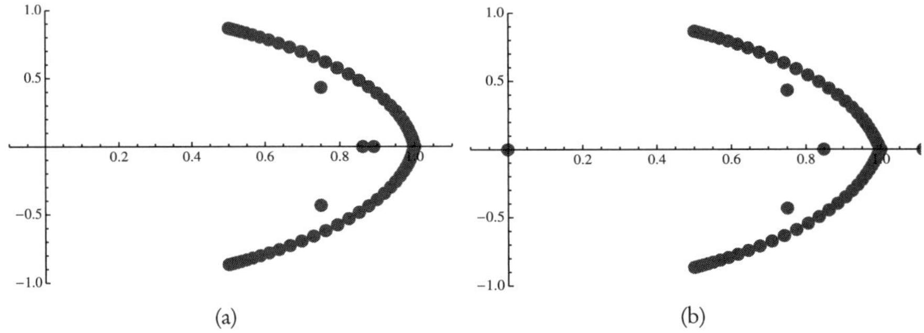

Figure 8.27. *A demonstration of a singular point for the operator* $C_n[G;\Gamma]$ *associated with the KdV equation when solitons are present in the solution.* (a) *The spectrum of* $C_n[G;\Gamma]$ *when* $x = 0$. (b) *The spectrum of* $C_n[G;\Gamma]$ *when* $x = -0.64$. *It is clear that the operator is close to singular. Despite the important fact that* $\mathscr{C}[G;\Gamma]$ *may not be invertible at isolated points (for fixed* t*), it does not adversely affect the computation of the solution of the KdV equation with this method.*

is imposed when the soliton contours $A_j^{\pm}$ are present [60]. The fact that the matrix problem ($\Phi(\infty) = I$) fails to have a (unique) solution can be visualized by computing the eigenvalues of $C_n[G;\Gamma]$ as x varies for $t = 0$; see Figure 8.27. When $x = 0$ the spectrum is well separated from the origin, but for relatively small values of x, a real eigenvalue λ crosses the imaginary axis at $x \approx -0.64$ and the matrix is singular. This eigenvalue has a high velocity as x varies and $|\lambda| > .01$ for $|x + 0.65| > .01$. Therefore this singular point is not encountered, generically, when the solution is computed, as our numerical method is based on inverting $C_n[G;\Gamma]$, which does not impose the symmetry condition.

8.2.5 ▪ Miura transformation

Assume q satisfies the defocusing version of the modified KdV equation (8.2); then $u = -q^2 - q_x$ satisfies the KdV equation (8.1). This is the well-known *Miura transformation* [85]. The numerical approach used here allows for q_x to be computed in a straightforward way, by essentially differentiating the linear system resulting from the collocation method for RH problems [92]. In Figure 8.28 we use the Miura transformation to check the

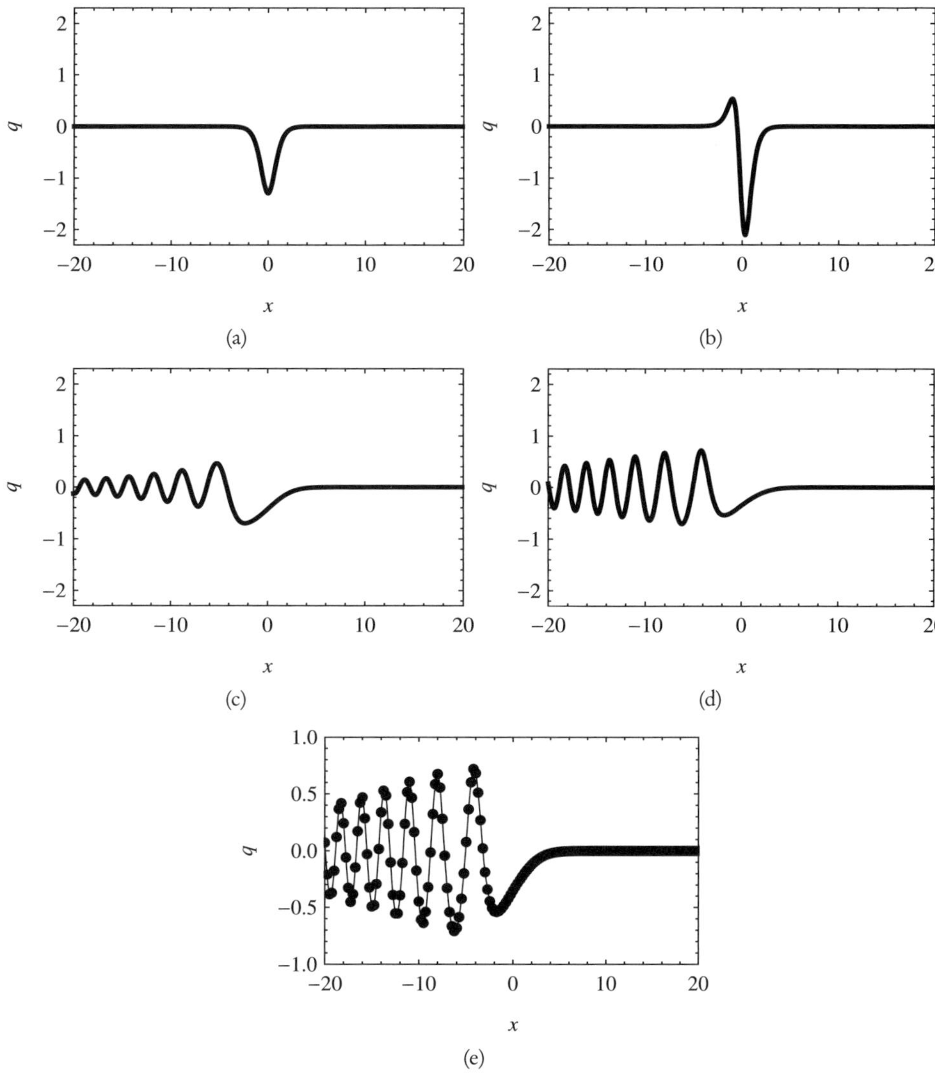

Figure 8.28. *Numerical demonstration of the Miura transformation.* (a) *Initial condition for the modified KdV equation,* $q(x,0) = q_0(x) = -1.3\,\text{sech}^2(x)$. (b) *Initial condition for the KdV equation,* $q(x,0) = -q_0^2(x) - q_0'(x)$. (c) *Evolution using the modified KdV equation at* $t = 0.75$. (d) *Evolution using the KdV equation at* $t = 0.75$. (e) *Solid: Evolution using the KdV equation at* $t = 0.75$, *Dots: Miura transformation of the evolution using the modified KdV equation at* $t = 0.75$. *Reprinted with permission from Elsevier* [118].

consistency of our numerics for $q(x,0) = -1.3\,\text{sech}^2(x)$. As expected, the evolution of the KdV equation and the Miura transformation of the modified KdV equation coincide.

8.3 ▪ Uniform approximation

In this section we prove uniform approximation for the numerical solution of the modified KdV equation (8.2) for $x < 0$. The analysis presented below demonstrates the ease with which the ideas in Chapter 6 can be applied.

Recall that the RH problem for the modified KdV equation is

$$\Phi^+(s) = \Phi^-(s)G(s), \quad s \in \mathbb{R}, \quad \Phi(\infty) = I,$$
$$G(z) = \begin{bmatrix} 1-\rho(z)\rho(-z) & -\rho(-z)e^{-\theta(z)} \\ \rho(z)e^{\theta(z)} & 1 \end{bmatrix}, \quad \theta(z) = 2izx + 8iz^3 t.$$

In the cases we consider, ρ is analytic in a strip that contains $\mathbb{R}$. If $x \ll -ct^{1/3}$, the deformation is similar to the case considered in a subsequent chapter for Painlevé II (see Section 10.2.2) and uniform approximation follows by the same arguments. We assume $x = -12c^2 t^{1/3}$ for some positive constant c. This deformation is found in [118]. We rewrite θ:

$$\theta(z) = -24ic^2(zt^{1/3}) + 8i(zt^{1/3})^3.$$

We note that $\theta'(z_0) = 0$ for $z_0 = \pm\sqrt{-x/(12t)} = \pm ct^{-1/3}$. We introduce a new variable $\zeta = zt^{1/3}/c$ so that

$$\theta(\zeta ct^{-1/3}) = -24ic^3\zeta + 8ic^3\zeta^3 = 8ic^3(\zeta^3 - 3\zeta).$$

For a function of $f(z)$ we use the new scaling operator: $\tilde{f}(\zeta) = f(\zeta ct^{-1/3})$. The functions $\tilde{\theta}$, $\tilde{G}$, and $\tilde{\rho}$ are identified similarly. After deformation and scaling, we obtain the following RH problem for $\tilde{\Phi}(\zeta)$:

$$\tilde{\Phi}^+(s) = \tilde{\Phi}^-(s)J(s), s \in \Sigma = [-1,1] \cup \Gamma_1 \cup \Gamma_2 \cup \Gamma_3 \cup \Gamma_4,$$
$$J(\zeta) = \begin{cases} \tilde{G}(\zeta) & \text{if } \zeta \in [-1,1], \\ \begin{bmatrix} 1 & 0 \\ \tilde{\rho}(\zeta)e^{\tilde{\theta}(\zeta)} & 1 \end{bmatrix} & \text{if } \zeta \in \Gamma_1 \cup \Gamma_2, \\ \begin{bmatrix} 1 & -\tilde{\rho}(-\zeta)e^{-\tilde{\theta}(\zeta)} \\ 0 & 1 \end{bmatrix} & \text{if } \zeta \in \Gamma_3 \cup \Gamma_4, \end{cases}$$

where Γ_i, $i = 1,2,3,4$, shown in Figure 8.29, are locally deformed along the path of steepest descent. To reconstruct the solution to the modified KdV equation we use the formula

$$q(x,t) = 2i\frac{c}{t^{1/3}} \lim_{\zeta \to \infty} \zeta \tilde{\Phi}_{12}(\zeta). \tag{8.26}$$

Remark 8.3.1. *We assume ρ decays rapidly at ∞ and is analytic in a strip that contains the real axis. This allows us to perform the initial deformation which requires modification of the contours at ∞. As t increases, the analyticity requirements on ρ are reduced; the width of the strip can be taken to be smaller if needed. We only require that each Γ_i lie in the domain of analyticity for $\tilde{\rho}$. More specifically, we assume t is large enough so that when we truncate the contours for numerical purposes using Proposition* 2.78*, they lie within the strip of analyticity for $\tilde{\rho}$.*

The parametrix derived in [35] is an approximate solution of $[J;\Sigma]$ and this is used to show that $\mathscr{C}[J;\Sigma]$ has an inverse that is uniformly bounded by appealing to Lemma 2.67:

$$\mathscr{C}[J;\Sigma]^{-1}u = \mathscr{C}_\Sigma^+[u(\Psi^+)^{-1}]\Psi^+ - \mathscr{C}_\Sigma^-[u(\Psi^+)^{-1}]\Psi^-,$$

where $\Psi_\pm^{-1} \in L^\infty(\Sigma)$ is the unique solution of $[J;\Sigma]$. Thus an approximate solution gives an approximate inverse operator and an estimate on the true inverse operator.

Figure 8.29. *Jump contours for the RH problem for the modified KdV equation.*

Repeated differentiation of $J(z)$ proves that this deformation yields a uniform numerical approximation by Theorem 7.11. Furthermore, replacing c by any smaller value yields the same conclusion. This proves the uniform approximation of the modified KdV equation in the Painlevé region

$$R_\epsilon \triangleq \{(x,t) : t \geq \epsilon, \;\; x \leq -\epsilon, x \geq -12c^2 t^{1/3}\}, \;\; \epsilon > 0,$$

where ϵ is determined by the analyticity of ρ. We assume the spectral data is computed exactly.

Theorem 8.2 (Uniform approximation). *If Assumption* 7.4.1 *holds, then for every* $\epsilon > 0$ *and* $k > 0$ *there exists* $C_{k,\epsilon} > 0$ *such that*

$$\sup_{(x,t)\in R_\epsilon} |q_n(x,t) - q(x,t)| \leq C_{k,\epsilon} n^{-k} + \epsilon.$$

Here ϵ *is the effect of contour truncation.*

Chapter 9
The Focusing and Defocusing Nonlinear Schrödinger Equations

In this chapter we consider the numerical solution of the initial-value problem on the whole line for the nonlinear Schrödinger (NLS) equations

$$\begin{aligned} iq_t + q_{xx} + 2\lambda|q|^2 q = 0, \quad \lambda = \pm 1, \\ q(\diamond, 0) = q_0 \in \mathscr{S}_\delta(\mathbb{R}). \end{aligned} \tag{9.1}$$

When $\lambda = 1$ we obtain the focusing NLS equation and, for $\lambda = -1$, the defocusing NLS equation. The majority of what we describe here was originally published in [117]. The NLS equations describe physical phenomena in optics [78], Bose–Einstein condensates [59, 100], and water waves [121]. Together with the KdV equation, these equations are the canonical examples of $(1+1)$-dimensional integrable PDEs. In this chapter we solve the NLS equation, both focusing and defocusing, numerically via the IST. We will see that solutions of NLS equations exhibit the same properties that make us consider the solutions of the KdV equation to be nonlinear special functions.

The presence of an oscillatory dispersive tail is seen for both the focusing and the defocusing NLS equations. Examination of the linear dispersion relationship for the linearization of the PDEs indicates that small amplitude waves will travel at a speed proportional to their wave number. Unlike the KdV equation, solitons in the solution do not separate from dispersion asymptotically. These factors make traditional numerics inefficient to capture the solution for large time. The computational cost of computing the solution using time-stepping methods grows rapidly with time; see Appendix E.1.

The method presented in this chapter, in the same way as the method in Chapter 8, has a computational cost of computing the solution at given x and t values that is seen to be independent of x and t. In Chapter 8 this claim was verified with numerical tests for the KdV and modified KdV equations and proved in specific cases. In this chapter, we prove this fact using results from Section 7.1. It is also worth noting that we do not present asymptotic regions in this chapter like we did in the previous chapter. This is because the NLS equations, generically, have only one asymptotic region. This is evidenced by the fact that (3.30) gives a full description of the long-time behavior in the defocusing case. Furthermore, below we derive only two deformations for numerical evaluation: small-time and long-time.

In addition to solving the problem on the whole line, we use symmetries to solve specific boundary-value problems on $\mathbb{R}^+$. We also compute unbounded solutions to the defocusing NLS equation which have poles. Lastly, we prove that our approximation, under a specific assumption on the underlying numerical scheme, approximates solutions of

(9.1) uniformly away from possible poles in the entire (x,t)-plane. It should also be noted that the scattering problem (see Section 9.1) for the focusing NLS equation is a non-self-adjoint spectral problem and this complicates the asymptotic analysis of the problem [28]. Because of our assumptions on initial data, the numerical method outlined in Chapter 6 is not adversely affected by this additional complication.

9.1 ▪ Integrability and Riemann–Hilbert problems

The focusing and defocusing NLS equations are both completely integrable [2, p. 110]. We look for Lax pairs of the form

$$\mu_x = L(z,q)\mu, \quad \mu_t = M(z,q)\mu,$$

through the modified Zakharov–Shabat scattering problem (see also Section 8.0.1) given by

$$\mu_x = \begin{bmatrix} -\mathrm{i}z & q \\ r & \mathrm{i}z \end{bmatrix}\mu, \quad \mu_t = \begin{bmatrix} A & B \\ C & D \end{bmatrix}\mu. \tag{9.2}$$

In this case, choosing

$$A = -4\mathrm{i}z^2 + \mathrm{i}qr, \qquad B = 2qz + \mathrm{i}q_x, \qquad C = -2rz + \mathrm{i}r_x, \qquad D = -A, \tag{9.3}$$

and $r = -\lambda\bar{q}$, we obtain Lax pairs for both the focusing and the defocusing NLS equations.

We briefly describe the IST. Some conventions used here differ from those in Chapter 3, but the unification is easily seen. For the reasoning behind the difference in approach we refer the reader to Remark 8.0.1. For simplicity, we assume $q \in \mathcal{S}_\delta(\mathbb{R})$. We define two matrix solutions of (9.2) by their corresponding asymptotic behavior:

$$\mu^-(x;z) \backsim \begin{bmatrix} e^{-\mathrm{i}zx} & 0 \\ 0 & -e^{\mathrm{i}zx} \end{bmatrix} \text{ as } x \to -\infty, \quad \mu^+(x;z) \backsim \begin{bmatrix} e^{-\mathrm{i}zx} & 0 \\ 0 & e^{\mathrm{i}zx} \end{bmatrix} \text{ as } x \to \infty. \tag{9.4}$$

Liouville's formula implies that the determinants of these solutions are constant in x. Since the determinants are nonzero at $\pm\infty$, the columns of these matrix solutions define a linearly independent set which spans the solution space. There must exist a scattering matrix

$$S(z) = \begin{bmatrix} a(z) & \mathcal{B}(z) \\ b(z) & \mathcal{A}(z) \end{bmatrix}$$

such that $\mu^+(x;z) = \mu^-(x;z)S(z)$. Define $\rho(z) = b(z)/a(z)$ to be the reflection coefficient. Symmetry properties of S follow (see (3.14))

$$a(z) = -\overline{\mathcal{A}(\bar{z})}, \qquad b(z) = -\lambda\overline{\mathcal{B}(\bar{z})}. \tag{9.5}$$

For the focusing NLS equation ($\lambda = 1$) there may exist values $z_j \notin \mathbb{R}$ and $\operatorname{Im} z_j > 0$ such that

$$\mu_1^-(x;z_j) = S(z_j)\mu_2^+(x;z_j), \; a(z_j) = 0,$$

where the subscripts refer to columns. This implies that $\mu_1^-(x;z_j)$ decays at both $\pm\infty$, exponentially. Thus, $\mu_1^-(x;z_j)$ is an $L^2(\mathbb{R})$ eigenfunction of (9.2). From the symmetries

(9.5) $\bar{z}_j$ is also an $L^2(\mathbb{R})$ eigenvalue. Each such value z_j corresponds to a soliton in the solution. For these values of z we define the norming constants

$$c_j = \frac{b(z_j)}{a'(z_j)}.$$

For the defocusing NLS equation ($\lambda = -1$) it is known that there are no such eigenfunctions [2]. This implies there are no smooth soliton solutions with spatial decay. As in the case of the KdV equation, we define the set

$$\left\{\rho, \{z_j\}_{j=1}^n, \{c_j\}_{j=1}^n\right\} \tag{9.6}$$

to be the scattering data, noting that the sets of eigenvalues and norming constants could be empty. As in Chapter 8, the process of finding the scattering data is called direct scattering, a forward spectral problem.

Remark 9.1.1. *We assume $q_0 \in \mathscr{S}_\delta(\mathbb{R})$ for $\delta > 0$. Theorem 3.10 demonstrates that $S(z)$ can be analytically continued to a neighborhood of the real axis. But it may happen that $S(z)$ (in particular $b(z)$) cannot be extended above z_j. Thus $b(z_j)$ is really an abuse of notation.*

As we see from (3.17) the RH problem associated with the NLS equations is

$$\begin{aligned} &\Phi^+(s) = \Phi^-(s)\begin{bmatrix} 1+\lambda\rho(s)\overline{\rho(s)} & \lambda\overline{\rho(s)}e^{-\theta(s)} \\ \rho(s)e^{\theta(s)} & 1 \end{bmatrix}, \quad s \in \mathbb{R}, \\ &\Phi(\infty) = I, \qquad \theta(z) = 2i(xz + 2tz^2), \end{aligned} \tag{9.7}$$

with the residue conditions (when $\lambda = 1$)

$$\begin{aligned} \operatorname{Res}_{z=z_j} \Phi(z) &= \lim_{z\to z_j} \Phi(z)\begin{bmatrix} 0 & 0 \\ c_j e^{\theta(z_j)} & 0 \end{bmatrix}, \\ \operatorname{Res}_{z=\bar{z}_j} \Phi(z) &= \lim_{z\to \bar{z}_j} \Phi(z)\begin{bmatrix} 0 & -\bar{c}_j e^{-\theta(\bar{z}_j)} \\ 0 & 0 \end{bmatrix}. \end{aligned}$$

These conditions follow in analogy with (1.12).

Once the solution of the RH problem is known, the solution to the corresponding NLS equation is given by the expression

$$q(x,t) = 2i \lim_{|z|\to\infty} z\Phi(z)_{12},$$

where the subscript denotes the $(1,2)$ component of the matrix. The process of solving the RH problem and reconstructing the solution is called inverse scattering.

We follow the standard procedure (see, e.g., Chapter 8) to turn the residue conditions into jump conditions. Fix $0 < \epsilon$ so that the circles $A_j^+ = \{z \in \mathbb{C} : |z - z_j| = \epsilon\}$ and $A_j^- = \{z \in \mathbb{C} : |z - \bar{z}_j| = \epsilon\}$ do not intersect each other or the real axis. We define $\hat{\Phi}$ by

$$\hat{\Phi}(z;x,t) = \begin{cases} \Phi(z;x,t)\begin{bmatrix} 1 & 0 \\ -c_j e^{\theta(z_j)}/(z - z_j) & 1 \end{bmatrix} & \text{if } |z - z_j| < \epsilon,\ j = 1,\dots,n, \\ \Phi(z;x,t)\begin{bmatrix} 1 & 0 \\ \bar{c}_j e^{-\theta(\bar{z}_j)}/(z - \bar{z}_j) & 1 \end{bmatrix} & \text{if } |z - \bar{z}_j| < \epsilon,\ j = 1,\dots,n, \\ \Phi(z;x,t) & \text{otherwise.} \end{cases} \tag{9.8}$$

It is straightforward to show that $\hat{\Phi}$ solves the RH problem:

$$\hat{\Phi}^+(z) = \begin{cases} \hat{\Phi}^-(z)G(z) & \text{if } z \in \mathbb{R}, \\ \hat{\Phi}^-(z)\begin{bmatrix} 1 & 0 \\ -c_j e^{\theta(z_j)}/(z-z_j) & 1 \end{bmatrix} & \text{if } z \in A_j^+, \\ \hat{\Phi}^-(z)\begin{bmatrix} 1 & -\bar{c}_j e^{-\theta(\bar{z}_j)}/(z-\bar{z}_j) \\ 0 & 1 \end{bmatrix} & \text{if } z \in A_j^-, \end{cases} \qquad \hat{\Phi}(\infty) = I,$$

where $A_j^-(A_j^+)$ has clockwise (counterclockwise) orientation. In addition to $q_0 \in \mathscr{S}_\delta(\mathbb{R})$, we assume that the set $\{z_j\}$ is bounded away from the real axis. This is sufficient to ensure that all RH problems we address in this chapter have unique solutions.

9.2 ▪ Numerical direct scattering

We describe a procedure to compute the scattering data (9.6). This follows Chapter 8. We look for solutions of the form (9.4) of (9.2). As before, define

$$\sigma_3 = \begin{bmatrix} 1 & 0 \\ 0 & -1 \end{bmatrix}, \quad Q = \begin{bmatrix} 0 & q \\ \lambda\bar{q} & 0 \end{bmatrix}, \quad \sigma_1 = \begin{bmatrix} 0 & 1 \\ 1 & 0 \end{bmatrix},$$

and two functions

$$J(z;x,t) = \mu^-(z;x,t)\sigma_3 e^{izx\sigma_3} - I, \qquad K(z;x,t) = \mu^+(z;x,t)e^{izx\sigma_3} - I. \tag{9.9}$$

Therefore $J \to 0$ as $x \to -\infty$ and $K \to 0$ as $x \to \infty$. Rewriting (9.2),

$$\mu_x = Q\mu - iz\sigma_3\mu, \tag{9.10}$$

we find that K and J both solve

$$N_x - iz[N,\sigma_3] - Q\sigma_1 N = Q\sigma_1.$$

For each z, this can be solved with the Chebyshev collocation method (see, for example, the methodology of [9]) on $(-L,0]$ for J and on $[0,L)$ for K using vanishing boundary conditions at $\pm L$ for L sufficiently large. If we use n collocation points, this gives two approximate solutions J_n and K_n for J and K, respectively. From J_n and K_n we obtain μ_n^- and μ_n^+, approximations of μ^- and μ^+, respectively, by inverting the transformations in (9.9). Furthermore, μ_n^- and μ_n^+ share the point $x=0$ in their domain of definition. Define

$$S_n(z) = (\mu_n^-)^{-1}(0;z)\mu_n^+(0;z).$$

This is an approximation of the scattering matrix, from which we extract an approximation of the reflection coefficient.

This procedure works well for z in a neighborhood of the real axis. The solutions which decay at both $\pm\infty$ are all that is needed to obtain $b(z)$ when $a(z)=0$. Furthermore, from the analyticity properties of a we have from (3.16)

$$a(z) - 1 = \int_{-\infty}^{\infty} \frac{a(s)-1}{s-z} đs, \qquad a'(z) = \int_{-\infty}^{\infty} \frac{a'(s)}{s-z} đs.$$

See Lemma 2.49 to justify the interchange of order of integration and differentiation because $a-1 \in H^1(\mathbb{R})$. Thus knowing $a(z)$ on the real axis and $b(z)$ when $a(z)=0$ allows us to compute $c_j = b(z_j)/a'(z_j)$. In practice we use the framework [96] discussed in Section 7.4 to compute these Cauchy integrals. Also, $a'(z)$ can be obtained accurately using spectral differentiation. The unit circle is mapped to the real axis with a Möbius transformation $M : \mathbb{U} \to \mathbb{R}$; see Section 4.3.1. Then $F_a(z) = a(M(z))$ is approximated using the Laurent series with the FFT so that $F_a'(z)$ is approximated by differentiating the Laurent series. The ratio $a'(M(z)) = F_a'(z)/M'(z)$ is re-expanded in a Laurent series using the FFT, and the map M is inverted to give an approximation of $a'(z)$.

The remaining problem is that of computing z_j. We consider (9.10):

$$\mu_x - Q\mu = -\mathrm{i}z\sigma_3\mu, \quad \Rightarrow \quad \mathrm{i}\sigma_3\mu_x - \mathrm{i}\sigma_3 Q\mu = z\mu. \tag{9.11}$$

Making the change of variables $x \mapsto \tan(s/2)$, $U(s) = \mu(\tan(s/2))$, $H(s) = Q(\tan(s/2))$ we obtain

$$2\mathrm{i}\cos^2(s/2)\sigma_3 U_s(s) - \mathrm{i}\sigma_3 H(s)U(s) = zU(s).$$

We use Hill's method [24] to compute the eigenvalues of the operator

$$2\mathrm{i}\cos^2(s/2)\sigma_3\frac{\mathrm{d}}{\mathrm{d}s} - \mathrm{i}\sigma_3 H(s) \tag{9.12}$$

in the space $L^2([-\pi,\pi])$. Following the arguments in Lemma 8.1 and using the convergence of Hill's method, even for non-self-adjoint operators [22, 68], the only eigenvalues we obtain off the real axis are those associated with (9.11). This allows us to compute the discrete spectrum $\{z_j\}_{j=1}^n$ with spectral accuracy. We may test to make sure all eigenvalues are captured by computing the IST at $t=0$ and ensuring that we recover the initial condition.

9.2.1 ▪ Numerical results

In this section we present numerical results for direct scattering. First, we compare the result of our method with a reflection coefficient that is known analytically. Next, we present numerically computed reflection coefficients, which we use later. For the focusing NLS equation ($\lambda = 1$) the authors in [108] present an explicit reflection coefficient for initial conditions of the form

$$q_0(x) = -\mathrm{i}A\,\mathrm{sech}(x)\exp(-\mathrm{i}\mu A \log\cosh(x)), \quad \mu, A \geq 0. \tag{9.13}$$

The components of the scattering matrix take the form

$$a(z) = \frac{\Gamma(w(z))\Gamma(w(z)-w_- - w_+)}{\Gamma(w-w_+)\Gamma(w-w_-)}, \qquad b(z) = \mathrm{i}A2^{-\frac{\mathrm{i}\mu}{2}}\frac{\Gamma(w(z))\Gamma(1-w(z)+w_+ + w_-)}{\Gamma(w_+)\Gamma(w_-)},$$

where

$$w(z) = -\mathrm{i}z - A\mu\frac{\mathrm{i}}{2} + \frac{1}{2}, \quad w_+ = -\mathrm{i}A\left(T + \frac{\mu}{2}\right), \quad w_- = \mathrm{i}A\left(T - \frac{\mu}{2}\right), \quad \text{and } T = \sqrt{\frac{\mu^2}{4} - 1}.$$

Here Γ is the Gamma function [91]. The set $\{z_j\}$ is nonempty for $0 \leq \mu < 2$. Its elements are

$$z_j = AT - \mathrm{i}(j-1/2), \ j \in \mathbb{N}, \text{ and } j < 1/2 + A|T|.$$

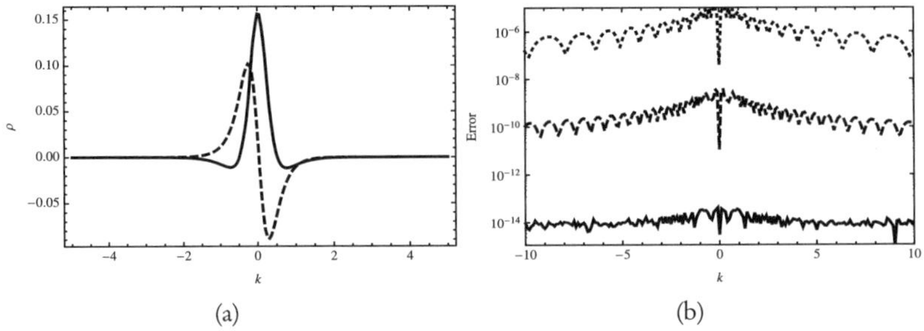

Figure 9.1. (a) *Plot of the known analytical formula for the reflection coefficient with* $A = 1$ *and* $\mu = 0.1$. *Solid: real part. Dashed: imaginary part.* (b) *Demonstration of spectral convergence for the reflection coefficient. Dotted:* 40 *collocation points. Dashed:* 80 *collocation points. Solid:* 120 *collocation points. Reprinted with permission from Royal Society Publishing* [117].

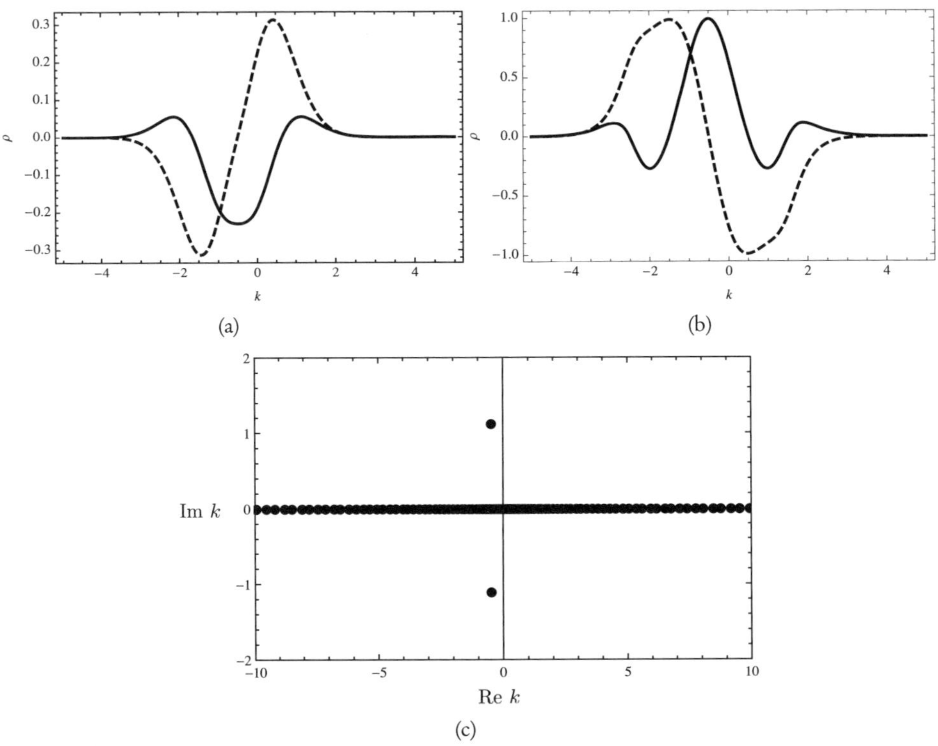

Figure 9.2. (a) *Plot of the computed reflection coefficient for the focusing NLS equation* ($\lambda = 1$) *when* $q_0(x) = 1.9e^{-x^2+ix}$. *Solid: real part. Dashed: imaginary part.* (b) *Plot of known reflection coefficient for the defocusing NLS equation* ($\lambda = -1$) *when* $q_0(x) = 1.9e^{-x^2+ix}$. *Solid: real part. Dashed: imaginary part.* (c) *The spectrum of* (9.12) *found using Hill's method when* $\lambda = 1$. *Reprinted with permission from Royal Society Publishing* [117].

In Figure 9.1 we plot the reflection coefficient for $A = 1$ and $\mu = 0.1$. These plots demonstrate spectral convergence.

In Figure 9.2 we show the computed reflection coefficient for $q_0(x) = 1.9\exp(-x^2+ix)$ for both focusing and defocusing NLS. For focusing NLS we find $z_1 \approx -0.5 + 1.11151i$; see Figure 9.2(c) for a plot of the spectrum of (9.12) found using Hill's method.

9.3 ▪ Numerical inverse scattering

As in the case of the KdV equation, numerical inverse scattering has two major components. The first is the use of a Chebyshev collocation method developed in Chapter 6 for solving RH problems, and the second is the deformation of contours in the spirit of the method of nonlinear steepest descent. As before, the use of nonlinear steepest descent is essential since the jump for the RH problem (9.7) is oscillatory for large values of x and t. Again we prove that the computational cost of computing the solution at a given point, accurate to within a given tolerance, is independent of x and t (see Section 9.6). Additionally, we demonstrate the deformation of the RH problem. The method proceeds in much the same way as in Section 3.2, and we include the full details so that this chapter may be read independently of Chapter 3.

9.3.1 ▪ Small time

When both x and t are small, the RH problem needs no deformation. When t is small, but x is large, the RH problem needs to be deformed. We introduce factorizations of the jump matrix. Define

$$\begin{aligned}
G(z) &= \begin{bmatrix} 1+\lambda\overline{\rho(\bar{z})}\rho(z) & \lambda\overline{\rho(\bar{z})}e^{-\theta(z)} \\ \rho(z)e^{\theta(z)} & 1 \end{bmatrix}, \\
M(z) &= \begin{bmatrix} 1 & \lambda\overline{\rho(\bar{z})}e^{-\theta(z)} \\ 0 & 1 \end{bmatrix}, \quad P(z) = \begin{bmatrix} 1 & 0 \\ \rho(z)e^{\theta(z)} & 1 \end{bmatrix}, \\
L(z) &= \begin{bmatrix} 1 & 0 \\ \frac{\rho(z)}{\tau(z)}e^{\theta(z)} & 1 \end{bmatrix}, \quad D(z) = \begin{bmatrix} \tau(z) & 0 \\ 0 & 1/\tau(z) \end{bmatrix}, \\
U(z) &= \begin{bmatrix} 1 & \frac{\lambda\overline{\rho(\bar{z})}}{\tau(z)}e^{-\theta(z)} \\ 0 & 1 \end{bmatrix}, \quad \tau(z) = 1+\lambda\rho(z)\overline{\rho(\bar{z})}.
\end{aligned} \tag{9.14}$$

Note that $G(z) = L(z)D(z)U(z) = M(z)P(z)$. We assume the sets $\{z_j\}_{j=1}^n$ and $\{c_j\}_{j=1}^n$ are empty. Later, we make the proper modifications to incorporate the extra contours required.

We deform contours of (9.7) off the real axis so that oscillations are turned to exponential decay. The matrix G contains the two factors $\exp(\pm\theta(z))$, and if one decays, the other must grow. This motivates separating these factors using the process of lensing; see Section 2.8.2.

Since $q_0 \in \mathscr{S}_\delta(\mathbb{R})$, we know that (see Lemma 3.4) ρ is analytic in the strip $S_\gamma = \{z \in \mathbb{C} : |\operatorname{Im} z| < \gamma\}$ for $\delta/2 > \gamma > 0$. The factors in (9.14) allow lensing, but we need to determine where to lens. We look for saddle points of the oscillator: $\theta'(z_0) = 0$ for $z_0 = -x/(4t)$. We use the LDU factorization for $z < z_0$ and MP for $z > z_0$. See Figure 9.3(b) for this deformation and note that the contours are locally deformed along the path of steepest descent, that is, the direction along which the jump matrix tends to the identity matrix most rapidly. We denote the solution of this lensed RH problem by Φ_1.

This RH problem can be solved numerically provided t is not too large. As t increases, the solution v on the contour $(-\infty, z_0)$ is increasingly oscillatory and is not well resolved using Chebyshev polynomials.

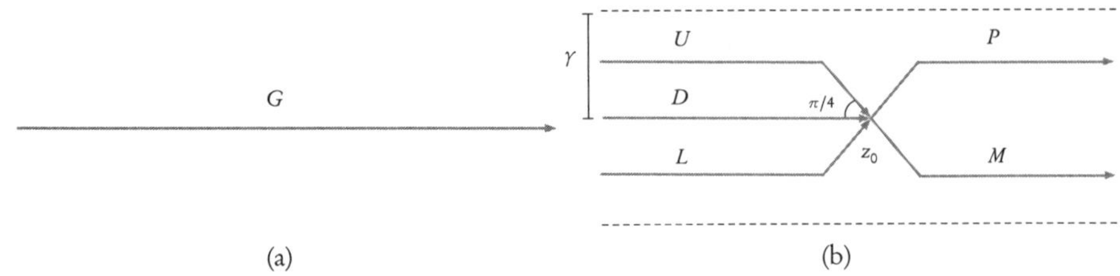

Figure 9.3. (a) *Jump contour for the initial RH problem for* Φ. (b) *Jump contours after lensing for the RH problem for* Φ_1.

9.3.2 ▪ Long time

Next, we provide a deformation that leads to a numerical method that is accurate for arbitrarily large x and t. In light of the results in Chapter 7 we need to remove the contour D from the RH problem so that all jumps decay to the identity matrix away from z_0 as x and t become large. The matrix-valued function

$$\Delta(z;z_0)=\begin{bmatrix} \delta(z;z_0) & 0 \\ 0 & \delta^{-1}(z;z_0) \end{bmatrix}, \qquad \delta(z;z_0)=\exp\left(\mathscr{C}_{(-\infty,z_0)}\log\tau(z)\right) \tag{9.15}$$

satisfies (see Problem 2.4.1)

$$\Delta^+(z;z_0)=\Delta^-(z;z_0)D(z), \quad z\in(-\infty,z_0), \qquad \Delta(\infty;z_0)=I.$$

We multiply the solution of the RH problem in Figure 9.3(b) by Δ^{-1} to remove the jump. To compute the solution to the new RH problem we use conjugation:

$$\Phi_1^+=\Phi_1^-J \Leftrightarrow \Phi_1^+\Delta_+^{-1}=\Phi_1^-J\Delta_+^{-1} \Leftrightarrow \Phi_1^+\Delta_+^{-1}=\Phi_1^-\Delta_-^{-1}\Delta_-J\Delta_+^{-1}.$$

Indeed, we see that if $J=D$, there is no jump. Define $\Phi_2=\Phi_1\Delta^{-1}$. See Figure 9.4 for a schematic overview of the new jumps. This deformation is not sufficient for a numerical solution, as $\Delta(z;z_0)$ has a singularity at $z=z_0$. We must perform one last deformation in a neighborhood of $z=z_0$ to bound contours away from this singularity. We use a piecewise definition of a function Φ_3 (see Figure 9.5(a)) and compute the jumps (Figure 9.5(b)). This is the final RH problem. It is used, after contour truncation and scaling, to compute solutions of the NLS equations for arbitrarily large time. We discuss the scaling of the RH problem in more detail in Section 9.6.

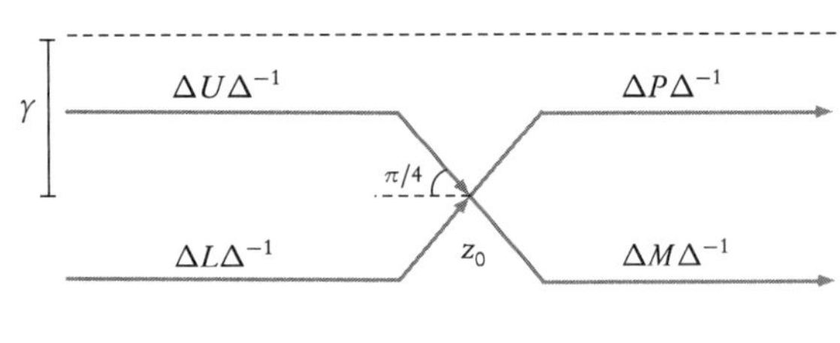

Figure 9.4. *Removal of the jump on* $(-\infty,z_0)$.

We use a combination of the deformation in Figure 9.3(b) for small time and the deformation in Figure 9.5(b) to obtain an approximation to focusing or defocusing NLS when

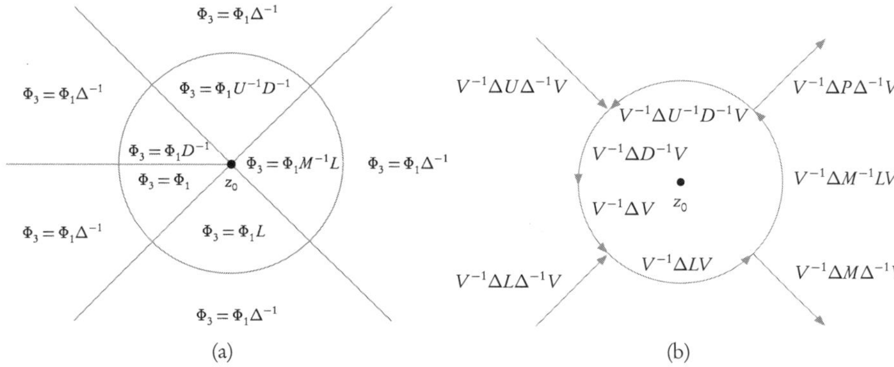

Figure 9.5. (a) *The piecewise definition of* Φ_3. (b) *Jump contours and jump matrices for the RH problem for* Φ_3.

no solitons are present in the solution. Lastly, we deal with the addition of solitons for the case of focusing NLS. There are additional jumps of the form

$$\Phi^+(z) = \begin{cases} \Phi^-(z)\begin{bmatrix} 1 & 0 \\ -c_j e^{\theta(z_j)}/(z-z_j) & 1 \end{bmatrix} & \text{if } z \in A_j^+, \\ \Phi^-(z)\begin{bmatrix} 1 & -\bar{c}_j e^{-\theta(\bar{z}_j)}/(z-\bar{z}_j) \\ 0 & 1 \end{bmatrix} & \text{if } z \in A_j^-. \end{cases}$$

We assume that $\operatorname{Re}(z_j) > \gamma$.

For small x and t, $|e^{\theta(z_j)}| = |e^{-\theta(\bar{z}_j)}|$ is close to unity and the contours and jumps need to be added to one of the deformations discussed above. This will not be the case for all x and t. When $|c_j e^{\theta(z_j)}| > 1$ we invert this factor through a deformation. Define the set $K_{x,t} = \{j : |c_j e^{\theta(z_j)}| > 1\}$. Note that the x and t dependence enters through $\theta(z_j)$. Next, define the functions

$$v(z) = \prod_{j \in K_{x,t}} \frac{z - z_j}{z - \bar{z}_j} \quad \text{and} \quad V(z) = \begin{bmatrix} v(z) & 0 \\ 0 & 1/v(z) \end{bmatrix}.$$

Define the piecewise-analytic matrix-valued function $\hat{\Phi}$:

$$\hat{\Phi}(z) = \Phi(z)\begin{cases} \begin{bmatrix} 1 & -(z-z_j)/(c_j e^{\theta(z_j)}) \\ c_j e^{\theta(z_j)}/(z-z_j) & 0 \end{bmatrix} V(z) & \text{if } |z - z_j| < \epsilon, \\ \begin{bmatrix} 0 & -\bar{c}_j e^{-\theta(\bar{z}_j)}/(z-\bar{z}_j) \\ (z-\bar{z}_j)/(\bar{c}_j e^{-\theta(\bar{z}_j)}) & 1 \end{bmatrix} V(z) & \text{if } |z - \bar{z}_j| < \epsilon, \\ V(z) & \text{otherwise.} \end{cases}$$

Computing the jumps that $\hat{\Phi}$ satisfies, we find, for $j \in K_{x,t}$,

$$\hat{\Phi}^+(z) = \hat{\Phi}^-(z) \begin{cases} V^{-1}(z)\begin{bmatrix} 1 & -(z-z_j)/(c_j e^{\theta(z_j)}) \\ 0 & 1 \end{bmatrix} V(z) & \text{if } z \in A_j^+, \\ V^{-1}(z)\begin{bmatrix} 1 & 0 \\ -(z-\bar{z}_j)/(\bar{c}_j e^{-\theta(\bar{z}_j)}) & 1 \end{bmatrix} V(z) & \text{if } z \in A_j^-. \end{cases}$$

This turns growth of the exponential into decay to the identity matrix. To simplify notation we define

$$T_{j,+}(z) = \begin{bmatrix} 1 & 0 \\ -c_j e^{\theta(z_j)}/(z-z_j) & 1 \end{bmatrix}, \qquad T_{j,-}(z) = \begin{bmatrix} 1 & -\bar{c}_j e^{-\theta(\bar{z}_j)}/(z-\bar{z}_j) \\ 0 & 1 \end{bmatrix}, \tag{9.16}$$

$$S_{j,+}(z) = \begin{bmatrix} 1 & -(z-z_j)/(c_j e^{\theta(z_j)}) \\ 0 & 1 \end{bmatrix}, \qquad S_{j,-}(z) = \begin{bmatrix} 1 & 0 \\ -(z-\bar{z}_j)/(\bar{c}_j e^{-\theta(\bar{z}_j)}) & 1 \end{bmatrix}. \tag{9.17}$$

In Figure 9.6 we present the full small-time and long-time RH problems. We use the notation $[J;\Sigma]$ to denote the RH problem in Figure 9.6(b).

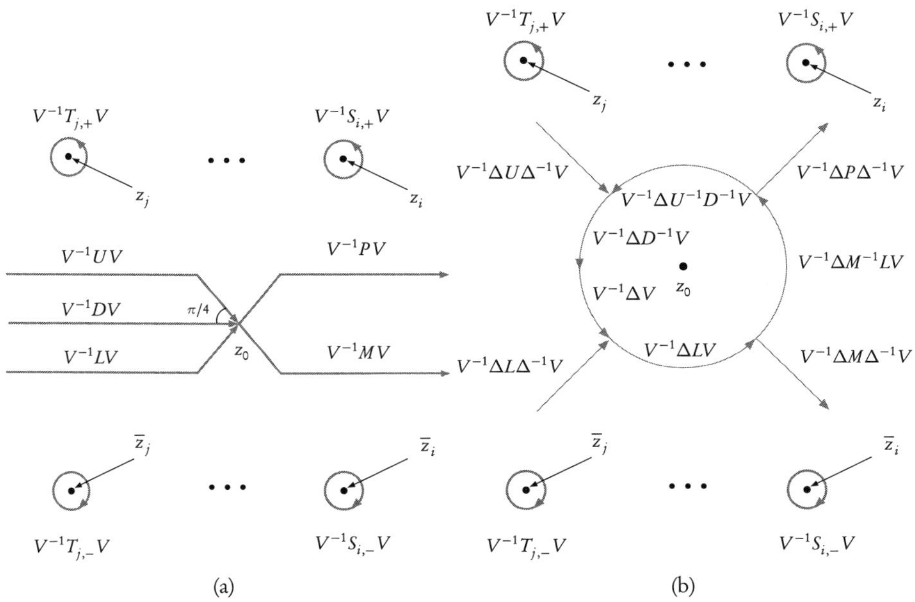

Figure 9.6. (a) *The jump contours and jump matrices for the final deformation for small time. In this schematic* $|c_j e^{\theta(z_j)}| \leq 1$ *and* $|c_i e^{\theta(z_i)}| > 1$. (b) *The jump contours and jump matrices for the final deformation for large time. Again, in this schematic* $|c_j e^{\theta(z_j)}| \leq 1$ *and* $|c_i e^{\theta(z_i)}| > 1$.

9.3.3 ▪ Numerical results

In Figure 9.7 we plot the solution of the focusing NLS equation with q_0 given by (9.13) with $A = 1$ and $\mu = 0.1$. The solution is nearly reflectionless, but Figure 9.7(d) shows the important dispersive aspect of the solution. Traditional numerical methods can fail to resolve this. In Figure 9.9 we plot the initial condition $q_0(x) = 1.9\exp(-x^2 + ix)$. The solutions of the focusing and defocusing NLS equations with this initial condition are computed. See Figure 9.10 for focusing and Figure 9.11 for defocusing. We also note that, when the initial condition is less localized, the corresponding reflection coefficient is more oscillatory. This makes it more difficult to resolve the solution of the corresponding RH problem. We restrict ourselves to initial data with rapid decay for this reason, i.e., in numerical examples we consider $q_0 \in \mathscr{S}_\delta(\mathbb{R})$ with δ large.

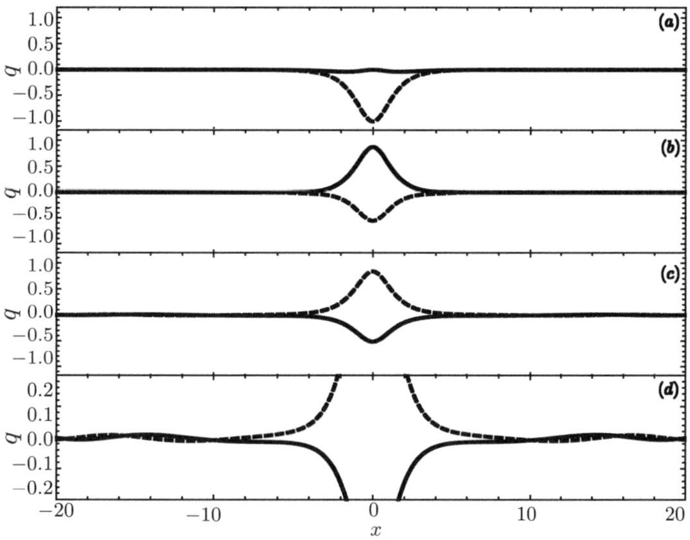

Figure 9.7. *The solution of the focusing NLS equation with q_0 given by* (9.13) *with $A = 1$ and $\mu = 0.1$. Solid: real part. Dashed: imaginary part.* (a) $q(x, 0)$. (b) $q(x, 1)$. (c) $q(x, 10)$. (d) *A scaled plot of $q(x, 10)$ showing the effects of dispersion. Traditional numerical methods can fail to resolve this. Reprinted with permission from Royal Society Publishing* [117].

We show numerical results that demonstrate spectral convergence. Let q_0 be given by (9.13) with $A = 1$ and $\mu = 0.1$ so that we can assume the reflection coefficient is computed to machine precision, i.e., $n > 80$ in Figure 9.1(b). Define $q_n(x, t)$ to be the approximate solution such that the number of collocation points per contour is proportional to n. In practice we set the number of collocation points to be n on shorter contours, like all contours in Figure 9.6(b). For larger contours, like the horizontal contours in Figure 9.6(a), we use $5n$ collocation points. To analyze the error we define

$$Q_m^n(x, t) = |q_n(x, t) - q_m(x, t)|. \tag{9.18}$$

Using this notation, see Figure 9.8 for a demonstration of spectral (Cauchy) convergence. Note that we choose x and t values to demonstrate spectral convergence in both the small-time and large-time regimes. As in the case of the KdV and modified KdV equations, implementational details are left to the software package [110].

9.4 ▪ Extension to homogeneous Robin boundary conditions on the half-line

Thus far, the results have been presented for the solution of the NLS equation posed on the whole line. We switch our attention to boundary-value problems on the half-line,

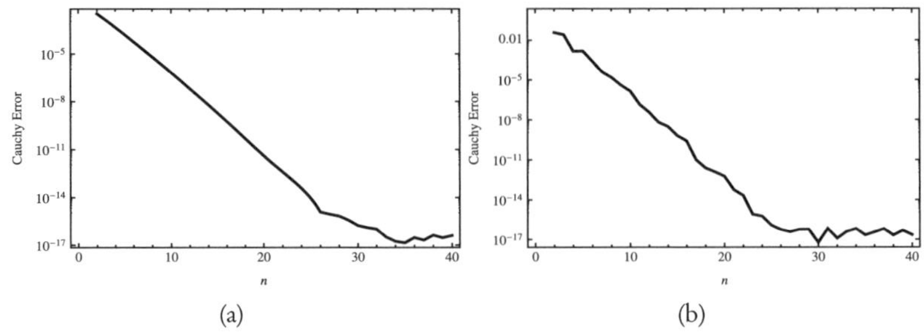

Figure 9.8. *The convergence of the numerical approximations of the solution of the focusing NLS equation with q_0 given by* (9.13) *with $A = 1$ and $\mu = 0.1$.* (a) $Q_n^{80}(2, 0.2)$ *as n ranges from 2 to 40.* (b) $Q_n^{80}(110, 110)$ *as n ranges from 2 to 40. Reprinted with permission from Royal Society Publishing* [117].

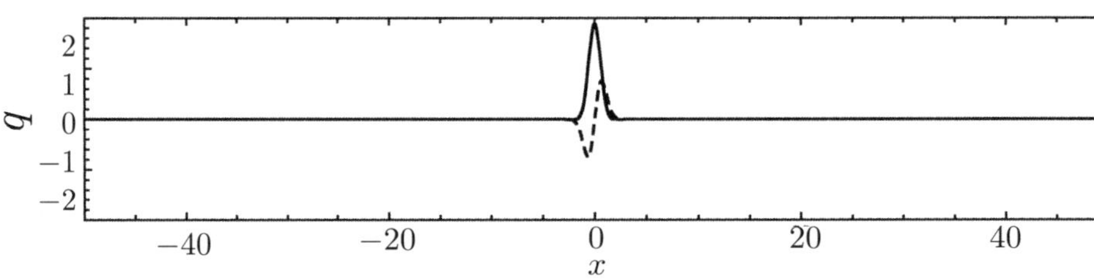

Figure 9.9. *The initial condition $q_0(x) = 1.9e^{-x^2+ix}$. Solid: real part. Dashed: imaginary part. Reprinted with permission from Royal Society Publishing* [117].

$x \geq 0$. Specifically, we extend the previous method to solve the following boundary-value problem:

$$\begin{aligned} iq_t + q_{xx} + 2\lambda|q|^2 q &= 0, \quad \lambda = \pm 1, \\ \alpha q(0,t) + q_x(0,t) &= 0, \quad \alpha \in \mathbb{R}, \\ q(x,0) = q_0(x) &\in \mathscr{S}_\delta(\mathbb{R}^+), \delta > 0. \end{aligned} \tag{9.19}$$

Here $\mathscr{S}_\delta(\mathbb{R}^+)$ is the space of smooth, rapidly decaying functions f on $[0, \infty)$ such that

$$\limsup_{x \to \infty} e^{\delta x} |f(x)| < \infty.$$

If we take $\alpha = 0$, we obtain a Neumann problem. Similarly, the limit $\alpha \to \infty$ effectively produces a Dirichlet problem. A method of images approach can be used to solve this problem. The approach of Biondini and Bui [13], first introduced in [12], takes the given initial condition on $[0, \infty)$ and produces an extension to $(-\infty, 0)$ using a *Darboux transformation*. For *Neumann boundary conditions* this results in an even extension and for *Dirichlet boundary conditions* the transformation produces an odd extension. Consider the system of ODEs

$$\begin{aligned} Y_1' &= Y_2, \\ Y_2' &= (4\lambda|q_0|^2 + \alpha^2) Y_1 - \lambda \bar{q}_0 Y_3 - q_0 Y_4, \\ Y_3' &= 2q_0' Y_1, \\ Y_4' &= 2\lambda \bar{q}_0' Y_1, \end{aligned} \tag{9.20}$$

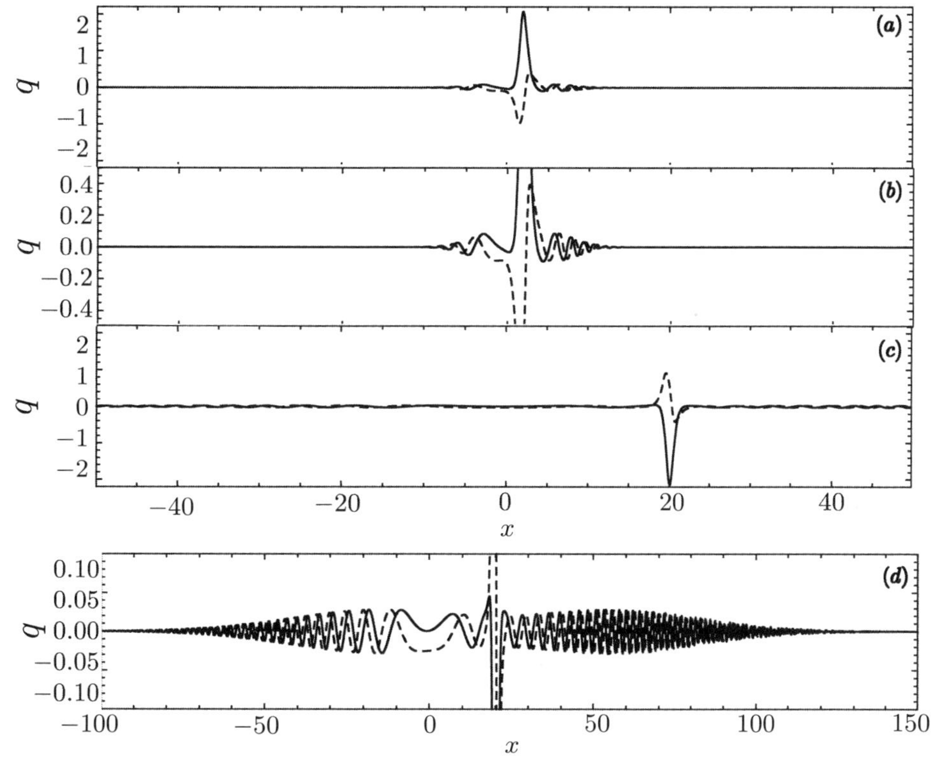

Figure 9.10. *The solution of the focusing NLS equation with q_0 shown in Figure 9.9. Solid: real part. Dashed: imaginary part.* (a) $q(x,1)$. (b) *A zoomed plot of* $q(x,1)$. (c) $q(x,10)$. (d) *A zoomed plot of* $q(x,10)$, *illustrating the dispersive effects in the solution. Reprinted with permission from Royal Society Publishing* [117].

with initial conditions

$$
\begin{aligned}
Y_1(0) &= 1,\\
Y_2(0) &= -\alpha,\\
Y_3(0) &= 2q_0(0),\\
Y_4(0) &= 2\lambda \bar{q}_0(0),
\end{aligned}
$$

and the function

$$
\tilde{q}(x) = \begin{cases} q_0(x) & \text{if } x \in [0,\infty),\\ -q_0(-x) + Y_3(x)/Y_1(x) & \text{if } x \in (-\infty, 0). \end{cases}
$$

It was shown in [13] that the solution of the Cauchy problem for the NLS equation on $\mathbb{R}$ with initial data $\tilde{q}$, restricted to $[0,\infty)$, is the unique solution of (9.19). To compute the extended initial data $\tilde{q}$ we first solve the system (9.20) numerically using a combination of Runge–Kutta 4 and 5 [76]. The IST for the extended potential can be used in the previous section's framework to numerically solve (9.19).

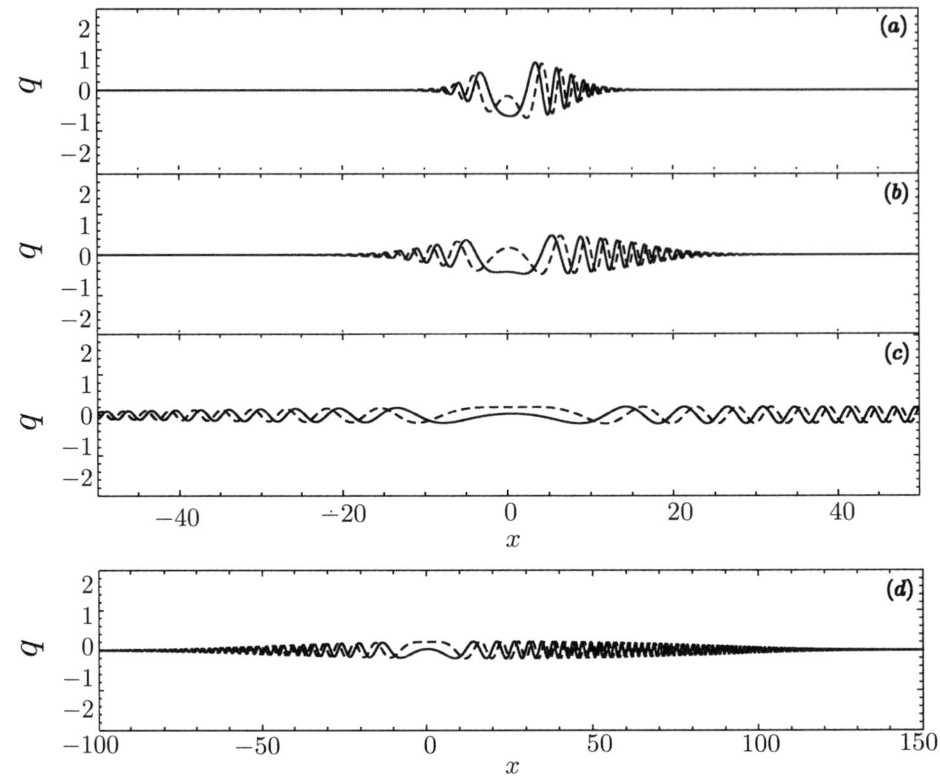

Figure 9.11. *The solution of the defocusing NLS equation with q_0 shown in Figure 9.9. Solid: real part. Dashed: imaginary part.* (a) $q(x,1)$. (b) $q(x,2)$. (c) $q(x,10)$. (d) *A scaled plot of* $q(x,10)$ *showing the dramatic effects of dispersion. Reprinted with permission from Royal Society Publishing* [117].

Remark 9.4.1. *The method of Bikbaev and Tarasov* [12] *was used to derive asymptotics by Deift and Park in* [30]. *Another approach would be to use the method of Fokas to compute solutions* [50, 67].

9.4.1 ▪ Numerical results

In this section we show numerical results for a Robin problem and a Neumann problem. As noted above, we could treat the Dirichlet problem by using an odd extension of our initial condition.

- **Robin boundary conditions**

  Here we show results for the case of the focusing NLS equation ($\lambda = 1$) with $\alpha = -1$ and with initial condition $q_0(x) = 1.9\exp(-x^2 + x)$. Note that the initial condition satisfies the boundary condition at $t = 0$. In Figure 9.12(a), we give the extended initial condition $\tilde{q}$ and in Figure 9.12(b) we show the corresponding reflection coefficient. The solution is shown in Figure 9.13. For this extended initial condition, we have four poles on the imaginary axis in the RH problem which corresponds to two stationary solitons:

$$z_1 \approx 1.84725\mathrm{i}, \quad c_1 \approx -14.4092\mathrm{i}, \quad z_2 \approx 1.21265\mathrm{i}, \quad c_2 \approx -8.17034\mathrm{i}.$$

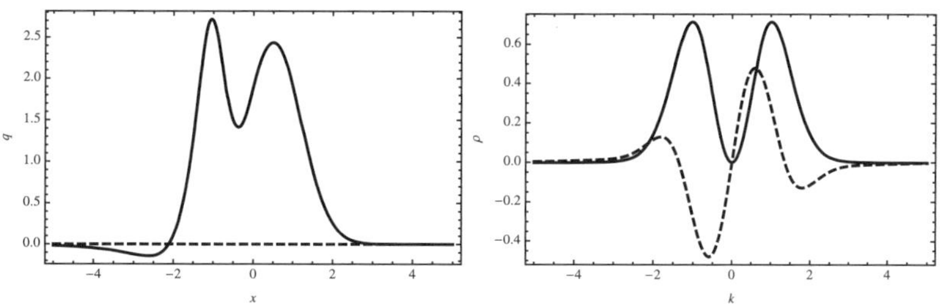

(a) (b)

Figure 9.12. (a) *The extended initial condition $\tilde{q}$. Solid: real part. Dotted: imaginary part.* (b) *The reflection coefficient for the extended initial condition $\tilde{q}$. Solid: real part. Dotted: imaginary part. Reprinted with permission from Royal Society Publishing* [117].

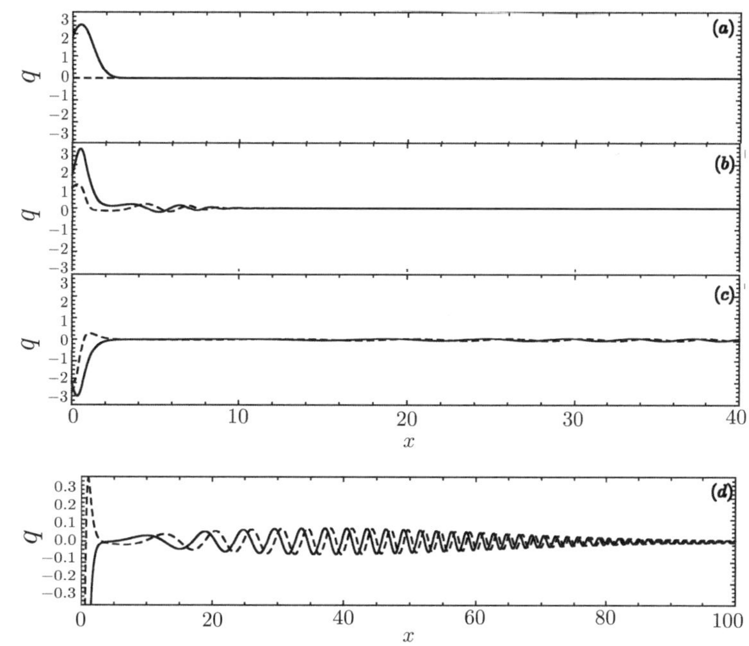

Figure 9.13. *The solution of the focusing NLS equation with Robin boundary conditions ($\alpha = 1$) and with q_0 as shown in Figure* 9.12(a). *Solid: real part. Dashed: imaginary part.* (a) $q(x, 0)$. (b) $q(x, 1)$. (c) $q(x, 10)$. (d) *A scaled plot of $q(x, 10)$ showing the extent of the dispersive tail. Reprinted with permission from Royal Society Publishing* [117].

- **Neumann boundary conditions**

  To show the reflection of a soliton off the boundary at $x = 0$ we solve a Neumann problem ($\alpha = 0$) with initial condition $q_0(x) = 1/2x^2 \exp(-.2x^2 - \mathrm{i}x)$. The extension $\tilde{q}$ of the initial condition can be seen in Figure 9.14(a) and the solution is shown in Figure 9.15. In this case it is just the even extension. The scattering data consists of

$$z_1 \approx 0.497613 + 0.371208\mathrm{i}, \quad c_1 \approx 0.110159 + 5.35099\mathrm{i},$$
$$z_2 \approx -0.497613 + 0.371208\mathrm{i}, \quad c_2 \approx -0.231104 - 0.0357421\mathrm{i}.$$

This shows that we have a pair of poles in the RH problem to the right of the imaginary axis and two to the left. This corresponds to one soliton moving to the left and one soliton moving to the right. The reflection coefficient is shown in Figure 9.14(b).

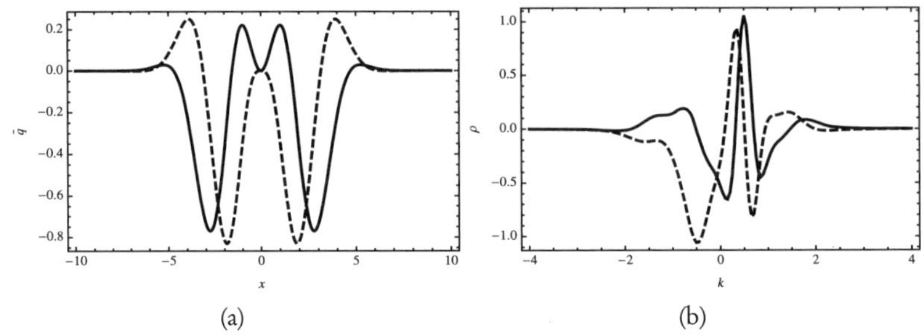

Figure 9.14. (a) *The extended initial condition $\tilde{q}$. Solid: real part. Dotted: imaginary part.* (b) *The reflection coefficient for the extended initial condition $\tilde{q}$. Solid: real part. Dotted: imaginary part. Reprinted with permission from Royal Society Publishing* [117].

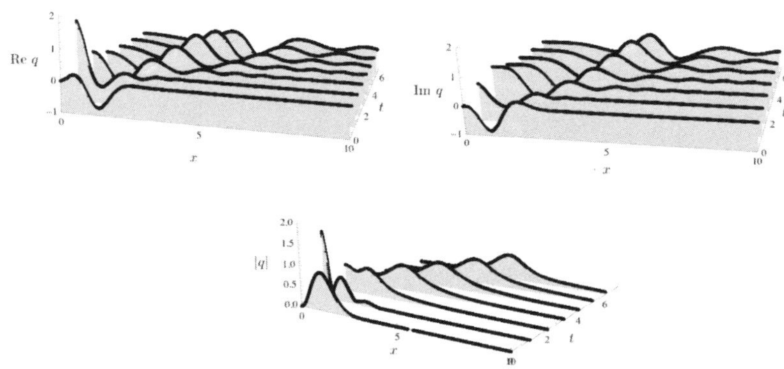

Figure 9.15. *The solution of the focusing NLS equation with Neumann boundary ($\alpha=-1$) conditions and q_0 as shown in Figure* 9.14(a). *The solution is shown up to $t=7$. Reprinted with permission from Royal Society Publishing* [117].

9.5 ▪ Singular solutions

As mentioned above, the defocusing NLS equation does not have soliton solutions that decay at infinity. We can insert the contours $A_j^{\pm}$ (see (9.8)) into the RH problem anyway. We introduce λ into (9.16) to obtain appropriate jump conditions for the defocusing NLS equations in the sense that the contours and jump matrices satisfy the hypotheses of the dressing method (Proposition 3.9). Define

$$T_{j,+}(z)=\begin{bmatrix} 1 & 0 \\ -c_j e^{\theta(z_j)}/(z-z_j) & 1 \end{bmatrix}, \quad T_{j,-}(z)=\begin{bmatrix} 1 & -\lambda\bar{c}_j e^{-\theta(\bar{z}_j)}/(z-\bar{z}_j) \\ 0 & 1 \end{bmatrix},$$

$$S_{j,+}(z)=\begin{bmatrix} 1 & -(z-z_j)/(c_j e^{\theta(z_j)}) \\ 0 & 1 \end{bmatrix}, \quad S_{j,-}(z)=\begin{bmatrix} 1 & 0 \\ -\lambda(z-\bar{z}_j)/(\bar{c}_j e^{-\theta(\bar{z}_j)}) & 1 \end{bmatrix}.$$

When $\lambda=1$ (focusing) this definition agrees with (9.16).

When $\lambda = -1$ (defocusing) we investigate how these additional contours will manifest themselves in the solution. Consider

$$u_1(x,t) = 2\eta e^{-4it(\xi^2-\eta^2)-2ix\xi} \operatorname{sech}(2\eta(4t\xi + x - x_0)),$$

which is the one-soliton solution of the focusing NLS equation [1]. A simple calculation shows that

$$u_2(x,t) = 2\eta e^{-4it(\xi^2-\eta^2)-2ix\xi} \operatorname{csch}(2\eta(4t\xi + x - x_0))$$

is a solution of the defocusing NLS. We are using the term "solution" loosely since this function has a pole when $4t\xi + x - x_0 = 0$. We call this a singular solution or singular soliton. These solutions are also called *positons*. Reference [43] contains a deeper discussion of these solutions with applications to rogue waves.

What we obtain when adding the above contours to the RH problem associated with the defocusing NLS equation is a nonlinear combination of these singular solutions in the presence of dispersion, as in the focusing case where the soliton was nonsingular. See Figure 9.16 for plots of a solution obtained using the reflection coefficient in Figure 9.2 along with

$$z_1 = 2 + 2i, \quad c_1 = 1000, \quad z_2 = -2 + 2i, \quad c_2 = 1/1000.$$

This corresponds to two of these singular solitons moving toward each other, until they interact with each other (the poles never cross). They interact in the presence of dispersion. We choose large and small norming constants to have the solitons away from $x = 0$ when $t = 0$. Not surprisingly, the relative accuracy of our numerical approximation breaks down near the poles. For points bounded away from the poles one should still expect uniform convergence, as discussed in the following section.

9.6 ▪ Uniform approximation

In this section we use the theory of Chapter 6 to prove the accuracy of the numerical method for arbitrarily large time when it is applied to $[J;\Sigma]$ in Figure 9.6(b). We assume the contours of the RH problem are truncated according to Proposition 2.78. Define $\Sigma_1 = \cup_j (A_j^+ \cup A_j^-)$ and $\Sigma_2 = \Sigma \setminus \Sigma_1$. Define the restrictions of J:

$$J_1(z) = J(z)|_{\Sigma_1}, \quad J_2(z) = J(z)|_{\Sigma_2}.$$

We introduce some x- and t-independent domains Ω_1 and Ω_2:

$$\Sigma_1 = \frac{\Omega_1}{\sqrt{t}} - \frac{x}{4t}, \quad \Sigma_2 = \Omega_2.$$

We use the change of variables

$$z = \frac{\zeta}{\sqrt{t}} - \frac{x}{4t}, \tag{9.21}$$

and the notation $\tilde{f}(\zeta) = f(\zeta/\sqrt{t} - x/(4t))$ for any function f, i.e., $\tilde{J}_1(\zeta) = J_1(\zeta/\sqrt{t} - x/(4t))$. Fix the trajectory in the (x,t)-plane: $x = 4ct$. We wish to use Algorithm 7.1. First, we numerically solve $[\tilde{J}_1, \Omega_1]$ with n collocation points to obtain a solution $\tilde{\Phi}_{1,n}$. The change of variables (9.21) is inverted, defining

$$\Phi_{1,n}(z) = \tilde{\Phi}_{1,n}\left(\sqrt{t}(z - z_0)\right).$$

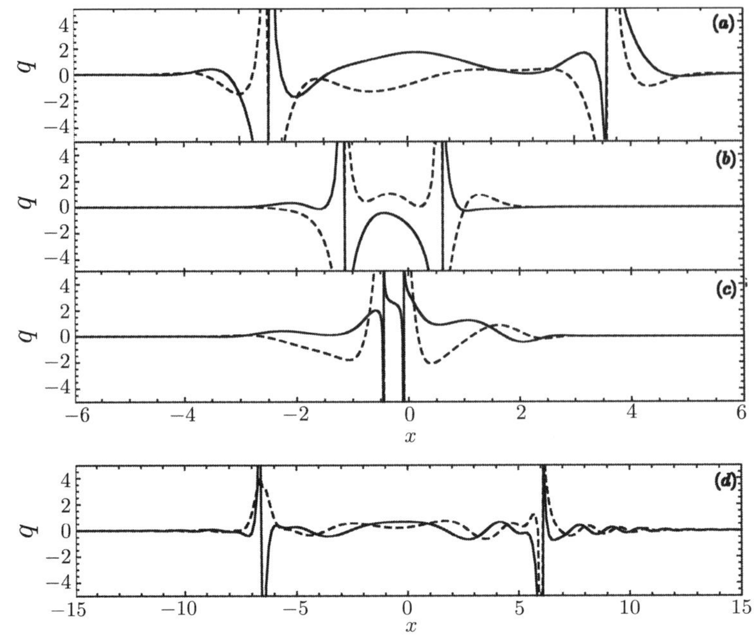

Figure 9.16. *A singular solution of the defocusing NLS equation. Note that the vertical lines are asymptotes. Solid: real part. Dashed: imaginary part.* (a) $q(x,0)$. (b) $q(x,0.1)$. (c) $q(x,0.2)$, *showing the interaction of the singular solutions.* (d) $q(x,1)$, *after the two poles have interacted with each other and a significant amount of dispersion is present. Reprinted with permission from Royal Society Publishing* [117].

Define $\tilde{J}_{2,n}(z) = \Phi_{1,n}(z)J_2(z)\Phi_{1,n}^{-1}(z)$. Then $[\tilde{J}_{2,n},\Omega_2]$ is solved numerically with n collocation points (for simplicity) to obtain a function $\Phi_{2,n}$. Note that there is no change of variables to invert for this RH problem. The function $\Phi_n = \Phi_{1,n}\Phi_{2,n}$ is an approximation of the solution to the full RH problem $[J;\Sigma]$. This is the scaled and shifted RH solver, Algorithm 7.1, applied in this context.

Since the arclength of Σ_1 tends to zero for large t, the conditions we check are the hypotheses of Theorem 7.11 (or Proposition 7.17):

- $\mathscr{C}[\tilde{J}_1;\Omega_1]^{-1}$ exists and is uniformly bounded in t,
- $\mathscr{C}[J_2;\Omega_2]^{-1}$ exists and is uniformly bounded in t, and
- all derivatives of $\tilde{J}_1(\zeta)$ and $J_2(\zeta)$, in the ζ variable, are uniformly bounded in t and ζ.

It is easy to see that all derivatives of $V^{-1}T_{j,\pm}V$ and $V^{-1}S_{j,\pm}V$ will be uniformly bounded. The transformation from $T_{j,\pm}$ to $S_{j,\pm}$ guarantees this.

The only possible singular behavior of $\tilde{J}_1$ will come from either the terms $\exp(\pm\theta(z))$ or from $\Delta(z;z_0)$. We proceed to show that under the chosen scaling, all derivatives of these two functions are bounded. From the definition of θ,

$$\theta(z) = 2\mathrm{i}(c4tz + 2tz^2) = -4\mathrm{i}c^2t + 4\mathrm{i}\zeta^2.$$

We see that all derivatives $\exp(\tilde{\theta}(\zeta))$ are bounded as long as ζ is bounded.

Now we consider $\Delta(z;z_0)$ in a neighborhood of z_0. We need to bound derivatives of $\exp(Y(z))$ on $\mathbb{C}\setminus(-\infty,z_0]$ where

$$Y(z)=\int_{-\infty}^{z_0}\frac{f(s)}{s-z}\mathrm{d}s,\quad f(s)=\log(1-\lambda\rho(s)\overline{\rho(s)}),$$

because it appears in Δ. We first note that $\exp(Y(z))$ is bounded for all z since $f(s)$ is real-valued; see (3.35) for a way to make this rigorous. Now, because the Cauchy integral is invariant under a Möbius change of variables that leaves infinity fixed (recall Lemma 5.4), we have

$$\tilde{Y}(\zeta)=\int_{-\infty}^{0}\frac{\tilde{f}(s)}{s-\zeta}\mathrm{d}s,\quad \tilde{f}(s)=f(s/\sqrt{t}-x/(4t)),\quad \tilde{Y}(\zeta)=Y(\zeta/\sqrt{t}-x/(4t)).$$

Therefore, from Lemma 2.49,

$$\tilde{Y}^{(j)}(\zeta)=\left(\int_{-\infty}^{0}\frac{\tilde{f}^{(j)}(s)}{s-\zeta}\mathrm{d}s-\sum_{i=1}^{j}\frac{\tilde{f}^{(j-i)}(0)}{(-\zeta)^{i+1}}\right),\quad \tilde{f}^{(j)}(0)=f^{(j)}\left(-\frac{x}{4t}\right)t^{-j/2}.$$

From the assumption that ρ is analytic and decays in a strip containing the real axis we see that all derivatives of $\tilde{Y}$ are uniformly bounded in t.

As stated, the analysis in Chapter 7 requires that the singular integral operators on Ω_i have uniformly bounded inverses as t becomes large. We describe how this follows from the asymptotic analysis of the RH problem [37, 39]. Again, a very useful fact is that once the solution Ψ of the RH problem $[G;\Gamma]$ is known, then the inverse of the operator is also known [27] (see also Lemma 2.67):

$$\mathscr{C}[G;\Gamma]^{-1}u=\mathscr{C}_\Gamma^+[u(\Psi^+)^{-1}]\Psi^+-\mathscr{C}_\Gamma^-[u(\Psi^+)^{-1}]\Psi^-. \tag{9.22}$$

If $\Psi\in L^\infty(\Gamma)$, then the inverse operator is bounded on $L^2(\Gamma)$. We show that $\mathscr{C}[\tilde{J}_1;\Omega_1]$ is close in operator norm (uniformly in t) to an operator with an explicit inverse that can be bounded uniformly in t using (9.22).

To construct the operator with an explicit inverse we follow the construction in Section 3.2. We factor off the singular behavior of $\Delta(z;z_0)$ ($\varphi(z)=f(z)/(4\pi\mathrm{i})$):

$$\begin{aligned}&\Delta(z;z_0)=\Delta_s(z;z_0)\Delta_r(z;z_0),\\&\Delta_s(z;z_0)=\operatorname{diag}((z-z_0)^{2\varphi(z_0)},(z-z_0)^{-2\varphi(z_0)}),\\&\Delta_r(z;z_0)\text{ is Hölder continuous at }z=z_0.\end{aligned}$$

Define (compare with (9.14))

$$\begin{aligned}&[M](z)=\begin{bmatrix}1 & \lambda\overline{\rho(\bar{z}_0)}e^{-\theta(z)}\\ 0 & 1\end{bmatrix}, && [P](z)=\begin{bmatrix}1 & 0\\ \rho(z_0)e^{\theta(z)} & 1\end{bmatrix},\\&[L](z)=\begin{bmatrix}1 & 0\\ \frac{\rho(z_0)}{\tau(z_0)}e^{\theta(z)} & 1\end{bmatrix}, && [D](z)=\begin{bmatrix}\tau(z_0) & 0\\ 0 & 1/\tau(z_0)\end{bmatrix},\\&[U](z)=\begin{bmatrix}1 & \frac{\lambda\overline{\rho(\bar{z}_0)}}{\tau(z_0)}e^{-\theta(z)}\\ 0 & 1\end{bmatrix}, && [\Delta](z;z_0)=\Delta_s(z;z_0)\Delta_r(z_0;z_0).\end{aligned} \tag{9.23}$$

We define an RH problem $[[\tilde{J}_1];\Omega_1]$ by replacing each appearance of M, P, L, D, and U in $\tilde{J}_1$ with $[M], [P], [L], [D]$, and $[U]$, respectively. If we assume $|z_0| \le C$, $C > 0$, the analyticity of ρ along with the Hölder continuity of Δ_r implies that $\|\tilde{J}_1 - [\tilde{J}_1]\|_\infty \to 0$, and therefore $\mathscr{C}[\tilde{J}_1;\Omega_1] - \mathscr{C}[[\tilde{J}_1];\Omega_1] \to 0$ in operator norm. If we extend Ω_1 using augmentation (Section 2.8.2) so that the outward rays are infinite, then $\mathscr{C}[[\tilde{J}_1];\Omega_1]^{-1}$ can be constructed explicitly out of parabolic cylinder functions. To see this, first note that the construction in Section 3.2.2 solves the associated RH problem explicitly. Then Lemma 2.67 gives the inverse operator explicitly. Furthermore, the inverse operator is uniformly bounded in t.

The estimates in Section 3.2.4 demonstrate that $\mathscr{C}[\tilde{J}_1;\Omega_1] - \mathscr{C}[[\tilde{J}_1];\Omega_1] \to 0$. To see this, one has to argue that any truncations/extensions have exponentially small contributions. Once convergence in operator norm is established, Theorem A.20 demonstrates that $\mathscr{C}[\tilde{J}_1;\Omega_1]^{-1}$ is uniformly bounded in t, for t sufficiently large.

The RH problem $[J_2, \Omega_2]$ has rational jump matrices and can be solved explicitly [2, p. 83]. The uniform boundedness of $\mathscr{C}[J_2,\Omega_2]^{-1}$ can be established by studying the explicit solution and again explicitly constructing its inverse in terms of the solution. If the set $\{z_j\}$ is large, then an asymptotic approach like [60] to show that the solitons separate may be more appropriate. Full details of this are beyond the scope of this chapter.

While we made the restriction $x = 4ct$ above, the bounds on derivatives and operators can be taken to be independent of c for $|c| \le C$. Define W^i to be the exact solution of the usual SIE (from (2.43)) on Ω_i and W^i_n to be its approximation with n collocation points per contour. From Theorem 7.11 in conjunction with Assumption 7.4.1 we have proved the following, which assumes that the reflection coefficient is obtained exactly.

Theorem 9.1 (Uniform approximation). *Assume that Assumption 7.4.1 holds. Fix $C > 0$. For every $\epsilon > 0$ and $k > 0$ there exists a constant $C_{k,\epsilon} > 0$ such that*

$$\sup_{|z_0|\le C} \|W^i_n - W^i\|_{L^2(\Omega_i)} < C_{k,\epsilon} n^{-k}, i = 1,2 \text{ if } |z_0| \le C,$$

and hence

$$|q_n(x,t) - q(x,t)| \le C_{k,\epsilon} \frac{|\Omega|^{1/2}}{\sqrt{\pi}} n^{-k} + \epsilon.$$

Here ϵ is the effect of truncation.

Since the arclength of the contours is bounded, we have (uniform in t) $L^1(\Gamma)$ convergence of W^i_n. It is then clear that (7.16) demonstrates that $q(x,t)$ is approximated uniformly.

Remark 9.6.1. *We emphasize that this theorem relies heavily on Assumption 7.4.1. If Assumption 7.4.1 fails to hold, the numerical method may not converge.*

Remark 9.6.2. *The results of [29, 37, 39] only apply to the defocusing NLS equation. The difficulty with the focusing NLS equation is the lack of information about $\{z_j\}$ and possible accumulation points on the real axis [28]. We are considering cases where we assume that $\{z_j\}$ are known (again see [28]), and the analysis proceeds in a way similar to the defocusing NLS equation.*

Remark 9.6.3. *Despite the fact that the theorem restricts to* $|z_0| < C$ *we still obtain uniform convergence. If* $q_0(x) \in \mathscr{S}_\delta(\mathbb{R})$*, for every* $\epsilon > 0$ *there exists* $C > 0$ *such that for* $|z_0| > C$*,* $|q(x,t)| < \epsilon$ [39]*. Thus we approximate the solution with zero when* $|z_0| > C$*.*

9.6.1 ▪ Numerical results

To demonstrate asymptotic accuracy we use the notation in (9.18) and fix n and m. We let x and t become large along a specific trajectory. For our purposes we use $x = 4t$. Note that along this trajectory q is on the order of $1/\sqrt{t}$ [122] (see also [39, 37, 36]). This allows us to estimate the relative error. See Figure 9.17 for a demonstration of the accuracy of the method for large x and t. We see that the relative error is bounded and small using relatively few collocation points.

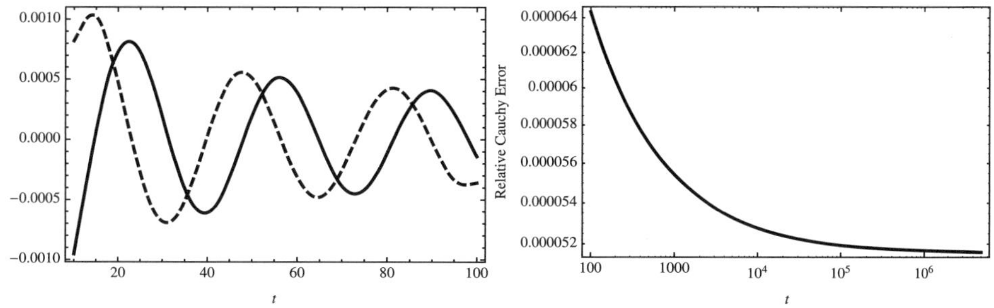

(a) (b)

Figure 9.17. *Asymptotic computations of solutions of the focusing NLS equation with* q_0 *given by* (9.13) *with* $A = 1$ *and* $\mu = 0.1$*. Solid: real part. Dashed: imaginary part.* (a) $q(8, 4t, t)$. (b) $Q_8^{16}(4t, t) \cdot t^{1/2}$ *for large values of* t*. Reprinted with permission from Royal Society Publishing* [117].

Chapter 10

The Painlevé II Transcendents

In this chapter we focus on the homogeneous Painlevé II ODE:

$$\frac{\mathrm{d}^2 u}{\mathrm{d}x^2} = xu + 2u^3. \tag{10.1}$$

(For brevity we refer to the *homogeneous Painlevé II equation* simply as *Painlevé II*.) The results of this section are compiled and expanded from [92, 97, 98]. There are many important applications of this equation: the Tracy–Widom distribution [109] from random matrix theory is written in terms of the Hastings–McLeod solution [64], and, as we have seen, asymptotic solutions to the KdV and modified KdV equations can be written in terms of the Ablowitz–Segur solutions [4] (see also (8.14)). The aim of this chapter is to demonstrate that the RH formulation can indeed be used effectively to compute solutions to Painlevé II. We remark that solutions of Painlevé II are commonly considered nonlinear special functions [19].

Solutions to differential equations such as (10.1) are typically defined by initial conditions; at a point x, we are given $u(x)$ and $u'(x)$. In the RH formulation, however, we do not specify initial conditions. Rather, the solution is specified by the *Stokes constants* $\{s_1, s_2, s_3\} \subset \mathbb{C}$, which satisfy the following condition:

$$s_1 - s_2 + s_3 + s_1 s_2 s_3 = 0. \tag{10.2}$$

We treat the Stokes constants as given, as in the aforementioned applications they arise naturally, while initial conditions do not. One can also think of the Stokes constants as being the analogous quantity to the scattering data in the case of PDEs. Given such constants, we denote the associated solution to (10.1) by

$$P_{\mathrm{II}}(s_1, s_2, s_3; x). \tag{10.3}$$

The solution P_{II} and its derivative can be viewed as the special functions which map Stokes constants to initial conditions.

Here we develop techniques to accurately and efficiently compute the *Ablowitz–Segur* and the *Hastings–McLeod solutions*. The method we describe in combination with other deformations [52] can be used to compute all solutions of (10.1), but for the sake of brevity we do not present more here.

The Ablowitz–Segur solutions satisfy $s_2 = 0$ and $s_1 = -s_3 \in \mathrm{i}\mathbb{R}$, with the special case of the Hastings–McLeod solution satisfying $s_1 = \pm 1$. Thus, as a special case, we are particularly interested in computing

$$P_{\mathrm{II}}(-\mathrm{i}\alpha, 0, \mathrm{i}\alpha; x), \quad \alpha \in \mathbb{R}.$$

Note that these are the solutions that appear in the asymptotic analysis of solutions of the KdV equation; see (8.11) and (8.14).

We divide the computation into five cases:

- Case 1: $1 - s_1 s_3 > 0$ and $x < 0$,
- Case 2: $1 - s_1 s_3 = 0$, $s_2 = 0$, and $x < 0$ (Hastings–McLeod solution),
- Case 3: $s_2 = 0$ and $x > 0$ (Ablowitz–Segur solutions),
- Case 4: $1 - s_1 s_3 = 0$, $s_2 \neq 0$, and $x < 0$, and
- Case 5: $s_2 \neq 0$ and $x > 0$.

For Case 1 we perform the deformation for $s_2 \in \mathbb{C}$, $1 - s_1 s_3 > 0$, with condition (10.2). Thus we calculate the Ablowitz–Segur solutions as a special case. One should compare Case 2 with the deformation for the collisionless shock region and $x > 0$ ($x < 0$) with that for the soliton (dispersive) regions in Chapter 8.

We omit Case 4, which can be obtained with only minor modifications using the method of Case 2, and Case 5, which can be obtained by modifying the deformations presented in [52]. All five cases are implemented in [93].

At first glance, computing the solutions to (10.1) appears elementary: given initial conditions, simply use one's favorite time-stepping algorithm, or better yet, input it into an ODE toolbox such as `ode45` in MATLAB® or `NDSolve` in *Mathematica*. Unfortunately, several difficulties immediately become apparent. In Figure 10.1, we plot several solutions to (10.1) (computed using the RH approach that we are advocating): the Hastings–McLeod solution and perturbations of the Hastings–McLeod solution. Note that the computation of the solution is inherently unstable, and small perturbations cause oscillations — which make standard ODE solvers inefficient — and poles — which completely break such ODE solvers (though this issue can be resolved using the methodology of [55]). This is examined more precisely in Section 10.4.1.

There are many other methods for computing the Tracy–Widom distribution itself as well as the Hastings–McLeod solution [16, 17], based on the Fredholm determinant formulation or solving a boundary-value problem. Moreover, accurate data values have been tabulated using high precision arithmetic with a Taylor series method [102]. However, we will see that there is a whole family of solutions to Painlevé II which exhibit similar sensitivity to initial conditions and the method presented here can capture these solutions.

We present the RH problem for the solution of Painlevé II (10.1) without any derivation. We refer the reader to [52] for a comprehensive treatment noting that many of the results we state originally appeared in [38]. Let $\Gamma = \Gamma_1 \cup \cdots \cup \Gamma_6$ with $\Gamma_j = \{s\mathrm{e}^{\mathrm{i}\pi(j/3-1/6)} : s \in \mathbb{R}^+\}$, i.e., Γ consists of six rays emanating from the origin; see Figure 10.2. The jump matrix is defined by $\check{G}(\lambda) = \check{G}_i(\lambda)$ for $\lambda \in \Gamma_i$, where

$$\check{G}_i(x;\lambda) = \check{G}_i(\lambda) = \begin{cases} \begin{bmatrix} 1 & s_i \mathrm{e}^{-\mathrm{i}8/3\lambda^3 - 2\mathrm{i}x\lambda} \\ 0 & 1 \end{bmatrix} & \text{if } i \text{ is even}, \\ \begin{bmatrix} 1 & 0 \\ s_i \mathrm{e}^{\mathrm{i}8/3\lambda^3 + 2\mathrm{i}x\lambda} & 1 \end{bmatrix} & \text{if } i \text{ is odd}. \end{cases}$$

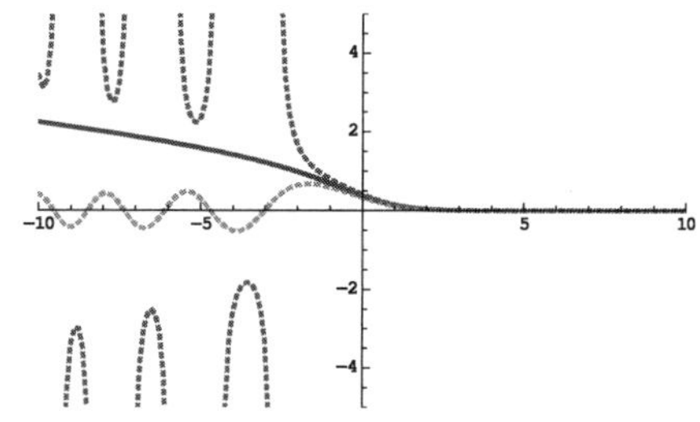

(a)

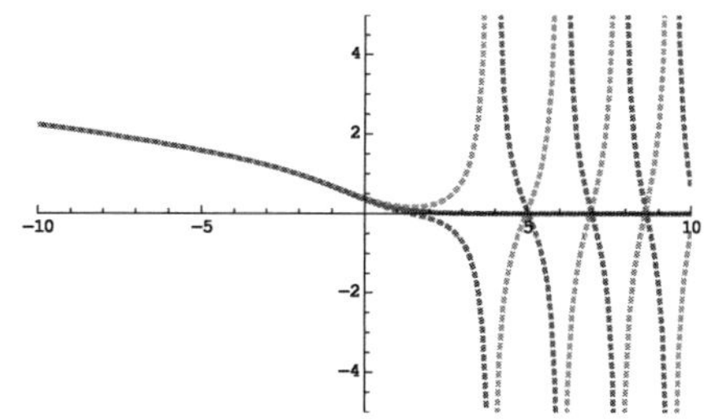

(b)

Figure 10.1. *Solutions to Painlevé II.* (a) *Radically different solutions for $x<0$.* (b) *Radically different solutions for $x>0$. Reprinted with permission from Springer Science+Business Media* [97].

From the solution $\check{\Phi}$ of $[\check{G};\Gamma]$, the Painlevé function is recovered by the formula

$$u(x)=\lim_{\lambda\to\infty}\lambda\check{\Phi}_{12}(\lambda),$$

where the subscripts denote the $(1,2)$ entry. This RH problem was solved numerically in [92] for small $|x|$.

For large $|x|$, the jump matrices $\check{G}$ off of the imaginary axis are increasingly oscillatory. We combat this issue by deforming the contour so that these oscillations turn into exponential decay. To simplify this procedure, and to start to mold the RH problem into the abstract form in Section 7.3, we first rescale the RH problem. If we let $\lambda=\sqrt{|x|}z$, then the jump contour Γ is unchanged, and

$$\begin{aligned}&\check{\Phi}^+(\lambda)=\check{\Phi}^+(\sqrt{|x|}z)=\check{\Phi}^-(\sqrt{|x|}z)\check{G}(\sqrt{|x|}z)=\check{\Phi}^-(\lambda)G(\lambda)\\ &\Leftrightarrow \Phi^+(z)=\Phi^-(z)G(z),\end{aligned}$$

where $\Phi(z)=\check{\Phi}(\sqrt{|x|}z)$ and $G(z)=G_i(z)$ on Γ_i with

$$G_i(z)=\begin{cases}\begin{bmatrix}1 & s_i e^{-\xi\theta(z)}\\ 0 & 1\end{bmatrix} & \text{if } i \text{ is even,}\\ \begin{bmatrix}1 & 0\\ s_i e^{\xi\theta(z)} & 1\end{bmatrix} & \text{if } i \text{ is odd,}\end{cases}$$

$\xi=|x|^{3/2}$, and

$$\theta(z)=\frac{2i}{3}\left(4z^3+2e^{i\arg x}z\right).$$

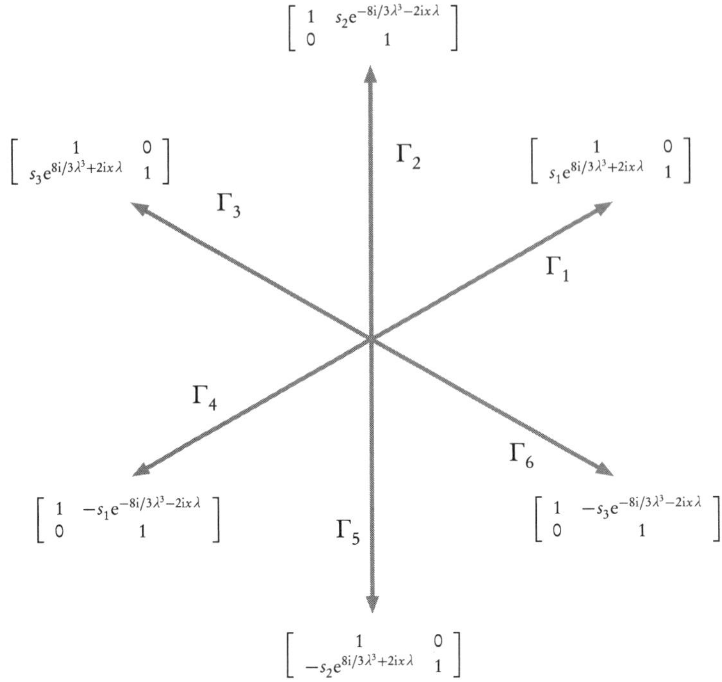

Figure 10.2. *The contour and jump matrix for the Painlevé II RH problem.*

Then

$$u(x) = \lim_{\lambda\to\infty} \lambda \check{\Phi}_{12}(x;\lambda) = \sqrt{|x|} \lim_{z\to\infty} z\Phi_{12}(z). \tag{10.4}$$

10.1 ▪ Positive x, $s_2 = 0$, and $0 \le 1 - s_1 s_3 \le 1$

We deform the RH problem for Painlevé II so that the numerics are asymptotically stable for positive x. We will see that the deformation is extremely simple under the following relaxed assumption.

Assumption 10.1.1. $x > 0$ *and* $s_2 = 0$.

We remark that, unlike other deformations we present, the following deformation can be easily extended to achieve uniform approximation for x in the complex plane such that $-\frac{\pi}{3} < \arg x < \frac{\pi}{3}$. This is primarily because the jumps on the contours Γ_2 and Γ_5 are the identity and the contours can be removed from the problem.

On the undeformed contour, the functions $e^{\pm i|x|^{3/2}\theta(z)}$ are oscillatory as $|x|$ becomes large. However, with the right choice of curve $h(t)$, $e^{\pm i\theta(h(t))}$ has no oscillations; instead, it decays exponentially as $t \to \infty$. But h is precisely the path of steepest descent, which passes through the stationary points of θ, i.e., the points where the derivative of θ vanishes. We readily find that

$$\theta'(z) = 2(4z^2 + 1),$$

and the stationary points are precisely $z_0 = \pm i/2$.

We note that, since $G_2 = I$, when we deform Γ_1 and Γ_3 through i/2 they become completely disjoint from Γ_4 and Γ_6, which we deform through $-$i/2. We point out that $G_3^{-1} = G_1$ and $G_6^{-1} = G_4$; thus we can reverse the orientation of Γ_3 and Γ_4, resulting in the jump G_1 on the curve $\Gamma_\uparrow$ and G_4 on $\Gamma_\downarrow$, as seen in Figure 10.3.

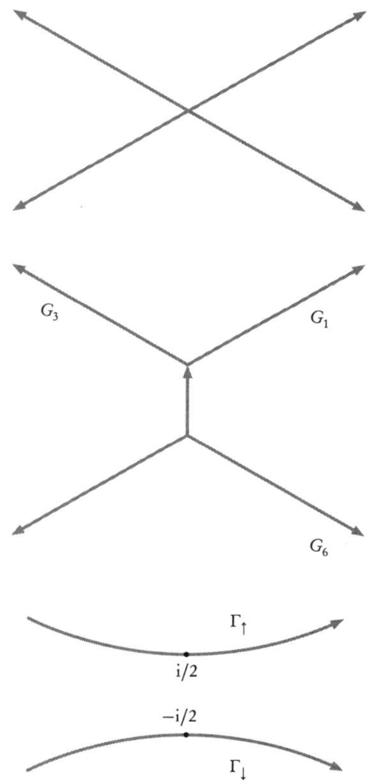

Figure 10.3. *Deforming the RH problem for positive x, with Assumption* 10.1.1.

We have

$$\theta\left(\pm\frac{\mathrm{i}}{2}\right) = \pm\frac{\mathrm{i}}{3}.$$

However, we now only have $\Gamma_\uparrow$ emanating from i/2, with jump matrix

$$G_1(z) = \begin{bmatrix} 1 & 0 \\ s_1 e^{\mathrm{i}|x|^{3/2}\theta(z)} & 1 \end{bmatrix}.$$

This matrix is exponentially decaying to the identity along $\Gamma_\uparrow$, as is G_6 along $\Gamma_\downarrow$. We employ the approach of Section 7.3 and Algorithm 7.1. We first use Proposition 2.78 to truncate the contours near the stationary point. What remains is to determine what "near" means. Because θ behaves like $\theta(\mathrm{i}/2) + \mathcal{O}(z \pm \mathrm{i}/2)^2$ near the stationary points, Assumption 7.0.1 implies that we should choose the shifting of $\beta_1 = \mathrm{i}/2$ and $\beta_2 = -\mathrm{i}/2$, the scalings $\alpha_1 = \alpha_2 = r|x|^{-3/4}$, and the canonical domains $\Omega_1 = \Omega_2 = [-1, 1]$. Here r is chosen so that what is truncated is negligible in the sense of Proposition 2.78. The treatment of G_6 is similar. The complete proof of asymptotic stability of the numerical method proceeds in a similar way as in Section 9.6.

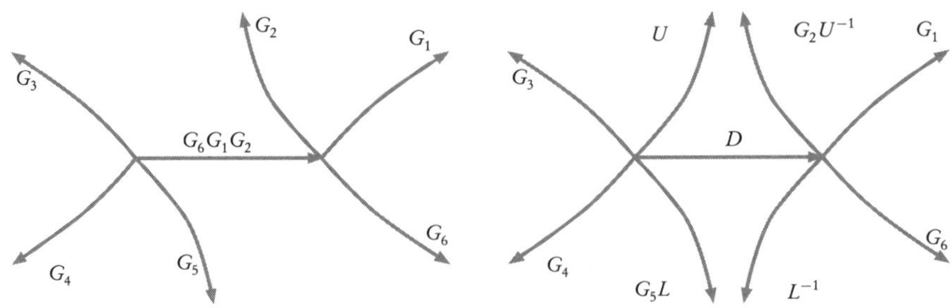

Figure 10.4. *Left: Initial deformation along the paths of steepest descent. Right: The deformed contour after lensing.*

10.2 ▪ Negative x, $s_2 = 0$, and $1 - s_1 s_3 > 0$

As mentioned above we perform this deformation under the following relaxed conditions.

Assumption 10.2.1. *$x < 0$, $s_2 \in \mathbb{C}$, and $1 - s_1 s_3 > 0$.*

We begin with the deformation of Γ to pass through the stationary points $\pm 1/2$, resulting in the RH problem on the left of Figure 10.4. The function

$$G_6(z)G_1(z)G_2(z) = \begin{bmatrix} 1 - s_1 s_3 & s_1 e^{-\xi\theta(z)} \\ s_2 e^{\xi\theta(z)} & 1 + s_1 s_2 \end{bmatrix}$$

has terms with $\exp(\pm\xi\theta(z))$. It cannot decay to the identity when deformed in the complex plane. We can resolve this issue by using lensing; see Section 2.8.2.

Consider the LDU factorization:

$$\begin{aligned} G_6(z)G_1(z)G_2(z) &= L(z)D(z)U(z) \\ &= \begin{bmatrix} 1 & 0 \\ e^{-\zeta\theta(z)}\frac{s_1}{1-s_1 s_3} & 1 \end{bmatrix} \begin{bmatrix} 1 - s_1 s_3 & 0 \\ 0 & \frac{1}{1-s_1 s_3} \end{bmatrix} \begin{bmatrix} 1 & e^{\zeta\theta(z)}\frac{s_1}{1-s_1 s_3} \\ 0 & 1 \end{bmatrix}. \end{aligned}$$

The entry U_{12} decays rapidly near $i\infty$, L_{21} decays rapidly near $-i\infty$, both L and U approach the identity matrix at infinity, and D is constant. Moreover, the oscillators in L and U are precisely those of the original G matrices. Therefore, we reuse the path of steepest descent and obtain the deformation on the right of Figure 10.4. The LDU decomposition is valid under the assumption $1 - s_1 s_3 \neq 0$.

10.2.1 ▪ Removing the connected contour

Although the jump matrix D is nonoscillatory (it is, in fact, constant), it is still incompatible with the theory presented in Section 7.3: we need the jump matrix to approach the identity matrix away from the stationary points. Therefore, it is necessary to remove this connecting contour. Since $D = \operatorname{diag}(d_1, d_2)$ is diagonal, we can solve $P^+ = P^- D$ with $P(\infty) = I$ on $(-1/2, 1/2)$ in closed form; see Problem 2.4.1:

$$P(z) = \begin{bmatrix} \left(\frac{2z+1}{2z-1}\right)^{i\log d_1/2\pi} & 0 \\ 0 & \left(\frac{2z+1}{2z-1}\right)^{i\log d_2/2\pi} \end{bmatrix}.$$

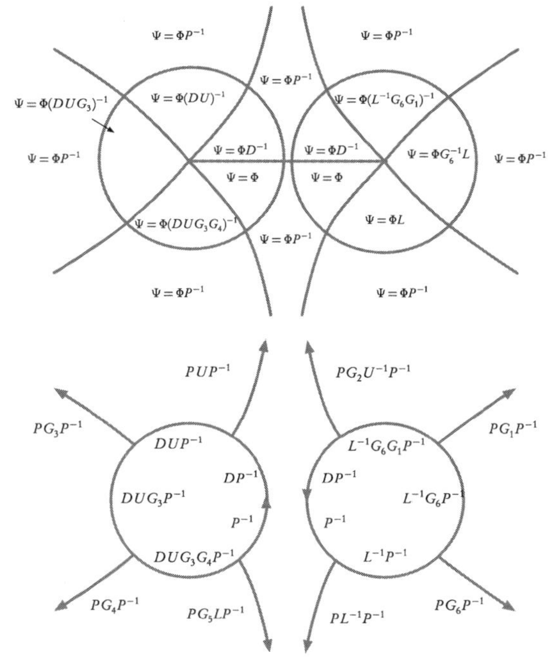

Figure 10.5. *Top: Definition of Φ in terms of Ψ. Bottom: Jump contour for Ψ.*

This local parametrix solves the desired RH problem for any choice of branch of the logarithm. However, we must choose a branch so that the singularity is locally integrable. This is accomplished by choosing the principal branch.

We write $\Phi = \Psi P$. Since P satisfies the required jump on $(-1/2, 1/2)$, Ψ has no jump there. Moreover, on each of the remaining curves we have

$$\Psi^+ = \Phi^+ P^{-1} = \Phi^- G P^{-1} = \Psi^{-1} P G P^{-1},$$

and our jump matrix becomes PGP^{-1}. Unfortunately, we have introduced singularities at $\pm 1/2$ and the theory of Section 7.3 requires smoothness of the jump matrix. This motivates alternate definitions for Ψ in circles around the stationary points similar to what was done in Chapters 8 and 9. In particular, we define Φ in terms of Ψ by the left panel of Figure 10.5, where Ψ has the jump matrix defined in the right panel of the figure. A quick check demonstrates that this definition of Φ satisfies the required jump relations.

We are ready to apply Algorithm 7.1. Define

$$\Omega = \{z : |z| = 1\} \cup \{r e^{i\pi/4} : r \in (1,2)\} \cup \{r e^{3i\pi/4} : r \in (1,2)\}$$
$$\cup \{r e^{-3i\pi/4} : r \in (1,2)\} \cup \{r e^{-i\pi/4} : r \in (1,2)\}.$$

In accordance with Assumption 7.0.1, we have

$$\Gamma_1 = \frac{1}{2} + \xi^{-1/2}\Omega \quad \text{and} \quad \Gamma_2 = -\frac{1}{2} + \xi^{-1/2}\Omega,$$

with the jump matrices defined according to Figure 10.5. Now, paths of steepest descent are local paths of steepest descent.

10.2.2 ▪ Uniform approximation

We have isolated the RH problem near the stationary points and constructed a numerical algorithm to solve the deformed RH problem. We show that this numerical algorithm approximates the true solution to the RH problem. In order to analyze the error, we introduce the local *model problem* for this RH problem following [52, Chapter 8]. The solution of the model problem is the parametrix used in asymptotic analysis. This section closely mirrors the approach in Section 9.6 for the NLS equations. It is important to note that knowledge of the parametrix is not needed to solve the RH problem numerically.

Let $\nu = \mathrm{i}\log d_2/(2\pi)$ and define the Wronskian matrix of parabolic cylinder functions $D_\nu(\zeta)$ [91],

$$Z_0(\zeta) = 2^{-\sigma_3/2}\begin{bmatrix} D_{-\nu-1}(\mathrm{i}\zeta) & D_\nu(\zeta) \\ \frac{\mathrm{d}}{\mathrm{d}\zeta}D_{-\nu-1}(\mathrm{i}\zeta) & \frac{\mathrm{d}}{\mathrm{d}\zeta}D_\nu(\zeta) \end{bmatrix}\begin{bmatrix} \mathrm{e}^{\mathrm{i}\pi/2(\nu+1)} & 0 \\ 0 & 1 \end{bmatrix}, \tag{10.5}$$

along with the constant matrices

$$H_{k+2} = \mathrm{e}^{\mathrm{i}\pi(\nu+1/2)\sigma_3}H_k\mathrm{e}^{\mathrm{i}\pi(\nu+1/2)\sigma_3},\quad H_0 = \begin{bmatrix} 1 & 0 \\ h_0 & 1 \end{bmatrix},\quad H_1 = \begin{bmatrix} 1 & h_1 \\ 0 & 1 \end{bmatrix},\quad \sigma_3 = \begin{bmatrix} 1 & 0 \\ 0 & -1 \end{bmatrix},$$

$$h_0 = -\mathrm{i}\frac{\sqrt{2\pi}}{\Gamma(\nu+1)},\quad h_1 = \frac{\sqrt{2\pi}}{\Gamma(-\nu)}\mathrm{e}^{\mathrm{i}\pi\nu}.$$

In addition, define

$$Z_{k+1}(\zeta) = Z_k(\zeta)H_k.$$

The sectionally holomorphic function $Z(\zeta)$ is defined as

$$Z(\zeta) = \begin{cases} Z_0(\zeta) & \text{if } \arg\zeta \in (-\pi/4, 0), \\ Z_1(\zeta) & \text{if } \arg\zeta \in (0, \pi/2), \\ Z_2(\zeta) & \text{if } \arg\zeta \in (\pi/2, \pi), \\ Z_3(\zeta) & \text{if } \arg\zeta \in (\pi, 3\pi/2), \\ Z_4(\zeta) & \text{if } \arg\zeta \in (3\pi/2, 7\pi/4). \end{cases}$$

This is used to construct the local solutions

$$\hat{\Psi}^r(z) = B(z)(-h_1/s_3)^{-\sigma_3/2}\mathrm{e}^{\mathrm{i}t\sigma_3/2}2^{-\sigma_3/2}\begin{bmatrix} \zeta(z) & 1 \\ 1 & 0 \end{bmatrix}Z(\zeta(z))(-h_1/s_3)^{\sigma_3/2},$$

$$\hat{\Psi}^l(z) = \sigma_2\hat{\Psi}^r(-z)\sigma_2,$$

where

$$\sigma_2 = \begin{bmatrix} 0 & -\mathrm{i} \\ \mathrm{i} & 0 \end{bmatrix},\quad B(z) = \left(\zeta(z)\frac{z+1/2}{z-1/2}\right)^{\nu\sigma_3},\quad \zeta(z) = 2\sqrt{-t\theta(z) + t\theta(1/2)}.$$

Consider the sectionally holomorphic matrix-valued function $\hat{\Psi}(z)$ defined by

$$\hat{\Psi}(z) = \begin{cases} P(z) & \text{if } |z \pm 1/2| > R, \\ \hat{\Psi}^r(z) & \text{if } |z - 1/2| < R, \\ \hat{\Psi}^l(z) & \text{if } |z + 1/2| < R. \end{cases}$$

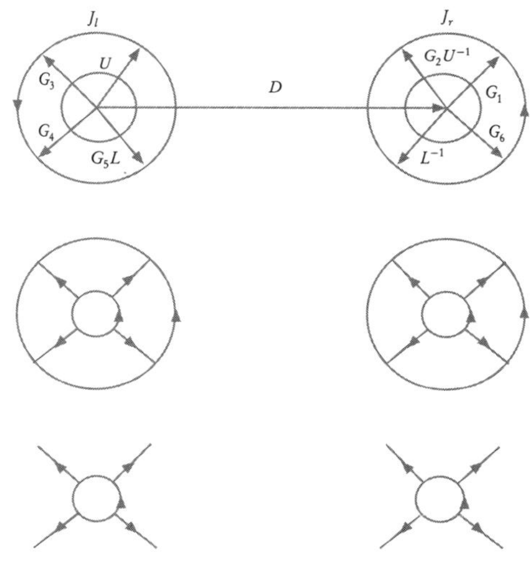

Figure 10.6. *Top: Jump contours for the model problem with solution $\hat{\Psi}$. Note that J_r and J_l are the jumps on the outside of the circles. They tend uniformly to the identity as $\xi \to \infty$. Center: The jump contours, $\hat{\Gamma}_1$, for the function $\hat{\Psi}_1$. The inner circle has radius r and the outer circle has radius R. Bottom: Contour on which $\hat{U}$ is nonzero. This can be matched up with the right contour in Figure* 10.5. *Reprinted with permission from John Wiley and Sons, Inc.* [98].

We use $[\hat{G};\hat{\Gamma}]$ to denote the RH problem solved by $\hat{\Psi}$. See the top panel of Figure 10.6 for $\hat{\Gamma}$. In [52, Chapter 4], it is shown that Ψ^r satisfies the RH problem for Φ exactly near $z = 1/2$ and for Ψ^l near $z = -1/2$. Notice that $\hat{\Psi}^r$ and $\hat{\Psi}^l$ are bounded near $z = \pm 1/2$. In the special case where $\log d_1 \in \mathbb{R}$, P remains bounded at $\pm 1/2$. Following the analysis in [52] we write

$$\Phi(z) = \chi(z)\hat{\Psi}(z),$$

where $\chi \to I$ as $\zeta \to \infty$.

We deform the RH problem for $\hat{\Psi}$ to open up a small circle of radius r near the origin as in Figure 10.5. We use $[\hat{G}_1;\hat{\Gamma}_1]$ to denote this deformed RH problem and solution $\hat{\Psi}_1$. See Figure 10.6 for $\hat{\Gamma}_1$. It follows that $\hat{\Psi}(z)P^{-1}(z)$ is uniformly bounded in z and ξ. Further, $\hat{\Psi}_1$ has the same properties. Since $\hat{\Psi}_1$ is uniformly bounded in both z and ξ, we use Lemma 2.67 to show that $\mathscr{C}[\hat{G}_1;\hat{\Gamma}_1]^{-1}$ has a uniformly bounded norm. We wish to use this to show the uniform boundedness of the inverse $\mathscr{C}[G;\Gamma]^{-1}$. To do so we extend the jump contours and jump matrices in the following way. Set $\Gamma_e = \Gamma \cup \hat{\Gamma}_1$ and define

$$G_e(z) = \begin{cases} G(z) & \text{if } z \in \Gamma, \\ I & \text{otherwise}, \end{cases}$$

$$\hat{G}_e(z) = \begin{cases} \hat{G}_1(z) & \text{if } z \in \hat{\Gamma}_1, \\ I & \text{otherwise}. \end{cases}$$

The estimates in [52] show that $G_e - \hat{G}_e \to 0$ uniformly as $\xi \to \infty$ on Γ_e. It follows that $\mathscr{C}[\hat{G}_e;\Gamma_e]^{-1}$ is uniformly bounded since the extended operator is the identity operator on $\Gamma \setminus \hat{\Gamma}_1$. Lemma 2.76 implies that $\mathscr{C}[G_e;\Gamma_e]^{-1}$ is uniformly bounded for sufficiently large ξ, which implies that $\mathscr{C}[G;\Gamma]^{-1}$ is uniformly bounded for ξ sufficiently large, noting that the extended operator is the identity operator on the added contours. We use this construction to prove the uniform convergence of the numerical method using both direct and indirect estimates.

Remark 10.2.1. *Solutions to Painlevé II often have poles on the real line, which correspond to the RH problems not having a solution. In other words, $\|\mathscr{C}[\Gamma,\Omega]^{-1}\|$ is not uniformly bounded, which means that the theory of Chapter 7 does not apply. However, the theorems can be adapted to the situation where x is restricted to a subdomain of the real line such that $\|\mathscr{C}[\Gamma,\Omega]^{-1}\|$ is uniformly bounded. This demonstrates asymptotic stability of the numerical method for solutions with poles, provided that x is bounded away from the poles, similar to the restriction of the asymptotic formulae in* [52].

10.2.3 ▪ Application of direct estimates

We see below that the RH problem for Ψ satisfies the properties of a numerical parametrix. This requires that the jump matrices have uniformly bounded Sobolev norms. The only singularities in the jump matrices are of the form

$$s(z) = \left(\frac{z-1/2}{z+1/2}\right)^{\mathrm{i}v}, v \in \mathbb{R}.$$

After transforming to a local coordinate η, $z = \xi^{-1/2}\eta - 1/2$, we see that

$$S(\eta) = s(\xi^{-1/2}\eta - 1/2) = \xi^{-\mathrm{i}v/2}\left(\frac{\xi^{-1/2}\eta + 1}{\eta}\right)^{\mathrm{i}v}.$$

The function $S(\eta)$ is smooth and has uniformly bounded derivatives provided η is bounded away from $\eta = 0$. The deformations applied thus far guarantee that η is bounded away from 0. To control the behavior of the solution for large η we look at the exponent which appears in the jump matrix

$$\theta(z) = \frac{2\mathrm{i}}{3} - 4\mathrm{i}\left(z + \frac{1}{2}\right)^2 + \frac{8\mathrm{i}}{3}\left(z + \frac{1}{2}\right)^3$$

and define

$$\tilde{\theta}(\eta) = \theta(\xi^{-1/2}\eta - 1/2) = \frac{2\mathrm{i}}{3} - 4\mathrm{i}\eta^2/\xi + \frac{8\mathrm{i}}{3}\eta^3/\xi^{3/2}.$$

If we assume that the contours are deformed along the local paths of steepest descent, all derivatives of $\mathrm{e}^{\xi\tilde{\theta}(\eta)}$ are exponentially decaying, uniformly in ξ. After applying the same procedure at $z = 1/2$ and after contour truncation, Proposition 2.78 implies the RH problem for Ψ satisfies the hypotheses of Theorem 7.11, proving uniform convergence. Let $u_n(x)$ be the approximation of P_{II} with n collocation points on each contour.

Theorem 10.1 (Uniform approximation). *We assume Assumption 7.4.1 holds. Assume $s_2 = 0$ and $1 - s_1 s_3 > 0$. For any $\epsilon > 0$ and $k > 0$ there exists a constant $C_{k,\epsilon} > 0$ such that for $x \in \mathbb{R}$*

$$|u_n(x) - P_{\mathrm{II}}(s_1, 0, s_3; x)| \le C_{k,\epsilon} n^{-k} + \epsilon \quad \textit{for all } k > 0.$$

Here ϵ is the effect of contour truncation.

10.2.4 ▪ Application of indirect estimates*

The second approach is to use the solution of the model problem to construct a numerical parametrix. Since we have already established strong uniform convergence, we proceed only to establish a deeper theoretical link with the method of nonlinear steepest descent, demonstrating that the success of nonlinear steepest descent implies the success of the numerical method, even though the numerical method does not depend on the details of nonlinear steepest descent. We start with the RH problem $[\hat{G}_1;\hat{\Gamma}_1]$ and its solution $\hat{\Psi}_1$. As before, see Figure 10.6 for $\hat{\Gamma}_1$. Define $\hat{u} = (\hat{\Psi}_1)^+ - (\hat{\Psi}_1)^-$ which is the solution of the associated SIE on $\hat{\Gamma}_1$. The issue here is that we cannot scale the deformed RH problem in Figure 10.5 so that it is posed on the same contour as $[G;\Gamma]$. We need to remove the larger circle.

Let us define a new function $\hat{U} = \hat{u}\phi$, where ϕ is a C^∞ function with support in $(B(-1/2,R)\cup B(1/2,R))\cap\hat{\Gamma}_1$ such that $\phi = 1$ on $(B(1/2,r)\cup B(-1/2,r))\cap\hat{\Gamma}_1$ for $r < R$. Let $\hat{\Gamma}_2$ be the support of $\hat{U}$ (see bottom contour in Figure 10.6). Define $\hat{\Psi}_2 = I + \mathscr{C}_{\hat{\Gamma}_2}\hat{U}$. From the estimates in [52, Chapter 8], it follows that

$$\hat{\Psi}_2^+ = \hat{\Psi}_2^- \hat{G}_2,$$

where $\hat{G}_2 - G$ tends uniformly to zero as $\xi \to \infty$. We have to establish the required smoothness of $\hat{U}$. We do this explicitly from the above expression for $\hat{\Psi}P^{-1}$ after using the scalings $z = \xi^{-1/2}k \pm 1/2$. The final step is to let ξ be sufficiently large so that we can truncate both $[G;\Gamma]$ and $[\hat{G}_2;\hat{\Gamma}_2]$ to the same contour. We use Proposition 7.17 to prove that this produces a numerical parametrix. Additionally, this shows how the local solution of RH problems can be tied to stable numerical computations of solutions.

Remark 10.2.2. *This analysis relies heavily on the boundedness of P. These arguments fail if we let P have unbounded singularities. In this case one approach would be to solve the RH problem for χ. The jump for this RH problem tends to the identity. To prove weak uniformity for this problem one needs to only consider the trivial RH problem with the jump being the identity matrix as a numerical parametrix. Another approach is to remove the growth in the parametrix through conjugation by constant matrices; see [97] for such an approach to rectify an unbounded global parametrix in the orthogonal polynomial RH problem.*

10.3 ▪ Negative x, $s_2 = 0$, and $s_1 s_3 = 1$

We develop deformations for the Hastings–McLeod Stokes constants. We realize uniform approximation in the aforementioned sense. The imposed conditions reduce to the following.

Assumption 10.3.1. $x < 0$, $s_2 = 0$, *and* $s_1 = -s_3 = \pm\mathrm{i}$.

We begin by deforming the RH problem (Figure 10.2) to the one shown in Figure 10.7. The horizontal contour extends from $-\alpha$ to α for $\alpha > 0$. We determine α below. Define

$$G_0 = G_6 G_1 = \begin{bmatrix} 0 & s_1 e^{-\mathrm{i}|x|^{3/2}\theta(z)} \\ s_1 e^{\mathrm{i}|x|^{3/2}\theta(z)} & 1 \end{bmatrix}.$$

Note that the assumption $s_2 = 0$ simplifies the form of the RH problem substantially; see Figure 10.7(b). We replace θ with a function possessing more desirable properties. Define

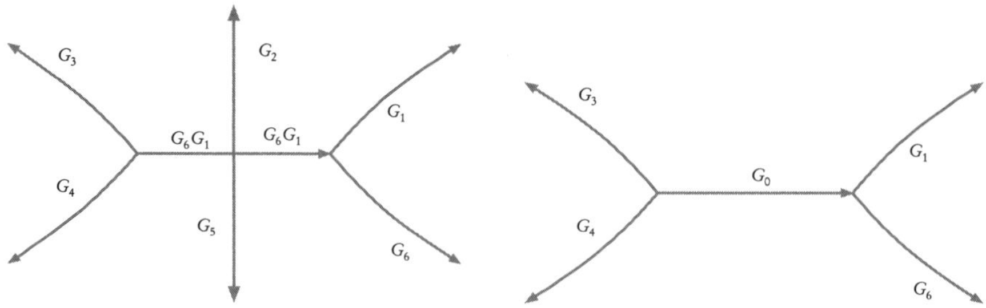

(a) (b)

Figure 10.7. *Deforming the RH problem for negative x, with Assumption* 10.3.1. *The black dots represent* $\pm\alpha$. (a) *Initial deformation.* (b) *Simplification stemming from Assumption* 10.3.1.

$$\Theta(z) = \mathrm{e}^{\mathrm{i}|x|^{3/2}\frac{g(z)-\theta(z)}{2}\sigma_3}, \quad g(z) = (z^2-\alpha^2)^{3/2}.$$

The branch cut for $g(z)$ is chosen along $[-\alpha,\alpha]$. If we equate $\alpha = 1/\sqrt{2}$, the branch of g can be chosen so that $g(z)-\theta(z) \sim \mathcal{O}(z^{-1})$. Furthermore, $g_+(z)+g_-(z) = 0$ and $\mathrm{Im}(g_-(z)-g_+(z)) > 0$ on $(-\alpha,\alpha)$. Define $\hat{G}_i = \Theta_-^{-1} G_i \Theta_+$ and note that

$$\hat{G}_0(z) = \begin{bmatrix} 0 & s_1 \mathrm{e}^{-\mathrm{i}|x|^{3/2}\frac{g_+(z)+g_-(z)}{2}} \\ s_1 \mathrm{e}^{\mathrm{i}|x|^{3/2}\frac{g_+(z)+g_-(z)}{2}} & \mathrm{e}^{\mathrm{i}|x|^{3/2}\frac{g_-(z)-g_+(z)}{2}} \end{bmatrix} = \begin{bmatrix} 0 & s_1 \\ s_1 & \mathrm{e}^{\mathrm{i}|x|^{3/2}\frac{g_-(z)-g_+(z)}{2}} \end{bmatrix}.$$

As $x \to -\infty$, G_0 tends to the matrix

$$J = \begin{bmatrix} 0 & s_1 \\ s_1 & 0 \end{bmatrix}.$$

The solution of the RH problem

$$\Psi^+(z) = \Psi^-(z) J(z), \quad z \in [-\alpha,\alpha], \quad \Psi(\infty) = I,$$

is given by

$$\Psi^{\mathrm{out}}_{\mathrm{HM}}(z) = \frac{1}{2}\begin{bmatrix} \beta(z)+\beta(z)^{-1} & -\mathrm{i}s_1(\beta(z)-\beta(z)^{-1}) \\ -\mathrm{i}s_1(\beta(z)-\beta(z)^{-1}) & \beta(z)+\beta(z)^{-1} \end{bmatrix}, \quad \beta(z) = \left(\frac{z-\alpha}{z+\alpha}\right)^{1/4}.$$

Here β has a branch cut on $[-\alpha,\alpha]$ and satisfies $\beta(z) \to 1$ as $z \to \infty$. It is clear that $(\Psi^{\mathrm{out}}_{\mathrm{HM}})_+ \hat{G}_0 (\Psi^{\mathrm{out}}_{\mathrm{HM}})_-^{-1} \to I$ uniformly on every closed subinterval of $(-\alpha,\alpha)$.

We define local parametrices near $\pm\alpha$:

$$\Psi^{\alpha}_{\mathrm{HM}}(z) = \begin{cases} I & \text{if } -\frac{\pi}{3} < \arg(z-a) < \frac{\pi}{3}, \\ \hat{G}_1^{-1}(z) & \text{if } \frac{\pi}{3} < \arg(z-a) < \pi, \\ \hat{G}_1(z) & \text{if } -\pi < \arg(z-a) < -\frac{\pi}{3}, \end{cases}$$

$$\Psi^{-\alpha}_{\mathrm{HM}}(z) = \begin{cases} I & \text{if } \frac{2\pi}{3} < \arg(z+a) < \pi \text{ or } -\pi < \arg(z+a) < -\frac{2\pi}{3}, \\ \hat{G}_1^{-1}(z) & \text{if } 0 < \arg(z+a) < \frac{2\pi}{3}, \\ \hat{G}_1(z) & \text{if } -\frac{2\pi}{3} < \arg(z+a) < \frac{2\pi}{3}. \end{cases}$$

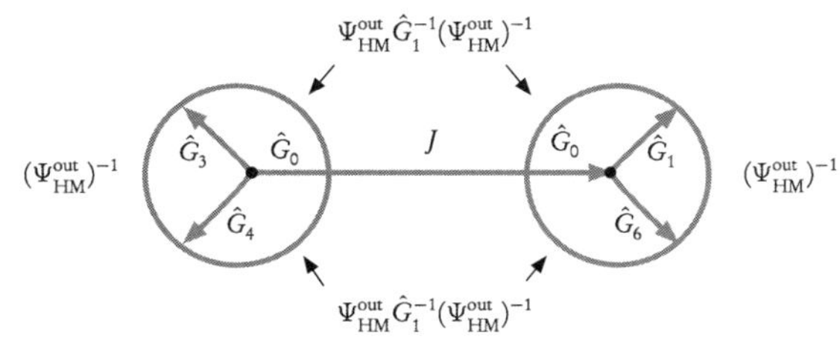

Figure 10.8. *The jump contours and jump matrices for the RH problem solved by Ψ_{HM}. The radius for the two circles is r.*

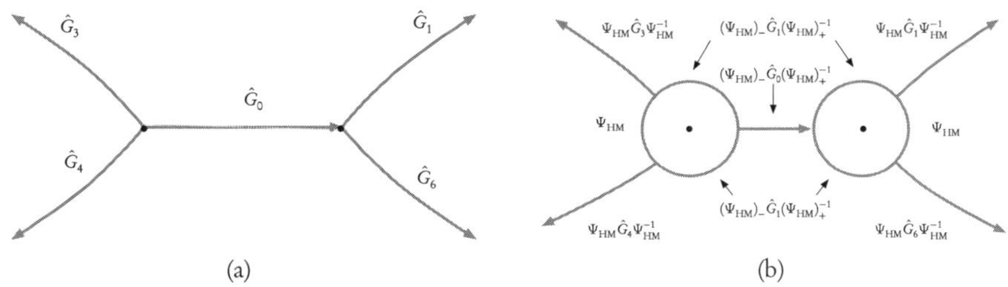

Figure 10.9. *The final deformation of the RH problem for negative x, with Assumption 10.3.1. The black dots represent $\pm\alpha$.* (a) *After conjugation by Θ.* (b) *Bounding the contours away from the singularities of g and β using Ψ_{HM}.*

We are ready to define the global parametrix. Given $r > 0$, define

$$\Psi_{\mathrm{HM}}(z) = \begin{cases} \Psi_{\mathrm{HM}}^{\alpha}(z) & \text{if } |z-a| < r, \\ \Psi_{\mathrm{HM}}^{-\alpha}(z) & \text{if } |z+a| < r, \\ \Psi_{\mathrm{HM}}^{\mathrm{out}}(z) & \text{if } |z+a| > r \text{ and } |z-a| > r. \end{cases}$$

It follows that $\Psi_{\mathrm{HM}}(z)$ satisfies the RH problem shown in Figure 10.9(b), after using the deformation in Figure 10.8.

Let Φ be the solution of the RH problem shown in Figure 10.9(a). It follows that $X \triangleq \Phi\Psi_{\mathrm{HM}}^{-1}$ solves the RH problem shown in Figure 10.9(b). The RH problem for X has jump matrices that decay to the identity away from $\pm\alpha$. We use Assumption 7.0.1 to determine that we should use $r = |x|^{-1}$. We solve the RH problem for X numerically. To compute the solution of Painlevé II used in the asymptotic analysis of the KdV equation (8.14) we use the formula

$$P_{\mathrm{II}}(\pm \mathrm{i}, 0, \mp \mathrm{i}; x) = 2\mathrm{i}\sqrt{|x|} \lim_{z\to\infty} zX(z)_{12}.$$

See Figure 10.11(a) below for a plot of the Hastings–McLeod solution with $s_1 = \mathrm{i}$.

10.4 ▪ Numerical results

In Figure 10.10 we plot the solution to Painlevé II with $(s_1, s_2, s_3) = (1, -2, 3)$ and demonstrate numerically that the computation remains accurate in the asymptotic regime. We use $u_n(x)$ to denote the approximate solution obtained with n collocation points per contour. Since we are using (10.4), we consider the estimated relative error by dividing the absolute error by $\sqrt{x}$. We see that we retain relative error as x becomes large. Implementational details can be found in [93].

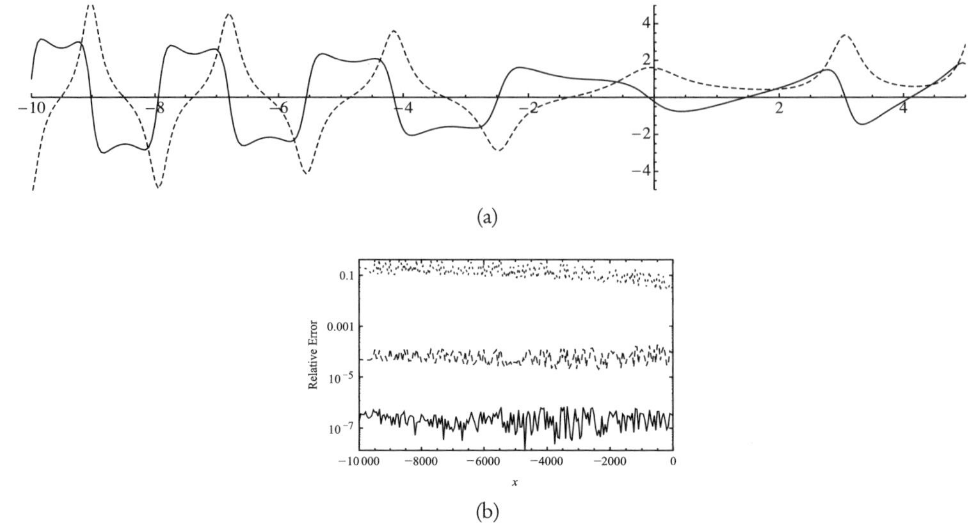

Figure 10.10. (a) *Plot of the approximation of* $P_{\mathrm{II}}(1,-2,3;x)$ *for small* x. *Solid: real part. Dashed: imaginary part.* (b) *Relative error. Solid:* $|x|^{-1/2}|u_{12}(x)-u_{36}(x)|$. *Dashed:* $|u_8(x)-u_{36}(x)|/\sqrt{|x|}$. *Dotted:* $|u_4(x)-u_{36}(x)|/\sqrt{|x|}$. *This plot demonstrates both uniform approximation and spectral convergence. Reprinted with permission from John Wiley and Sons, Inc.* [98].

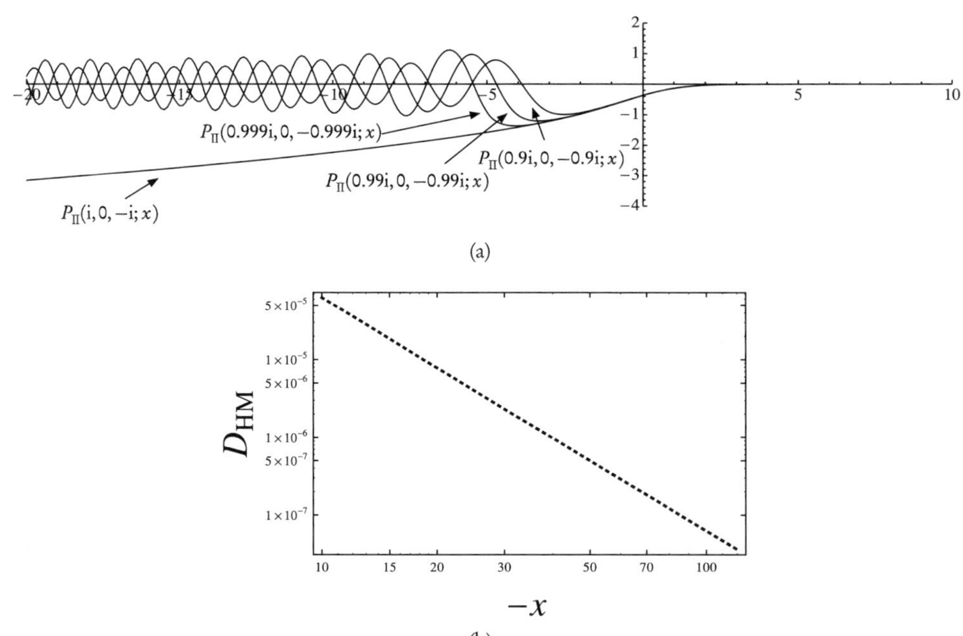

Figure 10.11. *Plotting and analysis of the numerical approximation of* $P_{\mathrm{II}}(\mathrm{i},0,-\mathrm{i};x)$ *along with some "nearby solutions." When comparing with Figure* 10.1 *note that* $-P_{\mathrm{II}}(\mathrm{i},0,-\mathrm{i};x)=P_{\mathrm{II}}(-\mathrm{i},0,\mathrm{i};x)$. (a) $P_{\mathrm{II}}(\alpha\mathrm{i},0,-\alpha\mathrm{i};x)$, $\alpha=0.9,0.99,0.999,1$ *for positive and negative* x. *For small* $|x|$ *we solve the undeformed RH problem.* (b) *A verification of the numerical approximation using the asymptotics when* $\alpha=1$ (10.6); *reprinted with permission from Springer Science+Business Media* [97].

10.4.1 ▪ The Hastings–McLeod solution

To verify our computations for the Hastings-McLeod solution $P_{\mathrm{II}}(\mathrm{i},0,-\mathrm{i};x)$ we may use the asymptotics [52]

$$P_{\mathrm{II}}(\mathrm{i},0,-\mathrm{i};x) \sim -\sqrt{\frac{-x}{2}} + \mathcal{O}\left(x^{-5/2}\right) \quad \text{as } x \to -\infty. \tag{10.6}$$

Define

$$D_{\mathrm{HM}}(x) = \left| \frac{P_{\mathrm{II}}(\mathrm{i},0,-\mathrm{i};x)}{\sqrt{\frac{-x}{2}}} + 1 \right|$$

to be an estimate of the *relative error* which should tend to a constant for x large and negative. We demonstrate this in Figure 10.11(b).

This solution, shown in Figure 10.11(a), is indeed a very special solution. First and foremost are its applications. We have seen that it arises in the asymptotic analysis of the KdV equation. Above we have mentioned that it arises in the cumulative distribution function for the famous Tracy–Widom distribution.

From a purely differential equation point of view, the solution is also remarkable. The solution has exponential decay to zero for $x > 0$, as is evidenced by the decay of the jump matrices to the identity in the deformations leading up to Figure 10.3. Additionally, the solution has monotonic behavior for all x, as seen in Figure 10.11(a). As was discussed in Figure 10.1, this may not be initially surprising until one considers nearby solutions. Typically, Ablowitz–Segur solutions with $|s_1| < 1$ are oscillatory for $x < 0$, as seen in Figure 10.11(a), and convergence is very nonuniform as the Stokes constants (s_1, s_2, s_3) approach $(\mathrm{i}, 0, -\mathrm{i})$.

As hinted at by Figure 10.1, additional singular behavior can occur for $x > 0$. To see this, consider the Stokes constants $(s_1, s_2, s_2) = (\mathrm{i}, 10^{-7}\mathrm{i}, -\mathrm{i})$. A symmetry in the RH problem is broken when $s_2 \neq 0$, and the problem does not have a solution for all $x \in \mathbb{R}$. This manifests itself in the solution of the Painlevé II equation as poles. See Figure 10.12(a) for this solution plotted in the complex plane. This solution is plotted alongside the Hastings–McLeod solution and another solution with poles in Figure 10.12(b). It is clear that the Hastings–McLeod solution strikes a delicate balance.

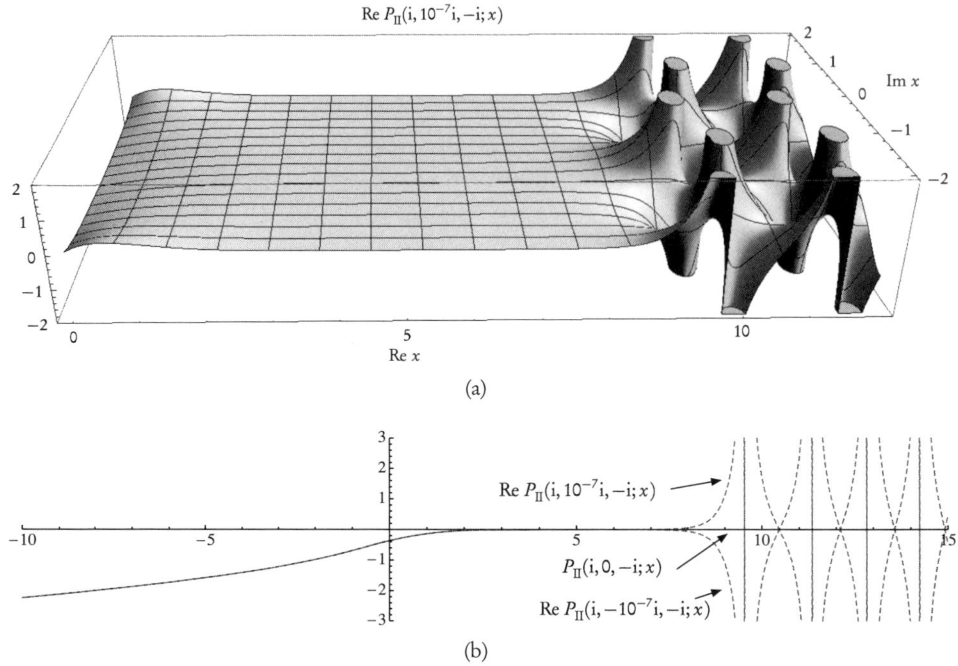

Figure 10.12. *Plotting the numerical approximation of* $P_{\text{II}}(s_1, s_2, s_3; x)$ *for Stokes constants near those for the Hastings–McLeod solution when* $\operatorname{Re} x > 0$*. These plots are made for relatively small* $|x|$ *and deformation is not necessary, i.e., the RH problem shown in Figure* 10.2 *can be solved directly.* (a) $P_{\text{II}}(\text{i}, 10^{-7}\text{i}, -\text{i}; x)$ *in the complex plane. Poles are clearly evident in the solution. Reprinted with permission from John Wiley and Sons, Inc.* [98]. (b) *Three solutions plotted on the real axis for* $x > 0$*. Poles are clearly present on the real axis.*

Chapter 11

The Finite-Genus Solutions of the Korteweg–de Vries Equation

This chapter presents a description (originally appearing in [113]) for the so-called *finite-genus* or *finite-gap solutions* of the KdV equation

$$q_t + 6qq_x + q_{xxx} = 0, \quad (x,t) \in \mathbb{R} \times \mathbb{R}, \tag{11.1}$$

and uses this description to compute them.

The finite-genus solutions arise in the spectral analysis of the Schrödinger operator with a periodic or quasi-periodic potential, where the spectrum has only a finite number g of finite-length bands separated by g gaps. The analysis presented in Section 1.6 is the analogous procedure for a potential with sufficient decay. The finite-genus solutions are explicitly described in terms of *Riemann theta functions.* The relevant theta functions are determined by hyperelliptic compact Riemann surfaces of genus g. In this chapter we reserve the symbol g for the genus of the relevant surface. In Section 1.6 and Chapter 8 we demonstrated how the spectral analysis of the Schrödinger operator can lead to a method (the inverse scattering transform) for solving the initial-value problem on the line. When one looks to solve (11.1) with periodic initial data, the finite-genus solutions play the same role that is played by trigonometric polynomials for the linear KdV equation $q_t + q_{xxx} = 0$, in the sense that the solution of the periodic problem in the space of square-integrable functions is approximated arbitrarily close by a finite-genus solution with sufficiently high g. An eloquent overview of the extensive literature on these solutions is found in McKean's review [79] of [81]. Of particular importance in the development of this literature are the pioneering works of Lax [74] and Novikov [89]. Excellent reviews are also found in Chapter 2 of [90], Dubrovin's oft-cited review article [45], and [11], parts of which focus specifically on the computation of these solutions.

The computation of finite-genus solutions is a nontrivial matter. Several approaches have appeared.

- Lax's original paper [74] includes an appendix by Hyman, where solutions of genus two were obtained through a variational principle.
- The now-standard approach of their computation is via their algebro-geometric description in terms of Riemann surfaces; see [23] or [56], for instance.
- Yet another approach is through the numerical solution of the so-called *Dubrovin equations*, a set of coupled ODEs that describe the dynamics of the zeros and poles of an auxiliary eigenfunction of the spectral problem, the *Baker–Akhiezer function*

[11, 44]. The finite-genus solution is recovered from the solution of the Dubrovin equations [90, 99].

One advantage of the last two methods over the variational method employed by Lax and Hyman is that periodic and quasi-periodic solutions are constructed with equal effort. The same is true for our approach, described below.

The main result of this chapter is the derivation of a Riemann–Hilbert representation of what is known as the Baker–Akhiezer function. We construct an RH problem whose solution is used to find the Baker–Akhiezer function. From this, one extracts the associated solution of the KdV equation. The x and t dependence of the solution appear in an explicit way so that no time stepping is required to obtain the value of the solution at a specific x and t. This should be contrasted with, for instance, the numerical solution of the Dubrovin equations [99]. Furthermore, just like for the method of inverse scattering (Chapters 8 and 9), the infinite-line counterpart of the problem under investigation, this dependence of the KdV solution on its independent variables appears linearly in an exponential function in the RH problem.

In order to solve this RH problem, we employ a regularization procedure using a g-function [33] which is related to the function that appeared in the collisionless shock region of the KdV equation. This simplifies the x and t dependence further. The resulting RH problem has piecewise-constant jumps. Straightforward modifications allow the RH problem to be numerically solved efficiently using the techniques in Chapter 6. This results in an approximation of the Baker–Akhiezer function that is uniformly valid on its associated Riemann surface. From this, we produce a uniform approximation of the associated solution of the KdV equation in the entire (x,t)-plane.

In this chapter, we begin by introducing the required fundamentals from the theory of Riemann surfaces. Next we use the methods of [90, Chapter 2] to describe how hyperelliptic Riemann surfaces are used to solve the KdV equation for a restricted class of initial conditions. The representation of the Baker–Akhiezer function in terms of an RH problem is derived in the following section. The modification of this RH problem is discussed in the two subsequent sections. The final form of the RH problem is presented in Section 11.5. In the final section the RH problem is solved numerically and the convergence of the method is verified. The method is illustrated there with many examples.

11.1 ▪ Riemann surfaces

We use this section to introduce the fundamental ideas from the theory of Riemann surfaces that are needed below. Most of these fundamental facts can be found in [11, 44]. The unfinished lecture notes by Dubrovin [46] provide an especially readable introduction, and most results stated below can also be found there. We include additional classical results on Riemann surfaces to give the reader some insight into the depth of the subject.

Definition 11.1. *Let*

$$F(\lambda,w)=w^2-P_{2g+2}(\lambda) \text{ or } F(\lambda,w)=w^2-P_{2g+1}(\lambda),$$

where P_m is a polynomial[43] *of degree m. The algebraic curve associated with this function is the solution set in $\mathbb{C}^2$ of the equation $F(\lambda,w)=0$. The desingularization and compactification of this curve is a* Riemann surface, Γ. *For this restricted class of polynomials the associated Riemann surface Γ is said to be hyperelliptic.*

[43] We will only consider polynomials with simple roots.

Note that in this chapter Γ is no longer a contour: it refers to a Riemann surface, and we only consider hyperelliptic surfaces. A *local parameter* on the surface Γ at a point $Q \in \Gamma$ is a locally analytic parameterization $z(P)$ that has a simple zero at Q.

Example 11.2. In the case $F(\lambda, w) = w^2 - (\lambda - a)(\lambda - b)$, a local parameter at $Q = (\mu, \sqrt{(\mu - a)(\mu - b)})$ for $\mu \neq a, b$ is $z(P) = \lambda - \mu$ for $P = (\lambda, w)$. Then

$$\lambda = z + \mu,$$
$$w = w(\lambda) = w(z + \mu)$$

are both analytic functions of z near $z = 0$. If $\mu = a$, then a local parameter is $z(P) = \sqrt{\lambda - a}$ for $P = (\lambda, w)$. Then

$$\lambda = z^2 + a,$$
$$w = w(\lambda) = w(z^2 + a) = z\sqrt{z^2 - b + a}$$

are both analytic functions of z near $z = 0$. ∎

It is well known that the hyperelliptic surfaces in Definition 11.1 are of genus g; they can be identified with a sphere with g handles. Define the a cycles $\{a_j\}_{j=1}^{g}$ and the b cycles $\{b_j\}_{j=1}^{g}$ on the Riemann surface as in Figure 11.1. The set $\cup_j \{a_j, b_j\}$ is a basis for the *homology* of the Riemann surface. It is also well known that a genus g surface has g linearly independent *holomorphic differentials*, denoted by $\omega_1, \ldots \omega_g$. We choose the normalization for this basis as such:

$$\oint_{a_j} \omega_k = 2\pi i \delta_{jk}, \quad j, k = 1, \ldots, g.$$

The matrix

$$B = \left(B_{jk}\right)_{1 \leq j,k \leq g}, \quad B_{jk} = \oint_{b_j} \omega_k, \tag{11.2}$$

is known as a *Riemann matrix*. Although this matrix has important properties and is necessary for computing the theta function representation of the finite-genus solutions, we do not need it directly.

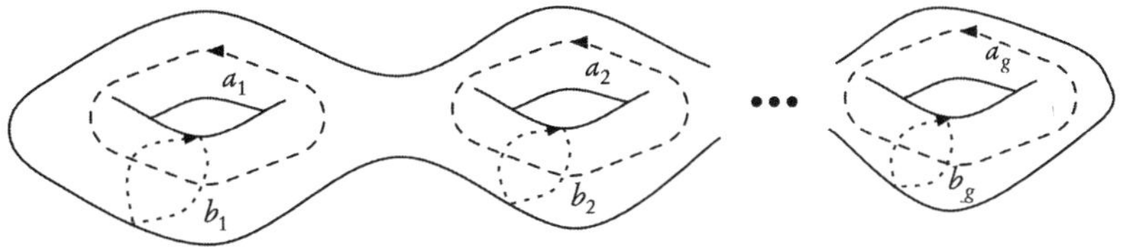

Figure 11.1. *An idealization of a hyperelliptic Riemann surface with a choice for the a and b cycles. Reprinted with permission from Elsevier* [113].

Lemma 11.3 (See [45]). *Let ω be a holomorphic differential on a Riemann surface of genus g. If*

$$\oint_{a_j} \omega = 0, \; j = 1, \ldots, g,$$

then $\omega = 0$.

Lemma 11.4 (See [45]). *Every holomorphic differential ω on a genus g hyperelliptic Riemann surface $w^2 - P_{2g+1}(\lambda) = 0$ can be expressed locally as*

$$\omega = \frac{q(\lambda)}{w} \mathrm{d}\lambda,$$

where q is a polynomial of degree at most $g-1$. Conversely, any differential of this form is holomorphic.

This lemma also holds for $2g+1$ replaced with $2g+2$, but we will not need this in what follows.

A *divisor* is a formal sum

$$D = \sum_{j=1}^{k} n_j Q_j, \quad n_j \in \mathbb{Z},$$

of points Q_j on the Riemann surface. Given a meromorphic function f on the Riemann surface with poles at Q_j of multiplicity n_j and zeros at R_j with multiplicity m_j, we define the associated divisor

$$(f) \triangleq \sum_{j=1}^{l} m_j R_j - \sum_{j=1}^{k} n_j Q_j.$$

The degree of a divisor is

$$\deg D \triangleq \sum_{j=1}^{k} n_j \text{ so that } \deg(f) = \sum_{j=1}^{l} m_j - \sum_{j=1}^{k} n_j.$$

A divisor is said to be positive if each n_j is positive, and $D > D'$ holds if $D - D'$ is positive. We use $l(D)$ to denote the dimension of the space of meromorphic functions f such that $(f) \geq D$.

Lemma 11.5 (Riemann inequality [45]). *For a genus g surface, if $\deg D \geq g$, then*

$$l(D) \geq 1 + \deg D - g.$$

A divisor D is said to be *nonspecial* if the Riemann inequality is an equality. Fix a point Q_0 on the Riemann surface and define the *Abel mapping* for points on the Riemann surface by

$$A(Q) \triangleq \left[\int_{Q_0}^{Q} \omega_1, \ldots, \int_{Q_0}^{Q} \omega_g \right], \tag{11.3}$$

where the path of integration is taken to be the same for all integrals. Note that this is well-defined for the appropriately normalized differentials. We extend this map to divisors $D = \sum_{j=1}^{k} n_j Q_j$ by

$$A(D) = \sum_{j=1}^{k} n_j A(Q_j).$$

Theorem 11.6 (See [45]). *The Abel map A maps divisors to the associated* Jacobi variety $J(\Gamma)=\mathbb{C}^g/\{2\pi M+BN\}$ *for* $M,N\in\mathbb{Z}^g$*, where B is defined in* (11.2). *Furthermore, if the divisor* $D=Q_1+\cdots+Q_g$ *is nonspecial, then A has a single-valued inverse in a neighborhood of* $A(D)$.

We do not make use of this theorem directly but include it for completeness. A *meromorphic differential* is a differential ω such that for every $Q\in\Gamma$ there exists a local parameter such that $\omega=f(z)\mathrm{d}z$, where $f(z)$ is (locally) a meromorphic function in the classical sense. Next, we describe properties of Abelian differentials of the second kind that are needed below.

Definition 11.7. *Given a point Q on the Riemann surface and a positive integer n, an Abelian differential of the second kind is a meromorphic differential that has a single pole of order* $n+1$ *so that its local representation is*

$$\nu_Q^n=\left(z^{-n-1}+\mathcal{O}(1)\right)\mathrm{d}z$$

with respect to a local parameter z, $z(Q)=0$.

When Q is the point at infinity we construct these differentials explicitly. As a local parameter we take $z^2=1/\lambda$ since Q is a branch point. If n is even, we set

$$\nu_\infty^n=-\frac{1}{2}\lambda^{n/2-1}\mathrm{d}\lambda.$$

When n is odd, there is more to be done. First, it follows that

$$\frac{\lambda^j}{w}\mathrm{d}\lambda=-2\frac{z^{-2j-3}}{\sqrt{P(z^{-2})}}\mathrm{d}z.$$

Then

$$P(z^{-2})=z^{-4g-2}(1-z^2\alpha_{g+1})\prod_{j=1}^{g}(1-z^2\alpha_j)(1-z^2\beta_j).$$

Thus

$$\begin{aligned}\frac{\lambda^j}{w}\mathrm{d}\lambda&=-2z^{-2j-2+2g}\left((1-z^2\alpha_{g+1})\prod_{j=1}^{g}(1-z^2\alpha_j)(1-z^2\beta_j)\right)^{-1/2}\mathrm{d}z\\&=(-2z^{-2j-2+2g}+\mathcal{O}(1))\mathrm{d}z.\end{aligned}$$

We choose $j=g+(n-1)/2$ so that

$$\nu_\infty^n=-\frac{1}{2}\frac{\lambda^{g+(n-1)/2}}{w}\mathrm{d}\lambda.$$

Let μ_∞^n be the differential obtained from ν_∞^n by adding holomorphic differentials so that it has zero a cycles. We state a lemma concerning the b periods of these differentials.

Lemma 11.8 (See [46]). *Define* $y_k(z)$ *through the equalities* $\omega_k=y_k(z)\mathrm{d}z$ *and* $z^2=1/\lambda$. *Then*

$$\oint_{b_k}\mu_\infty^n=\frac{1}{n!}\frac{\mathrm{d}^{n-1}}{\mathrm{d}z^{n-1}}y_k(z)\bigg|_{z=0},\quad k=1,\ldots,g.$$

11.2 ▪ The finite-genus solutions of the KdV equation

We turn to the consideration of the scattering problem associated with the KdV equation. The time-independent Schrödinger equation

$$-\Psi_{xx} - q_0(x)\Psi = \lambda\Psi \tag{11.4}$$

is solved for eigenfunctions $\Psi(x,\lambda)$ bounded for all x. We define the *Bloch spectrum*

$$\sigma_{\mathrm{B}}(q_0) = \left\{ \lambda \in \mathbb{C} : \text{ there exists a solution of (11.4) such that } \sup_{x\in\mathbb{R}} |\Psi(x,\lambda)| < \infty \right\}.$$

It is well known that for $q_0(x)$ smooth and periodic the Bloch spectrum consists of a countable collection of real intervals (see, for example, [80]):

$$\sigma_{\mathrm{B}}(q_0) = \bigcup_{j=1}^{\infty} [\alpha_j, \beta_j], \quad \alpha_j < \beta_j < \alpha_{j+1} < \beta_{j+1}.$$

If there are only $n+1$ intervals, then $\beta_{n+1} = \infty$. We refer to the intervals $[\alpha_j, \beta_j]$ as *bands* and the intervals $[\beta_j, \alpha_{j+1}]$ as *gaps*.

We make the following assumption in what follows.

Assumption 11.2.1. $\sigma_{\mathrm{B}}(q_0) \subset [0,\infty)$ *consists of a finite number of intervals. In this case we say that* q_0 *is a finite-gap potential.*

Define Γ to be the hyperelliptic Riemann surface associated with the function

$$F(\lambda, w) = w^2 - P(\lambda), \quad P(\lambda) = (\lambda - \alpha_{g+1}) \prod_{j=1}^{g} (\lambda - \alpha_j)(\lambda - \beta_j).$$

See Figure 11.2 for a cartoon. We divide this surface into two sheets. Choose the branch cuts for the function $\sqrt{P(\lambda)}$ along $\sigma_{\mathrm{B}}(q_0)$. We fix the branch by the requirement that $\sqrt{P(\lambda)} \sim (-)^g \mathrm{i} |\lambda|^{g+1/2}$ as $\lambda \to -\infty$. Define $\sqrt{P(\lambda)}^+$ to be the value $\lim_{\epsilon \downarrow 0} \sqrt{P(\lambda + \mathrm{i}\epsilon)}$. This allows us to define

$$\Gamma_\pm = \{(\lambda, \pm\sqrt{P(\lambda)}^+) : \lambda \in \mathbb{C}\}.$$

When considering a function f defined on Γ we use the notation $f_\pm$ so that $f_+ (f_-)$ denotes the function restricted to $\Gamma_+ (\Gamma_-)$. In this way we can consider $f_\pm$ as a function of only λ. We need an explicit description of the a cycles since we take a computational approach below:

$$a_i = \{(\lambda, \sqrt{P(\lambda)}^+) : \lambda \in (\beta_i, \alpha_{i+1}]\} \cup \{(\lambda, -\sqrt{P(\lambda)}^+) : \lambda \in [\beta_i, \alpha_{i+1})\}.$$

The a_i component on $\Gamma_+ (\Gamma_-)$ is oriented in the direction of decreasing (increasing) λ. This description is also useful since we will consider poles and zeros lying on the a cycles.

Remark 11.2.1. *There is some inconsistency in the notation* $f_\pm$ *which is also present in the literature. In what follows, it will be clear from the context whether we are referring to a function defined on the Riemann surface or to* f_+ *and* f_- *separately. Furthermore, this should not be confused with the boundary values of an analytic function as discussed in Chapter 2.*

We introduce further notation that will be of use later. Given a point $Q=(\lambda,w)\in\Gamma$, we follow [52] and define the involution $*$ by $Q^*=(\lambda,-w)$. This is an isomorphism from one sheet of the Riemann surface to the other. The first sheet, with the a cycles removed, is isomorphic to the cut plane

$$\hat{\mathbb{C}} \triangleq \mathbb{C}\setminus\left\{[\alpha_{g+1},\infty)\cup\bigcup_{j=1}^{g}[\alpha_j,\beta_j]\right\},$$

through the mapping $\hat{}:\Gamma\to\mathbb{C}$ defined by $\hat{Q}=\lambda$. The mapping $\check{}:\mathbb{C}\to\Gamma_+$ defined by $\check{\lambda}=(\lambda,\sqrt{P(\lambda)}^+)$ is the inverse of the $\hat{}$ mapping restricted to Γ_+.

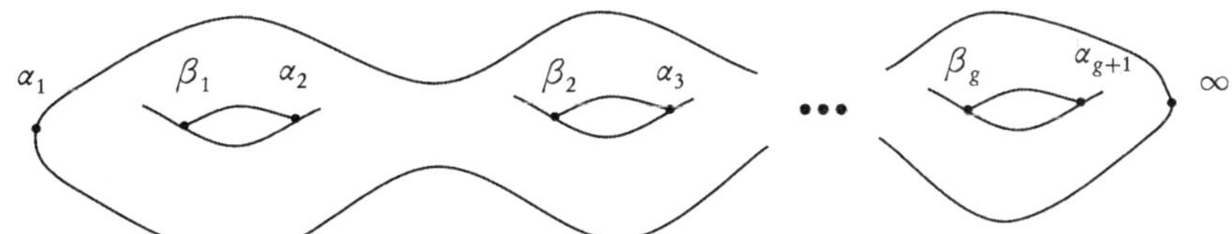

Figure 11.2. *A cartoon of the Riemann surface associated with the finite-gap potential* q_0. *Reprinted with permission from Elsevier* [113].

Lemma 11.9 (See [90]). *For every* $x_0\in\mathbb{R}$ *there exists two solutions* $\Psi_\pm$ *of* (11.4) *such that*

- $\Psi_\pm(x,\lambda)$ *is meromorphic with respect to* λ *on* $\Gamma\setminus\{\infty\}$ *with* g *poles and* g *zeros such that*

$$(\Psi_\pm)=D-D',$$

where

$$D=\sum_{i=1}^{g}Q_i,\quad Q_i\in a_i,\quad D'=\sum_{i=1}^{g}R_i,\quad R_i\in a_i.$$

- *Define* $\gamma_i=\hat{R}_i$. *Then* $\{\gamma_j\}_{j=1}^{g}$ *as functions of* x *satisfy the Dubrovin equations*

$$\gamma_j(x_0)\in[\beta_j,\alpha_{j+1}], \tag{11.5}$$

$$\frac{d\gamma_j}{dx}=-\frac{2i\sqrt{P(\gamma_j)}}{\prod_{k\neq j}(\gamma_j-\gamma_k)}. \tag{11.6}$$

- *The solutions are uniquely specified by* $\Psi_\pm(x,\lambda)=e^{\pm i\sqrt{\lambda}(x-x_0)}(1+\mathcal{O}(\lambda^{-1/2}))$ *as* $\lambda\to\infty$.

For simplicity we take $x_0=0$ below. We will always take the branch cut for $\lambda^{1/2}$ to be along $[0,\infty)$ and fix the branch by $\lambda^{1/2}\sim i|\lambda|^{1/2}$ as $\lambda\to-\infty$. If the potential $q_0(x)$ is taken as an initial condition for the KdV equation, then these zeros have both time and space dependence $\gamma_j=\gamma_j(x,t)$. This dependence is given by [44]

$$\dot{\gamma}_j=-\frac{8i(\gamma_j+q_0/2)\sqrt{P(\gamma_j)}}{\prod_{k\neq j}(\gamma_j-\gamma_k)}, \tag{11.7}$$

and the solution to the KdV equation can be reconstructed through

$$q(x,t) = -2\sum_{j=1}^{g}\gamma_j(x,t) + \alpha_{g+1} + \sum_{j=1}^{g}(\alpha_j + \beta_j).$$

The (now time-dependent) function $\Psi_\pm(x,t,\lambda)$ is known as a Baker–Akhiezer (BA) function. From the general theory of BA functions [90] it is known that it is uniquely determined by a nonspecial divisor

$$D = \sum_{i=1}^{g} Q_i$$

for the poles and the asymptotic behavior [90]. The following lemma shows that all divisors we consider are nonspecial.

Lemma 11.10. *On the hyperelliptic surface $w^2 = P(\lambda)$ the divisor $D = \sum_{i=1}^{g} R_i$, where $R_i \in a_i$ is nonspecial.*

Proof. Assume f is meromorphic with $D \leq (f)a$ and we show that f must be constant so that $l(D) = 1$ and the Riemann inequality is an equality. The differential $\omega = \mathrm{d}f$ has double poles with zero residues at the points R_i. We have the following representation:

$$\omega = \sum_{i=1}^{g} y_i + \eta.$$

Here y_i are Abelian differentials of the second kind, normalized so that they have zero periods along the a cycles and second-order poles at the points R_i. Since f is single-valued on the Riemann surface,

$$\oint_{a_k}\omega = 0, \quad \oint_{b_k}\omega = 0, \quad k = 1,\ldots,g.$$

Since the a periods vanish, we conclude that η has zero a periods and must be zero. From the b period condition we obtain

$$\oint_{b_k}\omega = \sum_{j=1}^{g} c_j \psi_{kj}(z_j(0)) = 0, \tag{11.8}$$

where z_j is a local parameter near $R_j = z_j(0)$ and ψ_{kj} is determined from the equality $\omega_k = \psi_{kj}(z_j)\mathrm{d}z_j$ near R_j. We know that ω_k can be expressed uniquely as the sum of differentials of the form

$$u_l = \frac{\lambda^{l-1}}{w}\mathrm{d}\lambda \text{ for } l = 1,\ldots,g,$$

with coefficients d_{kl}. If $R_j = (z_j, \pm\sqrt{P(z_j)})$ is not a branch point, we obtain

$$\psi_{kj}(z_j) = \sum_{l=1}^{g} d_{kl}\frac{z_j^{l-1}}{\sqrt{P(z_j)}}.$$

If it is a branch point $R_j = (z_j, 0)$, we use the local parameter $s = \sqrt{\lambda - z_j}$ so that

$$\psi_{kj}(z_j) = \sum_{l=1}^{g} d_{kl} \frac{2z_j^{l-1}}{\sqrt{P'(z_j)}}.$$

Since the matrix $d = (d_{kl})_{1\leq k,l\leq g}$ is invertible, the condition (11.8) is reduced to the study of the matrix

$$Z = (z_j^{l-1})_{1\leq j,l\leq g},$$

after multiplying rows by suitable constants. This is a Vandermonde system and thus is invertible. This shows that $c_j = 0$, $j = 1,2,\ldots,g$, and thus $w = 0$ and $f = C \in \mathbb{C}$. This proves the result. □

Remark 11.2.2. *We have shown that the Abel map is invertible from the Jacobi variety to the symmetrized Riemann surface in a neighborhood of $A(D)$ for every divisor we consider.*

Being precise, we obtain the following unique characterization of the function $\Psi_\pm(\lambda) = \Psi_\pm(x,t,\lambda)$ [90].

Definition 11.11. *The BA function for the solution of the KdV with initial condition $q_0(x)$ is the unique function such that the following hold:*

1. *$\Psi_\pm$ solves* (11.4).
2. *$\Psi_\pm$ is meromorphic on $\Gamma \setminus \{\infty\}$ with poles at*
$$D = \sum_{i=1}^{g} Q_i, \quad Q_i \in a_i, \quad \hat{Q}_i = \gamma_i(0,0). \tag{11.9}$$
3. $\Psi_\pm(\lambda) = \mathrm{e}^{\pm \mathrm{i}\lambda^{1/2}x \pm 4\mathrm{i}\lambda^{3/2}t}(1 + \mathcal{O}(\lambda^{-1/2}))$ *as* $\lambda \to \infty$.
4. $q_0(x) = -2\sum_{j=1}^{g}\gamma_j(x,0) + \alpha_{g+1} + \sum_{j=1}^{g}(\alpha_j + \beta_j)$.

We note that 1 and 3 are sufficient to uniquely specify the function $\Psi_\pm$.

Instead of computing the zeros of the BA function we derive a Riemann–Hilbert formulation of the BA function to compute the function itself. The main benefit of this approach is that the roles of x and t in the problem are reduced to that of parameters. This gives an approximation to the solution of the KdV equation that is uniformly convergent in the (x,t)-plane. In this sense our method is comparable to the theta function approach which can also achieve uniform convergence [56]. On the other hand, no time stepping is required, as for the direct numerical solution of the PDE or the numerical solution of (11.5) and (11.7).

In what follows we assume without loss of generality that $\alpha_1 = 0$. If $\alpha_1 \neq 0$, we define $\tau = \lambda - \alpha_1$ and consider a modified scattering problem

$$\begin{aligned} -\Psi_{xx} - q_0(x)\Psi &= (\tau + \alpha_1)\Psi, \\ -\Psi_{xx} - \tilde{q}_0(x)\Psi &= \tau\Psi, \quad \tilde{q}_0(x) = q_0(x) + \alpha_1. \end{aligned} \tag{11.10}$$

Let $q(x,t)$ and $\tilde{q}(x,t)$ be the solutions of the KdV equation with $q_0(x)$ and $\tilde{q}_0(x)$, respectively, as initial conditions. If $\tilde{q}(x,t)$ satisfies the KdV equation, then so does $\tilde{q}(x - 6ct, t) + c$. Therefore, by uniqueness, $q(x,t) = \tilde{q}(x + 6\alpha_1 t, t) - \alpha_1$.

11.3 ▪ From a Riemann surface of genus g to the cut plane

Consider the hyperelliptic Riemann surface Γ from Section 11.2. We represent a function $f_\pm$ defined on Γ by a vector-valued function f on $\hat{\mathbb{C}}$ by

$$f(\lambda)=\left[f_+(\check{\lambda}),\ f_-((\check{\lambda})^*)\right].$$

Assume the function $f_\pm$ is continuous on all of Γ. Let $\lambda\in(\alpha_j,\beta_j)$ and define $\lambda_{\pm\epsilon}=\lambda\pm i\epsilon$. It follows that $\lim_{\epsilon\downarrow 0}\check{\lambda}_{\pm\epsilon}=\lim_{\epsilon\downarrow 0}(\check{\lambda}_{\mp\epsilon})^*$. From the continuity of $f_\pm$

$$\lim_{\epsilon\downarrow 0} f_+(\check{\lambda}_{\pm\epsilon})=\lim_{\epsilon\downarrow 0} f_-((\check{\lambda}_{\mp\epsilon})^*).$$

Define the boundary values $f^\pm(\lambda)=\lim_{\epsilon\downarrow 0} f(\lambda_{\pm\epsilon})$. Then

$$f^+(\lambda)=f^-(\lambda)\begin{bmatrix} 0 & 1 \\ 1 & 0 \end{bmatrix}.$$

We form a planar representation of the BA function

$$\Psi(\lambda)=\left[f_+(\check{\lambda}),\ f_-((\check{\lambda})^*)\right].$$

The function $\Psi(\lambda)=\Psi(x,t,\lambda)$ is analytic in $\hat{\mathbb{C}}$ and satisfies

$$\Psi^+(\lambda)=\Psi^-(\lambda)\begin{bmatrix} 0 & 1 \\ 1 & 0 \end{bmatrix},\ \lambda\in(\alpha_{n+1},\infty)\cup\bigcup_{j=1}^{g}(\alpha_j,\beta_j),$$
$$\Psi(\lambda)=\left[e^{i\lambda^{1/2}x+4i\lambda^{3/2}},\ e^{-i\lambda^{1/2}x-4i\lambda^{3/2}}\right](I+\mathcal{O}(\lambda^{-1/2}))$$

as $\lambda\to\infty$. The next step is to remove the oscillatory nature of Ψ for large λ. This procedure will affect the jumps; thus some care is in order. Define

$$R(\lambda)=R(x,t,\lambda)=\begin{bmatrix} e^{-\zeta(x,t,\lambda)/2} & 0 \\ 0 & e^{\zeta(x,t,\lambda)/2} \end{bmatrix},$$
$$\zeta(x,t,\lambda)=2ix\lambda^{1/2}+8it\lambda^{3/2}.$$

The function $\Phi(\lambda)=\Phi(x,t,\lambda)=\Psi(x,t,\lambda)R(x,t,\lambda)$ satisfies

$$\begin{aligned}\Phi^+(\lambda)&=\Phi^-(\lambda)\begin{bmatrix} 0 & 1 \\ 1 & 0 \end{bmatrix},\ \lambda\in(\alpha_{n+1},\infty)\cup\bigcup_{j=1}^{g}(\alpha_j,\beta_j),\\ \Phi^+(\lambda)&=\Phi^-(\lambda)\begin{bmatrix} e^{-\zeta(x,t,\lambda)} & 0 \\ 0 & e^{\zeta(x,t,\lambda)} \end{bmatrix},\ \lambda\in\bigcup_{j=1}^{g}(\beta_j,\alpha_{j+1}),\\ \Phi(\lambda)&=[1,\ 1](I+\mathcal{O}(\lambda^{-1/2})).\end{aligned} \tag{11.11}$$

This is an RH problem for Φ when the poles at $\gamma_j(0,0)$ coincide with α_j or β_j. The boundary values of the solution to the RH problem should be at least locally integrable. A pole at a band end (α_j or β_j) corresponds to a square root singularity. In general, we have poles in the intervals (β_j,α_{j+1}) where there are smooth jumps. In Section 11.4.1 we treat the case where $\gamma_j(0,0)=\beta_j$, $j=1,\ldots,g$, while enforcing that Φ remains bounded at $\{\alpha_j\}_{j=1}^g$. No such enforcement is made at $\{\beta_j\}_{j=1}^g$. The general case of poles on the *a* cycles is treated in Section 11.4.2.

11.4 ▪ Regularization

We show how the jump conditions in (11.11) can be reduced to piecewise constant jumps. As mentioned above, we first perform the calculations in the simpler case when the poles are located at $(\beta_j, 0)$ on Γ. In the general case, we use an additional BA function as a parametrix to move the poles to the band ends, thus reducing the problem to the first case.

11.4.1 ▪ All poles at the band ends

We assume $\gamma_j(0,0) = \beta_j$. Define the g-function

$$\mathcal{G}(\lambda) \triangleq \mathcal{G}(x,t,\lambda) \triangleq \sqrt{P(\lambda)} \sum_{j=1}^{g} \int_{\beta_j}^{\alpha_{j+1}} \frac{-\zeta(x,t,s) + \mathrm{i}\Omega_j(x,t)}{\sqrt{P(s)}^{+}} \frac{\mathrm{d}s}{s-\lambda}, \tag{11.12}$$

where $\Omega_j(x,t)$ is constant in λ and will be determined below.

Lemma 11.12. *The g-function satisfies*

- $\mathcal{G}^+(\lambda) + \mathcal{G}^-(\lambda) = 0$ *for* $\lambda \in (\alpha_j, \beta_j)$,
- $\mathcal{G}^+(\lambda) - \mathcal{G}^-(\lambda) = -\zeta(x,t,\lambda) + \mathrm{i}\Omega_j(x,t)$ *for* $\lambda \in (\beta_j, \alpha_{j+1})$,
- $\mathcal{G}(\lambda)/\sqrt{P(\lambda)} = \sum_{k=1}^{g} m_k(x,t)\lambda^{-k} + \mathcal{O}(\lambda^{-g-1})$ *as* $\lambda \to \infty$, *where*

$$m_k(x,t) = -\sum_{j=1}^{g} \int_{\beta_j}^{\alpha_{j+1}} \frac{-\zeta(x,t,s) + \mathrm{i}\Omega_j(x,t)}{\sqrt{P(s)}^{+}} s^{k-1} \mathrm{d}s.$$

Proof. The first two properties follow from the branching properties of $\sqrt{P(\lambda)}$ and the Plemelj lemma, Lemma 2.7. The last property follows from Lemma 2.12. □

Define the matrix function

$$G(\lambda) \triangleq G(x,t,\lambda) \triangleq \begin{bmatrix} \mathrm{e}^{-\mathcal{G}(x,t,\lambda)} & 0 \\ 0 & \mathrm{e}^{\mathcal{G}(x,t,\lambda)} \end{bmatrix}$$

and consider the function

$$\Lambda(\lambda) \triangleq \Lambda(x,t,\lambda) \triangleq \Phi(x,t,\lambda) G(x,t,\lambda).$$

Using Lemma 11.12 we compute the jumps of Λ:

$$\Lambda^+(\lambda) = \Lambda^-(\lambda) \begin{bmatrix} 0 & 1 \\ 1 & 0 \end{bmatrix}, \quad \lambda \in (\alpha_{g+1}, \infty) \cup \bigcup_{j=1}^{g} (\alpha_j, \beta_j), \tag{11.13}$$

$$\Lambda^+(\lambda) = \Lambda^-(\lambda) \begin{bmatrix} \mathrm{e}^{\mathrm{i}\Omega_j(x,t)} & 0 \\ 0 & \mathrm{e}^{-\mathrm{i}\Omega_j(x,t)} \end{bmatrix}, \quad \lambda \in \bigcup_{j=1}^{g} (\beta_j, \alpha_{j+1}). \tag{11.14}$$

Since $\sqrt{P(\lambda)} = \mathcal{O}(|\lambda|^{g+1/2})$, G has growth in λ at ∞ unless $m_k(x,t) = 0$ for $k = 1, \ldots, g$. We wish to determine $\{\Omega_j\}_{j=1}^{g}$ so that Λ has the same asymptotic behavior as Φ as $\lambda \to \infty$;

see (11.11). Thus, we must solve the following problem, which we put in slightly more abstract terms since we make use of it again below.

Problem 11.4.1. *Given continuous functions*

$$f_j : [\beta_j, \alpha_{j+1}] \to \mathbb{C}, \quad j = 1, \ldots, g,$$

we seek constants Ω_j *satisfying the moment conditions*

$$0 = \sum_{j=1}^{g} \int_{\beta_j}^{\alpha_{j+1}} \frac{-f_j(\lambda) + \mathrm{i}\Omega_j}{\sqrt{P(\lambda)}^+} \lambda^{k-1} \mathrm{d}\lambda, \quad k = 1, \ldots, g.$$

Theorem 11.13. *Problem* 11.4.1 *has a unique solution. Further, if each* f_j *takes purely imaginary values, then each* Ω_j *is real-valued.*

Proof. The second claim follows from the fact that $\sqrt{P(\lambda)}^+$ takes purely imaginary values in the gaps, $[\beta_j, \alpha_{j+1}]$. To establish the first claim, notice that Problem 11.4.1 is equivalent to the linear system

$$M\Omega = V, \quad (M)_{kj} = \mathrm{i} \int_{\beta_j}^{\alpha_{j+1}} \frac{\lambda^{k-1}}{\sqrt{P(\lambda)}} \mathrm{d}\lambda,$$
$$(\Omega)_j = \Omega_j, \quad (V)_k = \sum_{j=1}^{g} \int_{\beta_j}^{\alpha_{j+1}} \frac{f_j(\lambda)}{\sqrt{P(\lambda)}} \lambda^{k-1} \mathrm{d}\lambda.$$

Assume the rows of M are linearly dependent. Then there exist constants $\{d_k\}$ such that

$$\sum_{k=1}^{g} d_k M_{kj} = 0 \text{ for } j = 1, \ldots, g.$$

Explicitly, this implies

$$0 = \sum_{k=1}^{g} \int_{\beta_j}^{\alpha_{j+1}} \frac{d_k}{\sqrt{P(\lambda)}} \lambda^{k-1} \mathrm{d}\lambda = \int_{\beta_j}^{\alpha_{j+1}} \left(\sum_{k=1}^{g} d_k \lambda^{k-1} \right) \frac{\mathrm{d}\lambda}{\sqrt{P(\lambda)}} \text{ for } j = 1, \ldots, g.$$

We show this implies that the holomorphic differential

$$f = \sum_{k=1}^{g} d_k \lambda^{k-1} \frac{\mathrm{d}\lambda}{\sqrt{P(\lambda)}} = \sum_{k=1}^{g} d_k \lambda^{k-1} \frac{\mathrm{d}\lambda}{w}$$

has zero a periods. Compute

$$\int_{\alpha_j}^{\beta_{j+1}} \frac{\lambda^{k-1}}{w} \mathrm{d}\lambda = \frac{1}{2} \left(\int_{\alpha_j}^{\beta_{j+1}} \frac{\lambda^{k-1}}{w} \mathrm{d}\lambda + \int_{\beta_{j+1}}^{\alpha_j} \frac{\lambda^{k-1}}{-w} \mathrm{d}\lambda \right) = \frac{1}{2} \oint_{a_j} \frac{\lambda^{k-1}}{w} \mathrm{d}\lambda.$$

Indeed, f integrates to zero around every a cycle, implying that f is the zero differential; see Lemma 11.3. But since each of $\lambda^{k-1} w^{-1} \mathrm{d}\lambda$ is linearly independent, we conclude that $d_k = 0$, $k = 1, \ldots, g$, and the rows of M are linearly independent. The linear system is uniquely solvable. □

If we select Ω_j to make all m_k vanish, we use the condition

$$\lim_{\lambda\to\infty} \Lambda(x,t,\lambda) = [1,\ 1]$$

in conjunction with (11.13) to obtain an RH problem for Λ. It is important that Ω in Problem 11.4.1 is real-valued. This implies the piecewise-constant jump matrix in (11.13) is bounded for all x and t.

11.4.2 ▪ Poles in the gaps

In this section we show how to use an additional BA function to, in effect, reduce the case where $\gamma_j(0,0) \in (\beta_j, \alpha_{j+1})$ to that of $\gamma_j(0,0) = \beta_j$. We assume that not all poles of $\Psi_\pm$ lie on the band ends $\{\beta_j\}_{j=1}^g$. Consider the planar representation of a BA function $(\Psi_p)_\pm$ which satisfies

$$\Psi_p^+(\lambda) = \Psi_p^-(\lambda)\begin{bmatrix} 0 & 1 \\ 1 & 0 \end{bmatrix},\ \lambda \in (\alpha_{g+1}, \infty) \cup \bigcup_{j=1}^{g} (\alpha_j, \beta_j),$$

$$\Psi_p(\lambda) = \left[e^{\kappa(\lambda)/2}, e^{-\kappa(\lambda)/2}\right](I + \mathcal{O}(\lambda^{-1/2})),\quad \kappa(\lambda) = \sum_{j=1}^{g} i t_j \lambda^{j-1/2},\ \ t_j \in \mathbb{R},$$

with poles at β_j. The goal is to choose $\{t_j\}_{j=1}^g$ so that $(\Psi_p)_\pm$ has zeros precisely at the poles of $\Psi_\pm$. Define $(\Psi_r)_\pm = \Psi_\pm(\Psi_p)_\pm$. The planar representation $\Psi_r = \Psi\Psi_p$ with entrywise multiplication will now have poles at β_j and zeros at the zeros of Ψ. We find $\Psi_\pm$ by first finding the two functions $(\Psi_p)_\pm$ and $(\Psi_r)_\pm$, both of which have poles at $\{(0,\beta_j)\}_{j=1}^g$, and dividing. Thus, the general case of poles in gaps is reduced to poles at band ends provided we can find the required $\{t_j\}_{j=1}^g$.

Remark 11.4.1. *We are using the term "poles" loosely. On $\hat{\mathbb{C}}$, Ψ_p has unbounded square root singularities, while on Γ, $(\Psi_p)_\pm$ has poles.*

We show that we can choose $\{t_j\}_{j=1}^g$ so that the zeros of $(\Psi_p)_\pm$ will be at an arbitrary divisor

$$D' = \sum_{j=1}^{g} R_j,\ \ R_j \in a_j.$$

We first state a lemma about the location of the zeros and poles of a BA function from [11, p. 44].

Lemma 11.14. *Let D' be the divisor of the zeros of the BA function and D be that of the poles. Assume*

$$\Psi_\pm(\lambda) = e^{\ell(z)}(1 + \mathcal{O}(z^{-1})),\ \ z \to \infty,\ \ z^2 = \lambda. \tag{11.15}$$

Then, on the Jacobi variety $J(\Gamma)$,

$$A(D') = A(D) - V, \tag{11.16}$$

where V is the vector of the b-periods of a normalized Abelian differential of the second kind ν that satisfies

$$\nu(Q) = d\ell(z) + \mathcal{O}(z^{-2})dz, \quad z = z(Q) \to \infty, \tag{11.17}$$

$$\oint_{a_l} \nu = 0, \quad (V)_l = \oint_{b_l} \nu, \quad l = 1, \ldots, g. \tag{11.18}$$

Conversely, if two divisors satisfy (11.16), *then they are the divisors of the poles and zeros of some BA function which satisfies* (11.15).

To determine Ψ_p we have D and D'. We need to show we can choose $\nu = d\kappa + \mathcal{O}(z^{-2})dz$ so that (11.16) holds. The following lemma provides this result.

Lemma 11.15. *Assume*

$$D = \sum_{j=1}^{g} Q_j, \quad D' = \sum_{j=1}^{g} R_j, \quad Q_j, R_j \in a_j.$$

Then there exist real constants $\{t_j\}_{j=1}^{g}$ so that the differential

$$\nu = \sum_{j=1}^{g} t_j \nu_j$$

satisfies the properties in (11.17) *with $\ell(z) = \kappa(z)$, and ν_j can be constructed explicitly.*

Proof. Recall that the terms with negative exponents in a Laurent expansion of a function $f(z)$ about a point are called the *principal part*. The principal part about infinity is found by finding the principal part of $f(1/z)$ about $z = 0$. Define τ_j to be the Abelian differential of the second kind with principal part (see Section 11.1)

$$\tau_j = (2j-1)\left(z^{2j-3} + \mathcal{O}(z^{-2})\right)dz, \quad z \to \infty,$$

where $1/z$ is a parameter in the neighborhood of ∞. For $j \geq 1$, we choose a path of integration that lies on one sheet. We have

$$\int_{\lambda_0}^{\lambda} \tau_j = \pm \lambda^{j-1/2}(1 + \mathcal{O}(\lambda^{j-3/2})) \text{ as } \lambda \to \infty.$$

Define

$$\nu = \sum_{j=1}^{g} \mathrm{i} t_j (\tau_j + \eta_j),$$

where η_j is a holomorphic differential chosen so that $\tau_j + \eta_j$ has vanishing a periods. We define $\nu_j = \mathrm{i}(\tau_j + \eta_j)$. Consider the system of equations

$$\oint_{b_k} \nu = (V)_k, \quad k = 1, \ldots, g.$$

It follows that (see Lemma 11.8)

$$\oint_{b_k} \nu = \sum_{j=1}^{g} \mathrm{i}t_j \frac{1}{(2j-2)!} \frac{\mathrm{d}^{2j-2}}{\mathrm{d}z^{2j-2}} r_k(z)\bigg|_{z=0}, \quad k=1,\ldots,g.$$

Here z is a local parameter in the neighborhood of ∞: $z(\infty)=0$ and $\omega_k = r_k(z)\mathrm{d}z$. To compute these derivatives we again use a convenient basis, not normalized, for the holomorphic differentials:

$$u_j = \frac{\lambda^{j-1}}{w}\mathrm{d}\lambda, \quad j=1,\ldots,g.$$

Set $\lambda = 1/z^2$ and compute

$$u_j = -2z^{2(g-j)}\left((1-\alpha_{j+1}z)\prod_{i=1}^{g}(1-\alpha_i z)(1-\beta_i z)\right)^{-1/2} \mathrm{d}z = s_j(z)\mathrm{d}z.$$

It is clear that the matrix

$$(A)_{kj} = \frac{\mathrm{d}^{2k-2}}{\mathrm{d}z^{2k-2}} s_j(z)$$

is triangular with nonvanishing diagonal entries. There exists an invertible linear transformation from $\{u_j\}_{j=1}^{g}$ to $\{\omega_k\}_{k=1}^{g}$, and since A is invertible, it follows that the system

$$\oint_{b_k} \nu = (V)_k \text{ for } k=1,\ldots,g \tag{11.19}$$

is uniquely solvable for $\{t_j\}_{j=1}^{g}$. □

This proves the existence of a BA function with asymptotic behavior (11.15) and one arbitrary zero on each a cycle. In summary, the BA function $(\Psi_r)_\pm = (\Psi_p)_\pm \Psi_\pm$ has poles located at $(\beta_j, 0)$ and one zero on each a cycle corresponding to the zeros of $\Psi_\pm$. We show below how to compute such a BA function. We use the approach of Section 11.3 to formulate an RH problem for $(\Psi_p)_\pm$:

$$\Lambda_p^+(\lambda) = \Lambda_p^-(\lambda)\begin{bmatrix} 0 & 1 \\ 1 & 0 \end{bmatrix}, \ \lambda \in (\alpha_{n+1}, \infty) \cup \bigcup_{j=1}^{g} (\alpha_j, \beta_j), \tag{11.20}$$

$$\Lambda_p^+(\lambda) = \Lambda_p^-(\lambda)\begin{bmatrix} e^{\mathrm{i}W_j} & 0 \\ 0 & e^{-\mathrm{i}W_j} \end{bmatrix}, \ \lambda \in \bigcup_{j=1}^{g} (\beta_j, \alpha_{j+1}), \tag{11.21}$$

where each of the $W_j \in \mathbb{R}$ is chosen so that the g-function

$$\mathcal{G}_p(\lambda) = \sqrt{P(\lambda)} \sum_{j=1}^{g} \int_{\beta_j}^{\alpha_{j+1}} \frac{-\kappa(s) + \mathrm{i}W_j}{\sqrt{P(s)}^{+}} \frac{\mathrm{d}s}{s-\lambda} \tag{11.22}$$

satisfies $\mathcal{G}_p(\lambda) = \mathcal{O}(\lambda^{-1/2})$ as $\lambda \to \infty$. Theorem 11.13 provides a well-defined map from $\{t_j\}_{j=1}^{g}$ to $\{W_j\}_{j=1}^{g}$. Furthermore, each W_j can be taken modulo 2π. The RH problem for $(\Psi_r)_\pm$ is similar, but $\kappa(\lambda)$ must be replaced with $\kappa(\lambda) + 2\mathrm{i}x\lambda^{1/2} + 8\mathrm{i}t\lambda^{3/2}$ to account for the x and t dependence in $\Psi_\pm$. In this case we write $W_j(x,t)$. We elaborate on this below.

11.5 ▪ A Riemann–Hilbert problem with smooth solutions

The numerical method described in Chapter 6 requires solutions of the RH problem to be smooth. We need to deform the RH problem to take into account the singularities explicitly if we wish to solve it numerically. One should compare the deformations that follow with those used for the monodromy problem in Section 6.4. In this section, we assume the divisor for the poles of the BA function is

$$D = \sum_{j=1}^{g} (\beta_j, 0)$$

and that the $\Omega_j(x,t)$ are chosen so that the moment conditions for $\mathcal{G}$ are satisfied. We replace $\Omega_j(x,t)$ with W_j and $W_j(x,t)$ in the case of $(\Psi_p)_\pm$ and $(\Psi_r)_\pm$, respectively. In light of the previous section, all other cases can be reduced to this. Define $\Delta(\lambda) = \Delta(x,t,\lambda)$ by

$$\Delta(x,t,\lambda) = \begin{bmatrix} \delta(x,t,\lambda) & 0 \\ 0 & 1/\delta(x,t,\lambda) \end{bmatrix}, \quad \delta(x,t,\lambda) = \prod_{j=1}^{g} \left(\frac{\lambda - \alpha_{j+1}}{\lambda - \beta_j} \right)^{\Omega_j(x,t)/(2\pi)}.$$

The branch cut for δ is to be along the intervals $[\beta_j, \alpha_{j+1}]$, and we assume $\Omega_j(x,t) \in [0, 2\pi)$. Note that Δ satisfies

$$\Delta^+(\lambda) = \Delta^-(\lambda) \begin{bmatrix} e^{i\Omega_j(x,t)} & 0 \\ 0 & e^{-i\Omega_j(x,t)} \end{bmatrix}, \quad \lambda \in (\beta_j, \alpha_{j+1}).$$

Define (see (6.6))

$$H(\lambda) = \frac{1}{2} \begin{bmatrix} 1 & 1 + \sqrt{\lambda - \alpha_{n+1}} \\ 1 & 1 - \sqrt{\lambda - \alpha_{n+1}} \end{bmatrix},$$

where the function $\sqrt{\lambda - \alpha_{n+1}}$ has its branch cut on $[\alpha_{n+1}, \infty)$ and satisfies $\sqrt{\lambda - \alpha_{n+1}} \sim i|\lambda|^{1/2}$ as $\lambda \to -\infty$ to fix the branch. The last function we need is the g-function matrix

$$G(\lambda) = G(x,t,\lambda) = \begin{bmatrix} e^{-\mathcal{G}(x,t,\lambda)} & 0 \\ 0 & e^{\mathcal{G}(x,t,\lambda)} \end{bmatrix}.$$

Note that if we were solving for $(\Psi_p)_\pm$ or $(\Psi_r)_\pm$, we would replace $\mathcal{G}$ with (11.22).

We introduce a local parametrix for what follows. Consider the RH problem

$$Y^+(\lambda) = Y^-(\lambda) \begin{bmatrix} 0 & c \\ 1/c & 0 \end{bmatrix}, \quad \lambda \in (a,b), \tag{11.23}$$

where we do not specify the asymptotic behavior since we wish to obtain multiple solutions. From Example 2.17 we find that

$$Y(\lambda; a, b, \alpha, \beta, c) = \begin{bmatrix} -i(\lambda - a)^\alpha (\lambda - b)^\beta / c & i(\lambda - a)^\alpha (\lambda - b)^\beta \\ 1/c & 1 \end{bmatrix}, \quad \alpha, \beta = \pm \frac{1}{2},$$

is a solution of (11.23). We choose the branch cut of $(\lambda - a)^\alpha (\lambda - b)^\beta$ to be along the interval $[a,b]$ with $(\lambda - a)^\alpha (\lambda - b)^\beta \sim |\lambda|^{\alpha+\beta}$ as $\lambda \to +\infty$. To simplify notation we

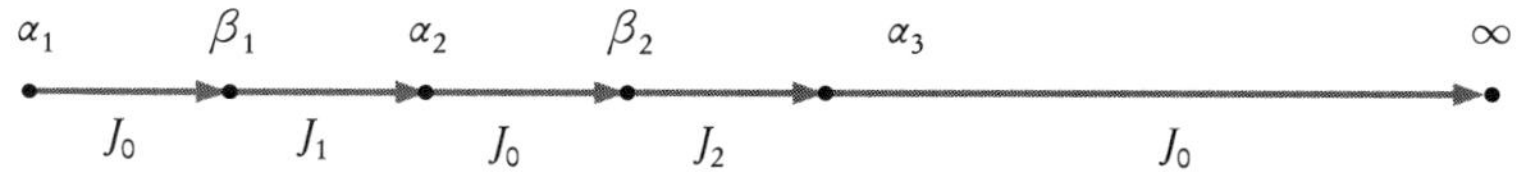

Figure 11.3. *The contours and jump matrices of the RH problem for* Λ.

define $J_j(x,t) = \operatorname{diag}(e^{-i\Omega_j(x,t)}, e^{i\Omega_j(x,t)})$ and

$$J_0 = \begin{bmatrix} 0 & 1 \\ 1 & 0 \end{bmatrix}.$$

We need a local parametrix at each point α_j or β_j because the RH problem does not satisfy the product condition. This motivates the definition, suppressing (x,t) dependence,

$$\begin{aligned} A_1(\lambda) &= Y(\lambda;\alpha_1,\beta_1,1/2,-1/2,1), \\ A_j(\lambda) &= Y(\lambda;\alpha_j,\beta_j,1/2,-1/2,e^{-i\Omega_{j-1}(x,t)}), \quad j=2,\ldots,g+1, \\ B_j(\lambda) &= Y(\lambda;\alpha_j,\beta_j,1/2,-1/2,e^{-i\Omega_j(x,t)}), \quad j=1,\ldots,g. \end{aligned}$$

This allows us to enforce boundedness at each α_j with a possibly unbounded singularity at β_j. The matrices A_j are used locally at α_j and B_j at β_j.

Consider the following example. The general case can be inferred from this.

Example 11.16 (Genus two). Our initial RH problem is (11.13) with the condition

$$\lim_{\lambda\to\infty} \Lambda(\lambda) = [1,\ 1];$$

see Figure 11.3. First, we introduce a circle around $\alpha_3 = \alpha_{g+1}$. In addition we place a large circle around all the gaps; see Figure 11.4. Now, we redefine our function Λ in various regions. Define Λ_1 by the piecewise definition in Figure 11.5(a). We compute the jumps satisfied by Λ_1; see Figure 11.5(b). An important calculation is that if $\Lambda_1(\lambda) = \Lambda_1(x,t,\lambda) = [1,\ 1] + \mathcal{O}(\lambda^{-1})$, then

$$\Lambda_1(\lambda)H(\lambda) = \frac{1}{2}([1,\ 1] + \mathcal{O}(\lambda^{-1}))\begin{bmatrix} 1 & 1+\sqrt{\lambda-\alpha_{n+1}} \\ 1 & 1-\sqrt{\lambda-\alpha_{n+1}} \end{bmatrix} = [1,\ 1] + \mathcal{O}(\lambda^{-1/2}).$$

This allows us to obtain functions with the correct asymptotic behavior.

We present the deformation in the interior of the large circle in Figure 11.5(a). See Figure 11.6(a) for the piecewise definition of Λ_2 and Figure 11.6(b) for the jumps and jump contours for Λ_2. While this RH problem can be solved numerically, we make a final deformation to reduce the number of contours present. Define $\mathcal{D}$ to be the region inside the large outer circle but outside each of the smaller circles around α_j, β_j. Then define

$$\Lambda_3(\lambda) = \Lambda_3(x,t,\lambda) = \begin{cases} \Lambda_2(x,t,\lambda)\Delta^{-1}(x,t,\lambda) & \text{if } \lambda \in \mathcal{D}, \\ \Lambda_2(x,t,\lambda) & \text{otherwise.} \end{cases}$$

See Figure 11.7 for the jumps and jump contours of the RH problem for Λ_3. We refer to this as the deformed and regularized RH problem associated with $\Psi_\pm$.

This resulting RH problem has smooth solutions by the theory in Chapter 2: the RH problem is k-regular for all k. Furthermore, the uniqueness of the BA function gives

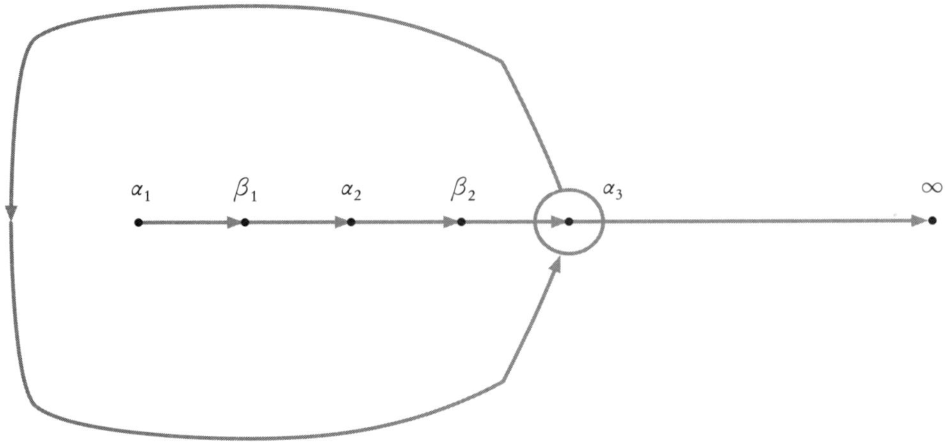

Figure 11.4. *Introducing a large circle around α_j and β_j.*

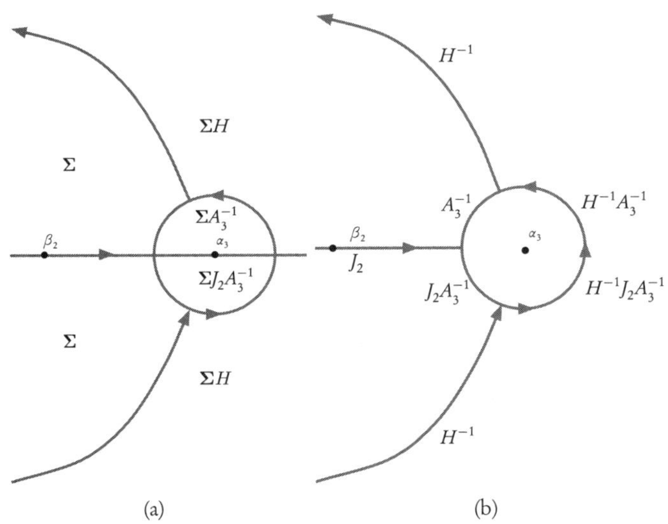

Figure 11.5. (a) *The piecewise definition of Λ_1.* (b) *The jump contours and jump matrices for the RH problem for Λ_1.*

us existence and uniqueness of the solution of this RH problem. See Section 11.7 for a more detailed discussion of the solvability of the RH problem. This justifies solving for Λ_3 numerically. ∎

11.5.1 ▪ Reconstruction of the solution to the KdV equation

Once the function Λ_3 above is known (at least numerically) we want to extract from it the solution of the KdV equation. We use that Λ_3 is analytic at infinity and that each component of Ψ satisfies (11.4). For λ large we write

$$\Lambda_3(\lambda) = \Psi(\lambda)R(\lambda)G(\lambda)H(\lambda). \tag{11.24}$$

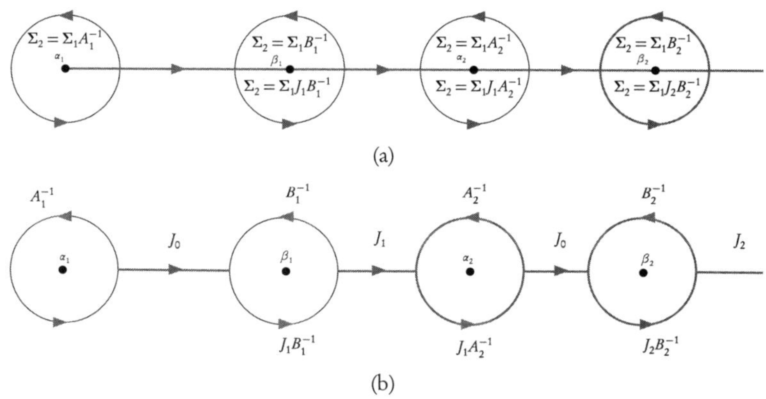

Figure 11.6. (a) *The piecewise definition of Λ_2 inside the outer circle.* (b) *The jump contours and jump matrices for the RH problem for Λ_2.*

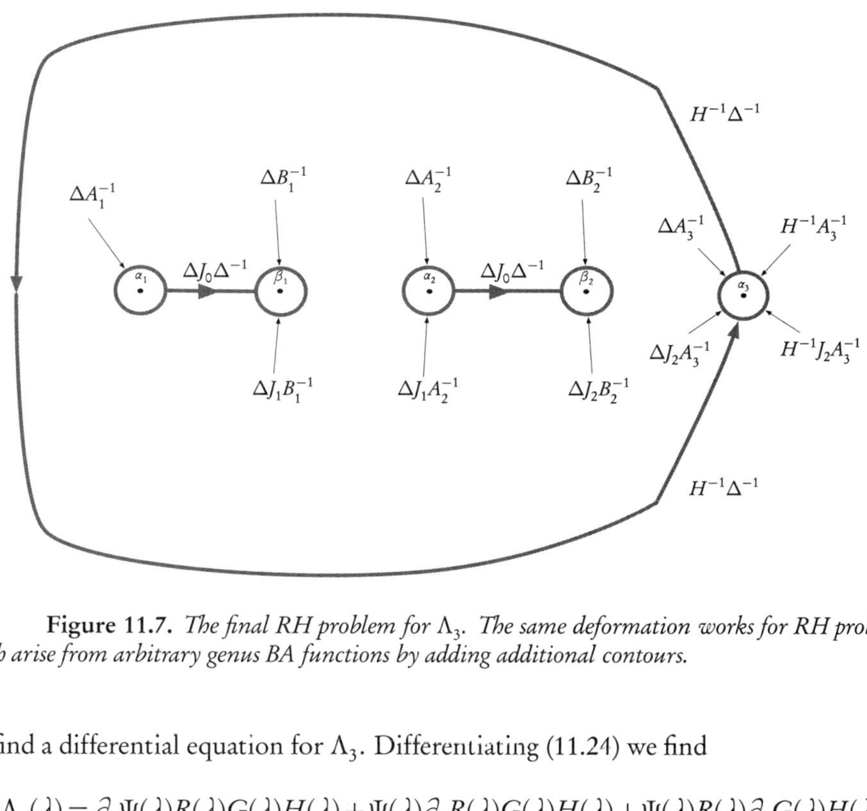

Figure 11.7. *The final RH problem for Λ_3. The same deformation works for RH problems which arise from arbitrary genus BA functions by adding additional contours.*

We find a differential equation for Λ_3. Differentiating (11.24) we find

$$\begin{aligned}\partial_x \Lambda_3(\lambda) &= \partial_x \Psi(\lambda)R(\lambda)G(\lambda)H(\lambda) + \Psi(\lambda)\partial_x R(\lambda)G(\lambda)H(\lambda) + \Psi(\lambda)R(\lambda)\partial_x G(\lambda)H(\lambda).\\ \partial_x^2 \Lambda_3(\lambda) &= \partial_x^2 \Psi(\lambda)R(\lambda)G(\lambda)H(\lambda) + \Psi(\lambda)\partial_x^2 R(\lambda)G(\lambda)H(\lambda) + \Psi(\lambda)R(\lambda)\partial_x^2 G(\lambda)H(\lambda)\\ &\quad + 2\Psi(\lambda)\partial_x R(\lambda)\partial_x G(\lambda)H(\lambda) + 2\partial_x \Psi(\lambda)\partial_x R(\lambda)G(\lambda)H(\lambda)\\ &\quad + 2\partial_x \Psi(\lambda)R(\lambda)\partial_x G(\lambda)H(\lambda).\end{aligned}$$

We seek to simplify this formula. Define $r(\lambda) = \operatorname{diag}(2i\lambda^{1/2}, -2i\lambda^{1/2})$, then

$$\partial_x R(\lambda) = r(\lambda)R(\lambda), \quad \partial_x^2 R(\lambda) = r^2(\lambda)R(\lambda).$$

It follows that each $\Omega_j(x,t)$ depends linearly on x. Define $g(\lambda)=\mathrm{diag}(-\partial_x\mathcal{G}(\lambda),\partial_x\mathcal{G}(\lambda))$; therefore

$$\partial_x G(\lambda)=g(\lambda)G(\lambda),\quad \partial_x^2 G(\lambda)=g^2(\lambda)G(\lambda).$$

Also, R,G,r, and g are diagonal and mutually commute. We write

$$\begin{aligned}\partial_x\Lambda_3(\lambda)&=\partial_x\Psi(\lambda)R(\lambda)G(\lambda)H(\lambda)+\Psi(\lambda)R(\lambda)G(\lambda)H(\lambda)H^{-1}(\lambda)[r(\lambda)+g(\lambda)]H(\lambda),\\ \partial_x^2\Lambda_3(\lambda)&=\partial_x^2\Psi(\lambda)R(\lambda)G(\lambda)H(\lambda)+2\partial_x\Psi(\lambda)R(\lambda)G(\lambda)H(\lambda)H^{-1}(\lambda)[r(\lambda)+g(\lambda)]H(\lambda)\\ &\quad+\Psi(\lambda)R(\lambda)G(\lambda)H(\lambda)H^{-1}(\lambda)[g^2(\lambda)+2g(\lambda)r(\lambda)+r^2(\lambda)]H(\lambda).\end{aligned}$$

We proceed to eliminate Ψ. Since $\partial_x^2\Psi=-\lambda\Psi-q(x,t)\Psi$, we obtain

$$\begin{aligned}\partial_x^2\Lambda_3(\lambda)=&[-\lambda-q(x,t)]\Lambda_3(\lambda)+2\partial_x\Lambda_3(\lambda)H^{-1}(\lambda)[g(\lambda)+r(\lambda)]H(\lambda)\\ &-\Lambda_3(\lambda)H^{-1}(\lambda)[g(\lambda)+r(\lambda)]^2H(\lambda).\end{aligned}\tag{11.25}$$

Set $\Lambda_3(\lambda)=[1,\ 1]+c_1(x,t)/\lambda+\mathcal{O}(\lambda^{-2})$ and substitute, assuming each derivative of Λ_3 has an induced asymptotic expansion,

$$\begin{aligned}\partial_x^2 c_1(x,t)/\lambda+\mathcal{O}(\lambda^{-2})=&[-\lambda-q(x,t)]([1,\ 1]+c_1(x,t)/\lambda+\mathcal{O}(\lambda^{-2}))\\ &+([1,\ 1]+\partial_x c_1(x,t)/\lambda+\mathcal{O}(\lambda^{-2}))H^{-1}(\lambda)[g(\lambda)+r(\lambda)]H(\lambda)\\ &+([1,\ 1]+c_1(x,t)/\lambda+\mathcal{O}(\lambda^{-2}))H^{-1}(\lambda)[g(\lambda)+r(\lambda)]^2H(\lambda).\end{aligned}$$

It can be shown that the $\mathcal{O}(\lambda)$ terms on each side of this equation cancel. Equating the $\mathcal{O}(1)$ terms we obtain

$$\begin{aligned}q(x,t)[1,\ 1]=&-\lim_{\lambda\to\infty}\partial_x c_1(x,t)/\lambda H^{-1}(\lambda)[g(\lambda)+r(\lambda)]H(\lambda)\\ &-\lim_{\lambda\to\infty}\left([-\lambda,\ -\lambda]+([1,\ 1]+c_1(x,t)/\lambda+\mathcal{O}(\lambda^{-2}))H^{-1}(\lambda)[g(\lambda)+r(\lambda)]^2H(\lambda)\right).\end{aligned}$$

Equating $\partial_x c_1(x,t)=[s_1(x,t),\ s_2(x,t)]$ and working this out explicitly, we find

$$q(x,t)=2\mathrm{i}(s_2(x,t)-s_1(x,t))+2\mathrm{i}E,\tag{11.26}$$
$$E=-\frac{1}{2\pi}\sum_{n=1}^{g}\int_{\beta_n}^{\alpha_{n+1}}\frac{\partial_x\Omega_n(x,t)-2\lambda^{1/2}}{\sqrt{P(\lambda)}^+}\lambda^g\,\mathrm{d}\lambda.$$

11.5.2 ▪ Regularization of the RH problem with poles in the gaps

In this section we deal with the case where the divisor for the poles of the BA function is of the form

$$D=\sum_{i=1}^{g}Q_i,\quad Q_i\in a_i.$$

We have proved the existence of a BA function with one arbitrary zero on each a cycle. We consider the BA function $(\Psi_r)_\pm=(\Psi_p)_\pm\Psi_\pm$ which has poles located at $\{(0,\beta_j)\}_{j=1}^g$ and one zero on each a cycle. In this section we assume we know $t_1,t_2,\ldots$, which are required to find $(\Psi_p)_\pm$. In the next section we discuss computing $\{t_j\}_{j=1}^g$. It follows that

$$(\Psi_r)_\pm\sim\mathrm{e}^{\pm Z(x,t,\lambda)/2},\tag{11.27}$$

where

$$Z(x,t,\lambda)=\kappa(\lambda)+2\mathrm{i}x\lambda^{1/2}+8\mathrm{i}t\lambda^{3/2}=2\mathrm{i}(x+t_1)\lambda^{1/2}+2\mathrm{i}(4t+t_2)\lambda^{3/2}+2\mathrm{i}\sum_{j=3}^{g}t_j\lambda^{(2j-1)/2}.$$

Using the techniques in Section 11.5.1 we see that this is all the information needed to set up a solvable RH problem for $(\Psi_r)_\pm$ with smooth solutions.

We have to extract the solution to the KdV equation from $(\Psi_r)_\pm$. We solve for a function $\Lambda_3(\lambda)=\Lambda_3(x,t,\lambda)$, the deformation of Ψ_r, that satisfies

$$\Lambda_3(x,t,\lambda)=\Psi(x,t,\lambda)R(x,t,\lambda)G(x,t,\lambda)\Psi_p'(\lambda)H(\lambda),\quad \Psi_p'(\lambda)=\operatorname{diag}\Psi_p(\lambda), \tag{11.28}$$

for large λ. If we perform the same calculations which result in (11.25), we obtain

$$\begin{aligned}\partial_x^2\Lambda_3(\lambda)&=[-\lambda-q(x,t)]\Lambda_3(\lambda)\\&\quad+2\partial_x\Lambda_3(\lambda)H^{-1}(\lambda)(\Psi_p'(\lambda))^{-1}[g(\lambda)+r(\lambda)]\Psi_p'(\lambda)H(\lambda)\\&\quad-\Lambda_3(\lambda)H^{-1}(\lambda)(\Psi_p'(\lambda))^{-1}[g(\lambda)+r(\lambda)]^2\Psi_p'(\lambda)H(\lambda).\end{aligned} \tag{11.29}$$

But Ψ_p' is diagonal and commutes with g and h. Therefore, all Ψ_p' dependence cancels out. We see that (11.26) is invariant under multiplication by $(\Psi_p)_\pm$. Thus, the solution $q(x,t)$ to the KdV equation is extracted from Λ_3 by (11.26). We summarize our results in the following theorem.

Theorem 11.17. *If $\Lambda_3(\lambda)$ is the solution of the deformed and regularized RH problem associated with $(\Psi_r)_\pm$ and*

$$\Lambda_3(\lambda)=[1,\ 1]+c_1(x,t)\lambda^{-1}+\mathcal{O}(\lambda^{-2}),\quad c_1(x,t)=[s_1(x,t),\ s_2(x,t)],$$

then the corresponding solution of the KdV equation is found through

$$q(x,t)=2\mathrm{i}(s_2(x,t)-s_1(x,t))+2\mathrm{i}E,$$

$$E=-\frac{1}{2\pi}\sum_{j=1}^{g}\int_{\beta_j}^{\alpha_{j+1}}\frac{\partial_x W_j(x,t)-2\lambda^{1/2}}{\sqrt{P(\lambda)}^{+}}\lambda^g\,\mathrm{d}\lambda,$$

where $\{W_j(x,t)\}_{j=1}^{\infty}$ are defined by the moment conditions for (11.22) with $\kappa(\lambda)$ replaced with $Z(x,t,\lambda)$.

This theorem states that despite the theoretical use of the function $(\Psi_p)_\pm$, the computation of the solution to the KdV equation does not require the computation of $(\Psi_p)_\pm$.

11.6 ▪ Numerical computation

In this section we discuss the computation of all the components of the theory. These components are

1. evaluating contour integrals used in the Abel map and Problem 11.4.1,
2. computing the singular integrals used in the representation of the g-function,
3. solving the deformed and regularized RH problem for the Baker–Akhiezer function, and
4. extracting the solution to the KdV equation from the Baker–Akhiezer function.

11.6.1 ▪ Computing contour integrals

The developments above require the computation of integrals of the form

$$I_j(f) = \int_{\beta_j}^{\alpha_{j+1}} \frac{f(\lambda)}{\sqrt{P(\lambda)}^+} \mathrm{d}\lambda, \tag{11.30}$$

to determine the g-function and compute Ω_j or W_j. Note that in the cases we consider f is analytic near the contour of integration. Also, we compute the Abel map of divisors whose points lie in gaps. We always choose $Q_0 = (\alpha_1, 0)$ in (11.3) and integrate along Γ_+ across the bands and gaps. Thus computing the Abel map of a point in a gap requires computation of integrals of the form (11.30) along with integrals of the form

$$K_j(f) = \int_{\alpha_j}^{\beta_j} \frac{f(\lambda)}{\sqrt{P(\lambda)}^+} \mathrm{d}\lambda, \quad I_j(f,\lambda) = \int_{\beta_j}^{\lambda} \frac{f(s)}{\sqrt{P(s)}^+} \mathrm{d}s. \tag{11.31}$$

While numerical integration packages can handle such integrals, as we have seen in Part II it is beneficial to use Chebyshev polynomials. For example, define

$$s = m(\lambda) = \frac{2}{\beta_j - \alpha_j}\lambda - \frac{\beta_j + \alpha_j}{\beta_j - \alpha_j}.$$

We change (11.30) to

$$I_j(f) = \int_{-1}^{1} \frac{f(m^{-1}(s))}{\sqrt{P(m^{-1}(s))}^+} \mathrm{d}m^{-1}(s).$$

The function

$$w(s) = \frac{\sqrt{1-s^2}}{\sqrt{P(m^{-1}(s))}}$$

is analytic in a neighborhood of the interval $[-1,1]$. We write

$$I_j(f) = \int_{-1}^{1} \left(\frac{f(m^{-1}(s))}{\sqrt{P(m^{-1}(s))}^+} w(s) \frac{\mathrm{d}}{\mathrm{d}s} m^{-1}(s) \right) \frac{\mathrm{d}s}{\sqrt{1-s^2}}.$$

The Chebyshev series approximation of the function in parentheses converges exponentially since it is analytic in a neighborhood of $[-1,1]$. A DCT is used to approximate the series, and the first coefficient in the series gives a very good approximation to $I_j(f)$. Alternatively, with one more change of variables, the integral can be mapped to $\mathbb{T}$ and the trapezoidal rule applied. Similar ideas work for $K_j(f)$, but we must modify our approach for $I_j(f,\lambda)$. Consider the integral

$$F_n(\lambda) = \int_{-1}^{\lambda} T_n(x) \frac{\mathrm{d}x}{\sqrt{1-x^2}}, \quad \lambda \in (1,1).$$

Using the standard change of variables $x = \cos\theta$,

$$F_n(\lambda) = -\int_{\pi}^{\arccos\lambda} T_n(\cos\theta)\mathrm{d}\theta = -\int_{\pi}^{\arccos\lambda} \cos(n\theta)\mathrm{d}\theta.$$

Therefore

$$F_n(\lambda)=\begin{cases} -\frac{\sin(n\arccos\lambda)}{n} & \text{if } n>0,\\ \pi-\arccos\lambda & \text{if } n=0. \end{cases}$$

Using the change of variables $m(\lambda)$ and the DCT we can compute each $I_j(f,\lambda)$ with this formula.

We need to compute the b periods. The b cycles have a more complicated relationship. Consider the $\tilde{b}_j$ cycles in Figure 11.8. Given any holomorphic differential $\omega=f(\lambda)\mathrm{d}\lambda$, we compute

$$\oint_{\tilde{b}_j}\omega=2\int_{\alpha_j}^{\beta_j}f(\lambda)\mathrm{d}\lambda.$$

From Figure 11.8, we see that $b_1=\tilde{b}_1$ and $b_i=\tilde{b}_i+b_{i-1}$. This gives a recursion relationship for b cycles.

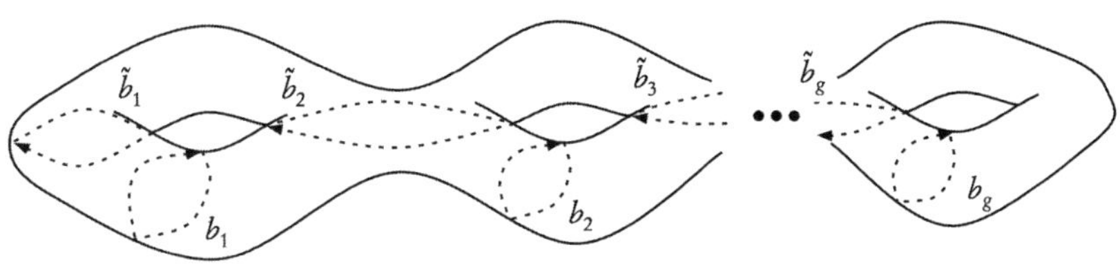

Figure 11.8. *The $\tilde{b}_j$ cycles on a schematic of the Riemann surface. Reprinted with permission from Elsevier* [113].

We must know ω_k before computing the Abel map. We describe how to compute the normalized differentials. Let $\omega=f(\lambda)\mathrm{d}\lambda$ be any holomorphic differential we showed in the proof of Theorem 11.13 that

$$\oint_{a_j}\omega=-2\int_{\beta_j}^{\alpha_{j+1}}f(\lambda)\mathrm{d}\lambda.$$

Given the branch point α_j,β_j, $j=1,\dots,g+1$, where $\beta_{g+1}=\infty$, we use the basis of unnormalized differentials

$$u_n=\frac{\lambda^{n-1}}{w}\mathrm{d}\lambda,\quad n=1,\dots,g,$$

and compute their a and b periods. This allows us to construct the basis ω_k of normalized differentials and gives us access to the Abel map.

Assume $Q=(\lambda,\sigma\sqrt{P(\lambda)}^{+})\in a_j$ for $\sigma=\pm1$. Then the kth component of the Abel map is computed by

$$(A(Q))_k=\sum_{l=1}^{j-1}(I_l(f_k)+K_l(f_k))+K_j(f_k)+\sigma I_j(f_k,\lambda),$$

where $f_k(\lambda)/w$ is the principal part of ω_k, the kth normalized holomorphic differential.

11.6.2 ▪ Computing the g-function

The g-function is defined by

$$\mathcal{G}(x,t,\lambda)=\sqrt{P(\lambda)}\sum_{j=1}^{g}\int_{\beta_j}^{\alpha_{j+1}}\frac{-\zeta(x,t,s)+\mathrm{i}\Omega_j(x,t)}{\sqrt{P(s)}^{+}}\frac{\mathrm{d}s}{s-\lambda}; \tag{11.32}$$

see (11.12). After mapping each interval of integration in (11.32) using a linear change of variables $z=m_j(s)$ ($m_j:[\beta_j,\alpha_{j+1}]\to[-1,1]$) we have the expression

$$\mathcal{G}(x,t,\lambda)=\frac{\sqrt{P(\lambda)}}{2\pi\mathrm{i}}\sum_{j=1}^{g}\int_{-1}^{1}H_j(z)\frac{\mathrm{d}z}{z-m_j(\lambda)},$$

where

$$H_j(z)=\frac{-\zeta(x,t,m_j^{-1}(z))+\mathrm{i}\Omega_j(x,t)}{\sqrt{P(m_j^{-1}(z))}^{+}}.$$

Note that $F_j(z)=H_j(z)\sqrt{1-z^2}$ is analytic in a neighborhood of $[-1,1]$. We use

$$\mathcal{G}(x,t,\lambda)=\sqrt{P(\lambda)}\sum_{j=1}^{g}\int_{-1}^{1}\frac{F_j(z)}{z-m_j(\lambda)}\frac{\mathrm{d}z}{\sqrt{1-z^2}}.$$

This reduces the problem of computing the g-function to that of computing integrals of the form

$$C(\lambda)=\int_{-1}^{1}\frac{f(s)}{s-\lambda}\frac{1}{\sqrt{1-s^2}}\mathrm{d}s,\quad \lambda\notin[-1,1],$$

where f is a smooth function on $[-1,1]$. The expressions in Lemma 5.10 provide a straightforward approach for this. We may apply the DCT and the transformation to vanishing basis coefficients from Section 4.4 to compute

$$f(s)\approx\sum_{j=0}^{n-1}c_jT_j(s)=\sum_{j=0}^{n-1}\tilde{c}_jT_j^z(s),\quad T_0^z(s)=1,\ T_1^z(s)=T_1(s).$$

Then Lemma 5.10 implies

$$C(\lambda)\approx\sum_{j=0}^{n-1}\tilde{c}_je_j(\lambda),\quad e_j(\lambda)=\begin{cases}\dfrac{\mathrm{i}}{2\sqrt{z^2-1}} & \text{if } j=0,\\[2ex] \dfrac{\mathrm{i}z}{2\sqrt{z^2-1}}-\dfrac{\mathrm{i}}{2} & \text{if } j=1,\\[2ex] -\mathrm{i}J_+^{-1}(z)^{j-1} & \text{otherwise.}\end{cases}$$

Although it is not important for our purposes in this chapter, one may wish to compute the limiting values $\mathcal{G}^{\pm}$ as λ approaches a gap from above or below. We use Corollary 5.11 with Lemma 2.7 so that the relation

$$\mathcal{C}_\Gamma^{\pm}=\pm\frac{1}{2}I-\frac{\mathrm{i}}{2}\mathcal{H}_\Gamma$$

allows for effective computation of $\mathcal{G}^{\pm}$.

11.6.3 ▪ Computing the Baker–Akhiezer function

This section is concerned with computing $(\Psi_r)_\pm$. Let D' be the divisor for the desired zeros of the BA function and D be the divisor for the poles. We compute the vector (see (11.16))

$$V = A(D' - D),$$

using the method for computing integrals described above. Next, consider the differentials

$$\nu_j = \mathrm{i}\frac{\lambda^{g+j-1}}{w}\mathrm{d}\lambda, \quad j = 1, \ldots, g,$$

which satisfy

$$\int_{\lambda_0}^{\lambda} \nu_j = \mathcal{O}(\lambda^{-1/2+j}) \text{ as } \lambda \to \infty.$$

We accurately compute the a periods of ν_j. We construct $\{\tilde{\nu}_j\}_{j=1}^{g}$ which each have vanishing a periods by adding an appropriate linear combination holomorphic differentials. We compute the matrix

$$(S)_{kj} = \oint_{b_k} \tilde{\nu}_j.$$

The system $SX = V$ is solved for the real-valued vector X, giving a differential

$$l = \sum_{j=1}^{g} (X)_j \tilde{x}_j,$$

that has b periods equal to the vector V. The final step is to compute the coefficients $\{t_j\}_{j=1}^{g}$ in the expansion

$$\int_{\lambda_0}^{\lambda} l = \sum_{n=1}^{g} \mathrm{i}t_n \lambda^{n-1/2} + \mathcal{O}(\lambda^{-1/2}) = \kappa(\lambda)/2 + \mathcal{O}(\lambda^{-1/2}).$$

The BA function with asymptotic behavior $(\Psi_p)_\pm \sim \mathrm{e}^{\pm\kappa(\lambda)/2}$ as $\lambda \to \infty$ has zeros at the points of D'. Theorem 11.17 tells us to seek $(\Psi_r)_\pm \sim \mathrm{e}^{\pm Z(x,t,\lambda)/2}$ as $\lambda \to \infty$. We construct the deformed and regularized RH problem for $(\Psi_r)_\pm$; see Section 11.5. This RH problem is solved numerically.

To demonstrate the method we use $\alpha_1 = 0$, $\beta_1 = 0.25$, $\alpha_2 = 1$, $\beta_2 = 1.5$, and $\alpha_3 = 2$. Thus we have a genus-two surface. We choose zeros to be at the points $(0.5, \sqrt{P(0.5)}^{+})$ and $(1.75, \sqrt{P(1.75)}^{+})$. See Figure 11.9 for a surface plot showing both the zeros and the poles of the BA function on a single sheet. See Figures 11.10 and 11.11 for contour plots of the real part, imaginary part, and modulus of the BA function on each sheet. Note that producing this plot requires the computation of the g-function. These plots are all produced in the genus-two case, but higher genus BA functions can also be plotted.

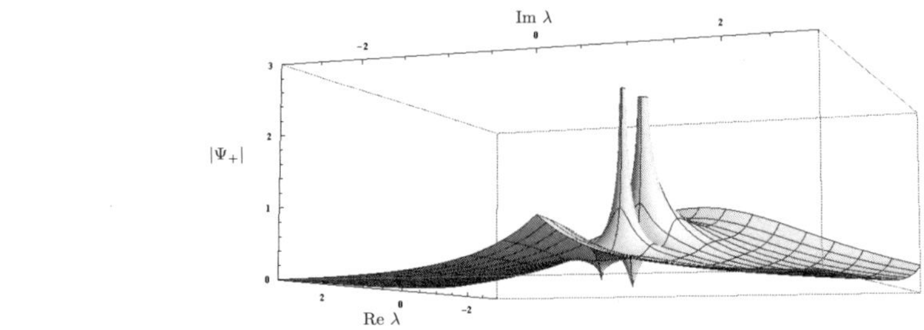

Figure 11.9. *A three-dimensional plot of the modulus of the BA function on one sheet of the Riemann surface. We see that two poles and two zeros are clearly present. Reprinted with permission from Elsevier* [113].

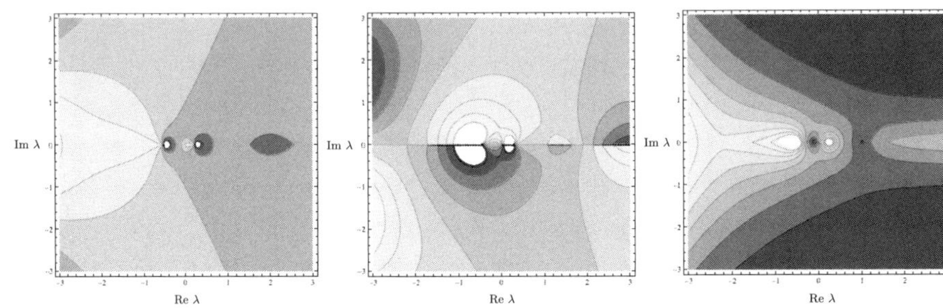

(a) (b) (c)

Figure 11.10. *A genus-two Baker–Akhiezer function. Darker shades indicate smaller values. Two poles and two zeros are clearly present.* (a) *The real part of* Ψ_+. (b) *The imaginary part of* Ψ_+. (c) *The modulus of* Ψ_+. *Reprinted with permission from Elsevier* [113].

11.6.4 ▪ Numerical solutions of the KdV equation

Before we move on to numerical results for the KdV equation, let us review the solution process. The constants α_j ($j = 1,\ldots,g+1$) and β_j ($j = 1,\ldots,g$) are chosen, all positive. This determines the polynomial $P(\lambda)$ and the unnormalized differentials u_k. The a periods of these differentials are computed using Chebyshev polynomials, and the normalized basis ω_k is constructed. Next, one point in each a cycle is chosen to be a pole of the BA function. These points make up the divisor for the poles of the BA function. The Abel map of this divisor is computed, along with the Abel map of the divisor

$$D = \sum_{j=1}^{g} (\beta_j, 0).$$

Through the process just outlined the constants t_j, $j = 1,\ldots,g$, are computed. The Riemann–Hilbert formulation is used to compute the function $(\Psi_r)_\pm$ by noting that its asymptotic behavior is (11.27). The function Λ_3 is found and $q(x,t)$ is computed using Theorem 11.17.

In this section we plot numerical solutions of the KdV equation. In the genus-two case we use numerical tests to demonstrate uniform spectral convergence.

Genus one: For a genus-one solution we set $\alpha_1 = 0$, $\beta_1 = 0.25$, and $\alpha_2 = 1$ with the zero of the BA function at $(0.5, \sqrt{P(0.5)}^+)$ at $t = 0$. See Figure 11.12 for plots of

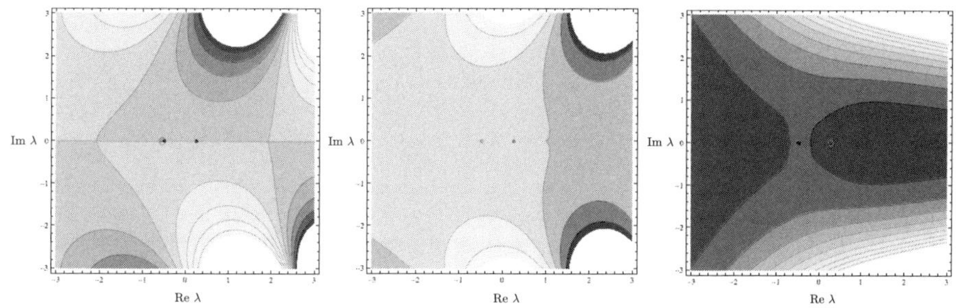

(a) (b) (c)

Figure 11.11. *A genus-two Baker–Akhiezer function. Darker shades indicate smaller values.* (a) *The real part of* Ψ_-. (b) *The imaginary part of* Ψ_-. (c) *The modulus of* Ψ_-. *Reprinted with permission from Elsevier* [113].

the corresponding solution of the KdV equation. This solution is an elliptic function. Explicitly [25],

$$q(x,t) = -\alpha_2 - \beta_1 + 2\,\mathrm{cn}^2(x - K(1-\beta_1) + 1.0768 \\ - (8(1-\beta_1)^2 - 4 - \alpha_2 - \beta_1)t, 1-\beta_1),$$

where $K(s)$ is the complete elliptic integral defined in (1.2) and cn is the Jacobi cn function [91]. The shift inside the cn function is computed numerically. See Figure 11.12 for another solution.

Genus two: For a genus-two solution we set $\alpha_1 = 0$, $\beta_1 = 0.25$, $\alpha_2 = 1$, $\beta_2 = 1.5$, and $\alpha_3 = 2$ with the zeros of the BA function at $(0.5, \sqrt{P(0.5)}^+)$ and $(1.75, \sqrt{P(1.75)}^+)$ at $t = 0$. See Figure 11.14 for plots of the corresponding solution of the KdV equation.

For this solution we numerically discuss convergence. We use $q_n(x,t)$ to denote the approximate solution of the KdV equation obtained with n collocation points per contour of the RH problem. We define the Cauchy error

$$Q_m^n(x,t) = |q_n(x,t) - q_m(x,t)|.$$

We fix $m = 80$ and let n vary: $n = 10, 20, 40$. See Figure 11.13 for plots of $Q_m^n(x,t)$ for various values of x and t. This figure demonstrates uniform spectral Cauchy convergence of the function $q_n(x,t)$ to $q(x,t)$, the solution of the KdV equation.

We plot another genus-two solution in Figure 11.14. If we shrink the widths of the bands, we can obtain solutions which are closer to the soliton limit. See Figure 11.15 for a solution demonstrating a soliton-like interaction. Note that our finite-genus solutions are often traveling to the left as opposed to the soliton solutions in Chapter 8 that travel to the right. This can be rectified with the discussion below (11.10) because we assume $\alpha_1 = 0$.

Genus three: For a genus-three solution we set $\alpha_1 = 0$, $\beta_1 = 0.25$, $\alpha_2 = 1$, $\beta_2 = 2$, $\alpha_3 = 2.5$, $\beta_3 = 3$, and $\alpha_4 = 3.5$ with the zeros of the BA function at $(0.5, \sqrt{P(0.5)}^+)$, $(1.75, \sqrt{P(1.75)}^+)$, and $(2.75, \sqrt{P(2.75)}^+)$ at $t = 0$. In Figure 11.16 we show the jump contours for the RH problem which are used in practice to compute the BA function. See Figure 11.18 for plots of the corresponding solution of the KdV equation and Figures 11.17 and 11.18 for another genus-three solution. We show the dynamics of the zeros of the BA function in Figure 11.17.

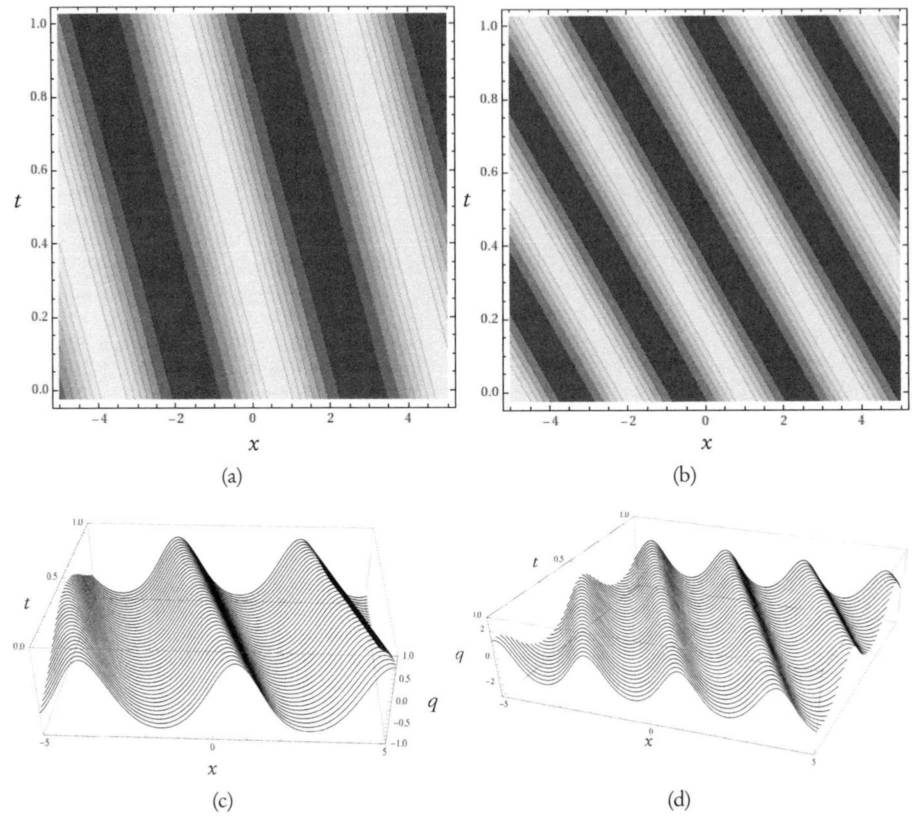

Figure 11.12. (a) *A contour plot of the genus-one solution with* $\alpha_1 = 0$, $\beta_1 = 0.64$, *and* $\alpha_2 = 1$ *with the zero of the BA function at* $(0.75, \sqrt{P(0.75)}^+)$ *at* $t = 0$. *Darker shades represent troughs.* (b) *A contour plot of the genus-one solution with* $\alpha_1 = 0$, $\beta_1 = 0.64$, *and* $\alpha_2 = 1$ *with the zero of the BA function at* $(0.75, \sqrt{P(0.75)}^+)$ *at* $t = 0$. *Again, darker shades represent troughs.* (c) *A three-dimensional plot of the solution in* (a) *showing the time evolution.* (d) *A three-dimensional plot of the solution in* (b) *showing the time evolution. Reprinted with permission from Elsevier* [113].

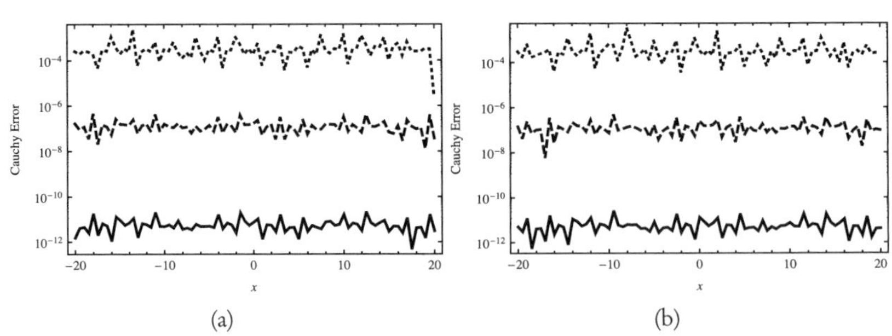

Figure 11.13. (a) *A logarithmically scaled plot of* $Q^n_{80}(x,0)$ *for* $n = 10$ *(dotted),* $n = 20$ *(dashed), and* $n = 40$ *(solid).* (b) *A logarithmically scaled plot of* $Q^n_{80}(x,25)$ *for* $n = 10$ *(dotted),* $n = 20$ *(dashed), and* $n = 40$ *(solid). This figure demonstrates uniform spectral convergence. Reprinted with permission from Elsevier* [113].

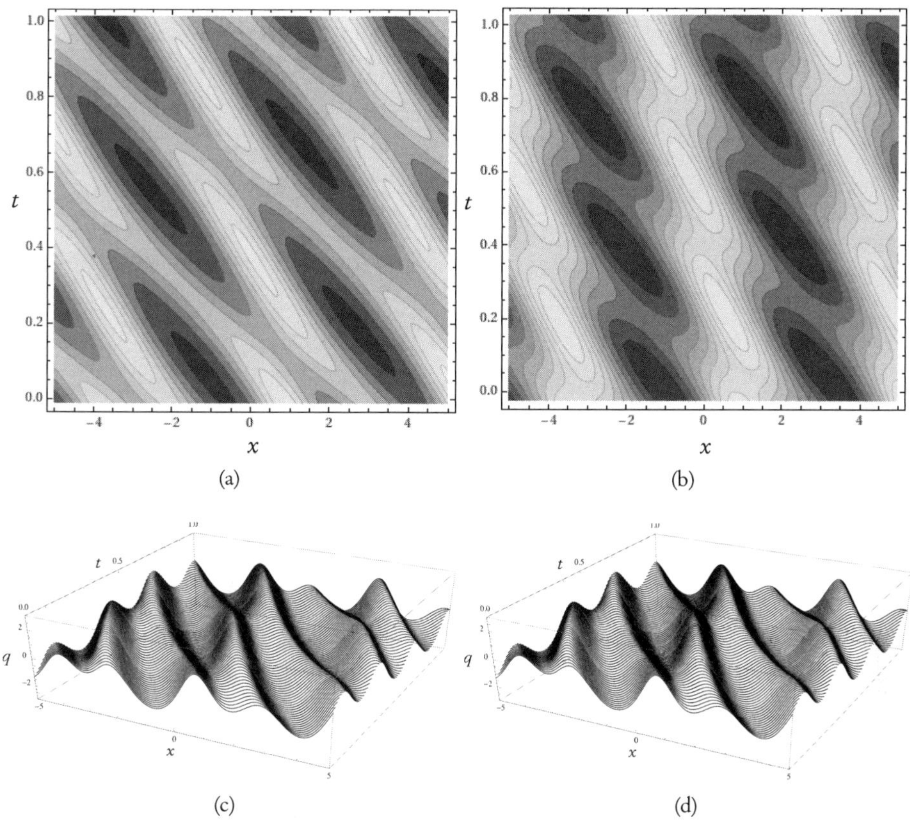

Figure 11.14. (a) *A contour plot of the genus-two solution with* $\alpha_1 = 0$, $\beta_1 = 0.25$, $\alpha_2 = 1$, $\beta_2 = 1.5$, *and* $\alpha_3 = 2$ *with the zeros of the BA function at* $(0.5, \sqrt{P(0.5)}^+)$ *and* $(1.75, \sqrt{P(1.75)}^+)$ *at* $t = 0$. *Darker shades represent troughs.* (b) *A contour plot of the genus-two solution with* $\alpha_1 = 0$, $\beta_1 = 0.25$, $\alpha_2 = 1$, $\beta_2 = 2$, *and* $\alpha_3 = 2.25$ *with the zeros of the BA function at* $(0.5, \sqrt{P(0.5)}^+)$ *and* $(2.2, \sqrt{P(2.2)}^+)$ *at* $t = 0$. *Again, darker shades represent troughs.* (c) *A three-dimensional plot of the solution in* (a) *showing the time evolution.* (d) *A three-dimensional plot of the solution in* (b) *showing the time evolution. Reprinted with permission from Elsevier* [113].

Genus five: Just to demonstrate the breadth of the method we compute a genus-five solution. We set $\alpha_1 = 0$, $\beta_1 = 0.25$, $\alpha_2 = 1$, $\beta_2 = 2$, $\alpha_3 = 2.5$, $\beta_3 = 3$, $\alpha_4 = 3.3$, $\beta_4 = 3.5$, $\alpha_5 = 4$, $\beta_5 = 5.1$, and $\alpha_6 = 6$ with the zeros of the BA function at $(0.5, \sqrt{P(0.5)}^+)$, $(2.2, \sqrt{P(2,2)}^+)$, $(3.2, \sqrt{P(3.2)}^+)$, $(3.6, \sqrt{P(3.6)}^+)$, and $(5.3, \sqrt{P(5.3)}^+)$ at $t = 0$. See Figure 11.19 for a plot of the corresponding solution of the KdV equation. This figure shows the time evolution.

11.7 ▪ Analysis of the deformed and regularized RH problem

In general, we consider an RH problem of the form

$$\Phi^+(s) = \Phi^-(s)G(s), \quad s \in \Sigma, \quad \Phi(\infty) = I, \tag{11.33}$$

where Σ is bounded and G depends on $\{\Omega_j(x,t)\}_{j=1}^g$, or alternatively $\{W_j(x,t)\}_{j=1}^g$. We use many of the results in Chapter 2. It is straightforward to check that G satisfies the

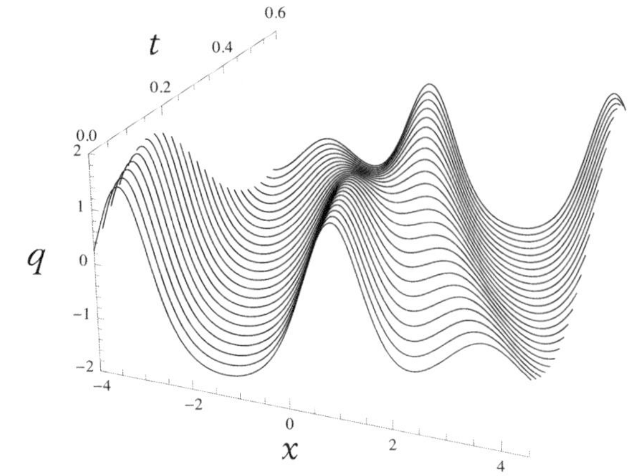

Figure 11.15. *A genus-two solution with* $\alpha_1 = 0$, $\beta_1 = 0.1$, $\alpha_2 = 1$, $\beta_2 = 1.05$, *and* $\alpha_3 = 1.75$ *with the zeros of the BA function at* $(0.5, \sqrt{P(0.5)}^+)$ *and* $(1.2, \sqrt{P(1.2)}^+)$ *at* $t = 0$. *This solution demonstrates a soliton-like interaction. Reprinted with permission from Elsevier* [113].

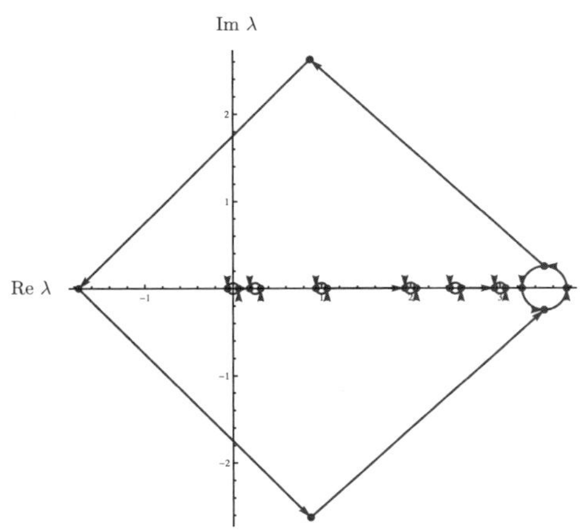

Figure 11.16. *The jump contours for the RH problem which are used in practice to compute the BA function. Here* $\alpha_1 = 0$, $\beta_1 = 0.25$, $\alpha_2 = 1$, $\beta_2 = 2$, $\alpha_3 = 2.5$, $\beta_3 = 3$, *and* $\alpha_4 = 3.5$. *Reprinted with permission from Elsevier* [113].

first-order product condition. Analyticity may be used to see that G satisfies the kth-order product condition for all $k > 0$.

We apply Theorem 2.69 to the RH problem derived in Section 11.5. We use G to denote the jump matrix. We note that when we augment the contour, $G = I$ on all added pieces, and these do not contribute to the integral. Also, $\det \Delta = 1$ away from α_j, β_j, and $\det J_0 = -1$. Both of these do not influence the index. We are left with

$$\begin{aligned}\operatorname{ind} \mathscr{C}[G;\Sigma] = &-\frac{1}{\pi \mathrm{i}} \sum_{l=1}^{g} \left(\int_{C_{\alpha_l}} \mathrm{d} \log \det A_l(s) + \int_{C_{\beta_l}} \mathrm{d} \log \det B_l(s) \right) \\ &- \frac{1}{\pi \mathrm{i}} \int_{\partial \mathscr{D}} \mathrm{d} \log \det H(s) - \frac{1}{\pi \mathrm{i}} \int_{C_{\alpha_{g+1}}} \mathrm{d} \log \det A_{g+1}(s).\end{aligned}$$

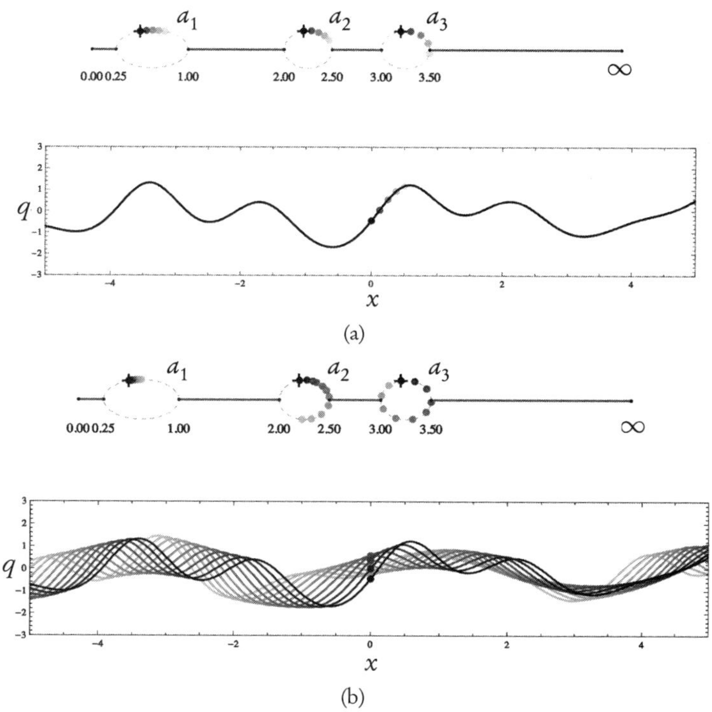

Figure 11.17. *A genus-three solution with* $\alpha_1 = 0$, $\beta_1 = 0.25$, $\alpha_2 = 1$, $\beta_2 = 2$, $\alpha_3 = 2.5$, $\beta_3 = 3$, *and* $\alpha_4 = 3.5$ *with the zeros of the BA function at* $(0.5, \sqrt{P(0.5)}^+), (2.2, \sqrt{P(2.2)}^+)$, *and* $(3.2, \sqrt{P(3.2)}^+)$ *at* $t = 0$. *These plots show the dynamics of the zeros of the BA function. The top plot in each panel gives a schematic of the Riemann surface with the a cycles labeled. Dots of the same shade across the panels are in correspondence. The* $+$ *on the plots represents where the pole of the BA function is located on the Riemann surface. These points are also the locations of the zeros at* $t = 0$. (a) *The solution at* $t = 0$. *We vary* x *from* $x = 0$ *up to* $x = 0.25$ *and plot how the zeros* $\{\gamma_1(x,0), \gamma_2(x,0), \gamma_3(x,0)\}$ *move on the Riemann surface.* (b) *The evolution of the same solution up to* $t = 0.125$. *We fix* $x = 0$ *and plot how the zeros* $\{\gamma_1(0,t), \gamma_2(0,t), \gamma_3(0,t)\}$ *move on the Riemann surface. Reprinted with permission from Elsevier* [113].

Here $C_{\alpha_j}, C_{\beta_j}$ are the circles around α_j, β_j, and $\mathcal{D}$ is again the region inside the large outer circle but outside each of the smaller circles, as before. Straightforward contour integration produces

$$\int_{C_{\alpha_l}} \mathrm{d}\log\det A_l(s) = \pi \mathrm{i}, \quad \int_{C_{\beta_l}} \mathrm{d}\log\det B_l(s) = -\pi \mathrm{i}, \quad \int_{\partial\mathcal{D}} \mathrm{d}\log\det H(s) = -\pi \mathrm{i}.$$

This proves that $\operatorname{ind}\mathscr{C}[G;\Sigma] = 0$. Every element in the kernel of $\operatorname{ind}\mathscr{C}[G;\Sigma]$ corresponds to a solution of the RH problem that vanishes at infinity; see the remarks in the proof of Theorem 2.73. Given a matrix-valued solution Φ, we sum the rows to get the vector representation of the BA function. If we have a vanishing solution, we zero out the second row and assume the first is nonzero. Call the new function Ψ. This is still a vanishing solution. Then $\Phi + c\Psi$ is a solution of the RH problem for any c. Summing the rows of $\Phi + c\Psi$ we obtain a function different from Φ for every c. This contradicts the uniqueness of the BA function and gives that $\mathscr{C}[G;\Sigma]$ must be boundedly invertible by the open mapping theorem. This shows that all RH problems considered here

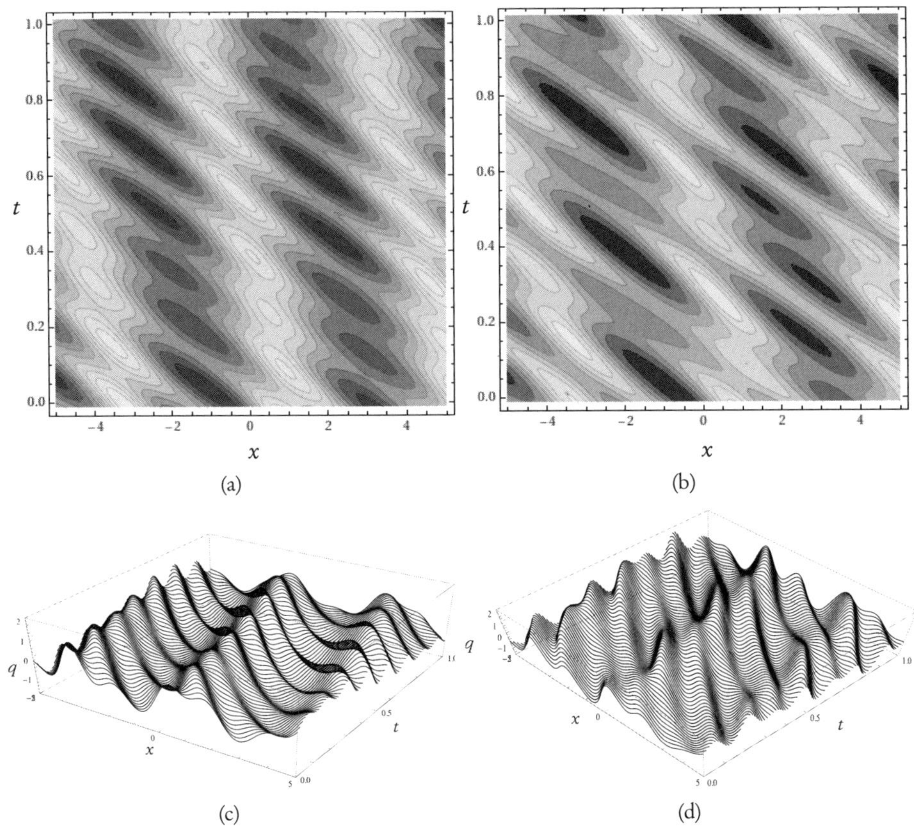

Figure 11.18. (a) *A contour plot of the genus-three solution with* $\alpha_1 = 0$, $\beta_1 = 0.25$, $\alpha_2 = 1$, $\beta_2 = 2$, $\alpha_3 = 2.5$, $\beta_3 = 3$, *and* $\alpha_4 = 3.5$ *with the zeros of the BA function at* $(0.5, \sqrt{P(0.5)}^+), (2.2, \sqrt{P(2.2)}^+)$ *and* $(3.2, \sqrt{P(3.2)}^+)$ *at* $t = 0$. *Darker shades represent troughs.* (b) *A contour plot of the genus-three solution in Figure* 11.17. *Again, darker shades represent troughs.* (c) *A three-dimensional plot of the solution in* (a) *showing the time evolution.* (d) *A three-dimensional plot of the solution in Figure* 11.17 *showing the time evolution. Reprinted with permission from Elsevier* [113].

are uniquely solvable with smooth solutions. This is the justification needed to use the numerical method for RH problems in Chapter 6.

11.8 ▪ Uniform approximation

We consider the RH problem (11.33). In this section we explain how our approximation of the BA function changes with x and t. We use the results from Section 7.1. As before, we consider the operator $\mathscr{C}[G;\Sigma]$ defined by

$$\mathscr{C}[G;\Sigma]U = U - (\mathscr{C}_\Sigma^- U)(G - I). \tag{11.34}$$

The operator equation

$$\mathscr{C}[G;\Sigma]U = [1,\ 1](G - I)$$

is discretized using the method in Section 7.4. Once an approximation U_n to U is known, an approximate solution $\Phi_n(\lambda) = \mathscr{C}_\Sigma U_n(\lambda) + [1,\ 1]$ of Φ is obtained. The residue of a

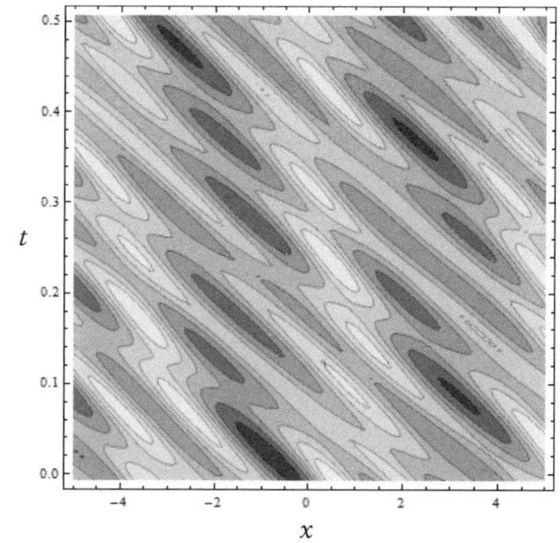

(a)

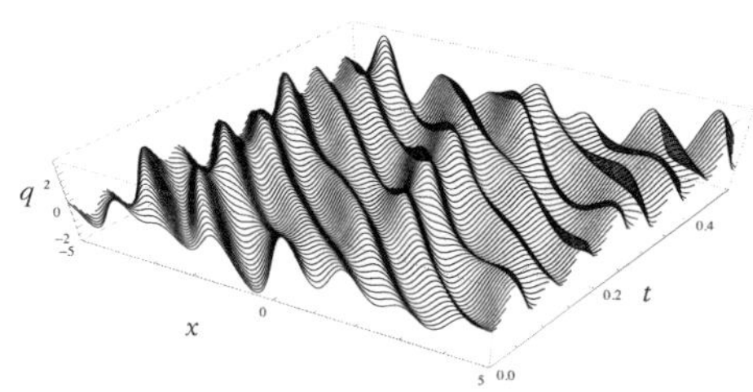

(b)

Figure 11.19. (a) *A contour plot of the genus-five solution with* $\alpha_1 = 0$, $\beta_1 = 0.25$, $\alpha_2 = 1$, $\beta_2 = 2$, $\alpha_3 = 2.5$, $\beta_3 = 3$, $\alpha_4 = 3.3$, $\beta_4 = 3.5$, $\alpha_5 = 4$, $\beta_5 = 5.1$, *and* $\alpha_6 = 6$ *with the zeros of the BA function at* $(0.5, \sqrt{P(0.5)}^+)$, $(2.2, \sqrt{P(2,2)}^+)$, $(3.2, \sqrt{P(3.2)}^+)$, $(3.6, \sqrt{P(3.6)}^+)$, *and* $(5.3, \sqrt{P(5.3)}^+)$ *at* $t = 0$. *Darker shades represent troughs.* (b) *A three-dimensional plot of the solution same solution showing the time evolution. Reprinted with permission from Elsevier* [113].

function at ∞ is computed by Lemma 2.13:

$$\lim_{\lambda\to\infty} \lambda(\Phi(\lambda) - [1,\ 1]) = -\int_\Sigma U(s)\text{đ}s.$$

This is what is used to compute s_1 and s_2 in (11.26). We make the fundamental assumption, Assumption 7.4.1, and establish two claims:

- There exists a constant, depending only on Γ, such that $||\mathscr{C}[G;\Sigma]^{-1}||_{\mathscr{L}(L^2(\Sigma))} \leq C$
- There exist constants D_k, depending only on Γ and k, such that $||G - I||_{W^{k,\infty}(\Sigma)} \leq D_k$ for each $k > 0$.

We emphasize that the operators $\mathscr{C}[G;\Sigma], \mathscr{C}[G;\Sigma]^{-1}$ depend on g constants $\Omega_j \in [0, 2\pi)$, $j = 1, \ldots, g$, in an analytic way.

It follows that the mapping

$$\Omega = (\Omega_1, \ldots, \Omega_g) \mapsto \mathscr{C}[G;\Sigma]$$

is continuous from $[0,2\pi)^g$ to $\mathscr{L}(L^2(\Gamma))$. Since the operator is always invertible, the same statement holds for the inverse operator. This implies

$$\sup_{\Omega} \|\mathscr{C}[G;\Sigma]^{-1}\|_{\mathscr{L}(L^2(\Gamma))} \leq C.$$

The second claim can be established by differentiating the jump matrix G. It is clear that all derivatives of G are bounded and that this bound can be made independent of Ω. This leads to the following theorem which shows that we expect uniform spectral convergence of all needed functions. We ignore the error from the approximation of $\{\Omega_j\}$.

Theorem 11.18 (Uniform approximation). *If Assumption* 7.4.1 *holds, then* $\Phi_n = I + \mathscr{C}_\Gamma u_n$, *the numerical approximation of* $\Phi = I + \mathscr{C}_\Gamma u$, *satisfies*

$$\sup_{\Omega} |\Phi_n(\lambda) - \Phi(\lambda)| < C_k \epsilon^{-1} n^{-k} \quad \textit{for every } k \geq 0, \ \inf_{s\in\Sigma} |\lambda - s| > \epsilon,$$

$$\sup_{\Omega} \|u_n - u\|_{L^2(\Sigma)} < L_k n^{-k} \quad \textit{for every } k \geq 0.$$

As a consequence (see (7.16)*), the approximate solution* $q_n(x,t)$ *of the KdV equation satisfies*

$$\sup_{\Omega} |q(x,t) - q_n(x,t)| < S_k n^{-k} \quad \textit{for every } k \geq 0.$$

Chapter 12

The Dressing Method and Nonlinear Superposition

In this chapter the computation of solutions of the KdV equation (8.1) with a particular class of step-like finite-gap initial data is considered. The material here concerning the KdV equation first appeared in [114, 115]. For our purposes, $q_0(x)$ is said to be a *step-like finite-gap function* if

$$\left| \int_0^{\pm\infty} \left| \frac{\mathrm{d}^n}{\mathrm{d}x^n}(q_0(x) - q_\pm(x)) \right| (1+|x|^m)\mathrm{d}x \right| < \infty$$

for all non-negative integers n and m and some finite-gap potentials $q_\pm(x)$. The existence and uniqueness of solutions for the KdV equation with this type of initial data is a highly nontrivial matter, as these functions do not fit into classical function spaces. The theory was discussed in [48] when the Bloch spectra associated with $q_\pm(x)$ either agree or are completely disjoint [48]. The solution of the KdV equation was shown to satisfy

$$\left| \int_0^{\pm\infty} \left| \frac{\mathrm{d}^n}{\mathrm{d}x^n}(q(x,t) - q_\pm(x,t)) \right| (1+|x|^m)\mathrm{d}x \right| < \infty \tag{12.1}$$

for all time.

Remark 12.0.1. *The analysis of Egorova, Grunert, and Teschl [48] allows for more general solutions than the numerical method that is discussed in this chapter. Particularly, we only concentrate on the case when the Bloch spectra of $q_\pm(x)$ coincide and we use the term "combination solutions" for this reason.*

The results of Egorova, Grunert, and Teschl are a significant step forward in the analysis of the KdV equation. Traditionally, the analysis proceeds in the Schwartz space ($q_\pm(x) = 0$) or towards the construction of finite-genus solutions ($q_+(x) = q_-(x)$ and $q_0(x) = q_+(x)$). Thus, their results are a generalization of both the inverse scattering transform for rapidly decaying initial data in Chapter 8 and of the analysis on Riemann surfaces for the construction of finite-genus solutions in Chapter 11. In the same spirit, the numerical methods in these chapters can be combined to compute a subclass of step-like finite gap solutions.

To combine these approaches we use the dressing method as applied to the KdV equation. The version of Proposition 3.9 that is applicable to the KdV equation is needed. It is important to note that there is a significant difference in presentation style between

Chapters 8 and 11. This is largely due to the fact that the analysis is performed in the z-plane in Chapter 8 and in the λ-plane in Chapter 11 ($z^2 = \lambda$). The reasoning for this is largely historical because we tried to mirror the typical presentation of such results. In Section 12.1.2 it is established that the finite-genus solutions may be computed in the z-plane. Therefore the dressing method may be applied directly in the z-plane, which is critical for the superposition of the solutions. The same methodology is also applicable to the defocusing NLS equation with the help of Proposition 3.9.

12.1 ▪ A numerical dressing method for the KdV equation

In this section, we discuss the construction of solutions of the KdV equation via the dressing method. It follows from the inverse scattering transform (essentially, by construction) that the solution Φ of the RH problem for the KdV equation (see (8.9)) is a solution of

$$-\Phi_{xx} + 2\mathrm{i}z\Phi_x\sigma_3 - q(x,t)\Phi = 0, \quad \sigma_3 = \begin{bmatrix} 1 & 0 \\ 0 & -1 \end{bmatrix}. \tag{12.2}$$

Furthermore, it is easy to check that Ψ in (11.11) also satisfies this equation with z replaced by $\lambda^{1/2}$. These functions satisfy a second equation determining their t dependence stemming from (8.3):

$$-\Phi_t + 4\mathrm{i}z^3\Phi\sigma_3 = (2q(x,t) - 4z^2)(\Phi_x - \mathrm{i}z\Phi\sigma_3) - q_x(x,t)\Phi. \tag{12.3}$$

Indeed, (12.2) and (12.3) essentially make up the Lax pair for the KdV equation. This is easily seen by writing $\Psi = \Phi e^{-(\mathrm{i}zx + 4\mathrm{i}z^3 t)\sigma_3}$ and finding the differential equations solved by Ψ. This produces the Lax pair derived from (8.3) (see also [2, p. 70]). This relationship is further explained by the dressing method. Recall the notation

$$e^{\alpha\hat{\sigma}_3}A = e^{\alpha\sigma_3}Ae^{-\alpha\sigma_3}.$$

We state the dressing method for the KdV equation as a proposition.

Proposition 12.1. *Let* $\Phi(z) = \Phi(x,t,z)$ *solve the RH problem*

$$\Phi^+(s) = \Phi^-(s)e^{-\theta(x,t,s)/2\hat{\sigma}_3}V(s), \quad s \in \Gamma, \quad \theta(x,t,s) = 2\mathrm{i}sx + 8\mathrm{i}s^3 t, \quad \Phi(\infty) = [1,\ 1],$$

where $\bar{\Gamma} = \Gamma$ *(with orientation),* $\det V(z) = 1$, $\overline{V(\bar{z})} = V(-z)$, *and* $V^{-1}(z) = \sigma_2\overline{V(\bar{z})}\sigma_2$ *with*

$$\sigma_2 = \begin{bmatrix} 0 & 1 \\ 1 & 0 \end{bmatrix}.$$

Assume that the L^2 2×2 *matrix RH problem has a unique solution that is sufficiently differentiable in* x *and* t *and that all existing derivatives are* $\mathcal{O}(1/z)$ *as* $z \to \infty$ *with* L^2 *boundary values. Define*

$$\left[Q(x,t),\ \tilde{Q}(x,t)\right] = 2\mathrm{i} \lim_{z\to\infty} z\partial_x\Phi(x,t,z)\sigma_3. \tag{12.4}$$

Then $Q = \tilde{Q}$, $\Phi(x,t,z)$ *solves*

$$\begin{aligned} -\Phi_{xx} + 2iz\Phi_x\sigma_3 - Q(x,t)\Phi &= 0, \\ -\Phi_t + 4iz^3\Phi\sigma_3 &= (2Q(x,t) - 4z^2)(\Phi_x - iz\Phi\sigma_3) - Q_x(x,t)\Phi, \end{aligned} \tag{12.5}$$

and Q *solves the KdV equation* (11.1).

Proof. We begin by establishing some symmetries of the solution. Let Φ be a matrix-valued function that tends to the identity matrix at infinity. We show that this matrix RH problem can be reduced to a vector RH problem.

Define $\hat{\Phi}(z) = \overline{\Phi(-\bar{z})}$. Note that $\hat{\Phi}^+(z) = \overline{\Phi^+(-\bar{z})}$ so that

$$\hat{\Phi}^+(z) = \hat{\Phi}^-(z)\overline{V(-\bar{z})} = \hat{\Phi}^-(z)V(z).$$

Therefore by uniqueness, $\Phi(z) = \overline{\Phi(-\bar{z})}$. Expand Φ near ∞ using this symmetry:

$$\begin{aligned} \Phi(z) &= I + \Phi_1 z^{-1} + \Phi_2 z^{-2} + \mathcal{O}(z^{-3}) \\ &= I - \bar{\Phi}_1 z^{-1} + \bar{\Phi}_2 z^{-2} + \mathcal{O}(z^{-3}). \end{aligned}$$

Thus Φ_1 is purely imaginary. Next, define $\tilde{\Phi}(z) = \sigma_2\overline{\Phi(\bar{z})}\sigma_2$ and note that $\tilde{\Phi}^+(z) = \sigma_2\overline{\Phi^-(\bar{z})}\sigma_2$. We obtain

$$\tilde{\Phi}^+(z) = \sigma_3\overline{\Phi^+(\bar{z})V^{-1}(\bar{z})}\sigma_2 = \tilde{\Phi}^-(z)\sigma_2\overline{V^{-1}(\bar{z})}\sigma_2 = \tilde{\Phi}^-(z)V(z).$$

Thus $\Phi(z) = \sigma_3\overline{\Phi(\bar{z})}\sigma_3$. Again, considering the series at infinity,

$$\begin{aligned} \Phi(z) &= I + \Phi_1 z^{-1} + \Phi_2 z^{-2} + \mathcal{O}(z^{-3}) \\ &= I + \sigma_2\bar{\Phi}_1\sigma_2 z^{-1} + \sigma_2\bar{\Phi}_2\sigma_2 z^{-2} + \mathcal{O}(z^{-3}). \end{aligned}$$

Therefore $\Phi_1 = \sigma_2\bar{\Phi}_1\sigma_2 = -\sigma_2\Phi_1\sigma_2$. If

$$\Phi = \begin{bmatrix} a & b \\ c & d \end{bmatrix},$$

then $a = -d$ and $c = -b$. Let ϕ be the vector consisting of the sum of the rows of Φ. It follows that

$$\phi(z) = [1,\ 1] + \phi_1 z^{-1} + \mathcal{O}(z^{-2}),$$

where $\phi_1\sigma_3 = Q(x,t)[1,\ 1]$ for some scalar-valued function Q. Thus the symmetries of the problem allow us to reduce it to a vector RH problem, justifying (12.4).

The fact that the RH problem has a unique solution implies that the only solution that decays at infinity is the zero solution. A straightforward but lengthy calculation shows that

$$\begin{aligned} &-\Phi_{xx} + 2iz\Phi_x\sigma_3 - Q(x,t)\Phi, \\ &\Phi_t - 4iz^3\Phi\sigma_3 + (2Q(x,t) - 4z^2)(\Phi_x - iz\Phi\sigma_3) - Q_x(x,t)\Phi \end{aligned}$$

are solutions of the RH problem that decay at infinity and therefore they must vanish identically. Hence, we obtain (12.5). The compatibility condition of (12.5) implies Q solves (11.1) as mentioned above. □

12.1.1 ▪ An RH problem on cuts

With the dressing method in hand, consider

$$\begin{aligned} &\varphi^+(s)=\varphi^-(s)\begin{bmatrix} 0 & -\mathrm{e}^{-\theta(x,t,s)} \\ \mathrm{e}^{\theta(x,t,s)} & 0 \end{bmatrix}, \quad s\in\bigcup_{j=1}^{g}\left[(-\alpha_{j+1},-\beta_j)\cup(\beta_j,\alpha_{j+1})\right], \\ &\varphi(\infty)=[1,1], \end{aligned} \tag{12.6}$$

where $0<\alpha_j<\beta_j<\alpha_{j+1}$. From Proposition 12.1 it follows that

$$q(x,t)=2\mathrm{i}\lim_{z\to\infty} z\partial_x\varphi_1(x,t,z)$$

must be a solution of the KdV equation. Below, we connect this solution to the finite-genus solutions of Chapter 11 and we superimpose this RH problem on the RH problem for the initial-value problem to obtain dispersive finite-genus solutions in Section 12.1.4. In the remainder of this section we discuss the numerical solution of this RH problem.

It is clear that (12.6) is an oscillatory RH problem. Solutions of the RH problem are more oscillatory as $|x|$ and t increase, and hence deformations must be performed in order to accurately approximate the solution. The g-function mechanism is used again to remove these oscillations. Consider the scalar RH problem for $j=1,2,\ldots,g$:

- $\mathcal{G}^+(s)+\mathcal{G}^-(s)=0$ for $s\in(-\alpha_{j+1},-\beta_j)\cup(\beta_j,\alpha_{j+1})$,
- $\mathcal{G}^+(s)-\mathcal{G}^-(s)=-\theta(x,t,s)+\mathrm{i}\Omega_{j+}(x,t)$ for $s\in(\beta_j,\alpha_{j+1})$,
- $\mathcal{G}^+(s)-\mathcal{G}^-(s)=-\theta(x,t,s)+\mathrm{i}\Omega_{j-}(x,t)$ for $s\in(-\alpha_{j+1},-\beta_j)$, and
- $\mathcal{G}(z)=\mathcal{O}(z^{-1})$ as $z\to\infty$.

Note the similarities with (11.22). Here $\{\Omega_{j\pm}(x,t)\}_{j=1}^{g}$ are constants (with respect to z) that must be determined. The methods of Chapter 2 allow us to find a function $\mathcal{G}$ that satisfies the first three properties:

$$\begin{aligned} \mathcal{G}(z)=\sqrt{P(z)}\sum_{j=1}^{g}\Bigg(&\int_{\beta_j}^{\alpha_{j+1}}\frac{-\theta(x,t,s)+\mathrm{i}\Omega_{j+}(x,t)}{\sqrt{P(s)}^+}\frac{\mathrm{d}s}{s-z} \\ &+\int_{-\alpha_{j+1}}^{-\beta_j}\frac{-\theta(x,t,s)+\mathrm{i}\Omega_{j-}(x,t)}{\sqrt{P(s)}^+}\frac{\mathrm{d}s}{s-z}\Bigg), \end{aligned}$$

where $P(z)=\prod_{j=1}^{g}\left[(z-\beta_j)(z-\alpha_{j+1})(z+\beta_j)(z+\alpha_{j+1})\right]$. To be precise, $\sqrt{P(z)}$ is taken to have branch cuts on the intervals (β_j,α_{j+1}) and $(-\alpha_{j+1},-\beta_j)$ and the behavior $\sqrt{P(z)}\sim z^{2g}$ as $z\to\infty$. Furthermore, we define $\sqrt{P(z)}^+=\lim_{\epsilon\to0^+}\sqrt{P(z+\mathrm{i}\epsilon)}$. The constants $\{\Omega_{j\pm}(x,t)\}_{j=1}^{g}$ are to enforce decay at infinity: $\mathcal{G}(z)=\mathcal{O}(z^{-1})$ as $z\to\infty$. Expanding $1/(s-z)$ in a geometric series (see Lemma 2.12) produces an expansion in powers of z^{-1}. Forcing the first $2g$ coefficients of this series to vanish results in $2g$ conditions:

$$\begin{aligned} &0=\sum_{j=1}^{g}\left(\int_{\beta_j}^{\alpha_{j+1}}\frac{-\theta(x,t,s)+\mathrm{i}\Omega_{j+}(x,t)}{\sqrt{P(s)}^+}s^m\mathrm{d}s+\int_{-\alpha_{j+1}}^{-\beta_j}\frac{-\theta(x,t,s)+\mathrm{i}\Omega_{j-}(x,t)}{\sqrt{P(s)}^+}s^m\mathrm{d}s\right), \\ &m=0,1,\ldots,2g-1. \end{aligned} \tag{12.7}$$

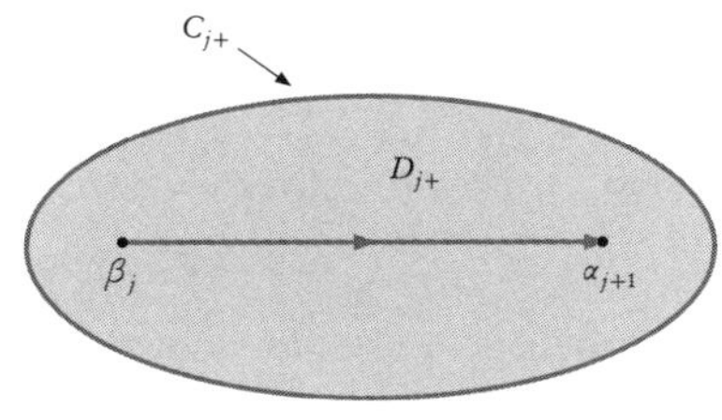

Figure 12.1. *The contour C_{j+} and the region D_{j+} surrounding (β_j, α_{j+1}).*

A linear system for $\{\Omega_{j\pm}(x,t)\}_{j=1}^{g}$ is found. The ideas from Theorem 11.13 may be adapted to show that this linear system is uniquely solvable. Furthermore, methods to compute all integrals that appear here are described in Section 11.6.

Define

$$G(z) \triangleq G(x,t,z) \triangleq \begin{bmatrix} e^{-\mathscr{G}(x,t,z)} & 0 \\ 0 & e^{\mathscr{G}(x,t,z)} \end{bmatrix}$$

and the vector-valued function

$$\Lambda(z) \triangleq \Lambda(x,t,z) \triangleq \varphi(x,t,z)G(x,t,z).$$

A direct calculation shows that Λ solves

$$\Lambda^+(s) = \Lambda^-(s) \begin{cases} \begin{bmatrix} 0 & -e^{-i\Omega_{j+}(x,t)} \\ e^{i\Omega_{j+}(x,t)} & 0 \end{bmatrix} & \text{if } s \in (\beta_j, \alpha_{j+1}), \\ \begin{bmatrix} 0 & -e^{-i\Omega_{j-}(x,t)} \\ e^{i\Omega_{j-}(x,t)} & 0 \end{bmatrix} & \text{if } s \in (-\alpha_{j+1}, -\beta_j) \end{cases}$$

for $j = 1, 2, \ldots, g$ with $\Lambda(x,t,\infty) = [1,\ 1]$. This is a piecewise-constant RH problem and we follow ideas from Chapter 11 to regularize it for numerical purposes. Define

$$R_{j\pm}(z) = \frac{1}{2}\begin{bmatrix} \beta_{j\pm}(z) + 1/\beta_{\pm j}(z) & ie^{-i\Omega(x,t)}(\beta_{j\pm}(z) - 1/\beta_{j\pm}) \\ -ie^{i\Omega(x,t)}(\beta_{j\pm}(z) - 1/\beta_{j\pm}(z)) & \beta_{j\pm}(z) + 1/\beta_{j\pm}(z) \end{bmatrix},$$

$$\beta_{j\pm}(z) = \left(\frac{z \mp \alpha_{j+1}}{z \mp \beta_j}\right)^{1/4}.$$

It follows that R_{j+} and R_{j-} satisfy the same jump as Λ in a neighborhood of (β_j, α_{j+1}) and $(-\alpha_{j+1}, -\beta_j)$, respectively. Let C_{j+} be a clockwise-oriented piecewise-smooth contour lying solely in the right-half plane surrounding (β_j, α_{j+1}) but not intersecting or surrounding (β_i, α_{i+1}) for $i \neq j$. Define C_{j-} in an analogous manner for $(-\alpha_{j+1}, -\beta_j)$, again with clockwise orientation. Define $D_{j\pm}$ to be the component of $\mathbb{C} \setminus C_{j\pm}$ that contains the interval $C_{j\pm}$ encloses; see Figure 12.1. Define

$$K(z) \triangleq K(x,t,z) \triangleq \begin{cases} \Lambda(x,t,z)R_{j\pm}^{-1}(x,t,z) & \text{if } z \in D_{j\pm}, \\ \Lambda(x,t,z) & \text{otherwise.} \end{cases}$$

Then $K(z)$ solves the following RH problem:

$$K^+(s) = K^-(s)R_{j\pm}(s), \quad s \in C_{j\pm}, \quad j = 1, 2, \ldots, g,$$
$$K(\infty) = [1,\ 1].$$

This RH problem can be solved numerically.

12.1.2 ▪ From the λ-plane to the z-plane

We describe a method to transform (11.11) to an RH problem in the z-plane so that it is straightforward to connect RH problems in the z-plane to finite-genus solutions of the KdV equation. First, notice that $\hat{\Phi}$ fails to be analytic on a subset of $(0,\infty)$. With $z^2=\lambda$, we write $\hat{\Phi}(\lambda)\triangleq\hat{\Phi}(x,t,\lambda)\triangleq\chi(x,t,\lambda^{1/2})$ and define

$$\tilde{\Phi}(z)\triangleq\tilde{\Phi}(x,t,z)\triangleq\begin{cases}\chi(x,t,z) & \text{if } \operatorname{Im} z>0,\\ \chi(x,t,-z) & \text{if } \operatorname{Im} z<0.\end{cases}$$

It is clear that $\tilde{\Phi}(z)$ fails to be analytic only on $\mathbb{R}$ and its jump matrices must be computed. For $z>0$,

$$\lim_{\epsilon\downarrow 0}\tilde{\Phi}(z\pm i\epsilon)=\lim_{\epsilon\downarrow 0}\chi(x,t,z\pm i\epsilon)=\chi^{\pm}(x,t,z).$$

For $z<0$,

$$\lim_{\epsilon\downarrow 0}\tilde{\Phi}(z\pm i\epsilon)=\lim_{\epsilon\downarrow 0}\chi(x,t,-z\mp i\epsilon)=\chi^{\mp}(x,t,-z).$$

For $\lambda>0$, if $\hat{\Phi}^+(\lambda)=\hat{\Phi}^-(\lambda)J(\lambda^{1/2})$, then $\tilde{\Phi}^+(z)=\tilde{\Phi}^-(z)J(z)$ for $z>0$, and $\tilde{\Phi}^-(z)=\tilde{\Phi}^+(z)J(-z)$ for $z<0$. Notice that all jumps in (11.11) satisfy $J(-z)=J^{-1}(z)$. For ease of notation, define

$$B_+=(\alpha_{g+1},\infty)\cup\bigcup_{j=1}^{g}(\alpha_j,\beta_j),\quad B_-=(-\infty,-\alpha_{g+1})\cup\bigcup_{j=1}^{g}(-\beta_j,-\alpha_j),$$
$$G_+=\bigcup_{j=1}^{g}(\beta_j,\alpha_{j+1}),\quad G_-=\bigcup_{j=1}^{g}(-\alpha_{j+1},-\beta_j).$$

One is led to consider an RH problem for $\tilde{\Phi}(z)$:

$$\begin{aligned}&\tilde{\Phi}^+(s)=\tilde{\Phi}^-(s)\begin{bmatrix}0&1\\1&0\end{bmatrix},\quad s\in B_+\cup B_-,\\ &\tilde{\Phi}^+(s)=\tilde{\Phi}^-(s)\begin{bmatrix}e^{-\theta(x,t,s)}&0\\0&e^{\theta(x,t,s)}\end{bmatrix},\quad s\in G_+\cup G_-,\\ &\tilde{\Phi}(\infty)=[1,\ 1].\end{aligned}\tag{12.8}$$

Due to its definition, $\tilde{\Phi}$ solves (12.2) in the upper-half plane and the same equation with $z\mapsto -z$ in the lower-half plane. This leads us to switch the entries of $\tilde{\Phi}$ in the lower-half plane. Define

$$\tilde{\Psi}(z)\triangleq\tilde{\Psi}(x,t,z)\triangleq\begin{cases}\tilde{\Phi}(x,t,z) & \text{if } \operatorname{Im} z>0,\\ \tilde{\Phi}(x,t,z)\begin{bmatrix}0&1\\1&0\end{bmatrix} & \text{if } \operatorname{Im} z<0.\end{cases}$$

Thus, $\tilde{\Psi}(z)$ satisfies

$$\begin{aligned} \tilde{\Psi}^+(s) &= \tilde{\Psi}^-(s)\begin{bmatrix} 0 & e^{-\theta(x,t,s)} \\ e^{\theta(x,t,s)} & 0 \end{bmatrix}, \quad s \in G_+ \cup G_-, \\ \tilde{\Psi}(\infty) &= [1,\ 1]. \end{aligned} \tag{12.9}$$

This differs from the RH problem for φ given above. The fundamental difference is that the determinant of the jumps for $\tilde{\Psi}$ is -1 instead of $+1$ in the case of φ. As discussed in Chapter 11 one column of $\tilde{\Psi}$ must have a pole in each connected component of $G_+ \cup G_-$. If the pole is at an endpoint of an interval, it is a pole on a Riemann surface corresponding to a square root singularity in the plane. Furthermore, in Chapter 11 it was established that if one point from each connected component of $G_+ \cup G_-$ is chosen, then there exists a solution of (12.9) that has a pole at each of these points. Momentarily, the presence of poles is ignored, although it highlights an important issue below.

It follows that one may consider (12.9) as a 2×2 RH problem normalized to the identity at infinity. Summing the rows allows the determination of a solution of the vector problem as was done in the proof of Proposition 12.1. Consider the auxiliary RH problem

$$\nu^+(s) = \nu^-(s)\begin{bmatrix} 0 & -1 \\ 1 & 0 \end{bmatrix}, \quad s \in G_+ \cup G_-, \quad \nu(\infty) = I.$$

Then for

$$\tilde{\Psi}(z) = \begin{bmatrix} \tilde{\Psi}_{11}(x,t,z) & \tilde{\Psi}_{12}(x,t,z) \\ \tilde{\Psi}_{21}(x,t,z) & \tilde{\Psi}_{22}(x,t,z) \end{bmatrix}$$

define

$$\tilde{\Psi}_\nu(z) = \tilde{\Psi}_\nu(x,t,z) = \begin{bmatrix} \nu_{11}(z)\tilde{\Psi}_{11}(x,t,z) & \nu_{12}(z)\tilde{\Psi}_{12}(x,t,z) \\ \nu_{21}(z)\tilde{\Psi}_{21}(x,t,z) & \nu_{22}(z)\tilde{\Psi}_{22}(x,t,z) \end{bmatrix}.$$

A calculation shows that $\tilde{\Psi}_\nu$ satisfies the same jumps as φ; see (12.6).

It follows that $\tilde{\Psi}_\nu$ has a pole in each interval $[\beta_j, \alpha_{j+1}]$ and $[-\alpha_{j+1}, -\beta_j]$, unless it is canceled out by an entry of ν. Thus if the RH problem for φ is solved and the transformation $\tilde{\Psi} \mapsto \tilde{\Psi}_\nu$ is inverted, the poles are at locations determined only by α_j and β_j: the zeros of the entries of ν. Thus this procedure is guaranteed to produce one solution of (12.9) despite the fact that there is a whole family of solutions. This family is described by the fact that for each $\gamma_j \in (\beta_j, \alpha_{j+1})$ and $\sigma_j \in \{1,2\}$ there exists a solution of (12.9) such that $\tilde{\Psi}_{\sigma_j}$ has a pole at $\pm\gamma_j$. This is a g parameter family of solutions, and it highlights the nonuniqueness of solutions of (12.9).

Remark 12.1.1. *It follows that ν can be found explicitly. Define*

$$\beta(z) = \left(\prod_{j=1}^{g} \frac{(z-\alpha_{j+1})(z+\beta_j)}{(z+\alpha_{j+1})(z-\beta_j)} \right)^{1/4};$$

then it is easy to check that

$$\nu(z) = \frac{1}{2}\begin{bmatrix} \beta(z)+\beta^{-1}(z) & -\mathrm{i}(\beta(z)-\beta^{-1}(z)) \\ \mathrm{i}(\beta(z)-\beta^{-1}(z)) & \beta(z)+\beta^{-1}(z) \end{bmatrix}$$

is the desired solution; see also Example 2.18. It can be shown that $\beta(z)-\beta^{-1}(z)$ has $2g$ zeros, located at $\pm u_j$ for $u_j \in (\beta_j, \alpha_{j+1})$ [124]. This justifies the construction above.

12.1.3 • Nonlinear superposition

Below, solutions of the initial-value problem are combined with finite-genus solutions using the following definition.

Definition 12.2. *Consider two L^2 RH problems for functions $\Phi_1(z) = \Phi_1(x,t,z)$ and $\Phi_2(z) = \Phi_2(x,t,z)$:*

$$\Phi_1^+(s) = \Phi_1^-(s) e^{-\theta(x,t,s)/2\hat{\sigma}_3} V_1(s), \quad s \in \Gamma_1, \quad \Phi_1(\infty) = [1,\ 1],$$
$$\Phi_2^+(s) = \Phi_2^-(s) e^{-\theta(x,t,s)/2\hat{\sigma}_3} V_2(s), \quad s \in \Gamma_2, \quad \Phi_2(\infty) = [1,\ 1],$$

such that V_1 and V_2 satisfy the hypothesis of Proposition 12.1. *In addition, assume V_1 and V_2 commute on $\Gamma_1 \cap \Gamma_2$. Thus $q_j(x,t) = 2\mathrm{i}\lim_{z\to\infty} z\partial_x \Phi_j(z)$, $j = 1,2$, is a solution of the KdV equation. Then $q_3(x,t) = 2\mathrm{i}\lim_{z\to\infty} z\partial_x \Phi_3(z)$ is called a nonlinear superposition of $q_1(x,t)$ and $q_2(x,t)$, where $\Phi_3(z) = \Phi_3(x,t,z)$ solves*

$$\Phi_3^+(s) = \Phi_3^-(s) e^{-\theta(x,t,s)/2\hat{\sigma}_3}(V_1(s)V_2(s)), \quad s \in \Gamma_1 \cup \Gamma_2 \quad \Phi_3(\infty) = [1,\ 1]. \tag{12.10}$$

Here V_1 and V_2 are extended to be the identity matrix outside their initial domain of definition.

Remark 12.1.2. *The condition that V_1 and V_2 commute is necessary so that V_1V_2 satisfies the hypotheses of Proposition* 3.9.

Consider the choice

$$V_1(z) = \begin{bmatrix} 1 - \overline{\rho(\bar{z})}\rho(z) & -\overline{\rho(\bar{z})} \\ \rho(z) & 1 \end{bmatrix}, \tag{12.11}$$

where ρ is as in (8.9). If $c = \{c_j\}$ and $Z = \{z_j\}$ are not empty, additional contours are added to the RH problem. Let

$$V_2(z) = \begin{cases} \begin{bmatrix} 0 & -1 \\ 1 & 0 \end{bmatrix} & \text{if } z \in G_+ \cup G_-, \\ I & \text{otherwise}. \end{cases} \tag{12.12}$$

Consider the numerical solution of (12.10), the nonlinear superposition of the solution of the initial-value problem and a finite-genus solution.

Assumption 12.1.1. *To simplify the computation of solutions, assume ρ is supported in a finite interval $[-\ell,\ell]$ that does not coincide with the spectral gaps: $[-\ell,\ell] \cap (G_+ \cup G_-) = \varnothing$.*

Thus, one is led to solve the following RH problem: Consider $\Phi_4(z) = \Phi_4(x,t,z)$, which satisfies

$$\Phi_4^+(s) = \Phi_4^-(s) \begin{cases} e^{-\theta(x,t,s)/2\hat{\sigma}_3} V_1(s) & \text{if } s \in [-\ell,\ell], \\ e^{-\theta(x,t,s)/2\hat{\sigma}_3} V_2(s) & \text{if } s \in G_+ \cup G_-, \end{cases} \quad \Phi_4(\infty) = [1,\ 1].$$

Remark 12.1.3. *If ρ has compact support, then it certainly cannot be analytic. In practice, one starts with a reflection coefficient ρ_a that is analytic in a strip that contains the real axis*

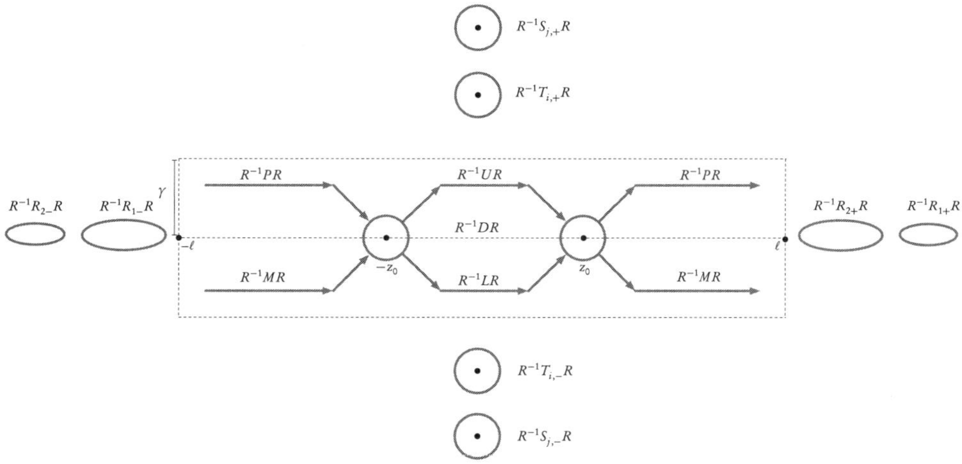

Figure 12.2. *The full RH problem that is solved to compute superposition solutions of the KdV equation. The elliptical contours represent $C_{1\pm}$ and $C_{2\pm}$. Note that these curves are located outside $[-\ell,\ell]$.*

as in Chapter 8. Then ρ is constructed from ρ_a by multiplying by smooth functions with compact support so that $\rho \sim \rho_a$ in an appropriate sense. This determines ℓ. It can be shown using Proposition 2.78 that the solution Φ_a of

$$\Phi_a^+(z) = \Phi_a^-(z) \begin{cases} e^{-\theta(x,t,z)/2\hat{\sigma}_3} \begin{bmatrix} 1-|\rho_a(z)|^2 & -\overline{\rho_a(\bar{z})} \\ \rho_a(z) & 1 \end{bmatrix} & \text{if } z \in \mathbb{C}\setminus(G_+\cup G_-), \\ e^{-\theta(x,t,z)/2\hat{\sigma}_3} \begin{bmatrix} -\overline{\rho_a(\bar{z})}-\rho_a(z) & -1 \\ 1 & 0 \end{bmatrix} & \text{if } z \in G_+\cup G_-, \end{cases}$$
$$\Phi_a(\infty) = [1,\ 1]$$

is close to Φ_4 in the sense that if $\|\diamond(\rho(\diamond)-\rho_a(\diamond))\|_{L^1\cap L^\infty(\mathbb{R})} < \epsilon$, then $|2\mathrm{i}\lim_{|z|\to\infty} \partial_x((\Phi_4)_1 - (\Phi_a)_1)| < C\epsilon$, i.e., $2\mathrm{i}\lim_{|z|\to\infty}\partial_x(\Phi_a)_1$ is a good approximation of the solution of the KdV equation. Importantly, all the matrix factorizations and contour deformations from Chapter 8 can be applied to the RH problem for Φ_a since

$$\begin{bmatrix} -\overline{\rho_a(\bar{z})}-\rho_a(z) & -1 \\ 1 & 0 \end{bmatrix} = \begin{bmatrix} 1 & -\overline{\rho_a(\bar{z})} \\ 0 & 1 \end{bmatrix} \begin{bmatrix} 0 & -1 \\ 1 & 0 \end{bmatrix} \begin{bmatrix} 1 & 0 \\ \rho_a(z) & 1 \end{bmatrix},$$

and the jump matrix factors in an appropriate way.

The nonlinear steepest descent method as described above transforms $[-\ell,\ell]$ to a contour $\Gamma(x,t)$ with jump $\tilde{V}_1$ that passes along appropriate paths of steepest descent. This process affects the jumps on $G_+\cup G_-$ but only by the multiplication of (to machine precision) analytic, diagonal matrix-valued function $R(x,t,z)$. The exact form of $R(x,t,z)$ can be inferred from the deformations in Chapter 8. In the dispersive region $R(x,t,z) = V^{-1}(z)\Delta(z;z_0)$ and $R(x,t,z) = V^{-1}(z)$ for the Painlevé and soliton regions (see (8.18) and (8.21) for definitions). This transforms $V_2(z)$ to $\tilde{V}_2(x,t,z) = R^{-1}(x,t,z)V_2(z)R(x,t,z)$. We display the full RH problem for the superposition solutions in Figure 12.2.

Remark 12.1.4. *We have highlighted a limitation of our approach. The contours $C_{j\pm}$ need to be in a location where the reflection coefficient is small. Furthermore, if $C_{j\pm}$ is near the origin, then the corresponding finite-genus solution of the KdV equation has a larger period. Thus, the decay rate of the reflection coefficient affects the periodicity/quasi-periodicity of the finite-genus solution that can be superimposed using this method.*

12.1.4 ▪ Numerical results

In this section we construct solutions of the KdV equation using the method described above. We choose a constant $\ell > 0$ and a reflection coefficient $\rho(z)$ for $z \in [-\ell, \ell]$, poles and norming constants ($Z = \{z_j\}_{j=1}^n$ and $c = \{c_j\}_{j=1}^n$), and gaps $0 < \ell < \beta_1 < \alpha_2 < \cdots < \alpha_{g+1}$.

A perturbed genus-two solution with no solitons

We choose ρ to be the reflection coefficient obtained from the initial condition $q_0(x) = -1.2\mathrm{e}^{-(x/4)^2}$ and $\ell = 2.4$. The sets c and Z are both empty. Finally, we equate $\beta_1 = 2.5$, $\alpha_2 = 2.54$, $\beta_2 = 4$, and $\alpha_3 = 4.013$. Recall that $q_1(x,t)$ is the solution of the KdV equation with initial condition $q_0(x)$, $q_2(x,t)$ is a genus-two solution, and $q_3(x,t)$ is the nonlinear superposition. We present the results in Figures 12.3, 12.4, and 12.5 below. We consider $\tilde{q}(x,t) = q_1(x,t) + q_2(x,t) - q_3(x,t)$ as a measure of nonlinearity. See Figure 12.6 for a plot of $\tilde{q}(x,t)$ at various times. We see that the nonlinear interaction is not local: as $x \to -\infty$ the genus-two solution experiences a phase shift. Thus the solution obtained from this method is a superposition function for all t in the sense that it satisfies (12.1).

We note that $q_\pm(x,t)$ in (12.1) can be computed. Assume there are n solitons in the solution, and for $z_0^2 = -x/(12t) > \ell$ let t and x be sufficiently large so that $K_{x,t} = \{1, 2, \ldots, n\}$; see (8.20). Then $R(x,t,z)$ is constant in x and t. Thus the RH problem created through the dressing method with $R^{-1}(x,t,z)V_2(z)R(x,t,z)$ defined on $G_+ \cup G_-$ produces a solution of the KdV equation. We change the definition of the g-function:

- $\mathcal{G}^+(s) - \mathcal{G}^-(s) = -\theta(x,t,s) - 2\log R_{11}(x,t,s) + \mathrm{i}\Omega_{j+}(x,t)$ for $s \in (\beta_j, \alpha_{j+1})$,
- $\mathcal{G}^+(s) - \mathcal{G}^-(s) = -\theta(x,t,s) - 2\log R_{11}(x,t,s) + \mathrm{i}\Omega_{j-}(x,t)$ for $s \in (-\alpha_{j+1}, -\beta_j)$.

When considering the analogue of (12.7) it is easy to see that the addition of the $\log R_{11}$ term contributes a constant to the right-hand side of the linear system for $\{\Omega_{j\pm}\}_{j=1}^g$. It is clear that this induces a phase shift. This modification is not needed for numerical purposes.

A perturbed genus-two solution with two solitons

We consider the addition of solitons and dispersion to a genus-two solution. Again, we let ρ be the reflection coefficient obtained from the initial condition $q_0(x) = -1.2\mathrm{e}^{-(x/4)^2}$. Also, we choose

$$Z = \{1.2589\mathrm{i}, 0.8571\mathrm{i}\}, \quad c = \{7604.0\mathrm{i}, 1206.3\mathrm{i}\}.$$

These are chosen by computing the eigenvalues and norming constants for a positive initial condition. Finally, to fix the genus-two solution we define $\beta_1 = 2.5$, $\alpha_2 = 2.52$, $\beta_2 = 4.1$, and $\alpha_3 = 4.105$. See Figure 12.7 for plots of this solution.

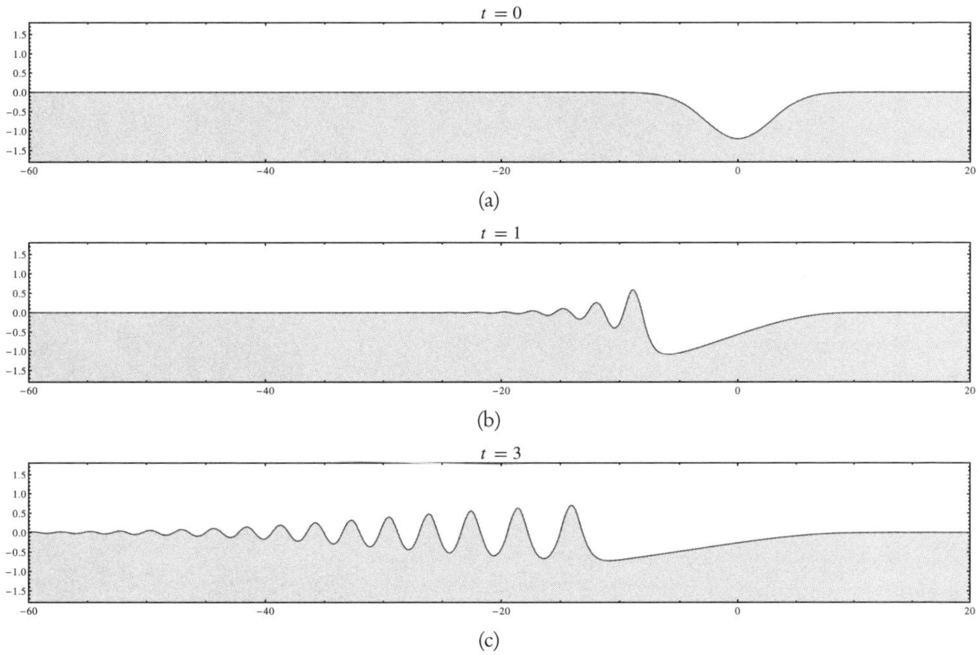

Figure 12.3. (a) *The initial condition for* $q_1(x,t)$. (b) *A plot of* $q_1(x,1)$. (c) *A plot of* $q_1(x,3)$. *Figure © IOP Publishing and London Mathematical Society. Reproduced with permission. All rights reserved* [114].

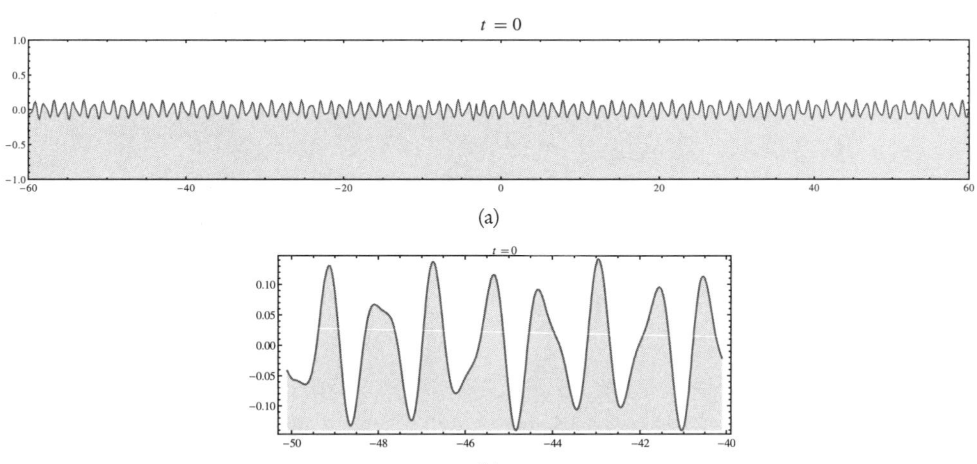

Figure 12.4. (a) *The initial condition for* $q_2(x,t)$. (b) *A zoomed plot of* $q_2(x,0)$. *Figure © IOP Publishing and London Mathematical Society. Reproduced with permission. All rights reserved* [114].

We examine the solution in different regions to demonstrate the phase shifts induced by $R(x,t,z)$ as discussed in the previous section. Specifically, we examine regions where the dispersive and solitonic effects decay rapidly in time and the solution of the KdV equation asymptotically looks like a phase-shifted finite-genus solution. As before, when $R(x,t,z)$ is constant with respect to z we expect the RH problem created through the

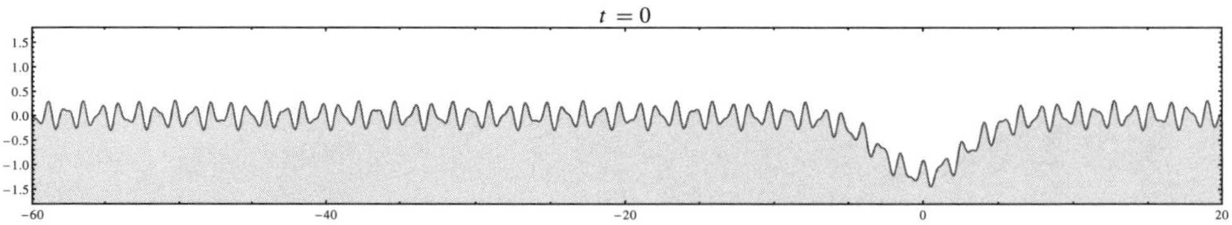

(a)

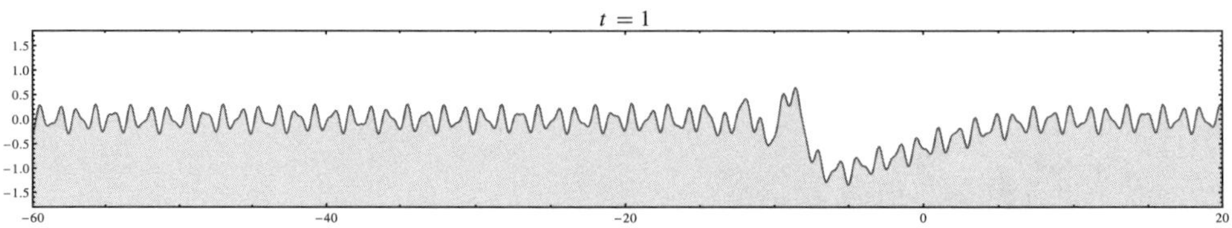

(b)

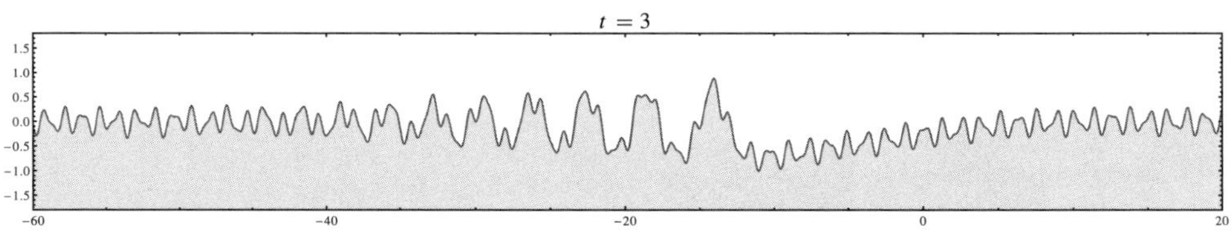

(c)

Figure 12.5. (a) *The initial condition for* $q_3(x,t)$. (b) *A plot of* $q_3(x,1)$. (c) *A plot of* $q_3(x,3)$. *Figure © IOP Publishing and London Mathematical Society. Reproduced with permission. All rights reserved* [114].

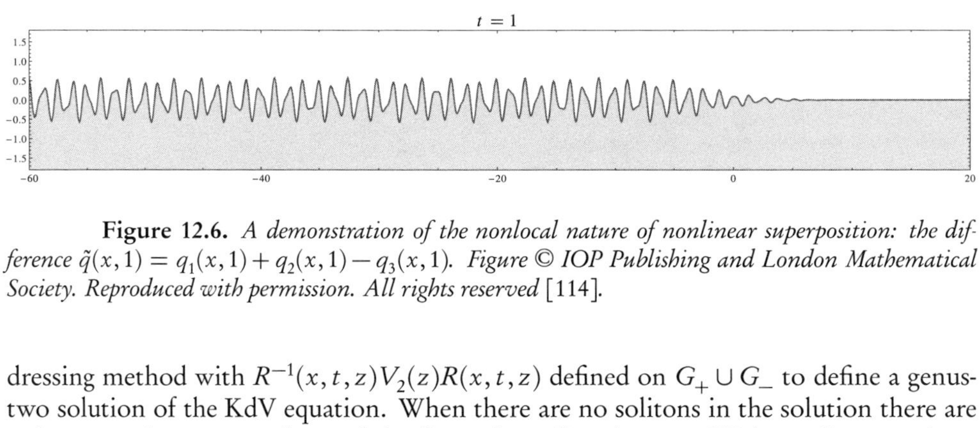

Figure 12.6. *A demonstration of the nonlocal nature of nonlinear superposition: the difference* $\tilde{q}(x,1) = q_1(x,1) + q_2(x,1) - q_3(x,1)$. *Figure © IOP Publishing and London Mathematical Society. Reproduced with permission. All rights reserved* [114].

dressing method with $R^{-1}(x,t,z)V_2(z)R(x,t,z)$ defined on $G_+ \cup G_-$ to define a genus-two solution of the KdV equation. When there are no solitons in the solution there are only two regions: $x \ll 0$ (beyond the dispersive tail) and $x \gg 0$. With n solitons we have $n+2$ regions:

- $x \gg 0$: in front of all solitons,
- the $n-1$ regions between solitons,
- the region between the trailing soliton and the dispersive tail, and
- $x \ll 0$: beyond the dispersive tail.

This is consistent with the results of [84]. In Figure 12.8 we demonstrate that using the definition of $R(x,t,z)$ we can compute these solutions.

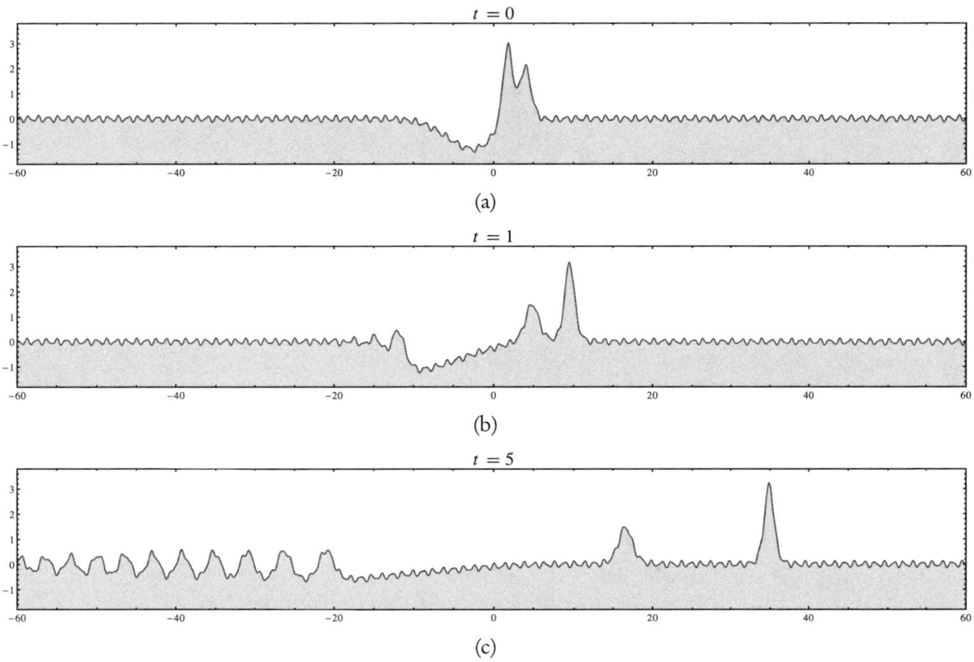

Figure 12.7. *The numerical dressing method applied to compute a solution of the KdV equation that contains two solitons, a genus-two solution, and dispersion.* (a) *The initial condition.* (b) *A plot of the solution at $t = 1$.* (c) *A plot of the solution at $t = 5$. Figure © IOP Publishing and London Mathematical Society. Reproduced with permission. All rights reserved* [114].

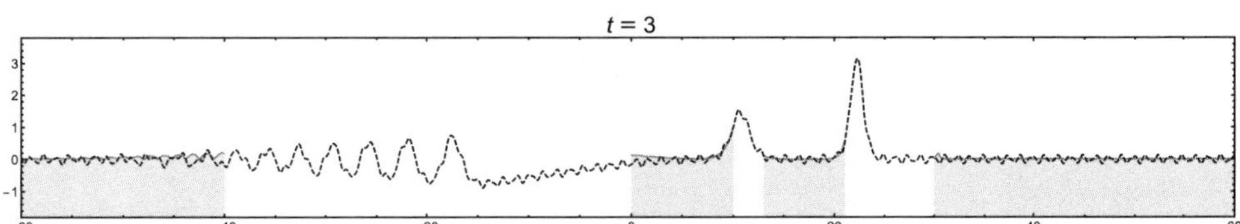

Figure 12.8. *A demonstration of the different regions in a two-gap, two-soliton solution. We numerically solve the RH problem created through the dressing method with $R^{-1}(x,t,z)V_2(z)R(x,t,z)$ defined on $G_+ \cup G_-$. The solution of the KdV equation obtained through this procedure is subtracted from the solution computed from the full RH problem. Dashed: solution from the full RH problem. Solid: the absolute difference of the two solutions. In this way we see that the solution limits to a different genus-two solution in each region. Figure © IOP Publishing and London Mathematical Society. Reproduced with permission. All rights reserved* [114].

12.2 ▪ A numerical dressing method for the defocusing NLS equation

With the ideas of the dressing method established for the KdV equation we turn to the defocusing NLS equation. Consider the RH problem

$$\varphi^+(s) = \varphi^-(s)\begin{bmatrix} 0 & -e^{-\theta(x,t,s)} \\ e^{\theta(x,t,s)} & 0 \end{bmatrix}, \quad s \in \bigcup_{j=1}^{g} (\alpha_j, \beta_j),$$

$$\varphi(\infty) = I, \quad \theta(x,t,s) = 2izx + 4iz^3 t,$$

where $\alpha_j < \beta_j < \alpha_{j+1}$. We use this definition of θ throughout this section. Similarly, it follows that $q(x,t) = 2\mathrm{i}\lim_{z\to\infty} z\varphi_{12}(x,t,z)$ must be a solution of the defocusing NLS equation because this RH problem has a unique solution. This follows from the theory of RH problems [123]. In [124] it is shown that $q(x,t)$ is a quasi-periodic (finite-genus) solution of the defocusing NLS equation.

It is clear that the methods of the previous section can be used to solve this RH problem. The only difference here, besides the quadratic phase, is that there is no symmetry of the contour across the imaginary axis. This provides no significant change from a computational standpoint. There is an associated g-function: for $j = 1,2,\ldots,g$,

- $\mathcal{G}^+(x,t,s) + \mathcal{G}^-(x,t,s) = 0$ for $s \in (\alpha_j, \beta_j)$,
- $\mathcal{G}^+(x,t,s) - \mathcal{G}^-(x,t,s) = -\theta(x,t,s) + \mathrm{i}\Omega_j(x,t)$ for $s \in (\alpha_j, \beta_j)$, and
- $\mathcal{G}(x,t,z) = \mathcal{O}(z^{-1})$ as $z \to \infty$,

where, again, $\{\Omega_j(x,t)\}_{j=1}^g$ are constants (with respect to z) to be determined. Again, it is straightforward to find a function $\mathcal{G}$ that satisfies the first three properties:

$$\mathcal{G}(x,t,z) \triangleq \sqrt{P(z)} \sum_{j=1}^{g} \left(\int_{\alpha_j}^{\beta_j} \frac{-\theta(x,t,s) + \mathrm{i}\Omega_j(x,t)}{\sqrt{P(s)}^+} \frac{\mathrm{d}s}{s-z} \right),$$

where $P(z) = \prod_{j=1}^{g} \left[(z-\beta_j)(z-\alpha_j) \right]$. Similar moment conditions apply:

$$0 = \sum_{j=1}^{g} \int_{\alpha_j}^{\beta_j} \frac{-\theta(x,t,s) + \mathrm{i}\Omega_j(x,t)}{\sqrt{P(s)}^+} s^m \mathrm{d}s, \qquad m = 0,1,\ldots,g-1. \tag{12.13}$$

This produces a linear system for $\{\Omega_j(x,t)\}_{j=1}^g$. In the same way as in the previous section, define

$$G(x,t,z) \triangleq \begin{bmatrix} \mathrm{e}^{-\mathcal{G}(x,t,z)} & 0 \\ 0 & \mathrm{e}^{\mathcal{G}(x,t,z)} \end{bmatrix}$$

and the matrix-valued function

$$\Lambda(x,t,z) \triangleq \varphi(x,t,z) G(x,t,z).$$

Thus

$$\Lambda^+(x,t,s) = \Lambda^-(x,t,s) \begin{bmatrix} 0 & -\mathrm{e}^{-\mathrm{i}\Omega_j(x,t)} \\ \mathrm{e}^{\mathrm{i}\Omega_j(x,t)} & 0 \end{bmatrix} \quad \text{if } s \in (\alpha_j, \beta_j)$$

for $j = 1,2,\ldots,g$ with $\Lambda(\infty) = I$. Define

$$R_j(z) = \frac{1}{2} \begin{bmatrix} \nu_j(z) + 1/\nu_j(z) & \mathrm{i}\mathrm{e}^{-\mathrm{i}\Omega(x,t)}(\nu_j(z) - 1/\nu_j(z)) \\ -\mathrm{i}\mathrm{e}^{\mathrm{i}\Omega(x,t)}(\nu_j(z) - 1/\nu_j(z)) & \nu_j(z) + 1/\nu_j(z) \end{bmatrix},$$

$$\nu_j(z) = \left(\frac{z-\alpha_j}{z-\beta_j} \right)^{1/4}.$$

It follows that R_j satisfies the same jump as Λ in a neighborhood of (α_j,β_j). Let C_j be a clockwise-oriented piecewise-smooth contour lying solely in the right-half plane surrounding (α_j,β_j) but not intersecting or surrounding (α_i,β_i) for $i \neq j$. Define D_j to be the component of $\mathbb{C}\setminus C_j$ that contains the interval C_j encloses, as in Figure 12.1.

Define

$$K(z) \triangleq K(x,t,z) \triangleq \begin{cases} \Lambda(x,t,z)R_j^{-1}(z) & \text{if } z \in D_j, \\ \Lambda(x,t,z) & \text{otherwise.} \end{cases}$$

Then $K(z)$ solves the following RH problem:

$$K^+(s) = K^-(s)R_j(s), \quad s \in C_j, \quad j = 1,2,\ldots,g, \quad K(\infty) = I.$$

This RH problem is solved numerically with the methodology of Chapter 6.

12.2.1 ▪ Superposition of solutions

Now, solutions of the initial-value problem and finite-genus solutions are combined using Definition 12.2 with Proposition 12.1 replaced by Proposition 3.9, the NLS analogue. Note that one less symmetry requirement is present in Proposition 3.9 because solutions of the NLS equations are complex-valued.

Consider the choice of V_1 as in (12.11) where $\rho(z)$ is a truncation (see Remark 12.1.3) of the reflection coefficient obtained from $q_0(x) = 1/2\,\mathrm{sech}(x/2)\mathrm{e}^{\mathrm{i}x}$. Then V_2 is chosen in a manner similar to (12.12):

$$V_2(z) = \begin{cases} \begin{bmatrix} 0 & -1 \\ 1 & 0 \end{bmatrix} & \text{if } z \in \mathcal{S}, \\ I & \text{otherwise,} \end{cases} \qquad \mathcal{S} = \bigcup_j (\alpha_j,\beta_j).$$

Consider the numerical solution of (12.10), the nonlinear superposition of the solution of the initial-value problem and a finite-genus solution.

Thus, the following RH problem is solved:

$$\Phi_4^+(z) = \Phi_4^-(z) \begin{cases} \mathrm{e}^{-\theta(x,t,z)/2\hat{\sigma}_3} V_1(z) & \text{if } z \in [-\ell,\ell], \\ \mathrm{e}^{-\theta(x,t,z)/2\hat{\sigma}_3} V_2(z) & \text{if } z \in \mathcal{S}, \end{cases} \qquad \Phi_4(\infty) = I.$$

We display the full RH problem for the superposition solutions in Figure 12.9. The deformations involve conjugation by a function as in the previous section. In the case of the defocusing NLS equation the conjugate is simplest and it only involves $\Delta(z;z_0)$ (see (9.15)).

Genus-one superposition: For this case let $(\alpha_1,\beta_1) = (6,6.1)$. A plot of the solution of the defocusing NLS equation when $V_1(z) = I$, i.e., a purely finite-genus solution, is shown in Figure 12.10(a). The purely dispersive solution with $V_2(z) = I$ is shown in Figure 12.10(b). The combination solution obtained by solving the RH problem with jump $V_2(z)V_1(z)$ is shown in Figure 12.11. When $t = 1$ we subtract the solution in Figure 12.10(a) and the purely dispersive solution in Figure 12.10(b) from the solution shown in Figure 12.11(b). If superposition acted linearly, we would obtain something identically (numerically) zero. This is not the case, and we see that the effect of superposition is a nonlocal phenomenon: the solution near $+\infty$ is a phase-shifted version of the solution near $-\infty$. See Figure 12.12 for a demonstration of this.

Genus-two superposition: We let $(\alpha_1,\beta_1) = (6,6.05)$ and $(\alpha_2,\beta_2) = (-7.4,-7.3)$. A plot of the solution of the defocusing NLS equation when $V_1(z) = I$, i.e., a purely finite-genus solution, is shown in Figure 12.13. The combination solution obtained by solving the RH problem with jump $V_2(z)V_1(z)$ is shown in Figure 12.14.

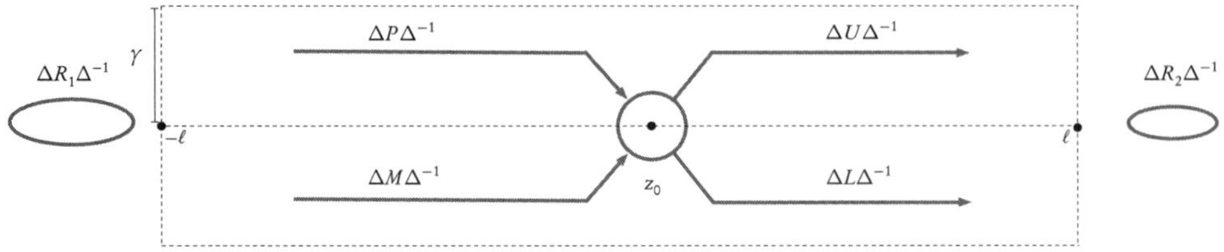

Figure 12.9. *The full RH problem that is solved to compute combination solutions of the defocusing NLS equation. The elliptical contours represent C_1 and C_2. Note that these curves are located outside $[-\ell,\ell]$.*

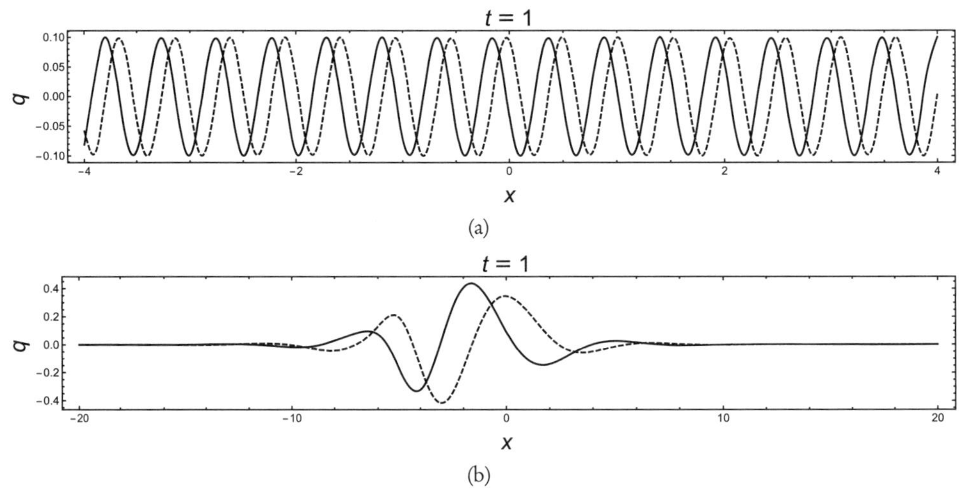

Figure 12.10. (a) *A genus-one solution of the defocusing NLS equation. Here $(\alpha_1,\beta_1) = (7,7.05)$ and $V_1(z) = I$.* (b) *A purely dispersive solution of the defocusing NLS equation with $q_0(x) = 1/2\,\mathrm{sech}(x/2)e^{ix}$.*

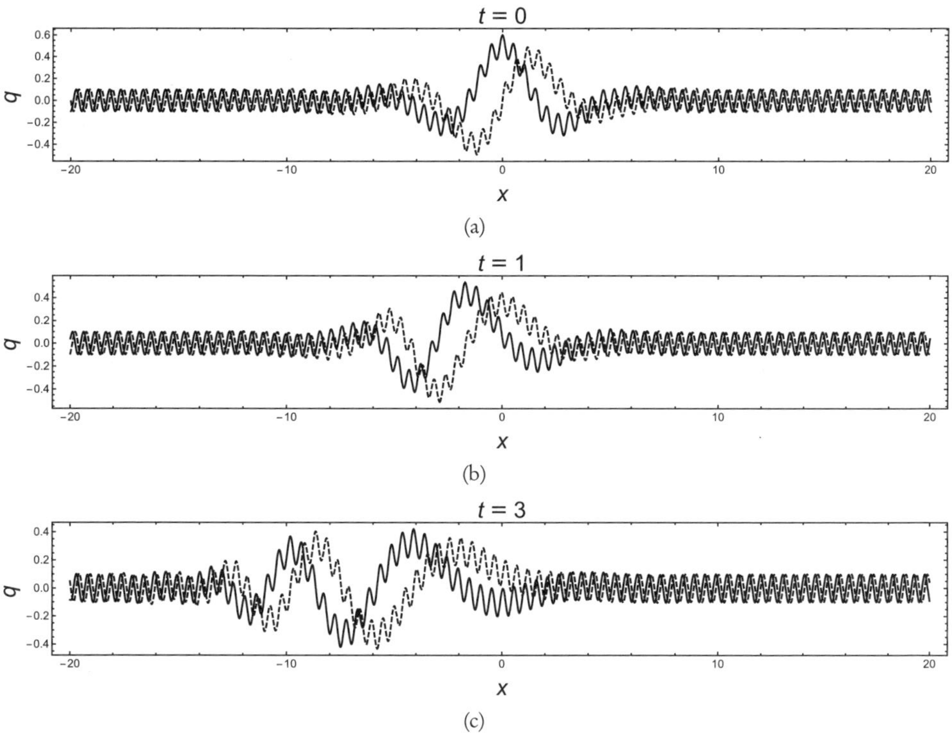

Figure 12.11. *Plot of the solution of the defocusing NLS equation that is a nonlinear superposition of the solution in Figures* 12.10(a) *and* 12.10(b). (a) $t = 0$. (b) $t = 1$. (c) $t = 3$.

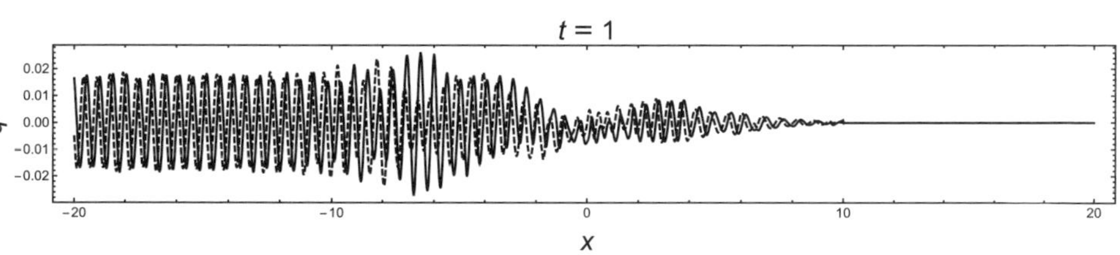

Figure 12.12. *The solution in Figure* 12.10(a) *subtracted from the solution shown in Figure* 12.10(b) *when* $t = 3$.

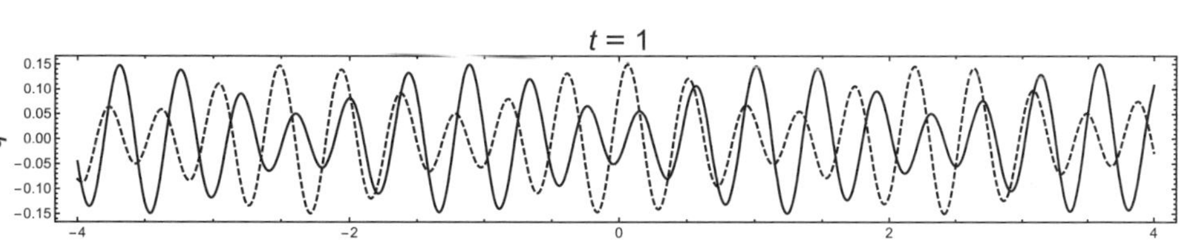

Figure 12.13. *A genus-two solution of the defocusing NLS equation. Here* $(\alpha_1, \beta_1) = (6, 6.05)$ *and* $(\alpha_2, \beta_2) = (-7.4, -7.3)$ *while* $V_1(z) = I$.

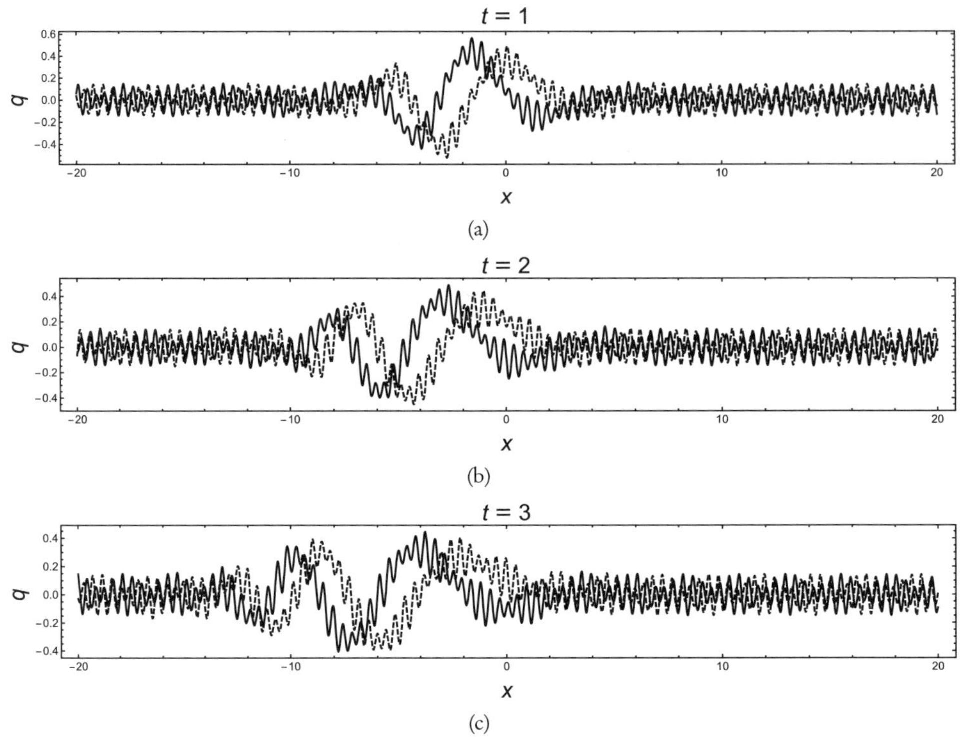

Figure 12.14. *Plot of the solution of the defocusing NLS equation that is a nonlinear superposition of the solution in Figures* 12.13 *and* 12.10(b). (a) $t = 1$. (b) $t = 2$. (c) $t = 3$.

Part IV

Appendices

Appendix A

Function Spaces and Functional Analysis

In this appendix we state (and restate) some results and definitions that are required in the text.

A.0.1 ▪ Bounded variation

Definition A.1. *A partition $P = \{x_1, \ldots, x_n\}$ over $[a, b]$, $b > a$, is a set of points that split the interval into subintervals:*

$$a = x_1 < x_2 < \cdots < x_n = b.$$

Define the variation over P by

$$\mathscr{V}_P[f] \triangleq \sum_{j=1}^{n-1} \left|\Delta_j f\right|,$$

where $\Delta_j f = f(x_{j+1}) - f(x_j)$. The total variation is then

$$\mathscr{V}[f] \triangleq \sup_P \mathscr{V}_P[f].$$

If $f(x) = \int_a^x g(s)\mathrm{d}s$ for $g \in L^1([a, b])$, then

$$\mathscr{V}[f] = \int_a^b |g(s)|\,\mathrm{d}s.$$

A.1 ▪ Banach spaces

We introduce a series of Banach spaces that are used in the text. Generally, we use $||\diamond||_X$ to denote the norm on the Banach space X. We may omit the subscript when the space is clear from the context.

A.1.1 ▪ Continuous functions

The Banach space of k-times continuously differentiable functions on a set A is denoted by $C^k(A)$. The norm on the space is given by

$$||f||_{C^k(A)} = \sum_{j=0}^{k} ||f^{(j)}||_u, \quad ||f||_u = \sup_{x \in A} |f(x)|.$$

An important (non-Banach) subspace of $C^k(A)$ is the space $C_c^k(A)$ of functions f with $\overline{\operatorname{supp} f} \subset B \subset A$ where B is an open, proper subset of A and where $\operatorname{supp} f \triangleq \{x \in A : f(x) \neq 0\}$ is compact. When considering the periodic interval $\mathbb{T} = [-\pi, \pi)$ we use periodic spaces:

$$C^k(\mathbb{T}) = \{f \in C^k([-\pi,\pi]) : f^{(j)}(\pi) = f^{(j)}(-\pi),\ j = 0,1,\ldots,k\},$$

equipped with the same norm as $C^k([-\pi,\pi])$. When $A = \Gamma$ is a self-intersecting contour with a decomposition $\Gamma = \Gamma_1 \cup \cdots \cup \Gamma_L$ where each Γ_j is a non-self-intersecting arc, we define

$$C^k(\Gamma) = \bigoplus_{j=1}^{L} C^k(\Gamma_j). \tag{A.1}$$

Note that this space consists of piecewise continuous functions and the locations of discontinuities depend on the decomposition. The choice of decomposition will always be clear from the context.

A.1.2 ▪ Lebesgue spaces

Unless otherwise stated, we assume that $\Gamma \subset \mathbb{C}$ is a piecewise-smooth oriented contour with at most a finite number of transverse self-intersections. Additionally we assume that any unbounded component of Γ tends to a straight line at infinity. Represent $\Gamma = \Gamma_1 \cup \cdots \cup \Gamma_L$ where each Γ_i is a non-self-intersecting arc.

Definition A.2.

$$L^p(\Gamma) \triangleq L^p(\Gamma, |\mathrm{d}s|) \triangleq \left\{ f \textit{ complex-valued on } \Gamma : f|_{\Gamma_i} \textit{ measurable}, \int_{\Gamma_i} |f(s)|^p |\mathrm{d}s| < \infty \forall i \right\},$$

where $|\mathrm{d}s|$ *represents the arclength measure.*

We make $L^p(\Gamma)$ into a Banach space by equipping it with the norm

$$\|f\|_{L^p(\Gamma)} \triangleq \left(\sum_{i=1}^{n} \int_{\Gamma_i} |f(s)|^p |\mathrm{d}s| \right)^{1/p}.$$

The notation $\|f\|_p \triangleq \|f\|_{L^p(\Gamma)}$ is used when Γ is clear from the context. We require some additional results concerning integration.

Theorem A.3 (Dominated convergence). *Let* $(f_n)_{n\geq 0}$ *be a sequence of functions in* $L^1(X,\mu)$ *such that*

1. $f_n(x) \to f(x)$ *for* μ*-a.e.* x*, and*
2. *there exists* $g \in L^1(X,\mu)$ *such that* $f_n(x) \leq g(x)$ *for* μ*-a.e.* x *and for every* n.

Then $\int_X f(x)\mathrm{d}\mu(x) = \lim_{n\to\infty} \int_X f_n(x)\mathrm{d}\mu(x)$.

The following result is a direct consequence of the dominated convergence theorem.

Theorem A.4 (See [54]). *Suppose* $f : X \times [a,b] \to \mathbb{C}$ *and that* $f(\diamond, t) : X \to \mathbb{C}$ *is integrable with respect to a measure* μ *for each fixed* $t \in [a,b]$. *Let* $F(t) = \int_X f(x,t)\mathrm{d}\mu(x)$.

1. *Suppose that there exists* $g \in L^1(X,\mu)$ *such that* $|f(x,t)| \le g(x)$ *for* μ*-a.e.* x *and for all* t. *If* $\lim_{t\to t_0} f(x,t) = f(x,t_0)$ *for* μ*-a.e.* x, *then* $\lim_{t\to t_0} F(t) = F(t_0)$.

2. *Suppose that* $\partial_t f(x,t)$ *exists and there is a* $g \in L^1(X,\mu)$ *such that* $|\partial_t f(x,t)| \le g(x)$ *for* μ*-a.e.* x *and for all* t. *Then* F *is differentiable and* $F'(t) = \int_X \partial_t f(x,t)\mathrm{d}\mu(x)$.

A.1.3 ▪ Sobolev spaces

Let Γ be a smooth contour and let $f : \Gamma \to \mathbb{C}$. We say f is Γ-differentiable if for each $s^* \in \Gamma$ there exists a value $f'(s^*)$ such that for $s \in \Gamma$

$$f(s) = f(s^*) + f'(s^*)(s-s^*) + (s-s^*)E_{s^*}(s-s^*),$$

where $|E_{s^*}(s-s^*)| \to 0$ as $|s-s^*| \to 0$. Note that this is weaker than complex differentiability since we restrict ourselves to $s \in \Gamma$.

Example A.5. The function $f(z) = |z|^2$ is Γ-differentiable for any smooth contour Γ, but it is nowhere analytic. ■

Our aim is to define D, the distributional differentiation operator for functions defined on Γ. For a function $\varphi \in C_c^\infty(\Gamma)$ we represent a linear functional f via the dual pairing $f(\varphi) \triangleq \langle f, \varphi \rangle_\Gamma$. To Df we associate the functional $-\langle f, \varphi' \rangle_\Gamma$. We are interested in the case where the distribution Df corresponds to a locally integrable function

$$\langle Df, \varphi \rangle_\Gamma = \int_\Gamma Df(s)\varphi(s)\mathrm{d}s = -\int_\Gamma f(s)\varphi'(s)\mathrm{d}s.$$

Definition A.6. *Define*

$$H^k(\Gamma) \triangleq \left\{ f \in L^2(\Gamma) : D^j f \in L^2(\Gamma),\ j = 0,\ldots,k \right\},$$

with norm

$$\|f\|_{H^k(\Gamma)}^2 \triangleq \sum_{j=0}^{k} \|D^j f\|_{L^2(\Gamma)}^2.$$

We write $W^{k,\infty}(\Gamma)$ for the Sobolev space with the L^2 norm replaced by the L^∞ norm. An important observation is that we are dealing with matrix-valued functions, and hence the definitions of all these spaces must be suitably extended. Since all finite-dimensional norms are equivalent, we can use the above definitions in conjunction with any matrix norm to define a norm for matrix-valued functions provided the norm is subadditive; see Appendix A.3. If the Hilbert space structure of $H^k(\Gamma)$ is needed, then a specific choice of the matrix norm is necessary so that it originates from an inner product.

It is clear that this construction works in any L^p space allowing for the definition of the space $W^{k,p}(\Gamma)$. Further development of Sobolev spaces is presented in Section 2.6.

Lemma A.7 (See [83]). *Let* Γ *be a smooth, nonclosed curve oriented from* $z = a$ *to* $z = b$. *If* $f \in H^1(\Gamma)$, *then*

$$D\int_\Gamma \frac{f(\zeta)}{\zeta - s}\mathrm{d}\zeta = \frac{f(a)}{a-s} - \frac{f(b)}{b-s} + \int_\Gamma \frac{Df(\zeta)}{\zeta - s}\mathrm{d}\zeta.$$

Proof. Let $g \in C_c^\infty(\Gamma)$. Let $s \in \Gamma$, $s \neq a, b$, and we choose the branch cut for $\log(z-s)$ to be a curve that passes from s to the point at infinity, along Γ. Define for ϵ sufficiently small

$$G_\epsilon(s) = \int_{\Gamma \setminus \Gamma_\epsilon} g'(\zeta) \log(\zeta - s) \mathrm{d}\zeta,$$

where $\Gamma_\epsilon = B(s,\epsilon) \cap \Gamma$ is an arc from s_1 to s_2. Integration by parts shows

$$\begin{aligned} G_\epsilon(s) = &-\int_{\Gamma\setminus\Gamma_\epsilon} \frac{g(\zeta)}{\zeta - s}\mathrm{d}\zeta \\ &+ g(b)\log(b-s) - g(a)\log(a-s) + g(s_1)\log(s_1 - s) - g(s_2)\log(s_2 - s). \end{aligned} \tag{A.2}$$

By definition

$$\begin{aligned} \mathscr{R}_\Gamma g(s) &= \fint_\Gamma \frac{g(\zeta)}{\zeta - s}\mathrm{d}\zeta = \lim_{\epsilon \downarrow 0} \int_{\Gamma\setminus\Gamma_\epsilon} \frac{g(\zeta)}{\zeta - s}\mathrm{d}\zeta, \\ \mathscr{R}_\Gamma &\triangleq \pi \mathrm{i}(\mathscr{C}_\Gamma^+ + \mathscr{C}_\Gamma^-). \end{aligned}$$

Next, write

$$\begin{aligned} g(s_1)\log(s_1 - s) - g(s_2)\log(s_2 - s) = &(g(s_1) - g(s))\log(s_1 - s) - (g(s_2) - g(s))\log(s_2 - s) \\ &+ g(s)\log\left(\frac{s_1 - s}{s_2 - s}\right). \end{aligned}$$

It is clear that this converges to $\mathrm{i}\pi g(s)$ as $\epsilon \downarrow 0$ from the fact that the contour is smooth so that

$$\lim_{\epsilon \downarrow 0} G_\epsilon(s) = \pi \mathrm{i} g(s) - \mathscr{R}_\Gamma g(s) + g(b)\log(b-s) - g(a)\log(a-s).$$

We compute $G_\epsilon'(s)$, noting that s_1 and s_2 depend on s:

$$\begin{aligned} G_\epsilon'(s) = &-\int_{\Gamma\setminus\Gamma_\epsilon} \frac{g'(\zeta)}{\zeta - s}\mathrm{d}\zeta \\ &+ g'(s_1)\log(s_1 - s)s_1'(s) - g'(s_2)\log(s_2 - s)s_2'(s). \end{aligned}$$

We claim that $s_1(s) = s + \epsilon e^{\mathrm{i}\theta(s)}$ for some smooth function $\theta(s)$ that has a bounded derivative as $\epsilon \downarrow 0$. Therefore, $s_1'(s) \to 1$ as $\epsilon \downarrow 0$. To see this more precisely, let $\gamma(x) : [-1,1] \to \Gamma$ be a smooth parameterization, and let $x_1(s,\epsilon)$ be such that $\gamma(x_1(s,\epsilon)) = s_1$ and $x(s) = \gamma^{-1}(s)$. We are looking to solve

$$|\gamma(x_1) - \gamma(x)| = \epsilon \quad \text{or} \quad (\gamma(x_1) - \gamma(x))(\overline{\gamma}(x_1) - \overline{\gamma}(x)) = \epsilon^2.$$

Express $x_1 = x + \epsilon\tau$, and we solve for τ as a function of the other variables. Examine

$$\frac{(\gamma(x + \epsilon\tau) - \gamma(x))(\overline{\gamma}(x + \epsilon\tau) - \overline{\gamma}(x))}{\epsilon^2} - 1 = F(x,\epsilon,\tau) = 0$$

by a Taylor expansion $\gamma(x + \epsilon\tau) = \gamma(x) + \gamma'(x)\epsilon\tau + \mathscr{O}(\epsilon^2\tau^2)$,

$$|\gamma'(x)|^2\tau^2 + \mathscr{O}(\epsilon^2\tau^4) - 1 = 0. \tag{A.3}$$

This clearly gives a positive and negative choice for τ, $\tau_{\pm} \approx \pm|\gamma'(x)|^{-1}$, and the implicit function theorem can be applied here. Making the negative choice, we have found $\tau_{-}(\epsilon, x)$, depending smoothly on both arguments. To examine the partial of τ_{-} with respect to x we have

$$0 = \frac{\mathrm{d}}{\mathrm{d}x} F(x, \epsilon, \tau_{-}(\epsilon, x)) \Rightarrow \partial_x \tau_{-}(\epsilon, x) = -\frac{F_x(x, \epsilon, \tau_{-}(\epsilon, x))}{F_\tau(x, \epsilon, \tau_{-}(\epsilon, x))}.$$

For sufficiently small ϵ, we know that F_τ does not vanish, the numerator must be bounded, and hence $|\partial_x \tau_{-}(\epsilon, x)|$ is uniformly bounded. We then have

$$\begin{aligned} s_1(s) &= \gamma(x(s) + \epsilon \tau_{-}(\epsilon, x(s))), \\ s_1'(s) &= \gamma'(x(s) + \epsilon \tau_{-}(\epsilon, x(s)))(x'(s) + \epsilon \partial_x \tau_{-}(\epsilon, x(s)) x'(s)) = 1 + \mathcal{O}(\epsilon). \end{aligned}$$

Similar calculations follow for s_2 and τ_{+}.

It is then seen that

$$\lim_{\epsilon \downarrow 0} G_\epsilon'(s) = \pi i g'(s) - \mathcal{R}_\Gamma[g'](s).$$

Now we show that $\lim_{\epsilon \downarrow 0} G_\epsilon'(s) = \frac{\mathrm{d}}{\mathrm{d}s} \lim_{\epsilon \downarrow 0} G_\epsilon(s)$, i.e., that the limit commutes with differentiation. A sufficient condition for this is that $G_\epsilon'(s)$ converges uniformly in a neighborhood of s as $\epsilon \downarrow 0$; see [105, Theorem 7.17]. This limit can be seen to be uniform because g' satisfies a uniform α-Hölder condition: $|g'(s_1) - g'(s_2)| \leq \Lambda |s_1 - s_2|^\alpha$. Consider

$$\begin{aligned} \int_{\Gamma \setminus \Gamma_\epsilon} \frac{g'(\zeta)}{\zeta - s} \mathrm{d}\zeta = \int_\Gamma \frac{g'(\zeta) - g'(s)}{\zeta - s} \mathrm{d}\zeta - \int_{\Gamma_\epsilon} \frac{g'(\zeta) - g'(s)}{\zeta - s} \mathrm{d}\zeta \\ + g'(s) \log\left(\frac{b - s}{a - s}\right) - g'(s) \log\left(\frac{s_2 - s}{s_1 - s}\right), \end{aligned} \tag{A.4}$$

$$\int_{\Gamma_\epsilon} \left| \frac{g'(\zeta) - g'(s)}{\zeta - s} \right| |\mathrm{d}\zeta| \leq \Lambda \int_{\Gamma_\epsilon} |\zeta - s|^{\alpha - 1} |\mathrm{d}\zeta| \leq 2C\Lambda\alpha^{-1}\epsilon^\alpha$$

from Lemma 2.5, which is clearly uniform in s. Now, examine

$$\begin{aligned} &\log\left(\frac{s_2 - s}{s_1 - s}\right), \quad \text{where} \\ &s_1(s) = s + \gamma'(x(s)) \epsilon \tau_{-}(\epsilon, x(s)) + \mathcal{O}(\epsilon^2), \\ &s_2(s) = s + \gamma'(x(s)) \epsilon \tau_{+}(\epsilon, x(s)) + \mathcal{O}(\epsilon^2). \end{aligned}$$

Then

$$\frac{s_2 - s}{s_1 - s} = \frac{\tau_{-}(\epsilon, x(s))}{\tau_{+}(\epsilon, x(s))} + \mathcal{O}(\epsilon) = -1 + \mathcal{O}(\epsilon)$$

from (A.3) because $\tau_{\pm}$ are bounded. This demonstrates that (A.4) converges uniformly in a neighborhood of s as $\epsilon \downarrow 0$. Thus $\frac{\mathrm{d}}{\mathrm{d}s} \lim_{\epsilon \downarrow 0} G_\epsilon(s) = \lim_{\epsilon \downarrow 0} G_\epsilon'(s)$ and (A.2) proves the lemma in the case that $g \in C_c^\infty(\Gamma)$.

For the general case we use density. Let $g_n \to f$ in $H^1(\Gamma)$, where $g_n \in C_c^\infty(\Gamma)$. Let Γ_0 be a smooth contour that either intersects Γ transversally, is disjoint from Γ, or coincides with a subarc of Γ. Further assume neither a nor b is in the interior of Γ_0. From the boundedness of $\mathscr{C}_\Gamma^\pm : L^2(\Gamma) \to L^2(\Gamma_0)$ (see Corollary 2.42) we have that

$$\mathscr{R}_\Gamma g_n \to \mathscr{R}_\Gamma f, \quad \mathscr{R}_\Gamma g_n' \to \mathscr{R}_\Gamma Df,$$
$$\frac{g_n(a)}{a-s} \to \frac{f(a)}{a-s}, \quad \frac{g_n(b)}{b-s} \to \frac{f(b)}{b-s},$$

in $L^2(\operatorname{supp}\varphi)$, for any $\varphi \in C_c^\infty(\Gamma_0)$. From the definition of the weak derivative we must show

$$\mathscr{I}[f] \triangleq \int_{\Gamma_0} \left(\mathscr{R}_\Gamma f(s)\varphi'(s) + \left(\frac{f(a)}{a-s} - \frac{f(b)}{b-s} + \mathscr{R}_\Gamma[Df](s) \right)\varphi(s) \right) \mathrm{d}s = 0$$

for all $\varphi \in C_c^\infty(\Gamma_0)$. This follows from the fact that $\mathscr{I}[g_n] = 0$ and $\mathscr{I}[g_n] \to \mathscr{I}[f]$ because all n-dependent functions in $\mathscr{I}[g_n]$ converge in $L^2(\operatorname{supp}\varphi)$. This proves the lemma for $f \in H^1(\Gamma)$. □

A.1.4 ▪ Hölder spaces

Definition A.8. *Given a domain $\Omega \subset \mathbb{C}$, a function $f : \Omega \to \mathbb{C}$ is α-Hölder continuous on Ω if for each $s \in \Omega$ there exists $\Lambda(s)$, $\delta(s) > 0$ such that*

$$|f(s) - f(s^*)| \leq \Lambda(s)|s - s^*|^\alpha \text{ for } |s - s^*| < \delta(s).$$

Note that this definition is useful when $\alpha \in (0,1]$. If $\alpha = 1$, f is Lipschitz and if $\alpha > 1$, f must be constant.

Definition A.9. *A function $f : \Gamma \to \mathbb{C}$ is uniformly α-Hölder continuous on a bounded contour Γ if Λ and δ can be chosen independently of s.*

Definition A.10. *For Γ smooth, bounded, and closed, the Banach space $C^{0,\alpha}(\Gamma)$ consists of uniformly α-Hölder continuous functions with norm*

$$\|f\|_{0,\alpha} \triangleq \sup_{s\in\Gamma} |f(s)| + |f|_{0,\alpha},$$

where

$$|f|_{0,\alpha} \triangleq \sup_{s_1 \neq s_2,\ s_1, s_2 \in \Gamma} \frac{|f(s_1) - f(s_2)|}{|s_1 - s_2|^\alpha}.$$

A.1.5 ▪ ℓ^p spaces

The numerical methods we develop hinge on representing solutions to linear equations in a chosen basis, in which case it is natural to consider spaces on the coefficients of a function in the basis. We use the following standard spaces.

Definition A.11. *For $S \subset \mathbb{Z}$, we define $\ell^p(S)$ as the Banach space with norms*

$$\|f\|_{\ell^p(S)} \triangleq \left(\sum_{k\in S} |f_k|^p\right)^{1/p} \qquad \text{and} \qquad \|f\|_{\ell^\infty(S)} \triangleq \sup_{k\in S} |f_k|,$$

where $f = [f_k]_{k\in S}$. We typically take $S = \mathbb{Z} = \{\ldots,-1,0,1,\ldots\}$, $S = \mathbb{N} = \{0,1,2,3,\ldots\}$, or $S = \mathbb{N}^+ = \{1,2,3,\ldots\}$. S is usually implied by the context, in which case we use the notation ℓ^p.

We mostly work in ℓ^1, ℓ^2, and ℓ^∞. The space $\ell^2(S)$ is distinguished because it is also a *Hilbert space* with inner product

$$\langle f, g\rangle_{\ell^2(S)} = \sum_{k\in S} \overline{f_k} g_k.$$

The space ℓ^1 is particularly important in the context of series expansion because it corresponds to evaluation being a bounded operator, provided the basis itself is uniformly bounded. In the context of the Fourier series, it also has the property that it defines the *Weiner algebra*: if the Fourier coefficients f and g both satisfy $\mathcal{F}f, \mathcal{F}g \in \ell^1$, then the Fourier coefficients of fg satisfy $\mathcal{F}[fg] \in \ell^1$. (See Definition B.1 for the definition of $\mathcal{F}$.)

A.1.6 ▪ Decaying coefficient spaces

Additional regularity in a function corresponds to faster decay in the coefficients. We introduce the following norms to capture algebraic decay.

Definition A.12. *$\ell^{\lambda,p}(S)$ is the Banach space with norm*

$$\|f\|_{\ell^{\lambda,p}(S)} \triangleq \left(\sum_{k\in S} \left|(|k|+1)^\lambda f_k\right|^p\right)^{1/p}.$$

In other words, λ imposes algebraic decay. We use $\ell^{\lambda,p}$ when S is implied by the context.

The $\ell^{\lambda,p}$ spaces are coefficient space analogues of Sobolev spaces $W^{\lambda,p}$. Indeed, $\mathcal{F}f \in \ell^{\lambda,2}$ is equivalent to $f \in H^\lambda(\mathbb{T})$. Analyticity in a function induces exponential decay in the coefficients, which we encapsulate via the following.

Definition A.13. *$\ell^{(\lambda,R),p}(S)$ is the Banach space with norm*

$$\|f\|_{\ell^{(\lambda,R),p}(S)} \triangleq \left(\sum_{k\in S} \left|R^{|k|}(|k|+1)^\lambda f_k\right|^p\right)^{1/p}.$$

We use $\ell^{(\lambda,R),p}$ when S is implied by the context.

A.1.7 ▪ Zero-sum spaces

The following space is used to encode when a function in a basis vanishes at a point.

Definition A.14. *The vanishing spaces $\ell^{(\lambda,R),p}_{\pm z}(S)$ consist of $f \in \ell^{(\lambda,R),p}(S)$ such that*

$$\sum_{k\in S} (\pm)^k f_k = 0,$$

and $\ell_{\mathrm{z}}^{(\lambda,R),p}(S)=\ell_{-\mathrm{z}}^{(\lambda,R),p}(S)\cap\ell_{+\mathrm{z}}^{(\lambda,R),p}(S)$. *We omit the dependency on* S *when it is implied by the context.*

A.2 ▪ Linear operators

Let X and Y be Banach spaces. Denote the set of bounded operators from X to Y by $\mathscr{L}(X,Y)$, and we equip the space with its standard induced operator norm, which makes the space into a Banach algebra if $X=Y$.

Definition A.15. *$\mathscr{K}\in\mathscr{L}(X,Y)$ is said to be compact if the image of a bounded set in X is precompact in Y. Or, equivalently, the image of the unit ball in X is precompact in Y. Let $\mathscr{K}(X,Y)$ denote the closed subspace of compact operators in $\mathscr{L}(X,Y)$.*

In a Hilbert space X a useful property of compact operators is that they are limits in the norm of a sequence of finite-rank, bounded operators. Explicitly, $\mathscr{K}$ is compact if and only if there exists a sequence $\mathscr{K}_n$ of finite-rank operators satisfying

$$||\mathscr{K}-\mathscr{K}_n||_{\mathscr{L}(X)}\to 0.$$

For discrete function spaces such as ℓ^2, the natural choice is $\mathscr{K}_n=\mathscr{I}_n\mathscr{K}\mathscr{I}_n$, where $\mathscr{I}_n$ is a rank n projection operator. These ideas allow us to define a (semi-)Fredholm operator.

Definition A.16. *$\mathscr{T}\in\mathscr{L}(X)$ is called left (right) semi-Fredholm if there exists $\mathscr{S}\in\mathscr{L}(X)$, called a regulator (or* Fredholm regulator*), such that $\mathscr{S}\mathscr{T}=I+\mathscr{K}$ ($\mathscr{T}\mathscr{S}=I+\mathscr{K}$) for some $\mathscr{K}\in\mathscr{K}(X)$. An operator is Fredholm if it is both left and right semi-Fredholm.*

Theorem A.17 (See [41]). *If X is a Banach space, then an operator $\mathscr{T}$ on X is a Fredholm operator if and only if* $\dim\ker\mathscr{T}$ *and* $\operatorname{codim}\operatorname{ran}T\triangleq\dim(X/\operatorname{ran}\mathscr{T})$ *are finite.*

Definition A.18. *The* Fredholm index *denoted by* $\operatorname{ind}\mathscr{T}$ *is defined by*

$$\operatorname{ind}\mathscr{T}=\dim\ker\mathscr{T}-\operatorname{codim}\operatorname{ran}\mathscr{T}.$$

We include three perturbation theorems from operator theory that are of great use.

Theorem A.19. *Let $\mathscr{T}$ be a Fredholm operator with index κ. There exists $\epsilon>0$ such that, for all operators $\mathscr{S}$, if $||\mathscr{S}-\mathscr{T}||<\epsilon$, then $\mathscr{S}$ is Fredholm with index κ.*

Theorem A.20 (See [7]). *Let X and Y be two normed linear spaces with at least one being a Banach space. Let $\mathscr{L}\in\mathscr{L}(X,Y)$ and $\mathscr{L}^{-1}\in\mathscr{L}(Y,X)$. Let $\mathscr{M}\in\mathscr{L}(X,Y)$ satisfy*

$$||\mathscr{M}-\mathscr{L}||<\frac{1}{||\mathscr{L}^{-1}||}.$$

Then $\mathscr{M}$ is invertible, $\mathscr{M}^{-1}\in\mathscr{L}(Y,X)$, and

$$||\mathscr{M}^{-1}||\leq\frac{||\mathscr{L}^{-1}||}{1-||\mathscr{L}^{-1}||||\mathscr{M}-\mathscr{L}||},$$

$$||\mathscr{M}^{-1}-\mathscr{L}^{-1}||\leq\frac{||\mathscr{L}^{-1}||^2||\mathscr{M}-\mathscr{L}||}{1-||\mathscr{L}^{-1}||||\mathscr{M}-\mathscr{L}||}.$$

Theorem A.21. *Let V be a Banach space, $\mathscr{L} \in \mathscr{L}(V)$. Assume that*

$$\sum_{n=0}^{\infty} \|\mathscr{L}^n\|_{\mathscr{L}(V)} < \infty.$$

Then $I - \mathscr{L}$ is invertible and

$$(I - \mathscr{L})^{-1} = \sum_{n=0}^{\infty} \mathscr{L}^n.$$

Proof. Consider

$$(I-\mathscr{L})\sum_{n=0}^{\infty}\mathscr{L}^n = \sum_{n=0}^{\infty}\mathscr{L}^n - \sum_{n=0}^{\infty}\mathscr{L}^{n+1} = I + \sum_{n=1}^{\infty}\mathscr{L}^n - \sum_{n=1}^{\infty}\mathscr{L}^n = I.$$

It is easily seen that same identity holds for right multiplication. □

A.2.1 ▪ Coefficient space operators

We often express (continuous) operators acting on $\ell^{(\lambda,R),p}$ spaces in matrix form. For example, here is an operator acting on $\ell^{(\lambda,R),p}(\mathbb{Z})$:

$$\mathscr{A} = \begin{bmatrix} \ddots & \vdots & \vdots & \vdots & \vdots & \vdots & \iddots \\ \cdots & A_{-2,-2} & A_{-2,-1} & A_{-2,0} & A_{-2,1} & A_{-2,2} & \cdots \\ \cdots & A_{-1,-2} & A_{-1,-1} & A_{-1,0} & A_{-1,1} & A_{-1,2} & \cdots \\ \cdots & A_{0,-2} & A_{0,-1} & \underline{A_{0,0}} & A_{0,1} & A_{0,2} & \cdots \\ \cdots & A_{1,-2} & A_{1,-1} & A_{1,0} & A_{1,1} & A_{1,1} & \cdots \\ \cdots & A_{2,-2} & A_{2,-1} & A_{2,0} & A_{2,1} & A_{2,2} & \cdots \\ \iddots & \vdots & \vdots & \vdots & \vdots & \vdots & \ddots \end{bmatrix}.$$

The operators for which we do this are always banded below or above in such a way that there is no ambiguity in the definition.

The following allows us to easily bound operators between different $\ell^{(\lambda,R),p}$ spaces.

Proposition A.22.

$$\|\mathscr{A}\|_{\mathscr{L}(\ell^{(\lambda,R),p},\ell^{(\alpha,r),q})} = \left\|\mathscr{D}_{\alpha,r}\mathscr{A}\mathscr{D}_{\lambda,R}^{-1}\right\|_{\mathscr{L}(\ell^p,\ell^q)},$$

where

$$\mathscr{D}_{\lambda,R} = \operatorname{diag}\left\{\ldots,3^\lambda R^2, 2^\lambda R, \underline{1}, 2^\lambda R, 3^\lambda R^2,\ldots\right\}.$$

As an illustrative example, if we take $\lambda = \alpha = 0$ and $R = r$, the norm on $\ell^{(0,R),p}(\mathbb{Z})$ is

$$\|\mathscr{A}\|_{\mathscr{L}(\ell^{(0,R),p})} = \left\| \begin{bmatrix} \ddots & \vdots & \vdots & \vdots & \vdots & \vdots & \iddots \\ \cdots & A_{-2,-2} & A_{-2,-1}R & A_{-2,0}R^2 & A_{-2,1}R & A_{-2,2} & \cdots \\ \cdots & \frac{A_{-1,-2}}{R} & A_{-1,-1} & A_{-1,0}R & A_{-1,1} & \frac{A_{-1,2}}{R} & \cdots \\ \cdots & \frac{A_{0,-2}}{R^2} & \frac{A_{0,-1}}{R} & \underline{A_{0,0}} & \frac{A_{0,1}}{R} & \frac{A_{0,2}}{R^2} & \cdots \\ \cdots & \frac{A_{1,-2}}{R} & A_{1,-1} & A_{1,0}R & A_{1,1} & \frac{A_{1,1}}{R} & \cdots \\ \cdots & A_{2,-2} & A_{2,-1}R & A_{2,0}R^2 & A_{2,1}R & A_{2,2} & \cdots \\ \iddots & \vdots & \vdots & \vdots & \vdots & \vdots & \ddots \end{bmatrix} \right\|_{\mathscr{L}(\ell^p)}.$$

A.3 ▪ Matrix-valued functions

Throughout this work we are interested in matrix-valued functions. For $M \in \mathbb{C}^{n\times m}$ we define the appropriate ℓ^p-based, Frobenius-type matrix norms:

$$|M|_p \triangleq \left(\sum_{j=1}^{n}\sum_{i=1}^{n}|M_{ij}|^p\right)^{1/p}.$$

Note that when $p=2$

$$|M|_2 = \sqrt{\operatorname{tr} M^* M},$$

where the * denotes conjugate transpose. This allows us to define the space, abusing notation,

$$L^p(\Gamma) \triangleq \left\{f : \Gamma \to \mathbb{C}^{n\times m}, \quad \text{measurable such that } \||f|_p\|_p < \infty\right\},$$

equipped with the obvious norm. For $p=2$ we set $W^{k,2}(\Gamma) \triangleq H^k(\Gamma)$ and this retains the Hilbert space structure with the inner product

$$\langle f, g\rangle = \operatorname{tr}\sum_{j=1}^{k}\int_\Gamma \overline{(f^{(j)}(z))}^\top g^{(j)}(z)|dz|.$$

Remark A.3.1. *Another important property of $H^k(\Gamma)$ that carries over for submultiplicative matrix norms is that for $k \geq 1$ there exists a constant $c > 0$ such that*

$$\|fg\|_{H^k(\Gamma)} \leq c\|f\|_{H^k(\Gamma)}\|g\|_{H^k(\Gamma)}.$$

This makes the space into an algebra, and, with a redefinition of the norm, the space is a Banach algebra ($c = 1$).

Appendix B
Fourier and Chebyshev Series

B.1 ▪ Fourier series

A function defined on the periodic interval $\mathbb{T} = [-\pi, \pi)$ can be formally represented by its *Fourier series*:

$$f(\theta) \sim \sum_{k=-\infty}^{\infty} \hat{f}_k \mathrm{e}^{\mathrm{i}k\theta},$$

where the *Fourier coefficients* are

$$\hat{f}_k \triangleq \frac{1}{2\pi} \int_{-\pi}^{\pi} f(\theta) \mathrm{e}^{-\mathrm{i}k\theta} \mathrm{d}\theta.$$

We denote the map from a function to its Fourier coefficients as follows.

Definition B.1. *For integrable f, define the* periodic Fourier transform

$$\mathcal{F} f \triangleq \left[\ldots, \hat{f}_{-1}, \hat{f}_0, \hat{f}_1, \ldots\right]^\top.$$

It is well known that $\mathcal{F} : L^2(\mathbb{T}) \to \ell^2$ and that this transformation is invertible: $\mathcal{F}^{-1} : \ell^2 \to L^2(\mathbb{T})$.

Definition B.2. *For $\mathbf{f} = \left[\ldots, f_{-1}, f_0, f_1, \ldots\right]^\top \in \ell^2$, define*

$$\mathcal{F}^{-1}\mathbf{f} = \sum_{k=-\infty}^{\infty} f_k \mathrm{e}^{\mathrm{i}k\theta},$$

where the sum converges in $L^2(\mathbb{T})$.

B.1.1 ▪ ℓ^1 and uniform convergence

For numerical purposes, we want to be able to evaluate the Fourier series pointwise, which requires the Fourier coefficients to converge absolutely:

$$\sum_{k=-\infty}^{\infty} \left|\hat{f}_k\right| < \infty.$$

In other words, we want $\mathcal{F} f \in \ell^1$. We note that this is sufficient to recover the function uniformly.

Proposition B.3. *If f is continuous and $\mathcal{F} f \in \ell^1$, then the Fourier series converges uniformly to f.*

Absolute convergence ensures that the Fourier series converges to a continuous function, i.e., $\mathcal{F}^{-1} : \ell^1 \to C(\mathbb{T})$. Unfortunately, the inverse is not true: continuous functions do not necessarily have Fourier coefficients that converge absolutely. Indeed, there exist continuous functions whose Fourier series do not converge *at all* at certain points.

Remark B.1.1. *While the Fourier series may not converge to f everywhere, a simple averaging of the Fourier coefficients*

$$\sum_{k=-n}^{n} \frac{n+1-|k|}{k+1} \hat{f}_k e^{ik\theta}$$

results in an expansion in trigonometric polynomials that does converge to $f \in C(\mathbb{T})$ pointwise [71, Theorem 1.5].

To ensure that $\mathcal{F} f \in \ell^1$, we describe several sufficient (but not necessary) conditions on the regularity of f. The simplest such condition is twice differentiability.

Proposition B.4. *Suppose $f \in C^2(\mathbb{T})$. Then $\mathcal{F} f \in \ell^1$.*

Proof. Integrate by parts twice:

$$\hat{f}_k = \frac{1}{2\pi} \int_{-\pi}^{\pi} f(\theta) e^{-ik\theta} d\theta = \frac{1}{2\pi i k} \int_{-\pi}^{\pi} f'(\theta) e^{-ik\theta} d\theta = \frac{1}{2\pi (ik)^2} \int_{-\pi}^{\pi} f''(\theta) e^{-ik\theta} d\theta.$$

The integral can be bounded for all k. Thus $\left|\hat{f}_k\right| \leq C\,|k|^{-2}$, and hence $\mathcal{F} f \in \ell^1$. □

We weaken this result slightly by using bounded variation (see Appendix A.0.1).

Lemma B.5. *If f has bounded variation, then*

$$\int_{-\pi}^{\pi} f(\theta) e^{-ik\theta} d\theta = \mathcal{O}(k^{-1}).$$

Proof. Because f is Riemann integrable, for sufficiently large n there exists a partition P satisfying

$$\left| \int_{-\pi}^{\pi} f(\theta) e^{-ik\theta} d\theta - \sum_{j=1}^{n} f(x_j) \int_{x_j}^{x_{j+1}} e^{-ik\theta} d\theta \right| \leq \frac{1}{k}.$$

But we can evaluate this approximation explicitly:

$$\begin{aligned} \sum_{j=1}^{n} f(x_j) \int_{x_j}^{x_{j+1}} e^{-ik\theta} d\theta &= \sum_{j=1}^{n-1} f(x_j) \frac{e^{-ikx_j} - e^{-ikx_{j+1}}}{ik} \\ &= \frac{1}{ik} \left(\left[f(-\pi) - f(x_{n-1}) \right] e^{ik\pi} + \sum_{j=2}^{n} \left[f(x_j) - f(x_{j-1}) \right] e^{-ikx_j} \right). \end{aligned}$$

The absolute value of this term is bounded by

$$\frac{2\|f\|_u + \mathcal{V}_P[f]}{k} \leq \frac{2\|f\|_u + \mathcal{V}[f]}{k}.$$

The lemma follows from the triangle inequality. □

We can now obtain a weaker condition for $\mathcal{F}f \in \ell^1$.

Theorem B.6. *If f is differentiable and f' has bounded variation, then $\mathcal{F}f \in \ell^1$.*

Proof. Integration by parts once gives the decay:

$$\int_{-\pi}^{\pi} f(\theta)e^{-ik\theta}d\theta = \frac{1}{ik}\int_{-\pi}^{\pi} f'(\theta)e^{-ik\theta}d\theta = \mathcal{O}(k^{-2}). \qquad \square$$

B.1.2 ▪ Smoothness implies fast decaying coefficients

For numerics, we exploit the fact that the functions with which we are concerned have high regularity: indeed, *smooth* functions — i.e., infinitely differentiable — are the norm, rather than the exception. Here, we see that smoothness of the function induces decay in the coefficients.

Lemma B.7. *Suppose $f \in C^\lambda(\mathbb{T})$ satisfies $f(x) = f(x+2\pi)$. If $f^{(\lambda)}$ has bounded variation, then $\mathcal{F}f \in \ell^{\lambda-1,1}$.*

Proof. Because of periodicity and differentiability, we can integrate by parts:

$$\hat{f}_k = \frac{1}{2\pi}\int_{-\pi}^{\pi} f(\theta)e^{-ik\theta}d\theta = \frac{1}{2\pi ik}\int_{-\pi}^{\pi} f'(\theta)e^{-ik\theta}d\theta = \cdots = \frac{1}{2\pi(ik)^\lambda}\int_{-\pi}^{\pi} f^{(\lambda)}(\theta)e^{-ik\theta}d\theta.$$

The theorem thus follows from Lemma B.5. □

We rephrase the preceding result in terms of Sobolev spaces.

Corollary B.8. *$\mathcal{F} : H^\lambda(\mathbb{T}) \to \ell^{\lambda,2}$ and $\mathcal{F} : H^{\lambda+1}(\mathbb{T}) \to \ell^{\lambda,1}$.*

Having a rough intuition of sufficient conditions allows us to put aside properties of f, in place of $\ell^{\lambda,p}$ spaces for $\mathcal{F}f$. This leads to simplification of the theory of discrete Fourier transforms in Chapter 4.

B.1.3 ▪ Fast decaying coefficients implies fast converging Fourier series

For numerics, the question is not only whether convergence is achieved but also the *rate of convergence*. It is clear that the convergence rate of the Fourier series is dictated by the decay in coefficients. We describe this decay in terms of $\ell^{(\lambda,R),p}$ spaces.

Proposition B.9. *If $\mathcal{F}f \in \ell^{(\lambda,R),2}$, then*

$$\left\| f(\theta) - \sum_{k=\alpha}^{\beta} \hat{f}_k e^{ik\theta} \right\|_{L^2(\mathbb{T})} \leq \|\mathcal{F}f\|_{\ell^{(\lambda,R),2}} (n+2)^{-\lambda} R^{-1-n},$$

where $n = \min\{|\alpha|,|\beta|\}$. *If* $\mathcal{F}f \in \ell^{(\lambda,R),1}$, *and then*

$$\left\| f(\theta) - \sum_{k=\alpha}^{\beta} \hat{f}_k e^{ik\theta} \right\|_u \leq \|\mathcal{F}f\|_{\ell^{(\lambda,R),1}} (n+2)^{-\lambda} R^{-1-n}.$$

Proof. The first norm is

$$\begin{aligned}
\left\| f(\theta) - \sum_{k=\alpha}^{\beta} \hat{f}_k e^{ik\theta} \right\|_{L^2(\mathbb{T})}^2 &= \sum_{k=-\infty, k\notin[\alpha,\beta]}^{\infty} \left|\hat{f}_k\right|^2 \\
&\leq (n+2)^{-2\lambda} R^{-2n-2} \sum_{k=-\infty, k\notin[-n,n]}^{\infty} \left|(|k|+1)^{\lambda} R^{|k|} \hat{f}_k\right|^2 \\
&\leq (n+2)^{-2\lambda} R^{-2n-2} \|\mathcal{F}f\|_{\ell^{(\lambda,R),2}}^2 .
\end{aligned}$$

Uniform convergence requires slightly more regularity:

$$\begin{aligned}
\left\| f - \sum_{k=\alpha}^{\beta} \hat{f}_k e^{ik\theta} \right\|_u &\leq \sum_{k=-\infty, k\notin[\alpha,\beta]}^{\infty} \left|\hat{f}_k\right| \\
&\leq (n+2)^{-\lambda} R^{-n-1} \|\mathcal{F}f\|_{\ell^{(\lambda,R),1}} . \qquad \square
\end{aligned}$$

B.1.4 ▪ Laurent series

The regularity of $f(e^{i\theta})$ is inherited from the regularity of $f(\zeta)$, and hence the convergence rate depends only on the regularity of f on the unit circle. On the unit circle *analyticity* is a natural phenomenon, and analyticity translates to exponential decay.

Theorem B.10. *Suppose* f *is analytic in an open annulus* $\left\{z : \frac{1}{\rho} < |z| < \rho\right\}$ *and is bounded on the closure. Then the Fourier coefficients have exponential decay:* $\mathcal{F}f \in \ell^{(0,\rho),\infty}$. *Therefore,* $\mathcal{F}f \in \ell^{(\lambda,R),p}$ *for all* p *and* λ *provided that* $R < \rho$.

Proof. We first rewrite the Fourier coefficient in the ζ plane:

$$\hat{f}_k = \frac{1}{2\pi} \int_{-\pi}^{\pi} f(e^{i\theta}) e^{-ik\theta} d\theta = \oint \frac{f(\zeta)}{\zeta^{-k-1}} d\zeta .$$

This is, of course, a specialization of Cauchy's integral formula. Now analyticity allows us to deform the integral, with the goal of making the integrand as small as possible. Thus for $k > 0$ we deform to the outer circle, obtaining (from the ML formula)

$$\left| \oint_{|\zeta|=1} \frac{f(\zeta)}{\zeta^{-k-1}} d\zeta \right| = \left| \oint_{|s|=R} \frac{f(s)}{s^{-k-1}} ds \right| \leq R^{-k-1} R \sup f \leq C R^{-k}.$$

Deforming to the inner circle proves the bound for negative k. $\square$

Remark B.1.2. *The regularity of* f *on the boundary of the annulus allows us to show that* $\mathcal{F}f \in \ell^{(\lambda,\rho),1}$ *or* $\mathcal{F}f \in \ell^{(\lambda,\rho),2}$, *in accordance with Lemma* B.7. *These sharper estimates are largely useless: we can perturb* ρ *to be slightly smaller, maintaining exponential decay.*

B.2 ▪ Chebyshev series

Consider now $f : [-1,1] \to \mathbb{C}$, i.e., a function defined on the unit interval. Unlike functions defined on $\mathbb{T}$, we no longer assume that there is inherent periodicity.

B.2.1 ▪ The Joukowsky map

The Joukowsky map is defined as

$$J(z) \triangleq \frac{1}{2}\left(z + \frac{1}{z}\right).$$

It is clearly analytic everywhere apart from the origin and infinity. Moreover, it is conformal apart from ± 1 since

$$J'(z) = \frac{1}{2}\left(1 - \frac{1}{z^2}\right)$$

does not vanish. It satisfies the property that it maps both z and $1/z$ to the same point. Furthermore, it maps the unit circle to the unit interval, twice:

$$J(e^{i\theta}) = \cos\theta.$$

In other words, it is a vertical projection from the unit circle to the unit interval; see Figure B.1.

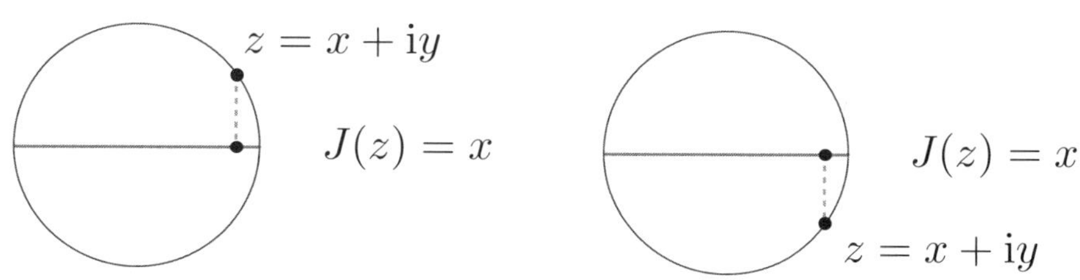

Figure B.1. *The Joukowsky map projects from the unit circle to the unit interval.*

In short, the Joukowsky map is a conformal map from both inside and outside the unit circle to the slit plane $\mathbb{C}\backslash[-1,1]$. For z in the slit plane $\mathbb{C}\backslash[-1,1]$, we define two right inverses of the Joukowsky map:

$$J_+^{-1}(z) \triangleq z - \sqrt{z-1}\sqrt{z+1} \quad : \quad \text{Map inside the circle.}$$
$$J_-^{-1}(z) \triangleq z + \sqrt{z-1}\sqrt{z+1} \quad : \quad \text{Map outside the circle.}$$

Proposition B.11. *J_+^{-1} and J_-^{-1} satisfy the following:*

1. *They are analytic on the slit plane $\mathbb{C}\backslash[-1,1]$.*
2. *For $x \in [-1,1]$, they satisfy the jump*
$$J_+^{-1}(x)^+ = J_-^{-1}(x)^- \quad \textit{and} \quad J_+^{-1}(x)^- = J_-^{-1}(x)^+.$$
3. *They are reciprocal: $\frac{1}{J_+^{-1}(z)} = J_-^{-1}(z)$.*

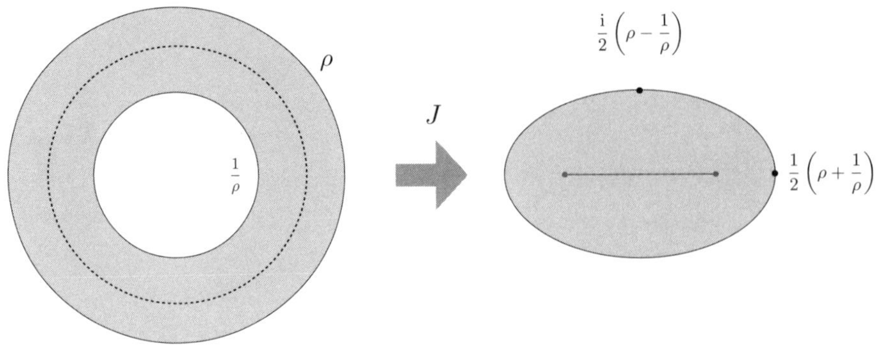

Figure B.2. *The behavior of an annulus under the Joukowsky map.*

4. J_+^{-1} (J_-^{-1}) *maps* $\mathbb{C}\backslash[-1,1]$ *to inside (outside) the unit circle.*

5. *They are right inverses of* J*:*

$$z = J(J_+^{-1}(z)) = J(J_-^{-1}(z)).$$

Proof. We first show that these are both analytic in $\mathbb{C}\backslash[-1,1]$. Analyticity off $(-\infty,1]$ is clear. Continuity on $(-\infty,-1)$ follows from $\sqrt{x}^+ = -\sqrt{x}^-$ for $x \in (-\infty,0)$:

$$J_+^{-1}(x)^+ = x - \sqrt{x-1}^+\sqrt{x+1}^+ = x - (-\sqrt{x-1}^-)(-\sqrt{x+1}^-) = J_+^{-1}(x)^-.$$

Analyticity follows since a function which is analytic off a curve Γ and continuous on Γ must be analytic on Γ (recall Theorem C.11).

The jumps on $[-1,1]$ also follow:

$$J_+^{-1}(x)^+ = x - \sqrt{x-1}^+\sqrt{x+1} = x + \sqrt{x-1}^-\sqrt{x+1} = J_-^{-1}(x)^-.$$

The reciprocal property follows from

$$J_+^{-1}(z)J_-^{-1}(z) = z^2 - (z-1)(z+1) = 1.$$

The fact that they are right inverses of J follows:

$$J(J_+^{-1}(z)) = \frac{1}{2}\left(J_+^{-1}(z) + J_-^{-1}(z)\right) = z.$$

Finally, assume z is real and $0 < z < 1$ and real, so that $J(z) > 1$. We have

$$\begin{aligned} J_+^{-1}(J(z)) &= \frac{1}{2}\left(z+\frac{1}{z}\right) - \sqrt{\frac{1}{2}\left(z+\frac{1}{z}\right)-1}\sqrt{\frac{1}{2}\left(z+\frac{1}{z}\right)+1} \\ &= \frac{1}{2}\left(z+\frac{1}{z}\right) - \sqrt{\frac{1}{4}z^2+\frac{1}{2}+\frac{1}{4z^2}} = \frac{1}{2}\left(z+\frac{1}{z}\right) - \frac{1}{2}\left(z-\frac{1}{z}\right) = z. \end{aligned}$$

Because $J(z)$ avoids the branch cut of J_+^{-1} for z inside the unit circle, both sides of the identity are analytic, and hence this identity holds true by analytic continuation. ∎

These two inverses have a branch cut on the unit interval; we therefore define two additional inverses from the unit circle:

$$J_{\uparrow}^{-1}(x) \triangleq x + \mathrm{i}\sqrt{1-x}\sqrt{1+x} = x + \mathrm{i}\sqrt{1-x^2} \quad : \quad \text{Map to upper-half circle.}$$
$$J_{\downarrow}^{-1}(x) \triangleq x - \mathrm{i}\sqrt{1-x}\sqrt{1+x} = x - \mathrm{i}\sqrt{1-x^2} \quad : \quad \text{Map to lower-half circle.}$$

We can use these inverses to describe the behavior of $J_{+}^{-1}(z)$ and $J_{-}^{-1}(z)$ as z approaches $(-1,1)$.

Proposition B.12.

$$[J_{+}^{-1}(x)]^{+} = J_{\downarrow}^{-1}(x), \quad [J_{+}^{-1}(x)]^{-} = J_{\uparrow}^{-1}(x),$$
$$[J_{-}^{-1}(x)]^{+} = J_{\uparrow}^{-1}(x), \quad [J_{-}^{-1}(x)]^{-} = J_{\downarrow}^{-1}(x).$$

Proof. These follow immediately from the limit of $\sqrt{x}$ on its branch cut. For example,

$$[J_{+}^{-1}(x)]^{+} = x - \sqrt{x-1}^{+}\sqrt{x+1} = x - \mathrm{i}\sqrt{1-x}\sqrt{x+1} = J_{\downarrow}^{-1}(x). \qquad \square$$

B.2.2 ▪ Chebyshev series

We use the Joukowsky map to approximate f on the unit interval. Our key observation is that $f(J(z)) = f\left(\frac{1}{2}\left(z + \frac{1}{z}\right)\right)$ is well-defined on the unit circle, and its smoothness is inherited from f. In other words, we immediately have

$$f(J(z)) = \sum_{k=-\infty}^{\infty} \hat{f}_k z^k.$$

We can recover f by either

$$f(x) = f(J(J_{\uparrow}^{-1}(x))) \quad \text{or} \quad f(x) = f(J(J_{\downarrow}^{-1}(x)))$$

since $f(J(z))$ has projected a copy of f to both the upper- and lower-half circles. Because of symmetry,

$$\hat{f}_k = \frac{1}{2\pi}\int_{-\pi}^{\pi} f(\cos\theta)\mathrm{e}^{-\mathrm{i}k\theta}\,\mathrm{d}\theta = \frac{1}{2\pi}\int_{-\pi}^{\pi} f(\cos\theta)\mathrm{e}^{\mathrm{i}k\theta}\,\mathrm{d}\theta = \hat{f}_{-k}.$$

Therefore

$$f(J(\zeta)) = \hat{f}_0 + \sum_{k=1}^{\infty} \hat{f}_k\left(\zeta^k + \zeta^{-k}\right).$$

Out of convention, we redefine the coefficients so that we can assemble this representation into one series: define

$$\check{f}_0 \triangleq \hat{f}_0 \qquad \text{and} \qquad \check{f}_k \triangleq 2\hat{f}_k = 2\hat{f}_{-k}$$

so that

$$f(J(\zeta)) = \sum_{k=0}^{\infty} \check{f}_k \frac{\zeta^k + \zeta^{-k}}{2}.$$

Transforming to the periodic interval we get a *cosine series*:

$$f(J(e^{i\theta})) = f(\cos\theta) = \sum_{k=0}^{\infty} \check{f}_k \frac{e^{ik\theta} + e^{-ik\theta}}{2} = \sum_{k=0}^{\infty} \check{f}_k \cos k\theta.$$

Transforming back to the $x = \cos\theta = J_{\uparrow}^{-1}(\zeta)$ variable, we obtain the *Chebyshev series*

$$f(x) = \sum_{k=0}^{\infty} \check{f}_k T_k(x),$$

where $T_k(x)$ are the Chebyshev polynomials.

Definition B.13. *Define the* Chebyshev polynomials *as*

$$T_k(x) \triangleq \cos k \arccos x = \frac{J_{\uparrow}^{-1}(x)^k + J_{\downarrow}^{-1}(x)^k}{2}$$

or equivalently as

$$T_k(\cos\theta) = \cos k\theta \qquad \textit{and} \qquad T_k(J(\zeta)) = \frac{\zeta^k + \zeta^{-k}}{2}.$$

Before continuing, we remark that $T_k(x)$ is indeed a polynomial.

Proposition B.14. *$T_k(x)$ is a degree k polynomial.*

Proof. We use the *binomial theorem*:

$$J_{\uparrow}^{-1}(x)^k = \left(x + i\sqrt{1-x^2}\right)^k = \sum_{j=0}^{k} i^j \binom{k}{j} x^{k-j} \sqrt{1-x^2}^j,$$

$$J_{\downarrow}^{-1}(x)^k = \left(x - i\sqrt{1-x^2}\right)^k = \sum_{j=0}^{k} (-)^j i^j \binom{k}{j} x^{k-j} \sqrt{1-x^2}^j.$$

In the sum, the odd terms cancel and the even terms are polynomial. □

We now define the Chebyshev transform and its inverse.

Definition B.15. *For integrable f, define the* Chebyshev transform

$$\mathcal{T} f \triangleq \left[\check{f}_0, \check{f}_1, \check{f}_2, \ldots\right]^\top.$$

For $\boldsymbol{f} = [f_0, f_1, \ldots]^\top \in \ell^1$, define the inverse Chebyshev transform

$$\mathcal{T}^{-1}\boldsymbol{f} = \sum_{k=0}^{\infty} \check{f}_k T_k(x).$$

B.2.3 ▪ Convergence of Chebyshev series

The decay in the Chebyshev coefficients $\mathcal{T} f$ is dictated by the regularity $f(J(\zeta))$ on the unit circle or equivalently $f(\cos\theta)$ on the periodic interval. This is *not* the same as the regularity of f on the unit interval. For example, differentiating $f(\cos\theta)$ we get

$\sin\theta f'(\cos\theta)$. The vanishing of $\sin\theta$ at 0 and $-\pi$ can potentially cancel singularities of f'. It is, however, convenient to use standard Sobolev spaces to determine the decay in coefficients.

Proposition B.16. *If $f \in H^{\lambda+1}(\mathbb{I})$, then $f(\cos\diamond) \in H^{\lambda}(\mathbb{T})$. In particular, $\mathcal{T} : H^{\lambda+1}(\mathbb{I}) \to \ell^{\lambda,2}$ and $\mathcal{T} : H^{\lambda+2}(\mathbb{I}) \to \mathcal{T} f \in \ell^{\lambda,1}$.*

In the case where f is analytic, the convergence rate of this expansion is dictated by the annulus of analyticity of $f(J(z))$. However, a circle of radius ρ (as well as a circle of radius $1/\rho$) is mapped to a *Bernstein ellipse*.

Definition B.17. *The Bernstein ellipse B_ρ is the open ellipse with major axis $\frac{1}{2}\left(\rho+\frac{1}{\rho}\right)$ and minor axis $\frac{1}{2}\left(\rho-\frac{1}{\rho}\right)$. Equivalently, B_ρ is the image of*

$$J(re^{i\theta}) = \frac{re^{i\theta}+r^{-1}e^{-i\theta}}{2} = \frac{r+r^{-1}}{2}\cos\theta + i\frac{r-r^{-1}}{2}\sin\theta$$

for $0 \le r < \rho$ and $-\pi \le \theta < \pi$.

Thus analyticity of f in the Bernstein ellipse dictates the decay rate of $\check{f}_k$.

Corollary B.18. *Suppose f is analytic in B_ρ and bounded on the closure. Then $\mathcal{T} f \in \ell^{(0,\rho),\infty}$ and $\mathcal{T} f \in \ell^{(\lambda,R),p}$ for all p and λ provided that $R < \rho$.*

Once we know the space that $\mathcal{T} f$ is in, however, the convergence properties of the series are identical to the Laurent and Fourier series.

B.2.4 ▪ Orthogonality of Chebyshev polynomials

Before continuing with the use of the Chebyshev series, we describe the standard theory of Chebyshev series, in particular the fact that $T_k(x)$ are orthogonal polynomials. We emphasize that for computational purposes, the connection with Laurent and Fourier series will prove a more powerful asset than orthogonality. Our first remark is that $T_k(x)$ are orthogonal with respect to a weighted inner product over the unit interval.

Proposition B.19.

$$\begin{aligned}
\int_{-1}^{1} \frac{1}{\sqrt{1-x^2}} dx &= \pi, \\
\int_{-1}^{1} \frac{T_k(x)^2}{\sqrt{1-x^2}} dx &= \frac{\pi}{2} \quad \textit{for} \quad k = 1, 2, \ldots, \\
\int_{-1}^{1} \frac{T_k(x)T_j(x)}{\sqrt{1-x^2}} dx &= 0 \quad \textit{for} \quad k \neq j.
\end{aligned}$$

Proof. The proof follows from the change of variables $x = \cos\theta$. ∎

As a result, we get direct expressions for the *Chebyshev coefficients*.

Proposition B.20.

$$\check{f}_0 = \frac{1}{\pi}\int_{-1}^{1} \frac{f(x)}{\sqrt{1-x^2}}\mathrm{d}x,$$
$$\check{f}_k = \frac{2}{\pi}\int_{-1}^{1} \frac{f(x)T_k(x)}{\sqrt{1-x^2}}\mathrm{d}x.$$

The Chebyshev polynomials also satisfy a three-term recurrence relationship.

Proposition B.21.

$$T_0(x) = 1,$$
$$T_1(x) = x,$$
$$T_{k+1}(x) = 2xT_k(x) - T_{k-1}(x).$$

Proof. Substitute $x = J(\zeta)$ in the above expressions. □

Of these properties, the three-term recurrence relationship is the most useful for computation. It allows one to avoid the evaluation of trigonometric functions or square roots via Clenshaw's algorithm [91, equation 3.11.15]. Furthermore, it can be used to reduce the problem of finding roots of polynomials to calculating eigenvalues of a *comrade matrix* [8].

B.2.5 ▪ Chebyshev U series

A sister family to the Chebyshev polynomials are the *Chebyshev U polynomials*,[44] which we define by

$$U_k(x) \triangleq \frac{T'_{k+1}(x)}{k+1}. \tag{B.1}$$

Expansions in this series are useful for the *integration* and *differentiation* of Chebyshev series, as well as calculation of Cauchy transforms for functions with square root singularities (see Section 5.4.1).

Fortunately, it is easy to convert between a Chebyshev series and Chebyshev U series. The starting point is the representation of U_k by mapping to the unit circle.

Proposition B.22.

$$U_k(J(\zeta)) = \frac{\zeta^{k+1} - \zeta^{-k-1}}{\zeta - \zeta^{-1}} = \zeta^{-k} + \cdots + \zeta^{k}$$

and

$$U_k(\cos\theta) = \frac{\sin(k+1)\theta}{\sin\theta}.$$

[44] Our Chebyshev polynomials are normally called *Chebyshev polynomials of the first kind* and Chebyshev U polynomials are called *Chebyshev polynomials of the second kind*.

Proof. We have $T_k(J(\zeta)) = \frac{\zeta^k+\zeta^{-k}}{2}$, and thus

$$\left(1-\frac{1}{\zeta^2}\right)T'_{k+1}(J(\zeta)) = (k+1)\left(\zeta^k - \zeta^{-k-2}\right),$$

or in other words,

$$U_k(J(\zeta)) = \frac{T'_{k+1}(J(\zeta))}{k+1} = \frac{\zeta^k - \zeta^{-k-2}}{1-\zeta^{-2}} = \frac{\zeta^{k+1} - \zeta^{-k-1}}{\zeta - \zeta^{-1}}.$$

A telescoping sum shows that $(\zeta - \zeta^{-1})(\zeta^{-k} + \cdots + \zeta^k) = \zeta^{k+1} - \zeta^{-k-1}$. Finally, the second equality follows from letting $\zeta = e^{i\theta}$. □

This leads to a *conversion relationship*.

Lemma B.23.

$$\begin{aligned} T_0(x) &= U_0(x), \\ T_1(x) &= \frac{U_1(x)}{2}, \\ T_k(x) &= \frac{U_k(x) - U_{k-2}(x)}{2}. \end{aligned}$$

Proof. Mapping to the unit circle, we have

$$\begin{aligned} U_k(J(\zeta)) - U_{k-2}(J(\zeta)) &= \frac{\zeta^{k+1} - \zeta^{-k-1}}{\zeta - \zeta^{-1}} - \frac{\zeta^{k-1} - \zeta^{1-k}}{\zeta - \zeta^{-1}} \\ &= \frac{\zeta^{k+1} - \zeta^{-k-1} - \zeta^{k-1} + \zeta^{1-k}}{\zeta - \zeta^{-1}} = \frac{(\zeta - \zeta^{-1})(\zeta^k + \zeta^{-k})}{\zeta - \zeta^{-1}} \\ &= 2T_k(J(\zeta)). \end{aligned}$$

□

Thus, for

$$f(x) = \sum_{k=0}^{\infty} \check{f}_k T_k(x) = \sum_{k=0}^{\infty} u_k U_k(x),$$

we have

$$\begin{bmatrix} u_0 \\ u_1 \\ \vdots \end{bmatrix} = \mathcal{S} \begin{bmatrix} \check{f}_0 \\ \check{f}_1 \\ \vdots \end{bmatrix}$$

for the operator

$$\mathcal{S} = \begin{bmatrix} 1 & 0 & -\frac{1}{2} & & & \\ & \frac{1}{2} & 0 & -\frac{1}{2} & & \\ & & \frac{1}{2} & 0 & -\frac{1}{2} & \\ & & & \ddots & \ddots & \ddots \end{bmatrix}.$$

If the vector of coefficients has only n nonzero coefficients (which is the case when we approximate the coefficients numerically as in Section 4.2.1), then we can clearly apply $\mathcal{S}$ in $\mathcal{O}(n)$ operations. We can also apply $\mathcal{S}^{-1}$ in $\mathcal{O}(n)$ operations by back substitution. This is despite its dense structure:

$$\mathcal{S}^{-1} = \begin{bmatrix} 1 & 0 & 1 & 0 & 1 & 0 & \cdots \\ & 2 & 0 & 2 & 0 & 2 & \cdots \\ & & 2 & 0 & 2 & 0 & \ddots \\ & & & \ddots & \ddots & \ddots & \end{bmatrix}.$$

Finally, we remark that $U_0(x), U_1(x), \ldots$ are orthogonal with respect to the inner product

$$\langle f, g \rangle = \int_{-1}^{1} f(x) g(x) \sqrt{1-x^2} \mathrm{d}x.$$

B.2.6 ▪ Differentiation and integration

Additional properties of Chebyshev polynomials lead to effective methods for differentiation and integration. The key formulae are the explicit derivatives (recall (B.1)) and the conversion relationship (Lemma B.23), leading to indefinite integrals (see also [104]):

$$\begin{aligned} \int T_0(x)\mathrm{d}x &= \int U_0(x)\mathrm{d}x = T_1(x), \\ \int T_1(x)\mathrm{d}x &= \int \frac{U_1(x)}{2}\mathrm{d}x = \frac{T_2(x)}{4}, \\ \int T_k(x)\mathrm{d}x &= \int \frac{U_k(x) - U_{k-2}(x)}{2}\mathrm{d}x = \frac{1}{2}\left(\frac{T_{k+1}(x)}{k+1} - \frac{T_{k-1}(x)}{k-1}\right). \end{aligned}$$

We therefore have, for $\mathcal{T} f \in \ell^1$,

$$\int f(x)\mathrm{d}x = \check{f}_0 T_1(x) + \check{f}_1\left(\frac{1}{4}T_2(x) - T_0(x)\right) + \frac{1}{2}\sum_{k=2}^{\infty} \check{f}_k \left(\frac{T_{k+1}(x)}{k+1} - \frac{T_{k-1}(x)}{k-1}\right).$$

Provided that f is sufficiently smooth, we also have

$$f'(x) = \sum_{k=1}^{\infty} \check{f}_k k U_{k-1}(x) = \sum_{k=1}^{\infty} \check{f}_k^D T_k(x),$$

where

$$f^D = \begin{bmatrix} f_0^D \\ f_1^D \\ \vdots \end{bmatrix} = \mathcal{S}^{-1} \begin{bmatrix} 0 & 1 & & & \\ & & 2 & & \\ & & & 3 & \\ & & & & \ddots \end{bmatrix} \mathcal{T} f.$$

It is clear (using Proposition A.22) that this conversion from Chebyshev coefficients of f to those of f' is bounded $\ell^{2,1} \to \ell^1$.

The expressions for integration and differentiation are amenable for numerics, where f is approximated by a finite-degree Chebyshev approximation. In the integration case, this forms the basis of Clenshaw–Curtis quadrature [20].

Appendix C
Complex Analysis

We collect here theorems and definitions from undergraduate complex analysis. A *curve* is the image of a one-to-one continuous function $\gamma : D \to \mathbb{C}$, where $D \subset \mathbb{R}$ is an interval. In the case that $D = [a,b)$ and $\lim_{x\to b} \gamma(x) = \gamma(a)$ we say that the curve is closed. When a curve is not closed, we say it is open and the points $\gamma(a)$ and $\gamma(b)$ are its *endpoints*. The term *arc* is also used to refer to an open curve. A *contour* is the finite union of curves and arcs. The differentiability of a curve, arc, or contour is determined by its smoothest parameterization.

Definition C.1. *Let γ_0 be the set of self-intersections of the contour Γ. We say a point $a \in \Gamma$ is an open endpoint if a is a limit point of only one of the connected components of $\Gamma \setminus (\gamma_0 \cup \{a\})$.*

For example, the interval $[-1,1]$ has ± 1 as open endpoints.

C.1 ▪ Inferred analyticity

Theorem C.2 (Liouville's theorem). *If $f(z)$ is entire (analytic for all $z \in \mathbb{C}$) and globally bounded, then f is a constant.*

Theorem C.3. *Suppose $D \subset \mathbb{C}$ is an open set and that $f : D \to \mathbb{C}$ is analytic in $D \setminus \{s\}$, has an isolated singularity at $s \in D$, and is bounded at s. Then f is analytic at s.*

Corollary C.4. *Suppose $f(z)$ has an isolated singularity at s and $\frac{f(z)}{(z-s)^a}$ is bounded for some a in a neighborhood of s. Then $f(z)(z-s)^{\lceil \operatorname{Re} a \rceil}$ is analytic at s.*

These results can be generalized with weaker hypotheses.

Definition C.5. *A function $f : D \to \mathbb{C}$ with $D \subset \mathbb{C}$ is locally integrable at a point $a \in D$ if*

$$\lim_{r \downarrow 0} \int_{\partial B(a,r)} |f(z)||dz| = 0.$$

Example C.6. The function

$$F(z) = \frac{f(z)}{(z-a)^r}$$

is locally integrable at $z = a$ when $f(z)$ is bounded and $r < 1$. ∎

While this differs from the usual definition of local integrability, it is useful because f is defined on a two-dimensional domain, but we want to capture singularities that are locally integrable in a one-dimensional sense.

Theorem C.7. *Suppose $f : D \to \mathbb{C}$ has an isolated singularity at $s \in D$ and is locally integrable. Then f is analytic at s.*

Proof. Let $A_{r,R}$ be the boundary of an annulus $A_{r,R} = \partial O_{r,R}$, $O_{r,R} \triangleq B(a,R) \setminus B(a,r)$ with $r < R$ both sufficiently small. With positive orientation, Cauchy's integral formula holds:

$$f(z) = \int_{A_{r,R}} \frac{f(s)}{s-z} \mathrm{d}s, \quad z \in O_{r,R}.$$

If the integral on $\partial B(a,r)$ vanishes as $r \downarrow 0$ for all z fixed, then f is analytic at a. This is guaranteed by local integrability:

$$\left| \int_{\partial B(a,r)} \frac{f(s)}{s-z} \mathrm{d}s \right| \leq \sup_{\partial B(a,r)} \frac{1}{|s-z|} \cdot \int_{\partial B(a,r)} |f(s)||\mathrm{d}s|. \qquad \square$$

Theorem C.8. *If f and g are analytic in a connected region D and agree on a set of points in D with a limit point in D, then $f = g$ in D.*

Corollary C.9. *Suppose f and g are analytic and agree on a smooth contour Γ in an open connected region D. Then $f = g$ in D.*

Theorem C.10 (Morera). *Let f be continuous in an open connected set $D \subset \mathbb{C}$. If*

$$\int_\Gamma f(z)\mathrm{d}z = 0$$

for every closed C^1 curve $\Gamma \subset D$, then f is analytic in D.

These results all combine to give an important theorem for the study of RH problems.

Theorem C.11. *Let $\Gamma \subset D$ be an arc and assume D is open. Suppose $f(z)$, analytic in $D \setminus \Gamma$, is continuous up to the interior of Γ and locally integrable for every point on Γ, and suppose*

$$f_+(s) = f_-(s).$$

Then f is analytic in a neighborhood of every point of Γ.

Proof. It follows from typical results on analytic continuation that f is analytic in $B(s,\delta_s)$ where s and $B(s,\delta_s) \cap \Gamma$ lie in the interior of Γ. Morera's theorem is used to prove this. Then f is analytic in $\cup_s B(s,\delta_s)$ and the only place it may fail to be analytic is at the endpoints of Γ where it could have isolated singularities. Local integrability demonstrates analyticity there, and the theorem follows. $\square$

Appendix D

Rational Approximation

We discuss the rational approximation of functions in appropriate L^p-based spaces. The results in this section are proved for $1 < p < \infty$, and we assume $1 < p < \infty$ throughout. It is instructive to see the generality of the methods of proof. We have defined the spaces $H^k_\pm(\Gamma)$ and $H^k_z(\Gamma)$ in Definition 2.48 and (2.41). The spaces $W^{k,p}_\pm(\Gamma)$ and $W^{k,p}_z(\Gamma)$ are defined in an analogous way, with the L^p norm. Note that the spaces $W^{k,p}_\pm(\Gamma)$ and $W^{k,p}_z(\Gamma)$ coincide when Γ is a piecewise-smooth Lipschitz graph.

D.1 ▪ Bounded contours

Theorem D.1 (See [47]). *If D is a bounded region such that ∂D is a piecewise differentiable Jordan curve, then polynomials in z are dense in $\mathcal{E}^p(D)$.*

We show for $f \in W^{k,p}_z(\partial D)$ that we can approximate it with rational functions. First, for $f \in W^{k,p}_z(\partial D) \cap \mathcal{E}^p(D)$ we approximate $D^k f$ by polynomials p_n. We have

$$\left| \int_a^s (D^k f(t) - p_n(t))\mathrm{d}t \right|^p \leq \|D^k f - p_n\|^p_{L^p(\partial D)} \cdot |\partial D|^{p/q}, \quad 1/p + 1/q = 1,$$

where $|\partial D|$ is the arclength of ∂D. Integrating both sides over Γ with respect to $|\mathrm{d}s|$ produces

$$\left\| \int_a^\diamond (D^k f(t) - p_n(t))\mathrm{d}t \right\|_{L^p(\partial D)} \leq \|D^k f - p_n\|_{L^p(\partial D)} \cdot |\partial D|.$$

Since $\int_a^s D^k f(t)\mathrm{d}t = D^{k-1} f + c$ for some c and $\int_a^s p_n(t)\mathrm{d}t - c$ is a polynomial that converges to $D^{k-1} f$ in $L^2(\partial D)$, this argument can be used again to construct a polynomial that converges to $D^{k-2} f$ in $L^2(\partial D)$.

We use that $f = \mathcal{C}^+_{\partial D} f - \mathcal{C}^-_{\partial D} f$ and we note that we may approximate $\mathcal{C}^+_{\partial D} f$ with polynomials. We must deal with the approximation of $\mathcal{C}^-_{\partial D} f$. We bootstrap from Theorem D.1. Without loss of generality, assume $0 \in D$. Note that $F(z) = \mathcal{C}^-_{\partial D} f(z) \in \mathcal{E}^p(D^{-1})$, where $D^{-1} = \{z : 1/z \in D\}$. If $f \in W^{k,p}_z$, then the kth derivative $G(z) = F^{(k)}(1/z) \in \mathcal{E}^p(D^{-1})$ may be approximated in $L^p(\partial D^{-1})$ by polynomials $p_n(z)$. It follows that G must have a $(k+1)$th-order zero at $z=0$. Thus

$$\left|\int_{\partial D^{-1}} \frac{p_n(s)}{s^j}\mathrm{d}s\right| \leq \left|\int_{\partial D^{-1}} \frac{p_n(s)-G(s)}{s^j}\mathrm{d}s\right| \leq C_j\|G-p_n\|_{L^p(\partial D^{-1})}, \quad j=1,2,\ldots,k+1.$$

Thus the first $(k+1)$ coefficients in p_n have magnitudes that are on the order of $\|G-p_n\|_{L^p(\partial D^{-1})}$. Define $\tilde{p}_n$ to be the polynomial obtained by dropping these terms. Since ∂D^{-1} has finite arclength, we find that $\|\tilde{p}_n - p_n\|_{L^p(\partial D^{-1})} \leq C\|G-p_n\|_{L^p(\partial D^{-1})}$ for some $C>0$, depending only on D and k. Therefore $\tilde{p}_n$ converges to G in $L^p(\partial D^{-1})$. A simple change of variables shows that $\tilde{p}_n(1/z)$ converges to F in $L^p(\partial D)$. Integrating $\tilde{p}_n(1/z)$ k times in the same way as for $\mathscr{C}^+_{\partial D} f$, we obtain a rational approximation of $\mathscr{C}^-_{\partial D} f$ in $W^{k,p}_z$. We conclude the following.

Theorem D.2. *If D is a bounded region such that ∂D is a piecewise differentiable Jordan curve, then rational functions of z are dense in $W^{k,p}_z(\partial D)$.*

D.2 ▪ Lipschitz graphs

We turn to the density of rational functions in $W^{k,p}_z(\Gamma)$ when Γ is a piecewise-smooth Lipschitz graph. It will be clear that the piecewise-smooth assumption can be relaxed in the $k=0$ or $L^p(\Gamma)$ case. We follow the ideas presented in [10] and complete their proof. We begin with some technical developments. Define the maximal function

$$M(f)(t) \triangleq \sup_{1\geq\delta>0} \frac{1}{|\Gamma\cap B(t,\delta)|}\int_{\Gamma\cap B(t,\delta)} |f(s)||\mathrm{d}s|.$$

Lemma D.3 (See [106]). *For $f\in L^p(\Gamma)$*

$$\|M(f)\|_{L^p(\Gamma)} \leq c(\Gamma)_p\|f\|_{L^p(\Gamma)}.$$

Lemma D.4. *Let $P_\epsilon(s,t)=P_\epsilon(t,s):\Gamma\times\Gamma\to\mathbb{C}$ satisfy*

1.

$$1-\int_\Gamma P_\epsilon(s,t)\mathrm{d}s \to 0 \text{ as } \epsilon\downarrow 0 \text{ for each fixed } t\in\Gamma,$$

2.

$$\sup_{t\in\Gamma}\int_\Gamma |P_\epsilon(s,t)||\mathrm{d}s| \leq C, \text{ and}$$

3. *for $\delta>0$, as $\epsilon\downarrow 0$*

$$\left\|\int_\Gamma |w_\delta(s,\diamond)||P_\epsilon(s,\diamond)||\mathrm{d}s|\right\|_\infty \to 0,$$

$$w_\delta(s,t)=\begin{cases} 0 & \text{if } |t-s|<\delta, \\ 1 & \text{otherwise.}\end{cases}$$

If Γ is a piecewise differentiable Lipschitz graph, then for $f \in W_+^{1,p}(\Gamma)$

$$\int_\Gamma f(t)P_\epsilon(t,s)\mathrm{d}t \to f(s)$$

in $L^p(\Gamma)$.

Proof. Consider

$$\int_\Gamma P_\epsilon(s,t)[f(t)-f(s)]\mathrm{d}t = \int_\Gamma P_\epsilon(s,t)w_\delta(s,t)f(t)\mathrm{d}t - \int_\Gamma P_\epsilon(s,t)w_\delta(s,t)f(s)\mathrm{d}t + \int_\Gamma (1-w_\delta(s,t))P_\epsilon(s,t)(f(t)-f(s))\mathrm{d}t.$$

We first show this function tends to zero in $L^p(\Gamma)$. Define $\Delta(t) = \{s \in \Gamma : |s-t| < \delta\}$ and consider

$$I_1 \triangleq \int_\Gamma \left| \int_{\Delta(t)} P_\epsilon(s,t)f(t)\mathrm{d}t \right|^p |\mathrm{d}s| = \int_\Gamma \left| \int_\Gamma P_\epsilon(s,t)w_\delta(s,t)f(t)\mathrm{d}t \right|^p |\mathrm{d}s|.$$

It follows that [54, Theorem 6.18]

$$I_1^{1/p} \leq K\|f\|_p, \quad K = \sup_{t\in\Gamma} \int_\Gamma |w_\delta(s,t)||P_\epsilon(s,t)||\mathrm{d}s| \to 0$$

as $\epsilon \to 0$ for any $\delta > 0$ by hypothesis 3. Now consider

$$I_2 \triangleq \int_\Gamma \left| \int_{\Delta(t)} P_\epsilon(s,t)f(s)\mathrm{d}t \right|^p |\mathrm{d}s| = \int_\Gamma |f(s)|^p \left| \int_\Gamma w_\delta(s,t)P_\epsilon(s,t)\mathrm{d}t \right|^p |\mathrm{d}s|.$$

Again, we obtain $I_2^{1/p} \leq K\|f\|_{L^p(\Gamma)}$. We are left estimating the $L^p(\Gamma)$ norm of

$$I_3(s) = \int_\Gamma (1-w_\delta(s,t))P_\epsilon(s,t)(f(t)-f(s))\mathrm{d}t.$$

A simple estimate gives

$$\begin{aligned}\left| \int_\Gamma (1-w_\delta(s,t))P_\epsilon(s,t)(f(t)-f(s))\mathrm{d}t \right| &\leq \sup_{s\in\Gamma} |(1-w_\delta(s,t))(f(t)-f(s))| \left| \int_\Gamma P_\epsilon(s,t)\mathrm{d}t \right|,\\ &\leq C \sup_{s\in\Delta(t)} |f(t)-f(s)|.\end{aligned}$$

We estimate

$$\begin{aligned}\sup_{s\in\Delta(t)} |f(t)-f(s)| &\leq \frac{|\Gamma\cap B(t,\delta)|}{|\Gamma\cap B(t,\delta)|} \int_{\Gamma\cap B(t,\delta)} |Df(x)||\mathrm{d}x|\\ &\leq |\Gamma\cap B(t,\delta)| M(Df).\end{aligned}$$

From the Lipschitz nature[45] of Γ, $|\Gamma\cap B(t,\delta)| \leq C\delta$, and thus

$$\|I_3\|_p \leq C\delta\|M(Df)\|_p \leq c_p C\delta\|Df\|_p,$$

[45]This estimate follows also for Carleson curves, where $\sup_{z\in\Gamma,R>0} \frac{|\Gamma\cap B(z,R)|}{R} < \infty$.

by Lemma D.3. We use all these facts to find

$$I \triangleq \left\| \int_\Gamma P_\epsilon(\diamond,t)f(t)\mathrm{d}t - f(\diamond)\right\|_p \leq \left\| \int_\Gamma P_\epsilon(\diamond,t)[f(t)-f(\diamond)]\mathrm{d}t\right\|_p + \left\| f(\diamond)-f(\diamond)\int_\Gamma P_\epsilon(\diamond,t)\mathrm{d}t\right\|_p.$$

The second term tends to zero as $\epsilon \to 0$ by hypotheses 1 and 2 and the dominated convergence theorem. Fix $\delta > 0$. We find

$$I \leq c_p C\delta ||Df||_p + H(\epsilon,\delta), \quad H(\epsilon,\delta) \to 0 \text{ as } \epsilon \to 0.$$

Letting $\epsilon \to 0$ followed by $\delta \to 0$ proves the lemma. □

Remark D.2.1. *Similar arguments show that if $f \in W^{1,\infty}$, then convergence in Lemma* D.4 *takes place in $L^\infty(\Gamma)$.*

Lemma D.5. *If Γ is a piecewise-smooth Lipschitz graph, then $W_z^{1,p}(\Gamma)$ is dense in $L^p(\Gamma)$.*

Proof. Let $\epsilon > 0$. On each smooth component of $\Gamma_i \subset \Gamma$ we approximate f with a smooth function f_n^j so that $\int_{\Gamma_j} |f(s)-f_n^j(s)|^p|\mathrm{d}s| \to 0$ as $n \to \infty$. The existence of such an approximation follows from the density of smooth functions in L^p spaces on the real axis. We multiply f_n^j by C^∞ functions ϕ_j that take values in $[0,1]$ and vanish in a neighborhood of each of the nonsmooth points of Γ to enforce the zero-sum condition. The functions ϕ_j are not equal to one only on a set E_j of arbitrarily small measure, say $\delta/2^j$. We find

$$||f-\phi_j f_n^j||_{L^p(\Gamma_j)} \leq ||f-f_n^j||_{L^p(\Gamma_j)} + ||f_n^j - \phi_j f_n^j||_{L^p(\Gamma_j)}.$$

We estimate the last term

$$||f_n^j - \phi_j f_n^j||^p_{L^p(\Gamma_j)} = \int_{E_j} |f_n^j(s) - \phi_j f_n^j(s)|^p|\mathrm{d}s| \leq \int_{E_j} |f_n^j(s)|^p|\mathrm{d}s|.$$

Therefore,

$$\limsup_{n\to\infty} ||f-\phi_j f_n^j||^p_{L^p(\Gamma_j)} \leq \int_{E_j} |f(s)|^p|\mathrm{d}s|.$$

Define $f_n = \phi_j f_n^j$ on Γ_j and

$$\limsup_{n\to\infty} ||f-f_n||^p_{L^p(\Gamma)} \leq \int_{\cup_j E_j} |f(s)|^p|\mathrm{d}s|.$$

Since $\int_{\cup_j E_j} |\mathrm{d}s| \to 0$ as $\delta \to 0$, density is proved. □

Theorem D.6. *Let Γ be a piecewise differentiable Lipschitz graph. Then*

$$K_\epsilon f(s) \triangleq \int_\Gamma f(t)\frac{\epsilon}{\pi}\frac{\mathrm{d}t}{(t-s)^2+\epsilon^2}, \quad \epsilon > 0,$$

converges to f in $L^p(\Gamma)$ as $\epsilon \to 0$ provided $f \in L^p(\Gamma)$.

Proof. First, we show this for $f \in W_+^{1,p}(\Gamma)$. We must check the three hypotheses in Lemma D.4. Let $\Gamma = \{x + \mathrm{i}\nu(x) : x \in \mathbb{R}\}$ and note that

$$P_\epsilon(s,t) = \frac{\epsilon}{\pi}\frac{1}{(t-s)^2+\epsilon^2} = \frac{1}{2\pi\mathrm{i}}\left(\frac{1}{t-(s+\mathrm{i}\epsilon)} - \frac{1}{t-(s-\mathrm{i}\epsilon)}\right).$$

Hypothesis 1 follows by a straightforward residue calculation. As will be seen below, hypotheses 2 and 3 follow in a straightforward way when $\Gamma = \mathbb{R}$. It suffices to show that

$$|P_\epsilon(x+\mathrm{i}\nu(x), y+\mathrm{i}\nu(y))||1+\mathrm{i}\nu'(x)| \le C|P_\epsilon(x,y)|, \quad x,y \in \mathbb{R}. \tag{D.1}$$

The sufficiency is clear for hypothesis 2 after a change of variables. For hypothesis 3, consider

$$I \triangleq \int_\Gamma |w(s,t)||P_\epsilon(s,t)||\mathrm{d}s| = \int_{\Gamma\setminus\Delta(t)} |P_\epsilon(s,t)||\mathrm{d}s|.$$

The set $\Gamma \cap \Delta(t)$ contains the set $\Delta'(t) = \{x+\mathrm{i}\nu(x) : x \in (t_1+t, t_2+t)\}$, where $t_1 + t + \mathrm{i}\nu(t_1+t), t_2+t+\mathrm{i}\nu(t_2+t)$ are the two points on $\Gamma \setminus \Delta(t)$ closest to t with respect to arclength. Thus

$$\begin{aligned} I &\le \int_{\Gamma\setminus\Delta'(t)} |P_\epsilon(s,t)||\mathrm{d}s| = \int_{\mathbb{R}\setminus(t_1+t,t_2+t)} |P_\epsilon(x+\mathrm{i}\nu(x), y+\mathrm{i}\nu(y))||1+\mathrm{i}\nu'(x)|\mathrm{d}x \\ &\le \int_{\mathbb{R}\setminus(t-t^*,t+t^*)} |P_\epsilon(x+\mathrm{i}\nu(x), y+\mathrm{i}\nu(y))||1+\mathrm{i}\nu'(x)|\mathrm{d}x \le C\int_{\mathbb{R}} |P_\epsilon(x,y)w_{t^*}(x,y)|\mathrm{d}x, \end{aligned}$$

where $t^* = \min\{-t_1, t_2\}$. Thus if hypothesis 3 holds for $P_\epsilon(x,y)$ and $\Gamma = \mathbb{R}$, then the right-hand side tends to zero uniformly.

Now we prove hypotheses 2 and 3 for $P_\epsilon(x,y)$ if $\Gamma = \mathbb{R}$. Since $P_\epsilon(x,y)$ is positive for $x,y \in \mathbb{R}$, the L^1 norm can be found to be unity by contour integration. For hypothesis 3 note that Cauchy's theorem implies

$$\int_{\mathbb{R}\setminus(t-\delta,t+\delta)} P_\epsilon(x,y)\mathrm{d}x + \int_{C_\delta} P_\epsilon(x,y)\mathrm{d}x = 0,$$

where C_δ is a half-circle in the upper-half plane connecting $t-\delta$ and $t+\delta$. We find that

$$I_2 = \int_{C_\delta} P_\epsilon(x,y)\mathrm{d}x = \frac{\epsilon}{\pi}\int_{-\pi}^{\pi} \frac{\mathrm{i}\delta \mathrm{e}^{\mathrm{i}\theta}\mathrm{d}\theta}{\delta^2\mathrm{e}^{2\mathrm{i}\theta}+\epsilon^2}.$$

We estimate, for $\epsilon < \delta$, $\delta > 0$,

$$|I_2| \le \frac{\epsilon}{\pi}\int_{-\pi}^{\pi}\frac{\delta\,\mathrm{d}\theta}{\delta^2-\epsilon^2} \to 0 \text{ as } \epsilon \to 0.$$

This establishes hypothesis 3 in this case.

Finally, we establish (D.1). Consider the ratio

$$r = \frac{\epsilon^2+(x-y)^2}{\epsilon^2+(x-y)^2-(\nu(x)-\nu(y))^2+2\mathrm{i}(x-y)(\nu(x)-\nu(y))}.$$

We rewrite r in terms of $X = \epsilon^2/(x-y)^2$ and $Y = |(\nu(x)-\nu(y))/(x-y)|$. Note that $0 \leq Y \leq M$ for a constant $M > 0$ and $0 \leq X \leq \infty$. Thus, we are led to

$$R(X,Y) = |r|^2 = \frac{(X+1)^2}{(X+1-Y^2)^2+4Y^2}, \quad (X,Y) \in [0,\infty] \times [0,M].$$

For $X+1 > 2M^2$ we find

$$R(X,Y) \leq \frac{4M^4}{M^4} \leq 4.$$

Since $R(X,Y)$ is continuous in $[0,2M^2] \times [0,M]$ and bounded on $[2M^2,\infty] \times [0,M]$, we obtain (D.1), establishing L^p convergence when $f \in W_+^{1,p}(\Gamma)$.

To prove the theorem we use the density of $W_+^{1,p}(\Gamma)$ in $L^p(\Gamma)$ and the fact that K_ϵ is a bounded operator on $L^p(\Gamma)$, $1 < p < \infty$, with a norm that is independent of ϵ (see Lemma 2.32). We have proved this fact in the case $p = 2$; see Lemma 2.32. Approximating f with a sequence $f_n \in W_+^{1,p}(\Gamma)$, we obtain

$$\begin{aligned} \|f - K_\epsilon f\|_p &\leq \|f - f_n\|_p + \|f_n - K_\epsilon f_n\|_p + \|K_\epsilon(f - f_n)\|_p \\ &\leq (1+C)\|f - f_n\|_p + \|f_n - K_\epsilon f_n\|_p. \end{aligned} \tag{D.2}$$

For $\delta > 0$ find N such that for $n > N$, $(1+C)\|f - f_n\|_p < \delta$. For such an n let $\epsilon \to 0$ in (D.2):

$$\limsup_{\epsilon \to 0} \|f - K_\epsilon f\|_p < \delta, \quad n > N.$$

This proves the lemma. □

Define the integral operator

$$K_\epsilon^j f(s) = \int_\Gamma f(t) \left(\frac{(-1)^j j!}{[t-(s+\mathrm{i}\epsilon)]^{j+1}} - \frac{(-1)^j j!}{[t-(s-\mathrm{i}\epsilon)]^{j+1}} \right) \mathrm{đ}t$$

whose kernel is the kth derivative of the Poisson kernel.

Corollary D.7. *If Γ is a piecewise-smooth Lipschitz graph and $f \in W_+^{k,p}(\Gamma)$, then $\|K_\epsilon^k f - D^k f\|_p \to 0$ as $\epsilon \downarrow 0$.*

Proof. Integration by parts and the zero-sum condition imply that $K_\epsilon^k f = K_\epsilon D^k f$. The conclusion follows from Theorem D.6. □

Theorem D.8. *If Γ is a piecewise-smooth Lipschitz graph, then $L^p(\Gamma)$ rational functions are dense in $W_z^{k,p}(\Gamma)$.*

Proof. Our proof is constructive. We assume $k \geq 1$. For $f \in W_z^{k,p}$ we approximate $K_\epsilon^k f$ with a Riemann sum. This is clearly a rational approximation of $D^k f$ for ϵ small. First, we show this Riemann sum approximation converges to $K_\epsilon^k f$ in $L^p(\Gamma)$. Define $\Gamma_R = \Gamma \cap B(0,R)$, where R is undetermined for now. It suffices to show the convergence of the Riemann sums on each smooth component of Γ_R, so we assume Γ_R is smooth.[46]

[46] If it happens that Γ_R is just piecewise smooth, then we decompose $\Gamma = \Gamma_1 \cup \cdots \cup \Gamma_L$, where Γ_i is smooth. Then $\mathscr{P}_n$ is taken to be the union of partitions of $\Gamma_\ell \cap B(0,R)$.

Let $\mathscr{P}_n$ be a partition of $\Gamma \cap B(0,R)$ with $\mathscr{P}_{n+1}$ being a refinement of $\mathscr{P}_n$. To simplify matters, we assume that for $x_i \in \mathscr{P}_n$, $\Delta x_i \triangleq x_i - x_{i-1}$ we have $\int_{x_{i-1}}^{x_i} |ds| = 2^{-n}$. Define the rational function

$$F(s;\mathscr{P}_n) = \sum_{x_i \in \mathscr{P}_n} P_{\epsilon,k}(s,x_i) f(x_i) \Delta x_i,$$

which is a Riemann sum approximation of $K_\epsilon^k f$. Here P_ϵ is the kernel of K_ϵ^k. We consider the difference of Riemann sums associated with $\mathscr{P}_m$ and $\mathscr{P}_n$ with $m > n$. With $x_i \in \mathscr{P}_m$, define $x_i' \in \mathscr{P}_n$ to be the next element of $\mathscr{P}_n$ (using the orientation of Γ). Notice that

$$F(s,\mathscr{P}_n) = \sum_{x_i \in \mathscr{P}_m} P_{\epsilon,k}(s,x_i') f(x_i') 2^{n-m} \Delta x_i',$$

because 2^{m-n} elements of $\mathscr{P}_m$ map to x_i' for each i. Also, consider for $x_i \in \mathscr{P}_m$ and let $m(t)$ be the arclength parameterization of Γ with $m(t_i) = x_i$. Then, using a Taylor expansion,

$$|\Delta x_i - 2^{n-m} \Delta x_i'| = 2^{-m} \left| m''(\zeta) \right| 2^{-n-1},$$

where $\zeta \in [t_{i-1}', t_i']$, $m(t_i') = x_i$. By our smoothness assumption, there exists $L > 0$ such that $\|m''\|_\infty \le L$. We also note that there exists $M > 0$ such that $|x_i - x_{i-1}| \le M 2^{-n}$, $x_i \in \mathscr{P}_n$. For fixed $\epsilon > 0$ because $k \ge 1$, there is a function $p(s) = p(s;\epsilon) > 0$ such that $|P_{\epsilon,k}(s,x) f(x)| \le p(s)$.

We obtain

$$\begin{aligned}
&|F(s;\mathscr{P}_n) - F(s;\mathscr{P}_m)| \\
&\le \left| \sum_{x_i \in \mathscr{P}_m} P_{\epsilon,k}(s,x_i') f(x_i') 2^{n-m} \Delta x_i' - \sum_{x_i \in \mathscr{P}_m} P_{\epsilon,k}(s,x_i) f(x_i) \Delta x_i \right| \\
&\le \left| \sum_{x_i \in \mathscr{P}_m} P_{\epsilon,k}(s,x_i') f(x_i') \Delta x_i - \sum_{x_i \in \mathscr{P}_m} P_{\epsilon,k}(s,x_i) f(x_i) \Delta x_i \right| \\
&+ \left| \sum_{x_i \in \mathscr{P}_m} P_{\epsilon,k}(s,x_i') f(x_i') (\Delta x_i - 2^{n-m} \Delta x_i') \right| \\
&\le 2^{-m} M \sum_{x_i \in \mathscr{P}_m} |P_{\epsilon,k}(s,x_i') f(x_i') - P_{\epsilon,k}(s,x_i) f(x_i)| + 2^{-n-1} L p(s).
\end{aligned} \tag{D.3}$$

It follows that $\mathscr{P}_m$ is the union of 2^{m-n-1} refinements of $\mathscr{P}_n$ and we may bound (D.3) using the total variation. Explicitly,

$$|F(s;\mathscr{P}_n) - F(s;\mathscr{P}_m)| \le 2^{-n-1} [M \mathscr{V}[P_{\epsilon,k}(s,\diamond) f(\diamond)] + L p(s)].$$

We use the well-known fact that for absolutely continuous functions

$$\mathscr{V}[F] \le \int_a^b |F'(x)| dx,$$

along with the smooth parameterization of Γ_R to find

$$|F(s;\mathscr{P}_n) - F(s;\mathscr{P}_m)| \le C 2^{-n-1} \int_{\Gamma_R} |\partial_t (P_{\epsilon,k}(s,t) f(t))| |dt|. \tag{D.4}$$

There exist positive constants $C_1(\epsilon)$ and $C_2(\epsilon)$ depending only on ϵ such that

$$\int_\Gamma |P_{\epsilon,k}(s,t)||\mathrm{d}t| \leq C_1(\epsilon), \quad \int_\Gamma |\partial_t P_{\epsilon,k}(s,t)||\mathrm{d}t| \leq C_2(\epsilon).$$

To see this, note that for fixed $\epsilon > 0$, these integrals are continuous functions that decay at infinity by the dominated convergence theorem. All this implies that (see [54, Theorem 6.18])

$$\left\| \int_{\Gamma_R} |P_{\epsilon,k}(\diamond,t)||Df(t))||\mathrm{d}t| \right\|_p \leq C_1(\epsilon)\|Df\|_p,$$

$$\left\| \int_{\Gamma_R} |\partial_t P_{\epsilon,k}(\diamond,t)||f(t))||\mathrm{d}t| \right\|_p \leq C_2(\epsilon)\|f\|_p.$$

There exists $C(\epsilon) > 0$ such that

$$\left\| \int_{\Gamma_R} |\partial_t (P_{\epsilon,k}(\diamond,t)f(t))||\mathrm{d}t| \right\|_p \leq C(\epsilon)\|f\|_{W^{1,p}(\Gamma)}.$$

Using this estimate after taking the $L^p(\Gamma)$ norm of (D.4) proves that our rational approximation is a Cauchy sequence in $L^p(\Gamma)$. By standard Riemann integration results, our rational approximation must converge pointwise to $\tilde{K}_\epsilon^k f$ where $\tilde{K}_\epsilon^k$ has integral kernel $P_{\epsilon,k}(s,t)\chi_{\Gamma_R}(t)$. This is the required L^p convergence for $k \geq 1$.

Continuing, we find for $f \in W_z^{k,p}(\Gamma)$, $f \neq 0$,

$$\begin{aligned} \|F(\diamond,\mathcal{P}_n) - D^k f\|_p &\leq \|K_\epsilon^k f - D^k f\|_p + \|K_\epsilon^k f - F(\diamond,\mathcal{P}_n)\|_p, \\ &\leq \|K_\epsilon^k f - D^k f\|_p + \|F(\diamond,\mathcal{P}_n) - \tilde{K}_\epsilon^k f\|_p + \|(\tilde{K}_\epsilon^k - K_\epsilon^k)f\|_p. \end{aligned}$$

For $\delta > 0$, fix $\epsilon > 0$ such that $\|K_\epsilon^k f - D^k f\|_p < \delta/3$. It follows that for R sufficiently large

$$\sup_{s\in\Gamma} \int_\Gamma |P_{\epsilon,k}(s,t) - P_{\epsilon,k}(s,t)\chi_{\Gamma_R}(t)||\mathrm{d}t| < \frac{\delta}{3\|f\|_p}.$$

This implies that $\|\tilde{K}_\epsilon^k f - K_\epsilon^k f\|_p < \delta/3$ [54, Theorem 6.18]. For such an R, we let N be large enough so that $\|F(\diamond,\mathcal{P}_n) - \tilde{K}_\epsilon^k f\|_p < \delta/3$ for $n > N$. We find

$$\|F(\diamond,\mathcal{P}_n) - D^k f\|_p \leq \delta, \quad n > N.$$

Because k is arbitrary in these calculations, the same L^p estimate holds with k replaced by j for $j = 0,\ldots,k$ and

$$\|F(\diamond,\mathcal{P}_n) - f\|_{W_z^{k,p}(\Gamma)} \leq C\delta.$$

This proves the density of rational functions for $k \geq 1$ by letting $\delta \downarrow 0$. To obtain the claim for $L^p(\Gamma)$ we use Lemma D.5. □

Corollary D.9. *$W_+^{l,p}(\Gamma)$ is dense in $W_+^{k,p}(\Gamma)$ for $l > k$.*

We arrive at our main rational approximation theorem. See Definition 2.57 for the definition of $R_\pm$.

Theorem D.10. *If Γ is admissible, then $R_\pm(\Gamma) \cap L^2(\Gamma)$ is dense in $H^k_\pm(\Gamma)$ and $L^2(\Gamma) \cap R_\pm(\Gamma) + \mathbb{C}^{n\times n}$ is dense in $\tilde{H}^k_\pm(\Gamma)$.*

Proof. For each $D_{i,\pm} \Subset \mathbb{C} \setminus \Gamma$, $\partial D_{i,\pm}$ is bounded or a Lipschitz graph, then the theorem follows from Theorems D.8 and D.2. $\square$

Remark D.2.2. *Unbounded contours need not be Lipschitz graphs for rational functions to be dense in the relevant function spaces. We present the results for this restricted class of contours because it suffices for our applied theory of RH problems. For general results see [82].*

Appendix E
Additional KdV Results

E.1 ▪ Comparison with existing numerical methods

The discussion that follows should be compared with the results of Chapter 8. The method we developed there requires $\mathcal{O}(1)$ operations to compute the solution of the KdV equation at any given point in the (x,t)-plane, even for arbitrarily large times.

It is clear that numerical aspects of dispersion are important, and they are studied in great detail in [56, 58]. In these studies, the authors invoke fourth-order time stepping and the fast fourier transform (FFT) in space. We describe a simplified version of their methodology.

From initial data q_0, define $\boldsymbol{q}_0 = (q_1,\ldots,q_n)^\top \triangleq q_0(\boldsymbol{x}^{n,L})$, a sampling of the initial condition with a uniform grid (with $2n$ points):

$$x^{n,L} = L\left[-1,\ \frac{1}{n}-1,\ \frac{2}{n}-1,\ldots,\ 1-\frac{1}{n}\right],$$

with L large. The DFT is applied to $\boldsymbol{q}_0$ producing coefficients $\{\hat{f}_k^n\}$ such that the series

$$q_{0,n-1}(x) = \sum_{k=-n+1}^{n} \hat{f}_k^n \mathrm{e}^{\pi i k x/L}$$

is an approximation of $q_0(x)$ on $[-L,L]$. This is differentiated explicitly, and the inverse DFT is applied to compute an approximation of the derivative $D_{n,L}$: $q_0'(\boldsymbol{x}^{n,L}) \approx D_{n,L}q_0(\boldsymbol{x}^{n,L})$. In this way, the right-hand side of the KdV equation is easily approximated at any fixed time:

$$-6q(x,t)q_x(x,t) - q_{xxx}(x,t) \approx -6\boldsymbol{q}^\top D_{n,L}\boldsymbol{q} - D_{n,L}^3\boldsymbol{q} \triangleq R_{n,L}(\boldsymbol{q}),\ \ \boldsymbol{q} = q(\boldsymbol{x}^{n,L},t).$$

This is coupled with an (autonomous) time-stepper (i.e., first-order Euler or fourth-order Runge–Kutta; see [76]) to evolve the system, which we abstractly represent as $F(\boldsymbol{q},\Delta t)$. With an acceptable choice of the time step Δt, order n; and interval length L, the sequence

$$\begin{aligned}\boldsymbol{q}_0 &= q_0(\boldsymbol{x}^{n,L}),\\ \boldsymbol{q}_{j+1} &= \boldsymbol{q}_j + F(R_{n,L}(\boldsymbol{q}_j),\Delta t)\end{aligned}$$

is an approximation of the solution of the KdV equation evaluated on the grid $\boldsymbol{x}^{n,L}$.

Remark E.1.1. *The third-order derivative can significantly complicate matters by creating a stiff system making small time steps necessary. The methodology in* [58] *deals with the linear term in the KdV right-hand side explicitly and allows a larger time step.*

This type of method is very efficient for approximating the solution of the Cauchy problem for small time and, when only a small-time solution is needed, the method in [58] is surely the method of choice. Approximation for larger time is complicated by the large velocity and increasingly oscillatory nature of the dispersive tail. If the dispersive tail reaches the boundary ($x = \pm L$), errors are immediately introduced into the approximation of the Cauchy problem because of the periodic nature of the approximation.

For a quantitative analysis we consider the asymptotic formula for the dispersive tail of KdV ([32]; see also (8.16)), which we write as

$$q(x,t) = t^{-1/2}\sqrt{\frac{\nu(k_0)}{3k_0}}\cos(-4/3k_0 x + \cdots) + \mathcal{O}(t^{-1}),$$

where ν is a function depending on q_0 which decays to zero, at a rate that depends on the regularity of q_0. To find the leading edge of the propagating tail, we choose the largest k' such that $\sqrt{\nu(k')/(3k')} > 10^{-10}$. Since k_0 depends only on the ratio x/t, we know the solution will be greater than machine precision for $t < 10^8$ in the neighborhood of the values of (x,t) such that $k' = k_0$. We estimate the number of operations required to compute the solution accurately. Assume the spatial computational domain is $[-L, L]$ with n equally spaced grid points. We solve from $t = 0$ to $t = T$ with a time step Δt. Since the solution behaves roughly like $\cos(-4/3k'x)$ with nonzero amplitude, we need to use a sufficient number of grid points to at least resolve this frequency, i.e., $\pi n/L \geq 4/3k'$ must be satisfied. Furthermore, using $k' = k_0 = \sqrt{-x/(12t)}$ we obtain that the point $(-12tk'^2, t)$ should be in our computational domain if $t \leq T$, requiring that $L > 12Tk'^2$. From these inequalities we obtain

$$n > \frac{16}{\pi}k'^3 T. \tag{E.1}$$

To ensure CFL stability of the method we require $\Delta t = c/n,\ c \leq 1$ [58]. Since the FFT costs on the order of $n \log n$ operations, the number of operations required to evaluate the spatial derivatives via an FFT is larger than $4n \log n$. The number of operations required for the fourth-order time stepper to evolve a point up to time T is greater than $4nTc^{-1}$. This combines to give a total number of operations larger than $16c^{-1}n^2T \log n$ which is $\mathcal{O}(T^3 \log T)$. For the initial condition used in Figure 8.2 we estimate $k' = 8$, which means that to evolve the solution to $T = 30$ requires more than 10^{13} flops, even with $c = 1$. The computation run time would be on the order of hours assuming it proceeded at one gigaflop (in serial). It becomes prohibitively expensive very quickly with increasing T, or for less regular initial conditions such that ν decays slower. For $T = 500$ the computation would run for nearly three years. These calculations are optimistic since convergence considerations and the presence of the nonlinearity can further increase the numerical effort required.

Other computational approaches which, just like our own, rely on the integrability of the equation being solved exist. For example, Osborne [99] and Boffetta and Osborne [14] compute the scattering data using an algorithm in the spirit of the FFT for which they expect second-order convergence, which is in stark contrast to the convergence of spectral methods.

Additionally, Hald [61] used a trapezoidal method to solve the Gel'fand–Levitan–Marchenko formulation of the inverse problem. This method lacked spectral accuracy and the flexibility to deform contours. Bornemann has also computed the solution of the inverse problem through the use of Fredholm determinants [17] for $t=0$.

E.2 ▪ The KdV g-function

We give the explicit form of g from Section 8.2.2. Restricting to $a^2+b^2=2$, the expression

$$s=24\int_a^b \sqrt{(p^2-a^2)(b^2-p^2)}\mathrm{d}p, \quad \tau=tz_0^3, \quad s=-\log z_0^2/\tau\in[0,8^{2/3}]$$

defines both $a(s)$ and $b(s)$ since it is a monotone function of a. Define the g-function to be

$$g(z)=12\int_{b(s)}^{z}\sqrt{(p^2-a^2(s))(p^2-b^2(s))}\mathrm{d}p+12\int_0^{a(s)}\sqrt{(p^2-a^2(s))(p^2-b^2(s))}\mathrm{d}p. \tag{E.2}$$

We choose the branch cuts for $\sqrt{(p^2-a^2(s))(p^2-b^2(s))}$ to be the straight-line segments along $[-b(s),-a(s)]$ and $[a(s),b(s)]$. In order for g to be single-valued it is necessary to add a branch cut on $[-a(s),a(s)]$.

Lemma E.1. *The g-function given by* (E.2) *satisfies that*

1. g *is bounded in the finite plane,*

2. $$g^+(z)-g^-(z)=\begin{cases} 24\int_{b(s)}^{z}\sqrt{(p^2-a^2(s))(p^2-b^2(s))}\mathrm{d}p & \text{if } z\in[a(s),b(s)],\\ 24\int_{b(s)}^{a(s)}\sqrt{(p^2-a^2(s))(p^2-b^2(s))}\mathrm{d}p & \text{if } z\in(-a(s),a(s)),\\ 24\int_{-b(s)}^{z}\sqrt{(p^2-a^2(s))(p^2-b^2(s))}\mathrm{d}p & \text{if } z\in[-b(s),-a(s)],\\ 0 & \text{otherwise,}\end{cases} \tag{E.3}$$

3. $$g^+(z)+g^-(z)=\begin{cases} 24\int_{0}^{a(s)}\sqrt{(p^2-a^2(s))(p^2-b^2(s))}\mathrm{d}p & \text{if } z\in[a(s),b(s)],\\ 24\int_{0}^{z}\sqrt{(p^2-a^2(s))(p^2-b^2(s))}\mathrm{d}p & \text{if } z\in(-a(s),a(s)),\\ 24\int_{0}^{-a(s)}\sqrt{(p^2-a^2(s))(p^2-b^2(s))}\mathrm{d}p & \text{if } z\in[-b(s),-a(s)],\\ 2g(z) & \text{otherwise,}\end{cases} \tag{E.4}$$

4. $$g(z)=4z^3-12z+\mathcal{O}(z^{-1}) \text{ as } z\to\infty, \text{ and}$$

5. $$z_0^2 e^{i\tau(g^+(z)-g^-(z))}=1, \text{ for } z\in(-a(s),a(s)).$$

Proof. 1 is clear from (E.2). 2 and 3 follow from contour integration and the fact that the integrand is invariant under $p \mapsto -p$. To prove 4 we look at the expansion of the integrand

$$12\sqrt{(p^2-a^2(s))(p^2-b^2(s))} = 12p^2\sqrt{(1-(a(s)/p)^2-(b(s)/p)^2+(a(s)/p)^2(b(s)/p)^2)}.$$

We use

$$\sqrt{1-y} = 1-\frac{1}{2}y-\frac{1}{8}y^2+\cdots,$$

to obtain (for large p)

$$12\sqrt{(p^2-a^2(s))(p^2-b^2(s))} = 12p^2\left(1-\frac{1}{2}(a(s)/p)^2-\frac{1}{2}(b(s)/p)^2\right)+\mathcal{O}(p^{-2}).$$

After integrating this we find

$$f(z) = \int_{a(s)}^{z}\sqrt{(p^2-a^2(s))(p^2-b^2(s))}\mathrm{d}p = 4z^3-6(a^2(s)+b^2(s))z+C+\mathcal{O}(z^{-1})$$

for $C \in \mathbb{C}$. To determine C, note that $f(z)$ has an expansion in odd powers of z plus some constant:

$$\begin{aligned}
f(z)+f(-z) &= 2C\\
&= \int_{a(s)}^{z}\sqrt{(p^2-a^2(s))(p^2-b^2(s))}\mathrm{d}p + \int_{a(s)}^{-z}\sqrt{(p^2-a^2(s))(p^2-b^2(s))}\mathrm{d}p\\
&= \int_{a(s)}^{z}\sqrt{(p^2-a^2(s))(p^2-b^2(s))}\mathrm{d}p + \int_{z}^{-a(s)}\sqrt{(p^2-a^2(s))(p^2-b^2(s))}\mathrm{d}p\\
&= -\int_{-a(s)}^{a(s)}\sqrt{(p^2-a^2(s))(p^2-b^2(s))}\mathrm{d}p\\
&= -2\int_{0}^{a(s)}\sqrt{(p^2-a^2(s))(p^2-b^2(s))}\mathrm{d}p.
\end{aligned}$$

Therefore $g(z) \sim 4z^3-12z+\mathcal{O}(z^{-1})$, where we used $a^2(s)+b^2(s) = 2$. Finally, for part 5 assume $z \in (-a(s), a(s))$,

$$\begin{aligned}
-\log z_0^2/\tau &= 24\int_{a(s)}^{b(s)}\sqrt{(p^2-a^2(s))(b^2(s)-p^2)}\mathrm{d}p,\\
-\log z_0^2/\tau &= 24\mathrm{i}\int_{a(s)}^{b(s)}\sqrt{(p^2-a^2(s))(p^2-b^2(s))}\mathrm{d}p = \mathrm{i}(g^+(z)-g^-(z)),\\
z_0^2 e^{\mathrm{i}\tau(g^+(z)-g^-(z))} &= 1. \qquad \square
\end{aligned}$$

These are all the properties mentioned in Section 8.2.2.

To compute $g(z)$ we rephrase it in terms of a solution of an RH problem as we did in Sections 1.2 and 5.5. What we really compute is $\mathfrak{g}'(x)$ where $\mathfrak{g}(z) = -\mathrm{i}\tau[4z^3-12z-g(z)]$, which as described above satisfies

$$\begin{aligned}
&\mathfrak{g}'_+(z)-\mathfrak{g}'_-(z) = 0 \quad \text{for } z \notin [-a(s),-b(s)]\cup[a(s),b(s)],\\
&\mathfrak{g}'_+(z)+\mathfrak{g}'_-(z) = r(z) \triangleq \mathrm{i}\tau 24(1-z^2) \quad \text{for } z \in [-a(s),-b(s)]\cup[a(s),b(s)],\\
&\mathfrak{g}'(z) = \mathcal{O}(z^{-1}) \quad \text{as } z \to \infty.
\end{aligned}$$

Recalling Lemma 5.8, it can be shown that

$$\mathfrak{g}'(z) = \mathscr{P}^{\infty}_{0,-b(s),-a(s)} f(z) + \mathscr{P}^{\infty}_{0,a(s),b(s)} h(z),$$

$$\mathscr{P}^{n}_{\alpha,a,b} f\left(\frac{a+b}{2} + \frac{b-a}{2} z\right) \triangleq \sum_{k=0}^{n-1} \frac{\check{f}_k}{2} J_+^{-1}(z)^k - \frac{\check{f}_0}{2} \frac{z}{\sqrt{z-1}\sqrt{z+1}} + \frac{\alpha}{\sqrt{z-1}\sqrt{z+1}}$$

for some undetermined functions f and h. Let $M_-(\mathbb{I}) = [-b(s), -a(s)]$ and $M_+(\mathbb{I}) = [a(s), b(s)]$ be the associated affine transformations and $\boldsymbol{x}^n_\pm = M_\pm(\boldsymbol{x}^n)$ be the mapped Chebyshev points. We can approximate the Chebyshev coefficients of $f \circ M_-$ and $h \circ M_+$ by collocation, i.e., solving the linear system

$$\mathscr{P}^{n}_{0,-b(s),-a(s)} f(\boldsymbol{x}^n_+) + \mathscr{P}^{n}_{0,a(s),b(s)} h(\boldsymbol{x}^n_+) = r(\boldsymbol{x}^n_+),$$
$$\mathscr{P}^{n}_{0,-b(s),-a(s)} f(\boldsymbol{x}^n_-) + \mathscr{P}^{n}_{0,a(s),b(s)} h(\boldsymbol{x}^n_-) = r(\boldsymbol{x}^n_-)$$

for the approximate first n Chebyshev coefficients of f and h. The choices made for $a(s)$ and $b(s)$ ensure that $\check{f}_0 = \check{h}_0 = 0$. Indefinite integration of this formula is straightforward from Section 5.4.4 to back out an approximation of g. This is a special case of the approach detailed in [94].

Bibliography

[1] M J Ablowitz. *Nonlinear Dispersive Waves*. Cambridge Texts in Applied Mathematics. Cambridge University Press, New York, 2011. (Cited on p. 247)

[2] M J Ablowitz and P A Clarkson. *Solitons, Nonlinear Evolution Equations and Inverse Scattering*. Cambridge University Press, Cambridge, UK, 1991. (Cited on pp. 16, 18, 212, 232, 233, 250, 304)

[3] M J Ablowitz and A S Fokas. *Complex Variables: Introduction and Applications*. Cambridge University Press, Cambridge, UK, second edition, 2003. (Cited on pp. 23, 33)

[4] M J Ablowitz and H Segur. Asymptotic solutions of the Korteweg–de Vries equation. *Stud. Appl. Math.*, 57:13–44, 1977. (Cited on pp. xii, 201, 210, 213, 216, 253)

[5] M J Ablowitz and H Segur. *Solitons and the Inverse Scattering Transform*. SIAM, Philadelphia, 1981. (Cited on p. 196)

[6] D V Anosov and A A Bolibrukh. *The Riemann–Hilbert Problem*. Springer, New York, 1994. (Cited on p. 13)

[7] K Atkinson and W Han. *Theoretical Numerical Analysis*. Springer, New York, 2009. (Cited on p. 330)

[8] S Barnett. A companion matrix analogue for orthogonal polynomials. *Linear Algebra Appl.*, 12:197–208, 1975. (Cited on p. 342)

[9] Z Battles and L N Trefethen. An extension of MATLAB to continuous functions and operators. *SIAM J. Sci. Comput.*, 25:1743–1770, 2004. (Cited on pp. 194, 202, 211, 234)

[10] R Beals and R R Coifman. Scattering and inverse scattering for first order systems. *Comm. Pure Appl. Math.*, 37(1):39–90, 1984. (Cited on pp. 64, 95, 96, 348)

[11] E D Belokolos, A I Bobenko, V Z Enol'skii, A R Its, and V B Matveev. *Algebro-Geometric Approach to Nonlinear Integrable Equations*. Springer, Berlin, 1994. (Cited on pp. 269, 270, 281)

[12] R F Bikbaev and V O Tarasov. Initial-boundary value problem for the nonlinear Schrödinger equation. *J. Phys. A*, 24(11):2507–2516, 1991. (Cited on pp. 242, 244)

[13] G Biondini and A Bui. On the nonlinear Schrödinger equation on the half line with homogeneous Robin boundary conditions. *Stud. Appl. Math.*, 129(3):249–271, 2012. (Cited on pp. 242, 243)

[14] G Boffetta and A R Osborne. Computation of the direct scattering transform for the nonlinear Schrödinger equation. *J. Comput. Phys*, 102:252–264, 1995. (Cited on p. 358)

[15] A A Bolibrukh. The Riemann–Hilbert problem. *Russ. Math. Surveys*, 45:1–58, 1990. (Cited on p. 13)

[16] F Bornemann. On the numerical evaluation of distributions in random matrix theory: A review. *Markov Process. Related Fields*, 16:803–866, 2010. (Cited on p. 254)

[17] F Bornemann. On the numerical evaluation of Fredholm determinants. *Math. Comp.*, 79:871–915, 2010. (Cited on pp. 254, 359)

[18] K F Clancey and I Gohberg. *Factorization of Matrix Functions and Singular Integral Operators*, volume 3 of *Operator Theory: Advances and Applications*. Birkhäuser Verlag, Basel, 1981. (Cited on pp. 23, 62, 70)

[19] P A Clarkson. Painlevé equations—nonlinear special functions. *J. Comput. Appl. Math.*, 153(1–2):127–140, 2003. (Cited on p. 253)

[20] C W Clenshaw and A R Curtis. A method for numerical integration on an automatic computer. *Numer. Math.*, 2(1):197–205, 1960. (Cited on p. 344)

[21] R R Coifman, P W Jones, and S Semmes. Two elementary proofs of the L^2 boundedness of Cauchy integrals on Lipschitz curves. *J. Amer. Math. Soc.*, 2(3):553–564, 1989. (Cited on pp. 49, 50)

[22] C W Curtis and B Deconinck. On the convergence of Hill's method. *Math. Comp.*, 79(269):169–187, 2010. (Cited on p. 235)

[23] B Deconinck, M Heil, A Bobenko, M van Hoeij, and M Schmies. Computing Riemann theta functions. *Math. Comp.*, 73:1417–1442, 2004. (Cited on p. 269)

[24] B Deconinck and J N Kutz. Computing spectra of linear operators using the Floquet–Fourier–Hill method. *J. Comput. Phys.*, 291:296–321, 2007. (Cited on pp. 194, 210, 211, 235)

[25] B Deconinck and D O Lovit. Data analysis and reduction using stationary solutions of the NLS equation. *Appl. Anal.*, 89:611–626, 2010. (Cited on p. 295)

[26] B Deconinck, T Trogdon, and V Vasan. The method of Fokas for solving linear partial differential equations. *SIAM Rev.*, 56(1):159–186, 2014. (Cited on p. 53)

[27] P Deift. *Orthogonal Polynomials and Random Matrices: A Riemann-Hilbert Approach*. American Mathematical Society, Providence, RI, 2008. (Cited on pp. 14, 16, 23, 97, 249)

[28] P Deift. Some open problems in random matrix theory and the theory of integrable systems. *Contemp. Math.*, 458:419–430, 2008. (Cited on pp. 232, 250)

[29] P Deift, A R Its, and X Zhou. Long-time asymptotics for integrable nonlinear wave equations. In *Important Developments in Soliton Theory*, Springer Series in Nonlinear Dynamics, A S Fokas and V E Zakharov, eds., pp. 181–204. Springer, Berlin, 1993. (Cited on pp. 102, 250)

[30] P Deift and J Park. Long-time asymptotics for solutions of the NLS equation with a delta potential and even initial data. *Int. Math. Res. Not.*, 2011:5505–5624, 2011. (Cited on p. 244)

[31] P Deift and E Trubowitz. Inverse scattering on the line. *Comm. Pure Appl. Math.*, 32(2):121–251, 1979. (Cited on pp. 16, 21)

[32] P Deift, S Venakides, and X Zhou. The collisionless shock region for the long-time behavior of solutions of the KdV equation. *Comm. Pure Appl. Math.*, 47:199–206, 1994. (Cited on pp. xii, 200, 201, 202, 210, 216, 358)

[33] P Deift, S Venakides, and X Zhou. An extension of the steepest descent method for Riemann–Hilbert problems: The small dispersion limit of the Korteweg–de Vries (KdV) equation. *Proc. Natl. Acad. Sci. USA*, 95:450–454, 1998. (Cited on pp. 216, 270)

[34] P Deift and X Zhou. A steepest descent method for oscillatory Riemann–Hilbert problems. *Bull. Amer. Math. Soc.*, 26:119–124, 1992. (Cited on pp. 64, 174, 202)

[35] P Deift and X Zhou. A steepest descent method for oscillatory Riemann–Hilbert problems. Asymptotics for the MKdV Equation. *Ann. Math.*, 137(2):295–368, 1993. (Cited on pp. xii, 97, 197, 199, 202, 228)

[36] P Deift and X Zhou. Long-time asymptotics for integrable systems. Higher order theory. *Comm. Math. Phys.*, 191:175–191, 1994. (Cited on pp. 97, 251)

[37] P Deift and X Zhou. *Long-Time Behavior of the Non-focusing Nonlinear Schrödinger Equation — A Case Study*, volume 1 of *Lectures in Mathematical Sciences*. University of Tokyo, Tokyo, 1994. (Cited on pp. 99, 249, 250, 251)

[38] P Deift and X Zhou. Asymptotics for the Painlevé II equation. *Comm. Pure Appl. Math.*, 48:277–337, 1995. (Cited on pp. 97, 254)

[39] P Deift and X Zhou. Long-time asymptotics for solutions of the NLS equation with initial data in a weighted Sobolev space. *Comm. Pure Appl. Math.*, 56:1029–1077, 2003. (Cited on pp. 97, 249, 250, 251)

[40] A Dienstfrey. *The Numerical Solution of a Riemann–Hilbert Problem Related to Random Matrices and the Painlevé V ODE*. PhD thesis, Courant Institute of Mathematical Sciences, New York, NY, 1998. (Cited on p. 155)

[41] R G Douglas. *Banach Algebra Techniques in Operator Theory*. Academic Press, New York, London, 1972. (Cited on p. 330)

[42] P G Drazin and R S Johnson. *Solitons: An Introduction*. Cambridge University Press, New York, 1996. (Cited on pp. 193, 223)

[43] P Dubard, P Gaillard, C Klein, and V B Matveev. On multi-rogue wave solutions of the NLS equation and positon solutions of the KdV equation. *Eur. Phys. J. Spec. Top.*, 185(1):247–258, 2010. (Cited on p. 247)

[44] B A Dubrovin. Inverse problem for periodic finite zoned potentials in the theory of scattering. *Funct. Anal. Appl.*, 9:61–62, 1975. (Cited on pp. 270, 275)

[45] B A Dubrovin. Theta functions and non-linear equations. *Russ. Math. Surveys*, 36:11–92, 1981. (Cited on pp. 269, 271, 272, 273)

[46] B A Dubrovin. *Integrable Systems and Riemann Surfaces Lecture Notes*. `http://people.sissa.it/~dubrovin/rsnleq_web.pdf`, 2009. (Cited on pp. 270, 273)

[47] P Duren. *Theory of H^p Spaces*. Academic Press, New York, London, 1970. (Cited on pp. 45, 46, 48, 347)

[48] I Egorova, K Grunert, and G Teschl. On the Cauchy problem for the Korteweg–de Vries equation with steplike finite-gap initial data: I. Schwartz-type perturbations. *Nonlinearity*, 22:1431–1457, 2009. (Cited on p. 303)

[49] L C Evans. *Partial Differential Equations*, volume 19 of *Graduate Studies in Mathematics*. American Mathematical Society, Providence, RI, second edition, 2010. (Cited on pp. 32, 59)

[50] A S Fokas. *A Unified Approach to Boundary Value Problems*. SIAM, Philadelphia, 2008. (Cited on pp. 53, 87, 88, 89, 97, 244)

[51] A S Fokas and I M Gel'fand. Integrability of linear and nonlinear evolution equations and the associated nonlinear Fourier transforms. *Lett. Math. Phys.*, 32:189–210, 1994. (Cited on p. 53)

[52] A S Fokas, A R Its, A A Kapaev, and V Y Novokshenov. *Painlevé Transcendents: The Riemann-Hilbert Approach*. American Mathematical Society, Providence, RI, 2006. (Cited on pp. xii, 13, 253, 254, 260, 261, 262, 263, 267, 275)

[53] A S Fokas and X Zhou. On the solvability of Painlevé II and IV. *Comm. Math. Phys.*, 144(3):601–622, 1992. (Cited on p. 77)

[54] G B Folland. *Real Analysis*. John Wiley & Sons, New York, 1999. (Cited on pp. 49, 93, 324, 349, 354)

[55] B Fornberg and J A C Weideman. A numerical methodology for the Painlevé equations. *J. Comput. Phys.*, 230:5957–5973, 2011. (Cited on p. 254)

[56] J Frauendiener and C Klein. Hyperelliptic theta-functions and spectral methods: KdV and KP solutions. *Lett. Math. Phys.*, 76:249–267, 2006. (Cited on pp. 269, 277, 357)

[57] C S Gardner, J M Greene, M D Kruskal, and R M Miura. Method for solving the Korteweg–de Vries equation. *Phys. Rev. Lett.*, 19:1095–1097, 1967. (Cited on p. 193)

[58] T Grava and C Klein. Numerical solution of the small dispersion limit of Korteweg–de Vries and Whitham equations. *Comm. Pure Appl. Math.*, 60:1623–1664, 2007. (Cited on pp. 357, 358)

[59] E P Gross. Structure of a quantized vortex in boson systems. *Nuovo Cim.*, 20:454–477, 1961. (Cited on p. 231)

[60] K Grunert and G Teschl. Long-time asymptotics for the Korteweg–de Vries equation via nonlinear steepest descent. *Math. Phys. Anal. Geom.*, 12(3):287–324, 2009. (Cited on pp. 198, 200, 212, 226, 250)

[61] O H Hald. Numerical solution of the Gel'fand-Levitan equation. *Linear Algebra Appl.*, 28:99–111, 1979. (Cited on p. 359)

[62] N Hale and A Townsend. A fast, simple, and stable Chebyshev–Legendre transform using an asymptotic formula. *SIAM J. Sci. Comput.*, 36(1):A148–A167, 2014. (Cited on p. 136)

[63] G H Hardy. *A Course of Pure Mathematics*. Cambridge University Press, Cambridge, UK, centenary edition, 2008. (Cited on p. xii)

[64] S P Hastings and J B McLeod. *Classical Methods in Ordinary Differential Equations*, volume 129 of *Graduate Studies in Mathematics*. American Mathematical Society, Providence, RI, 2012. (Cited on pp. 200, 253)

[65] J Helsing. A fast and stable solver for singular integral equations on piecewise smooth curves. *SIAM J. Sci. Comput.*, 33(1):153–174, 2011. (Cited on p. 155)

[66] A R Its. The Riemann–Hilbert problem and integrable systems. *Notices Amer. Math. Soc.*, 50(11):1389–1400, 2003. (Cited on p. 4)

[67] A R Its and D Shepelsky. Initial boundary value problem for the focusing NLS equation with Robin boundary conditions: Half-line approach. *Proc. R. Soc. Lond. Ser. A Math. Phys. Eng. Sci.*, 469:20120199, 2013. (Cited on p. 244)

[68] M A Johnson and K Zumbrun. Convergence of Hill's method for nonself-adjoint operators. *SIAM J. Numer. Anal.*, 50(1):64–78, 2012. (Cited on p. 235)

[69] T Kato. *Perturbation Theory for Linear Operators*. Springer, New York, 1995. (Cited on p. 16)

[70] A Klöckner, A Barnett, L Greengard, and M O'Neil. Quadrature by expansion: A new method for the evaluation of layer potentials. *J. Comput. Phys.*, 252:332–349, 2013. (Cited on p. 155)

[71] T W Körner. *Fourier Analysis*. Cambridge University Press, Cambridge, UK, 1989. (Cited on p. 334)

[72] D J Korteweg and G de Vries. On the change of form of long waves advancing in a rectangular canal, and on a new type of long stationary waves. *Philos. Mag.*, 39:422–443, 1895. (Cited on p. 193)

[73] R Kress and I H Sloan. On the numerical solution of a logarithmic integral equation of the first kind for the Helmholtz equation. *Numer. Math.*, 66(1):199–214, 1993. (Cited on p. 114)

[74] P D Lax. Periodic solutions of the KdV equation. *Comm. Pure Appl. Math.*, 28:141–188, 1975. (Cited on p. 269)

[75] J Lenells. Matrix Riemann–Hilbert problems with jumps across Carleson contours. *arXiv:1401.2506*, 2014. (Cited on pp. 23, 55)

[76] R J LeVeque. *Finite Difference Methods for Ordinary and Partial Differential Equations: Steady-State and Time-Dependent Problems*. SIAM, Philadelphia, 2007. (Cited on pp. 169, 243, 357)

[77] G S Litvinchuk and I M Spitkovskii. *Factorization of Measurable Matrix Functions*, volume 25 of *Operator Theory: Advances and Applications*. Birkhäuser, Basel, 1987. (Cited on p. 23)

[78] S V Manakov. On the theory of two-dimensional stationary self-focusing of electromagnetic waves. *Sov. Phys. JETP*, 38:248–253, 1974. (Cited on p. 231)

[79] H P McKean. Algebraic curves of infinite genus arising in the theory of nonlinear waves. In *Proc. International Congress of Mathematicians (Helsinki, 1978)*, pages 777–783, Helsinki, 1980. Acad. Sci. Fennica. (Cited on p. 269)

[80] H P McKean and E Trubowitz. Hill's operator and hyperelliptic function theory in the presence of infinitely many branch points. *Comm. Pure Appl. Math.*, 29(2):143–226, 1976. (Cited on p. 274)

[81] H P McKean and E Trubowitz. Hill's surfaces and their theta functions. *Bull. Amer. Math. Soc.*, 84:1042–1085, 1978. (Cited on p. 269)

[82] Y Meyer and R R Coifman. *Wavelets: Calderón-Zygmund and Multilinear Operators*. Cambridge University Press, Cambridge, UK, 1997. (Cited on pp. 49, 55, 355)

[83] S G Mikhlin and S Prössdorf. *Singular Integral Operators*. Springer, New York, 1980. (Cited on pp. 61, 325)

[84] A Mikikits-Leitner and G Teschl. Long-time asymptotics of perturbed finite-gap Korteweg-de Vries solutions. *J. Anal. Math.*, 116(1):163–218, 2012. (Cited on p. 314)

[85] R M Miura. Korteweg–de Vries equation and generalizations. I. A remarkable explicit nonlinear transformation. *J. Mathematical Phys.*, 9:1202–1204, 1968. (Cited on p. 226)

[86] N I Muskhelishvili. *Singular Integral Equations*. Noordoff, Groningen, The Netherlands, 1953. (Cited on pp. 33, 34)

[87] M J Narasimha and A M Peterson. On the computation of the discrete Cosine transform. *IEEE Trans. Commun.*, 26(6):934–936, 1978. (Cited on p. 117)

[88] M M S Nasser. Numerical solution of the Riemann–Hilbert problem. *Punjab Univ. J. Math.*, 40:9–29, 2008. (Cited on p. 155)

[89] S P Novikov. A periodic problem for the Korteweg–de Vries equation. I. *Funktional Anal. i Prilozhen.*, 8:54–66, 1974. (Cited on p. 269)

[90] S P Novikov, S V Manakov, L P Pitaevskii, and V E Zakharov. *Theory of Solitons*. Constants Bureau, New York, 1984. (Cited on pp. 269, 270, 275, 276, 277)

[91] F W J Olver, D W Lozier, R F Boisvert, and C W Clark. *NIST Handbook of Mathematical Functions*. Cambridge University Press, New York, 2010. (Cited on pp. 5, 8, 102, 103, 130, 140, 168, 189, 223, 235, 260, 295, 342)

[92] S Olver. Numerical solution of Riemann–Hilbert problems: Painlevé II. *Found. Comput. Math.*, 11(2):153–179, 2010. (Cited on pp. 146, 226, 253, 255)

[93] S Olver. RHPackage. `http://www.maths.usyd.edu.au/u/olver/projects/RHPackage.html`, 2010. (Cited on pp. 163, 166, 254, 265)

[94] S Olver. Computation of equilibrium measures. *J. Approx. Theory*, 163:1185–1207, 2011. (Cited on pp. 216, 361)

[95] S Olver. Computing the Hilbert transform and its inverse. *Math. Comp.*, 80:1745–1767, 2011. (Cited on p. 141)

[96] S Olver. A general framework for solving Riemann–Hilbert problems numerically. *Numer. Math.*, 122(2):305–340, 2012. (Cited on pp. 146, 163, 235)

[97] S Olver and T Trogdon. Numerical solution of Riemann–Hilbert problems: Random matrix theory and orthogonal polynomials. *Constr. Approx.*, 39(1):101–149, 2013. (Cited on pp. 16, 253, 255, 263, 266)

[98] S Olver and T Trogdon. Nonlinear steepest descent and numerical solution of Riemann–Hilbert problems. *Comm. Pure Appl. Math.*, 67:1353–1389, 2014. (Cited on pp. 173, 175, 253, 261, 266, 268)

[99] A R Osborne. *Nonlinear Ocean Waves and the Inverse Scattering Transform*, volume 97 of *International Geophysics Series*. Elsevier/Academic Press, Boston, MA, 2010. (Cited on pp. 270, 358)

[100] L P Pitaevskii. Vortex lines in an imperfect Bose gas. *Sov. Phys. JETP*, 13:451–454, 1961. (Cited on p. 231)

[101] J Plemelj. Riemannsche Funktionenscharen mit gegebener Monodromiegruppe. *Monatsh. Math.*, 1908. (Cited on p. 13)

[102] M Prähofer and H Spohn. Exact scaling functions for one-dimensional stationary KPZ growth. *J. Stat. Phys.*, 115:255–279, 2004. (Cited on p. 254)

[103] L G Reyna. L_2 estimates for Chebyshev collocation. *J. Sci. Comput.*, 3(1):1–23, 1988. (Cited on p. 189)

[104] T J Rivlin. *Chebyshev Polynomials*. Pure and Applied Mathematics (New York). John Wiley & Sons, New York, second edition, 1990. (Cited on p. 344)

[105] W Rudin. *Principles of Mathematical Analysis*. McGraw–Hill, New York, third edition, 1964. (Cited on p. 327)

[106] E M Stein. *Singular Integrals and Differentiability Properties of Functions*. Princeton Mathematical Series, No. 30. Princeton University Press, Princeton, NJ, 1970. (Cited on p. 348)

[107] E C Titchmarsh. *Introduction to the Theory of Fourier Integrals*. Oxford University Press, Oxford, UK, 1937. (Cited on p. 52)

[108] A Tovbis, S Venakides, and X Zhou. On semiclassical (zero dispersion limit) solutions of the focusing nonlinear Schrödinger equation. *Comm. Pure Appl. Math.*, 57:877–985, 2004. (Cited on p. 235)

[109] C A Tracy and H Widom. Level-spacing distributions and the Airy kernel. *Comm. Math. Phys.*, 159:151–174, 1994. (Cited on pp. xii, 253)

[110] T Trogdon. ISTPackage. `https://bitbucket.org/trogdon/istpackage`, 2013. (Cited on pp. 206, 222, 241)

[111] T Trogdon. *Riemann–Hilbert Problems, Their Numerical Solution and the Computation of Nonlinear Special Functions*. PhD thesis, University of Washington, Seattle, WA, 2013. (Cited on p. 173)

[112] T Trogdon. Rational approximation, oscillatory Cauchy integrals and Fourier transforms. To appear in Constr. Approx. doi: 10.1007/s00365-015-9294-2. (Cited on p. 119)

[113] T Trogdon and B Deconinck. A Riemann–Hilbert problem for the finite-genus solutions of the KdV equation and its numerical solution. *Phys. D*, 251:1–18, 2013. (Cited on pp. 269, 271, 275, 291, 294, 295, 296, 297, 298, 299, 300, 301)

[114] T Trogdon and B Deconinck. A numerical dressing method for the nonlinear superposition of solutions of the KdV equation. *Nonlinearity*, 27:67–86, 2014. (Cited on pp. 303, 313, 314, 315)

[115] T Trogdon and B Deconinck. Dispersive and soliton perturbations of finite-genus solutions of the KdV equation: Computational results. *Phys. Lett. A*, 378(7–8):617–622, 2014. (Cited on p. 303)

[116] T Trogdon and S Olver. A Riemann–Hilbert approach to Jacobi operators and Gaussian quadrature. To appear in IMA J. Numer. Anal. doi: 10.1093/imanum/dru061. (Cited on p. 16)

[117] T Trogdon and S Olver. Numerical inverse scattering for the focusing and defocusing nonlinear Schrödinger equations. *Proc. R. Soc. Lond. Ser. A Math. Phys. Eng. Sci.*, 469(2149):20120330, 2013. (Cited on pp. 231, 236, 241, 242, 243, 244, 245, 246, 248, 251)

[118] T Trogdon, S Olver, and B Deconinck. Numerical inverse scattering for the Korteweg–de Vries and modified Korteweg–de Vries equations. *Phys. D*, 241(11):1003–1025, 2012. (Cited on pp. 193, 194, 195, 199, 208, 209, 210, 223, 224, 225, 226, 227, 228)

[119] G Wechslberger and F Bornemann. Automatic deformation of Riemann-Hillbert problems with applications to the Painleve II transcendents, Constr. Approx. 39, pp. 152–171 (2014). (Cited on p. 173)

[120] T T Wu, B M McCoy, C A Tracy, and E Barouch. Spin-spin correlation functions for the two-dimensional Ising model: Exact theory in the scaling region. *Phys. Rev. B*, 13(1):316–374, 1976. (Cited on p. xii)

[121] V E Zakharov. Stability of periodic waves of finite amplitude on the surface of a deep fluid. *J. Appl. Mech. Tech. Phys.*, 9:190–194, 1968. (Cited on p. 231)

[122] V E Zakharov and S V Manakov. Asymptotic behavior of non-linear wave systems integrated by the inverse scattering method. *Sov. Phys. JETP*, 44:106–112, 1976. (Cited on p. 251)

[123] X Zhou. The Riemann–Hilbert problem and inverse scattering. *SIAM J. Math. Anal.*, 20(4):966–986, 1989. (Cited on pp. 23, 62, 63, 64, 70, 71, 77, 316)

[124] X Zhou. Riemann–Hilbert problems and integrable systems. *Lect. MSRI*, 1999. (Cited on pp. 46, 47, 63, 65, 309, 316)

Index